PROCEEDINGS OF THE XXII
INTERNATIONAL SYMPOSIUM

LEPTON AND PHOTON INTERACTIONS AT HIGH ENERGIES

Proceedings of the XXII
International Symposium

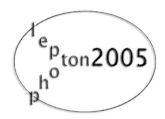

LEPTON AND PHOTON INTERACTIONS AT HIGH ENERGIES

Uppsala University, Sweden   30 June – 5 July 2005

EDITORS

Richard Brenner
Carlos P de los Heros
Johan Rathsman

Uppsala University, Sweden

World Scientific

NEW JERSEY · LONDON · SINGAPORE · BEIJING · SHANGHAI · HONG KONG · TAIPEI · CHENNAI

Published by

World Scientific Publishing Co. Pte. Ltd.

5 Toh Tuck Link, Singapore 596224

USA office: 27 Warren Street, Suite 401-402, Hackensack, NJ 07601

UK office: 57 Shelton Street, Covent Garden, London WC2H 9HE

British Library Cataloguing-in-Publication Data
A catalogue record for this book is available from the British Library.

ISBN 981-256-662-7

Printed in Singapore by World Scientific Printers (S) Pte Ltd

Editors' Foreword

The XXII International Symposium on Lepton-Photon Interactions at High Energy was held in Uppsala, between June 30th and July 5th 2005. The format of the symposium was the traditional in the Lepton-Photon series: plenary talks providing in-depth summaries of different topics. These proceedings follow the format of the conference, with chapters on Electroweak Physics and Beyond, Flavour Physics, QCD and Hadron Structure, Neutrino Physics, Astroparticle Physics and Cosmology and a chapter about Future Facilities. The accompanying DVD includes the video recordings of the talks and the papers submitted to the conference. About 300 papers and 30 posters were contributed by many collaborations and individuals, showing the interest that this symposium series arises among practitioners in the field. Two years before the Large Hadron Collider will come into operation, a majority of the contributed papers from experiments at colliders are submitted by LEP experiments at CERN, the electron-proton experiments at DESY and the dedicated flavour physics experiments at SLAC and KEK.

The symposium started with a session on electroweak physics and beyond. The electroweak measurements by LEP are close to final and they all agree with the Standard Model. This, together with recent improvements in the determination of the top mass at the Tevatron, has further constrained the Higgs mass. The session provided also insight on the latest results from the ongoing experiments at the Tevatron and HERA at DESY, and in-depth reviews of the status and planning for the upcoming LHC experiments at CERN. Theory was well covered by reports on the latest developments on the quest for understanding electroweak symmetry breaking and searching for a more fundamental theory replacing the Standard Model. Not only are the LHC-experiments set to search for signs of Supersymmetry, but there is a growing list of new theoretical possibilities, such as extra dimensions with a low string scale and split supersymmetry, to mention just a few, which the experimenters are gearing up to be able to confront with the physical reality.

The contributions to the session on Flavour Physics showed that the unitarity triangle is now determined with high precision in several complementary ways, and puts stringent constraints on the possible contributions allowed by different types of new physics scenarios. The experiments at b- and c-factories have reported many new discoveries and results on heavy quarkonium production and decays. With measurements that reach very high precision, predictions from the latest theoretical calculations are validated in great detail. Indirect searches of new physics in the decays of heavy flavours and searches for experimental evidence of long-standing predictions on exotic quark and gluon states were reported.

In the session on QCD we learned that the so called twistor space methods, which relate perturbative QCD to topological string theory, have developed tremendously in the last two years and that there is a steady progress in achieving higher precision QCD predictions for the LHC. There have also been new exiting experimental developments, exemplified by the accumulating evidence for a quark gluon plasma, some features of which may be understood using string theory, as well as on the still (at the time of writing) unsettled issue of possible pentaquark states. The session included a special talk to celebrate the 20th anniversary of the introduction of the idea of hard diffraction, which has proved so useful in the understanding of the strong interaction.

The progress in neutrino physics in the last few years has been enormous, as reflected by the results of the many experiments reviewed in the neutrino session. A consistent picture of the parameters of neutrino oscillations is emerging from atmospheric, long-baseline and reactor neutrino experiments. One must keep in mind that the observation of neutrino oscillations is an exciting evidence of physics beyond the Standard Model. But, as it is customary when a new observation is made, more questions than answers can be raised. We still have to understand why neutrino masses are so small and which is the hierarchy between the different masses. A wealth of precision measurements are being prepared, and we hope that the next Lepton-Photon symposium will have fresh news on these topics.

It is already several editions that the Lepton-Photon symposia include a session on Astroparticle Physics and Cosmology, reflecting the synergy between these fields and elementary particle physics. This year it was decided to have separate talks for neutrino astrophysics and cosmic rays, in order to be able to have detailed presentations about the latest experimental developments in these two rapidly evolving topics. The existing neutrino telescopes have already several years of accumulated exposure and are producing relevant physics results which constrain physics models of particle acceleration in cosmic sites. We also heard reports on the status of the next-generation neutrino telescopes and cosmic ray arrays which are designed to reach unprecedented sensitivity to study the issue of particle production and acceleration in cosmic objects.

The formation of the International Linear Collider project has increased the momentum for a future linear collider in the TeV energy range. In the session on Future Facilities the motivations for a linear collider were presented, along with interesting reviews on the status of acceleration technologies and ongoing R&D efforts on new particle detectors that will meet the challenges posed by the next generation accelerators.

One of the highlights of the symposium was the public lecture entitled "The Universe is a Strange Place", which was delivered by Frank Wilczek to a large audience of both participants and members of the general public that almost filled the Aula Magna. We are pleased to be able to include also a write-up of this public lecture in these proceedings.

We are grateful to all the contributors to these proceedings for their efforts in preparing their write-ups summarizing the status of the field. We know that the material from this conference series is a valuable tool in educating and stimulating young physicists, and we hope that this edition of the proceedings will also serve this purpose. We want also to thank the publisher that has allowed open access to the proceedings on the internet. This decision will ensure that these proceedings will reach a wide audience.

See you in Korea in 2007.

Richard Brenner,
Carlos P. de los Heros,
Johan Rathsman,
Editors

Symposium Organization and Acknowledgments

The successful organization of the 2005 Lepton Photon Symposium was made possible by the dedicated individual contributions of many people, in particular from members of the Organizing Committee and of the International Scientific Advisory Committee.

The Organizing Committee was composed of High Energy physicists from Uppsala University as well as colleagues from Stockholm University and the Royal Technical Institute in Stockholm (KTH). The International Scientific Advisory Committee was composed of High Energy physicists affiliated to research institutions in all parts of the world.

The principal lines of the symposium organization were decided by the Organizing Committee. Individual members of the committee were given executive responsibility for various specific organizational areas. Progress made in the various areas was reported at the meetings of the Organizing Committee and the actions in the different areas were discussed and coordinated.

The selection of titles and speakers for the plenary talks was made by the Organizing Committee on the basis of the advice given by International Scientific Advisory Committee. The Advisory Committee played an active and essential role in shaping the scientific program of the symposium. Advice was received from all its members and the draft program was iterated three times with the Advisory Committee.

Executive responsibilities were carried by the following members of the Organizing Committee; by Olga Botner for the local computers and, in collaboration with Göran Fäldt, for reviewing the submitted papers, by Adam Bouchta for the scientific web page, by Richard Brenner for the wireless computer link and the webcast, by Ulf Danielsson for contacts with public media, by Allan Hallgren for reviewing the contributed posters, by Carlos de los Heros for the symposium poster, graphic profile and the invitation of participants, by Gunnar Ingelman for the negotiations with different proceeding publishers and the administration of financial support to young participants from less favored regions and by Johan Rathsman for outreach and social activities. All members of the committee participated in the selection, based on the input from the Advisory Committee, of titles and speakers for the plenary talks. The younger scientists and research students of the Department of Radiation Sciences at Uppsala University acted as session secretaries and technical assistants during the conference. Richard Brenner, Carlos de los Heros and Johan Rathsman were responsible for the editing of these proceedings.

The conference organization bureau of Uppsala University, Akademikonferens, was contracted to carry out all administrative work, in particular the correspondence with, and registration of, the symposium participants, the organization of accommodation, of meals and of the social program.

There were in total 417 registered participants at the symposium. Of these, 21 participants from less favored regions received partial financial support in the form of fee waiver or accommodation support. The number of plenary speakers was 42 and the number of papers submitted was 299. Of these papers, 20 were submitted also in the form of posters. In total there were 31 contributed posters displayed permanently during the conference and presented during the poster sessions. In addition to the regular High Energy Physics sessions there was a reception hosted by the University the first day, a special Grid session with three introductory speakers the second day, a public lecture by Frank Wilczek and a concert in the Cathedral performed by the women's choir La Capella the third day, a one day excursion optionally by bus to Stockholm or by boat to Skokloster castle the fourth day and a conference banquet at Uppsala Castle the fifth day. After the end of the symposium there was the possibility to take part in a tour to the north of Sweden to visit Lappland and to experience the midnight sun.

The symposium was financially supported by The International Union of Pure and Applied Physics, The Royal Swedish Academy of Sciences through its Nobel Institute for physics, The Swedish Research Council, The City of Uppsala and Elsevier Publishers. These Proceedings are published by World Scientific who has, as a novel and important feature, agreed that the contents of the proceedings be made available on the symposium web site www.uu.se/LP2005 according to the principles of Open Access.

Sincere thanks are due to all speakers of the symposium for their hard and dedicated work in preparing excellent talks, to the members of the International Scientific Advisory Committee for their very active and constructive contributions to the shaping of the scientific program, to the members of the Organizing Committee, in particular those carrying executive responsibilities, for all the time, energy and good will put into the organizational work, to the session chair persons for their diligent guidance of the sessions, to the young scientists and research students for having managed the technical services for the sessions so well and to the administrative staff of Akademikonferens and of the Department of Radiation Sciences for carrying out the administrative ground-work so successfully. We also thank IUPAP for the confidence in charging us with the organization of this symposium and for providing experienced advice regarding its organization. The financial support of the sponsors is gratefully acknowledged.

Finally, we want to thank all the participants for coming to Uppsala and for making the symposium such a lively, seminal and memorable event.

Tord Ekelöf
Chair of the Organizing Committee

International Advisory Committee

Roger Blandford	Stanford	Paul Hoyer	Helsinki
Alexander Bondar	Budker INP	Karl Jakobs	Mainz
Lars Brink	Göteborg	Cecilia Jarlskog	Lund
Wilfried Buchmuller	DESY	Makoto Kobayashi	KEK
Per Carlson	KTH	Hugh Montgomery	FNAL
Janet Conrad	Columbia	Yorikiyo Nagashima	Osaka
Persis Drell	SLAC	Ritchie Patterson	Cornell
John Ellis	CERN	Ken Peach	RAL
Joel Feltesse	Saclay	Durga P. Roy	TIFR
Giorgio Giacomelli	Bologna	Alberto Santoro	Rio de Janeiro
Fabiola Gianotti	CERN	Dieter Schlatter	CERN
Igor Golutvin	JINR	Anthony Thomas	Jefferson lab
M. C. González-García	Stony Brook	Li Weiguo	IHEP Beijing
David Gross	Santa Barbara	Bruce Winstein	Chicago
Rolf Heuer	DESY	John Womersly	FNAL
Ian Hinchliffe	LBL		

Local Organizing Committee

Tord Ekelöf	Uppsala, chair	Göran Fäldt	Uppsala
Barbara Badelek	Uppsala	Allan Hallgren	Uppsala
Lars Bergström	Stockholm	Sten Hellman	Stockholm
Olga Botner	Uppsala	Gunnar Ingelman	Uppsala
Adam Bouchta	Uppsala	Kerstin Jon-And	Stockholm
Richard Brenner	Uppsala	Bengt Lund-Jensen	KTH
Ulf Danielsson	Uppsala	Mark Pearce	KTH
Inger Ericson	Uppsala, secretary	Johan Rathsman	Uppsala
Carlos de los Heros	Uppsala		

Session Chairs

Olga Botner	Uppsala	Vera Luth	SLAC
Hesheng Chen	IHEP, Beijing	Piermaria Oddone	Fermilab
Jonathan Dorfan	SLAC	Ken Peach	RAL
Tord Ekelöf	Uppsala	Alberto Santoro	Rio de Janeiro
Gösta Gustafson	Lund	Alexey Sissakian	JINR
Allan Hallgren	Uppsala	Yoji Totsuka	KEK
Boris Kayser	Fermilab	Albrecht Wagner	DESY
Sachio Komamiya	Tokio	Guy Wormser	Orsay
Sau Lan Wu	Wisconsin	Barbro Åsman	Stockholm

Speakers

Kazuo Abe	KEK	Aurelio Juste	FNAL
Ignatios Antoniadis	CERN	Igor Klebanov	Princeton U.
Marina Artuso	Syracuse	Stephan Lammel	FNAL
Laura Baudis	Florida	Jean-Marc Le Goff	CEA, Saclay
Ed Blucher	Chicago	Vera Luth	SLAC
Volker Burkert	Jefferson Lab	Klaus Mönig	DESY
Jon Butterworth	UCL, London	Ulrich Nierste	FNAL
Neil Calder	SLAC	René Ong	UCLA
Oliviero Cremonesi	INFN-Milano	Alan Poon	LBL
Ulf Dahlsten	European Comission	Luigi Rolandi	CERN
Michael Danilov	ITEP	Gavin Salam	Paris VI/VII
Sally Dawson	BNL	Xiaoyan Shen	IHEP, Beijing
Christinel Diaconu	CPPM, Marseille	Luca Silvestrini	INFN-Rome
Jonathan Dorfan	SLAC	Johanna Stachel	Heidelberg
Francesco Forti	INFN-Pisa	Iain Stewart	MIT
Fabiola Gianotti	CERN	Yoichiro Suzuki	Tokio
Srubabati Goswami	Harish-Chandra Institute	Eli Waxman	Weizmann Institute
Francis Halzen	Wisconsin	Christian Weinheimer	Munich
Steen Hannestead	NORDITA	Stefan Westerhoff	Columbia
Per Olof Hulth	Stockholm	Frank Wilczek	MIT
Gunnar Ingelman	Uppsala	Eric Zimmerman	Colorado
Rick Jesik	Imperial College, London	Frank Zimmermann	CERN

Scientic Secretaries

Johan Alwall	Uppsala	Mattias Lantz	Uppsala
Nils Bingefors	Uppsala	Inti Lehmann	Uppsala
Anna Davour	Uppsala	Johan Lundberg	Uppsala
David Duniec	Uppsala	Agnes Lundborg	Uppsala
Mattias Ellert	Uppsala	Bjarte Mohn	Uppsala
David Eriksson	Uppsala	Sophie Ohlsson	Uppsala
Robert Franke	Uppsala	Henrik Pettersson	Uppsala
Nils Gollub	Uppsala	Karin Schönning	Uppsala
Marek Jacewicz	Uppsala	Korinna Zapp	Heidelberg

Symposium Secretariat

Akademikonferens Uppsala:
Ulla Conti   Karin Hornay   Johanna Thyselius

Photo Credits

Teddy Thörnlund   Donald Griffiths   Uppsala University archive   Uppsala Kommun archive

Cover Page

The drawing on the cover is copyright by artist and illustrator Annika Giannini. The drawing highlights some of the buildings and landmarks by which Uppsala is best known. From botton-left and counterclock-wise: the Uppsala University library (Carolina Rediviva), the castle, the Gunilla bell tower, the three hills denoting ancient burial sites of Gamla Uppsala (Old Uppsala), the Wiks castle, the cathedral, the Museum Gustavianum with its characteristic cupola and the Old University building, the premises of this conference. Real photographs of these sites appear throughout this book

SYMPOSIUM PHOTOS

Contents

SUMMARY TALK

PUBLIC TALK

ELECTROWEAK PHYSICS AND BEYOND

Previous page:

Carolina Rediviva (above), the University central library, holds more than 5.000.000 books and publications in 132.133 m of shelves.

The Uppsala cathedral (below) is the largest religious building in Scandinavia. Building works started in 1287 and it was inaugurated in 1435

TOP QUARK MEASUREMENTS

AURELIO JUSTE

Fermi National Accelerator Laboratory, P.O. Box 500, MS 357, Batavia, IL 60510, USA
E-mail: juste@fnal.gov

Ten years after its discovery at the Tevatron collider, we still know little about the top quark. Its large mass suggests it may play a key role in the mechanism of Electroweak Symmetry Breaking (EWSB), or open a window of sensitivity to new physics related to EWSB and preferentially coupled to it. To determine whether this is the case, precision measurements of top quark properties are necessary. The high statistics samples being collected by the Tevatron experiments during Run II start to incisively probe the top quark sector. This report summarizes the experimental status of the top quark, focusing in particular on the recent measurements from the Tevatron Run II.

1 Introduction

The top quark vas discovered in 1995 by the CDF and DØ collaborations[1] during Run I of the Fermilab Tevatron collider. Like any discovery, this one caused a big excitement, although it did not really come as a surprise: the top quark existence was already required by self-consistency of the Standard Model (SM).

One of the most striking properties of the top quark is its large mass, comparable to the Electroweak Symmetry Breaking (EWSB) scale. Therefore, the top quark might be instrumental in helping resolve one of the most urgent problems in High Energy Physics: identifying the mechanism of EWSB and mass generation. In fact, the top quark may either play a key role in EWSB, or serve as a window to new physics related to EWSB and which, because of its large mass, might be preferentially coupled to it.

Ten years after its discovery, we still know little about the top quark. Existing indirect constraints on top quark properties from low-energy data, or the statistics-limited direct measurements at Tevatron Run I, are relatively poor and leave plenty of room for new physics. Precision measurements of top quark properties are crucial in order to unveil its true nature. Currently, the Tevatron collider is the world's only source of top quarks.

2 The Tevatron Accelerator

The Tevatron is a proton–antiproton collider operating at a center of mass energy of 1.96 TeV. With respect to Run I, the center of mass energy has been slightly increased (from 1.8 TeV) and the interbunch crossing reduced to 396 ns (from 3.6 μs). The latter and many other upgrades to Fermilab's accelerator complex have been made with the goal of achieving a significant increase in luminosity. Since the beginning of Run II in March 2001, the Tevatron has delivered an integrated luminosity of $L = 1$ fb^{-1}, and is currently operating at instantaneous luminosities $\mathcal{L} > 1 \times 10^{32}$ cm^{-2}s^{-1}. The goal is to reach $\mathcal{L} \sim 3 \times 10^{32}$ cm^{-2}s^{-1} by 2007, and $L \sim 4.1 - 8.2$ fb^{-1} by the end of 2009. This represents a $\times 40 - 80$ increase with respect to the Run I data set, which will allow the Tevatron experiments to make the transition from the discovery phase to a phase of precision measurements of top quark properties.

3 Top Quark Production and Decay

At the Tevatron, the dominant production mechanism for top quarks is in pairs, mediated by the strong interaction, with a predicted cross section at $\sqrt{s} = 1.96$ TeV of 6.77 ± 0.42 pb for $m_t = 175$ GeV [2]. Within the SM, top quarks can also be produced singly via the electroweak interaction, with $\sim 40\%$ of the top quark pair production rate. However, single top quark production has not been discovered yet. While the production rate of top quarks at the Tevatron is relatively high, ~ 2 $t\bar{t}$ events/hour at $\mathcal{L} = 1 \times 10^{32}$ cm^{-2}s^{-1}, this signal must be filtered out from the approximately seven million inelastic proton–antiproton collisions per second. This stresses the importance of highly efficient and selective triggers.

Since $m_t > M_W$, the top quark in the SM almost always decays to an on-shell W boson and a b quark. The dominance of the $t \to Wb$ decay mode results from the fact that, assuming a 3-generation and unitary CKM matrix[3], $|V_{ts}|, |V_{td}| << |V_{tb}| \simeq 1$ [4]. The large mass of the top quark also results in a large decay width, $\Gamma_t \simeq 1.4$ GeV for $m_t = 175$ GeV, which leads to a phenomenology radically different from that of lighter quarks. Because $\Gamma_t >> \Lambda_{QCD}$, the top quark decays before top-flavored hadrons or $t\bar{t}$-quarkonium bound-states have time to form[5]. As a result, the top quark provides a unique laboratory, both experimentally and theoretically, to study the interactions of a bare quak, not masked by non-perturbative QCD effects.

Thus, the final state signature of top quark events is completely determined by the W boson decay modes: $B(W \to q\bar{q}') \simeq 67\%$ and $B(W \to \ell\nu_\ell) \simeq 11\%$ per lepton (ℓ) flavor, with $\ell = e, \mu, \tau$. In the case of $t\bar{t}$ decay, the three main channels considered experimentally are referred to as *dilepton*, *lepton plus jets* and *all-hadronic*, depending on whether both, only one or none of the W bosons decayed leptonically. The *dilepton* channel has the smallest branching ratio, $\sim 5\%$, and is characterized by two charged leptons (e or μ), large transverse missing energy ($\not{E}_T$) because of the two undetected neutrinos, and at least two jets (additional jets may result from initial or final state radiation). The *lepton plus jets* channel has a branching ratio of $\sim 30\%$ and is characterized by one charged lepton (e

or μ), large $\not{E}_T$ and ≥ 4 jets. The largest branching ratio, $\sim 46\%$, corresponds to the *all-hadronic* channel, characterized by ≥ 6 jets. In all instances, two of the jets result from the hadronization of the b quarks and are referred to as b-jets. As it can be appreciated, the detection of top quark events requires a multipurpose detector with excellent lepton, jet and b identification capabilities, as well as hermetic calorimetry with good energy resolution.

4 The CDF and DØ detectors

The CDF and DØ detectors from Run I already satisfied many of the requirements for a successful top physics program. Nevertheless, they underwent significant upgrades in Run II in order to further improve acceptance and b identification capabilities, as well as to cope with the higher luminosities expected. CDF has retained its central calorimeter and part of the muon system, while it has replaced the central tracking system (drift chamber and silicon tracker). A new plug calorimeter and additional muon coverage extend lepton identification in the forward region. DØ has completely replaced the tracking system, installing a fiber tracker and silicon tracker, both immersed in a 2 T superconducting solenoid. DØ has also improved the muon system and installed new preshower detectors. Both CDF and DØ have upgraded their DAQ and trigger systems to accommodate the shorter interbunch time.

5 Top Quark Pair Production Cross Section

The precise measurement of the top quark pair production cross section is a key element of the top physics program. It provides a test of perturbative QCD and a sensitive probe for new physics effects affecting both top quark production and decay. Especially for the latter, the comparison of measurements in as many channels as possible is crucial. Also, by virtue of the detailed understanding required in terms of object identification and backgrounds, cross section analyses constitute the building blocks of any other top quark properties measurements. Finally, the precise knowledge of the top quark production cross section is an important input for searches for new physics having $t\bar{t}$ as a dominant background.

The measurements performed by CDF and DØ in Run I at $\sqrt{s} = 1.8$ TeV [6] were found to be in good agreement with the SM prediction[7], but limited in precision as a result of the low available statistics $((\Delta\sigma_{t\bar{t}}/\sigma_{t\bar{t}})_{stat} \sim 25\%)$. In Run II, the large expected increase in statistics will yield measurements *a priori* only limited by systematic uncertainties. These include jet energy calibration, signal/background modeling, luminosity determination (currently $\sim$ 6%), etc. However, it is also expected that such large data samples will allow to control/reduce many of these systematic uncertainties. One example is the use of large dedicated control samples to constrain parameters (e.g. gluon radiation) in the modeling of signal and background processes. The goal in Run II is to achieve a per-experiment uncertainty of $\Delta\sigma_{t\bar{t}}/\sigma_{t\bar{t}} \leq 10\%$ for $L \simeq 2$ fb^{-1}.

5.1 Dilepton Final States

Typical event selections require the presence of two high p_T isolated leptons (e, μ, τ or isolated track), large $\not{E}_T$ and ≥ 2 high p_T central jets. Physics backgrounds to this channel include processes with real leptons and $\not{E}_T$ in the final state such as $Z/\gamma^* \to \tau^+\tau^-$ ($\tau \to e, \mu$) and diboson production (WW, WZ, ZZ). The dominant instrumental backgrounds result from $Z/\gamma^* \to e^+e^-, \mu^+\mu^-$, with large $\not{E}_T$ arising from detector resolution effects, and processes where one or more jets fake the isolated lepton signature ($W+jets$ or QCD multijets). Additional kinematic or topolog-

ical cuts are usually applied to further reduce backgrounds, such as e.g on H_T (sum of p_T of jets in the event), exploiting the fact that jets from $t\bar{t}$ are energetic, whereas for backgrounds they typically arise from initial state radiation and have softer p_T spectra. CDF and DØ have developed different analysis techniques to exploit the potential of the sample. The *standard dilepton analysis* ($\ell\ell$), where two well identified leptons (e or μ) and at least two jets are required, has high purity ($S/B \geq 3$) but reduced statistics because of the stringent requirements on lepton identification and jet multiplicity. In order to improve the signal acceptance, the so-called *lepton+track analysis* ($\ell + track$) demands only one well identified lepton and an isolated track, and ≥ 2 jets (see Fig. 1). This analysis has increased acceptance for taus, in particular 1-prong hadronic decays. Finally, an *inclusive analysis* requiring two well identified leptons but placing no cuts on $\not{E}_T$ or jet multiplicity, shows the potential for the greatest statistical sensitivity. In this analysis, a simultaneous determination of $\sigma_{t\bar{t}}$ and σ_{WW} is performed from a fit to the two-dimensional distribution of $\not{E}_T$ vs jet multiplicity using templates from Monte Carlo (MC).

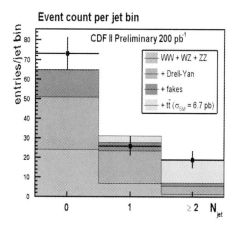

Figure 1. Jet multiplicity distribution for $t\bar{t}$ candidate events selected in the $\ell + track$ channel (CDF).

5.2 Lepton Plus Jets Final States

Typical event selections require one high p_T isolated lepton (e or μ), large $\not{E}_T$ and ≥ 3 high p_T central jets. The dominant background is $W{+}jets$, followed by QCD multijets with one of the jets faking a lepton. After selection the signal constitutes $\sim 10\%$ of the sample. Further signal-to-background discrimination can be achieved by exploiting the fact that all $t\bar{t}$ events contain two b quarks in the final state whereas only a few percent of background events do. CDF and DØ have developed b-tagging techniques able to achieve high efficiency and background rejection: *lifetime tagging* and *soft-lepton tagging*. *Lifetime tagging* techniques rely upon B mesons being massive and long-lived, traveling ~ 3 mm before decaying with high track multiplicity. The high resolution vertex detector allows to directly reconstruct secondary vertices significantly displaced from the event primary vertex (secondary vertex tagging, or SVT) or identify displaced tracks with large impact parameter significance. *Soft-lepton tagging* is based on the identification within a jet of a soft electron or muon resulting from a semileptonic B decay. Only soft-muon tagging (SMT) has been used so far, although soft-electron tagging is under development and should soon become available. The performance of the current algorithms can be quantified by comparing the event tagging probability for $t\bar{t}$ and the dominant $W{+}jets$ background. For instance, for events with ≥ 4 jets: $P_{\geq 1-tag}(t\bar{t}) \simeq 60\%(16\%)$ whereas $P_{\geq 1-tag}(W+jets) \simeq 4\%$, using SVT(SMT). These analyses are typically pure counting experiments and are performed as a function of jet multiplicity in the event (see Fig. 2). Events with 3 or ≥ 4 jets are expected to be enriched in $t\bar{t}$ signal, whereas events with only 1 or 2 jets are expected to be dominated by background. The former are used to estimate $\sigma_{t\bar{t}}$, and the latter to verify the background normalization pro-

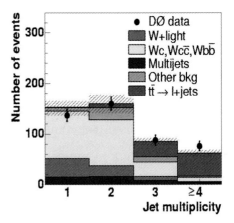

Figure 2. Jet multiplicity distribution for $t\bar{t}$ candidate events selected in the *lepton plus jets* channel, requiring at least one jet to be b-tagged by a secondary vertex algorithm (DØ).

cedure.

CDF and DØ have also developed analyses exploiting the kinematic and topological characteristics of $t\bar{t}$ events to discriminate against backgrounds: leptons and jets are more energetic and central and the events have a more spherical topology. The statistical sensitivity is maximized by combining several discriminant variables into a multivariate analysis (e.g. using neural networks), where the cross section is extracted from a fit to the discriminant distribution using templates from MC (see Fig. 3). Some of the dominant systematic uncertainties (e.g. jet energy calibration) can be reduced by making more inclusive selections (e.g. ≥ 3 jets instead of ≥ 4 jets). The combination of both approaches to improve statistical and systematic uncertainties have for the first time yielded measurements competitive with those using b-tagging (see Table 1).

5.3 All-Hadronic Final State

Despite its spectacular signature with ≥ 6 high p_T jets, the all-hadronic channel is extremely challenging because of the overwhelming QCD multijets background

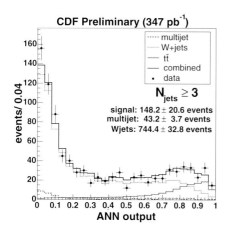

Figure 3. Neural network distribution for $t\bar{t}$ candidate events with ≥ 3 jets, selected in the *lepton plus jets* channel (CDF). This neural network exploits the kinematic and topological characteristics of $t\bar{t}$ events to discriminate against backgrounds.

$(S/B \sim 1/2500)$. Nevertheless, CDF and DØ successfully performed measurements of the production cross section and top quark mass in this channel in Run I. Current measurements by CDF and DØ focus on the *b*-tagged sample and make use of kinematic and topological information to further increase the signal-to-background ratio. CDF applies cuts on a set of four discriminant variables, whereas DØ builds an array of neural networks. In both cases, background is directly predicted from data.

5.4 Summary

Table 1 presents a summary of the best measurements in Run II in each of the different decay channels. Many more measurements have been produced by CDF and DØ and are available from their public webpages. So far, the different measurements are in agreement with each other and with the SM prediction. As precision continues to increase, the detailed comparison among channels will become sensitive to new physics effects. The single most precise measurement (*lepton plus jets*/SVT) has already

reached $\Delta\sigma_{t\bar{t}}/\sigma_{t\bar{t}} \sim 16\%$ and starts becoming systematics-limited. There is much work underway to further reduce systematic uncertainties as well as to combine the available measurements.

6 Top Quark Mass

The top quark mass (m_t) is a fundamental parameter of the SM, not predicted by the theory, and should be measured to the highest possible accuracy. In fact, it is an important ingredient in precision electroweak analyses, where some observables such as M_W receive loop corrections with a quadratic dependence on m_t. This fact was originally used to predict the value of m_t before the top quark discovery, which was ultimately found to be in good agreement with the experimental measurements and constituted a significant success of the SM. After the top quark discovery, the precise measurements of m_t and M_W can be used to constrain the value of the mass of the long-sought Higgs boson (M_H), since some of the electroweak precision observables also receive quantum corrections with a logarithmic dependence on M_H. The combined m_t from Run I measurements is $m_t = 178.0 \pm 4.3$ GeV[14], resulting on the preferred value of $M_H = 129^{+74}_{-49}$ GeV, or the upper limit $M_H < 285$ GeV at 95% C.L.. An uncertainty of $\Delta m_t \leq 2.0$ GeV would indirectly determine M_H to $\sim 30\%$ of its value.

Achieving such high precision is not an easy task, but the experience gained in Run I and the much improved detectors and novel ideas being developed in Run II provide a number of handles that seem to make this goal reachable. In Run I, the dominant systematic uncertainty on m_t was due to the jet energy scale calibration. The reason is that the top quark mass measurement requires a complicated correction procedure (accounting for detector, jet algorithm and physics effects) to provide a precise mapping between reconstructed jets and the original partons.

Table 1. Summary of the best $\sigma_{t\bar{t}}$ measurements at Tevatron Run II.

Channel	Method	$\sigma_{t\bar{t}}$ (pb)	L (pb^{-1})	Experiment
Dilepton	$\ell\ell, \ell + track$	$7.0^{+2.4}_{-2.1}$ (stat.)$^{+1.7}_{-1.2}$ (syst.)	200	CDF[8]
	$\ell\ell$	$8.6^{+3.2}_{-2.7}$ (stat.) ± 1.1 (syst.)	230	DØ[9]
Lepton plus Jets	SVT	8.1 ± 0.9 (stat.) ± 0.9 (syst.)	318	CDF[10]
	SVT	$8.6^{+1.2}_{-1.1}$ (stat.)$^{+1.1}_{-1.0}$ (syst.)	230	DØ[11]
	SMT	$5.2^{+2.9}_{-1.9}$ (stat.)$^{+1.3}_{-1.0}$ (syst.)	193	CDF[12]
	Kinematic	6.3 ± 0.8 (stat.) ± 1.0 (syst.)	347	CDF
	Kinematic	$6.7^{+1.4}_{-1.3}$ (stat.)$^{+1.6}_{-1.1}$ (syst.)	230	DØ[13]
All-Hadronic	SVT	7.8 ± 2.5 (stat.)$^{+4.7}_{-2.3}$ (syst.)	165	CDF
	SVT	$7.7^{+3.4}_{-3.3}$ (stat.)$^{+4.7}_{-3.8}$ (syst.)	162	DØ

To determine and/or validate the jet energy calibration procedure, data samples corresponding to *di-jet*, γ+jets and *Z+jets* production were extensively used. In addition to the above, the large $t\bar{t}$ samples in Run II allow for an *in situ* calibration of light jets making use of the W mass determination in $W \rightarrow jj$ from top quark decays, a measurement which is in principle expected to scale as $1/\sqrt{N}$. Also, dedicated triggers requiring displaced tracks will allow to directly observe $Z \rightarrow b\bar{b}$, which can be used to verify the energy calibration for b jets. Additional important requirements for the m_t measurement are: accurate detector modeling and state-of-the-art theoretical knowledge (gluon radiation, parton distribution functions, etc). The golden channel for a precise measurement is provided by the *lepton plus jets* final state, by virtue of its large branching ratio and moderate backgrounds, as well as the presence of only one neutrino, which leads to over-constrained kinematics. Powerful b-tagging algorithms are being used to reduce both physics and combinatorial backgrounds, and sophisticated mass extraction techniques are being developed, resulting in improvements in statistical as well as systematic uncertainties. An overview of the main analysis methods is given next.

6.1 Template Methods

These methods, traditionally used in Run I, start by constructing an event-by-event variable sensitive to m_t, e.g. the reconstructed top quark mass from a constrained kinematic fit in the *lepton plus jets* channel. The top quark mass is extracted by comparing data to templates on that particular variable built from MC for different values on m_t. Recent developments in this approach by CDF (see Fig. 4) have lead to the single most precise measurement to date: $m_t = 173.5^{+3.7}_{-3.6}$ (stat. + JES) ± 1.7 (syst.) GeV, exceeding in precision the current world average. The statistical uncertainty is minimized by separately performing the analysis in four subsamples with different b-tag multiplicity, thus each with a different background content and sensitivity to m_t. The dominant systematic uncertainty, jet energy calibration (JES), is reduced by using the *in situ* W mass determination from $W \rightarrow jj$ in a simultaneous fit of m_t and a jet energy calibration factor. The latter is also subjected to a constraint of $\sim 3\%$ from an external measurement in control samples. The remaining systematic uncertainties, amounting to $\Delta m_t = 1.7$ GeV, include contributions such as background shape, b-fragmentation, gluon radiation, parton distribution functions, etc,

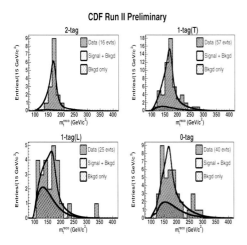

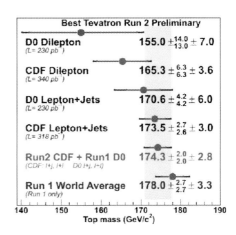

Figure 4. Reconstructed m_t distribution from a constrained kinematical fit in the *lepton plus jets* channel (CDF). The distribution is shown separately for the different subsamples defined based on the *b*-tag multiplicity.

Figure 5. Summary of the best m_t measurements at Tevatron Run II.

many of which are expected to be further reduced with larger data samples.

6.2 Dynamic Methods

The main objective of these methods is to make an optimal used of the statistical information of the sample. They are based on the calculation of the per-event probability density as function of m_t, taking into account resolution effects (better measured events contribute more) and summing over all permutations of jets as well as neutrino solutions. These methods typically include a complete or partial matrix element evaluation for the signal and dominant background processes. The so-called *Matrix Element Method* was pioneered by DØ and applied to the *lepton plus jets* Run I sample[15], leading to the single most precise measurement in Run I. In Run II, CDF has applied this method to the *b*-tagged *lepton plus jets* sample yielding a result competitive with the template method discussed above, and to the *lepton+track* sample, achieving the unprecedented accuracy in the *dilepton* channel of $m_t = 165.3 \pm 7.2$ (stat. + syst.) GeV.

6.3 Summary and Prospects

Fig. 5 summarizes the best Run II measurements for CDF and DØ in the different analysis channels. As it can be appreciated, some of the Run II individual measurements are already achieving uncertainties comparable or better than the Run I world average. The new preliminary combination of the DØ Run I and CDF Run II measurements in the *lepton plus jets* and *dilepton* channels yields: $m_t = 174.3 \pm 3.4$ GeV, $\chi^2/dof = 3.6/3$, improving upon the previous world average result. The resulting constraints on the Higgs boson mass are: $M_H = 98^{+52}_{-36}$ GeV or $M_H < 208$ GeV at 95% C.L.. Based on the current experience with Run II measurements, it is expected that an uncertainty of $\Delta m_t \leq 1.5$ GeV can be achieved at the Tevatron with 2 fb^{-1}, a precision which will probably be only matched by the LHC and will have to wait for the ILC to be exceeded.

7 Top Quark Couplings to the W boson

If the top quark is indeed playing a special role in the EWSB mechanism, it may have non-SM interactions to the weak gauge bosons. At the Tevatron, only the tWb vertex can be sensitively probed. The LHC will have in addition sensitivity to certain ttZ couplings[16].

Within the SM, the charge current interactions of the top quark are of the type V–A and completely dominated by the tWb vertex by virtue of the fact that $|V_{tb}| \simeq 1$. In fact, the tWb vertex defines most of the top quark phenomenology: it determines the rate of single top quark production and completely saturates the top quark decay rate. It is also responsible for the large top quark width, that makes it decay before hadronizing, thus efficiently transmitting its spin to the final state. The angular distributions of the top quark decay products also depend on the structure of the tWb vertex.

7.1 Single Top Quark Production

Within the SM, the main production mechanisms for single top quarks at the Tevatron involve the exchange of a timelike W boson (s-channel), $\sigma_s = 0.88 \pm 0.07$ pb, or a spacelike W boson (t-channel), $\sigma_t = 1.98 \pm 0.21$ pb[17]. Despite the relatively large expected rate, single top production has not been discovered yet. Upper limits on the production cross sections were obtained in Run I: $\sigma_s < 18$ pb, $\sigma_t < 13$ pb, $\sigma_{s+t} < 14$ pb (CDF) and $\sigma_s < 17$ pb, $\sigma_t < 22$ pb (DØ) at 95% C.L.. The experimental signature is almost identical to the *lepton plus jets* channel in $t\bar{t}$: high p_T isolated lepton, large $\not{E}_T$ and jets, but with lower jet multiplicity (typically 2 jets) in the final state, which dramatically increases the $W+jets$ background. In addition, $t\bar{t}$ production becomes a significant background with a very similar topology (e.g. if one lepton in the *dilepton* channel is not reconstructed).

Once it is discovered, the precise determination of the single top production cross section will probe, not only the Lorentz structure, but also the magnitude of the tWb vertex, thus providing the only direct measure-

ment of $|V_{tb}|$. The sensitivity to anomalous top quark interactions is enhanced by virtue of the fact that top quarks are produced with a high degree of polarization. In addition, the s- and t-channels are differently sensitive to new physics effects[18], so the independent measurement of σ_s and σ_t would allow to discriminate among new physics models should any deviations from the SM be observed.

In Run II the search for single top quark production continues with ever increasing data samples, improved detector performance, and increasingly more sophisticated analyses. The generic analysis starts by selecting b-tagged *lepton plus* $\geq 2jets$ candidate events. CDF considers one discriminant variable per channel (e.g. $Q(\ell) \times \eta(untagged\ jet)$ for the t-channel search) whereas DØ performs a multivariate analysis using using neural networks (see Fig. 6). The upper limit on σ is estimated exploiting the shape of the discriminant variable and using a Bayesian approach. From ~ 162 pb^{-1} data, CDF obtains the following observed (expected) 95% C.L. upper limits[19]: $\sigma_s < 13.6(12.1)$ pb, $\sigma_s < 10.1(11.2)$ pb and $\sigma_{s+t} < 17.8(13.6)$ pb. The world's best limits are obtained by DØ from ~ 230 pb^{-1} of data as a result of their more sophisticated analysis[20]: $\sigma_s < 6.4(5.8)$ pb and $\sigma_s < 5.0(4.5)$ pb. Both collaborations continue to add more data and improve their analyses and more sensitive results are expected soon.

7.2 W boson helicity in Top Quark Decays

While only single top quark production gives direct access to the magnitude of the tWb interaction, $t\bar{t}$ production can still be used to study its Lorentz structure. This is possible because the W boson polarization in top quark decays depends sensitively on the tWb vertex. Within the SM (V–A interaction), only two W boson helicity configurations, $\lambda_W = 0, -1$, are allowed. The frac-

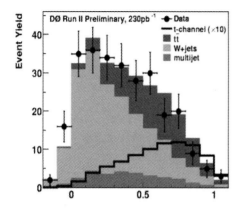

Figure 6. Neural network distribution for single top quark candidate events in the *b*-tagged *lepton plus* ≥ 2 *jets* sample (DØ). This neural network has been optimized to discriminate between *tb* (s-channel) and $W b \bar{b}$.

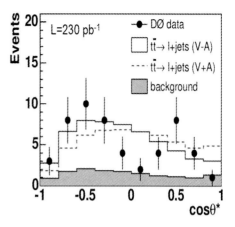

Figure 7. Lepton helicity angle distribution in the the *b*-tagged *lepton plus* ≥ 4 *jets* sample (DØ).

tion of longitudinal ($\lambda_W = 0$) and left-handed ($\lambda_W = -1$) W bosons are completely determined by the values of m_t, M_W and m_b and predicted to be: $F_0 \simeq 70\%$ and $F_- \simeq 30\%$, respectively (as a result, $F_+ \simeq 0\%$). The well-known quiral structure of the W interaction to leptons allows to use lepton kinematic distributions such as the p_T in the laboratory frame ($p_{T\ell}$) or the cosinus of the lepton decay angle in the W boson rest frame with respect to the W direction ($\cos\theta_\ell^*$) to measure the W helicity fractions. The $p_{T\ell}$ method can be applied to both *lepton plus jets* and *dilepton* final states. The $\cos\theta_\ell^*$ method can only be used in the *lepton plus jets* final state since explicit top quark reconstruction is required.

Current Run II measurements by CDF and DØ are based on $\sim 200 - 230$ pb^{-1} of data and, due to the still limited statistics, only consider the measurement of one W helicity fraction at a time, fixing the other one to the SM prediction. From the $p_{T\ell}$ method and using an unbinned likelihood, CDF has measured $F_0 = 0.27^{+0.35}_{-0.21}$ (stat. + syst.). DØ has instead focused on the $\cos\theta_\ell^*$ method to measure F_+ (see Fig. 7), using a binned likelihood[21]. The result from the combination of two analyses (*b*-tag and kinematic)

is $F_+ < 0.25$ at 95% C.L.. The best measurements in Run I yielded[22] $F_0 = 0.56 \pm 0.31$ (stat. + syst.) (DØ) and $F_+ < 0.18$ at 95% C.L. (CDF). All measurements, although still limited by statistics, are consistent with the SM prediction. The large expected samples in Run II should allow to make more sensitive measurements in the near future.

7.3 $B(t \to Wb)/B(t \to Wq)$

Assuming a 3-generation and unitary CKM matrix, $B(t \to Wb) = \Gamma(t \to Wb)/\Gamma_t \simeq 1$. An observation of $B(t \to Wb)$ significantly deviating from unity would be a clear indication of new physics such as e.g. a fourth fermion generation or a non-SM top quark decay mode. $\Gamma(t \to Wb)$ can be directly probed in single top quark production, via the cross section measurement. Top quark decays give access to $R \equiv B(t \to Wb)/B(t \to Wq)$, with $q = d, s, b$, which can be expressed as $R = \frac{|V_{tb}|^2}{|V_{td}|^2 + |V_{ts}|^2 + |V_{tb}|^2}$, and it's also predicted in the SM to be $R \simeq 1$.

R can be measured by comparing the number of $t\bar{t}$ candidates with 0, 1 and 2 *b*-tagged jets, since the tagging efficiencies for jets originating from light (d, s) and b quarks are very different. In Run I, CDF measured[23] $R = 0.94^{+0.31}_{-0.24}$ (stat. + syst.). In Run II, both CDF and DØ have performed this measurement using data samples of ~ 160 pb^{-1} and ~ 230 pb^{-1}, respectively. CDF considers events in both the *lepton plus jets* and *dilepton* channels and measures[24] $R = 1.12^{+0.27}_{-0.23}$ (stat. + syst.), whereas DØ only considers events in the *lepton plus jets* channel and measures $R = 1.03^{+0.19}_{-0.17}$ (stat. + syst.). All measurements are consistent with the SM prediction.

8 FCNC Couplings of the Top Quark

Within the SM, neutral current interactions are flavor-diagonal at tree level. Flavor Changing Neutral Current (FCNC) effects are loop-induced and thus heavily suppressed (e.g. $B(t \to cg) \simeq 10^{-10}, B(t \to c\gamma/Z) \simeq 10^{-12}$), so an observation would be a clear signal of new physics. Indeed, these effects can be significantly enhanced (by factors $\sim 10^3 - 10^4$) in particular extensions of the SM. Searches for FCNC interactions have been carried out in $p\bar{p}$, e^+e^- and $e^\pm p$ collisions. At Tevatron, FCNC couplings can manifest themselves both in the form of anomalous single top quark production ($qg \to t$, $q = u, c$) or anomalous top quark decays ($t \to qV$, $q = u, c$ and $V = g, \gamma, Z$). Only the latter has been experimentally explored so far, via the search for $t \to q\gamma/Z$ decays[25]. The same $tq\gamma/Z$ interaction would be responsible for anomalous single top quark production in e^+e^- ($e^+e^- \to \gamma^*/Z \to tq$) and $e^\pm p$ ($eq \to et$) collisions, and searches have been performed at LEP[26] and HERA[27,28], respectively. Fig. 8 shows the existing 95% upper limits on the magnitude of the tuZ and $tu\gamma$ couplings.

Recently, H1 has reported[28] a 2.2σ excess in their search for single top quark production in the leptonic channels. A total of 5 events were observed, compared to 1.31 ± 0.22 events expected. No excess was observed in

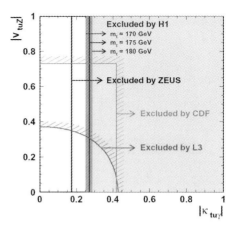

Figure 8. Exclusion limits at the 95% C.L. on the anomalous tuZ and $tu\gamma$ couplings obtained at the Tevatron, LEP (only L3 experiment shown) and HERA.

the hadronic channel. The combination of all channels yields a production cross section of $0.29^{+0.15}_{-0.14}$ pb. Interpreted as FCNC-mediated single top production, this measurement translates into $|\kappa_{tu\gamma}| = 0.20^{+0.05}_{-0.06}$. Higher statistics measurements at the Tevatron Run II and HERA-II should be able to confirm or exclude this measurement.

9 Searches for New Particles in Top Quark Production and Decay

Many models beyond the SM predict new particles preferentially coupled to the top quark: heavy vector gauge bosons (e.g. $q\bar{q} \to Z' \to t\bar{t}$ in Topcolor), charged scalars (e.g. $t \to H^+b$ in generic 2HDM), neutral scalars (e.g. $gg \to \eta_T \to t\bar{t}$ in Technicolor) or exotic quarks (e.g. $q\bar{q} \to W^* \to t\bar{b}'$ in E_6 GUT). Because of the large spectrum of theoretical predictions, experimentally it is very important to develop searches as model-independent as possible. These analyses usually look for deviations in kinematic properties (e.g. $t\bar{t}$ invariant mass or top p_T spectrum), compare cross section measurements in different decay channels, etc.

In Run I, a model-independent search for a narrow heavy resonance X decaying to $t\bar{t}$ in the *lepton plus jets* channel was performed[29]. The obtained experimental upper limits on $\sigma_X \times B(X \to t\bar{t})$ vs M_X were used to exclude a leptophobic X boson[30] with $M_X < 560$ GeV (DØ) and $M_X < 480$ GeV (CDF) at 95% C.L.. Similar searches are underway in Run II.

In Run II, CDF has performed a search for $t \to H^+ b$ decays in $t\bar{t}$ events. If $M_{H^+} < m_t - m_b$, $t \to H^+ b$ competes with $t \to W^+ b$ and results in $B(t \to Wb) < 1$. Since $H^\pm$ decays are different than $W^\pm$ decays, $\sigma_{t\bar{t}}$ measurements in the various channels would be differently affected. By performing a simultaneous fit to the observation in the *dilepton*, *lepton plus tau* and *lepton plus jets* channels, CDF has determined model-dependent exclusion regions in the $(\tan\beta, M_H^\pm)$ plane.

10   New Physics Contamination in Top Quark Samples

Top quark events constitute one of the major backgrounds to non-SM processes with similar final state signature. As a result, top quark samples could possibly contain an admixture of exotic processes. A number of model-independent searches have been performed at the Tevatron in Run I and Run II.

A slight excess over prediction in the *dilepton* channel (in particular in the $e\mu$ final state) was observed in Run I[31]. Furthermore, some of these events had anomalously large lepton p_T and $\not{E}_T$, which called into question their compatibility with SM $t\bar{t}$ production. In fact, it was suggested that these events would be more consistent with cascade decays from pair-produced heavy squarks[32]. In Run II, CDF and DØ continue to scrutinize the *dilepton* sample. To date, the event kinematics appears to be consistent with SM $t\bar{t}$ production[33,9]. Nevertheless, the flavor anomaly persists: the total number of events observed by both CDF and DØ in the

$e\mu(ee + \mu\mu)$ final state is 17(9), whereas the SM prediction is $10.2 \pm 1.0(9.4 \pm 1.0)$. More data is being analyzed and a definite conclusion on the consistency of the *dilepton* sample with the SM should be reached soon.

Also ongoing in Run II is the search for pair production of a heavy t' quark, with $t' \to Wq$. The final state signature would be identical to $t\bar{t}$, but the larger mass of the t' quark would cause the events to be more energetic than $t\bar{t}$. The current analysis is focused on the *lepton plus jets* channel and considers the H_T distribution as the observable to search for $t'\bar{t}'$ production. It is expected that with $L = 2$ fb^{-1}, $m_{t'} < 300$ GeV will be excluded at the 95% C.L..

11   Conclusions

Till the beginning of the LHC, the Tevatron will remain the world's only top quark factory and a comprehensive program of top quark measurements is well underway. The excellent performances of the accelerator and the CDF and DØ detectors open a new era of precision measurements in top quark physics, required to unravel the true nature of the top quark and possibly shed light on the EWSB mechanism. This is a largely unexplored territory, and thus it has the potential to reveal signs of new physics preferentially coupled to the top quark. Most existing measurements appear to be in agreement with the SM, but there are a number of tantalizing (although not statistically significant) anomalies, which should definitely be clarified with the large data samples expected from the Tevatron till the end of 2009. Furthermore, techniques developed at the Tevatron to carry out this rich program of precision top quark physics will be an invaluable experience for the LHC.

Acknowledgments

The author would like to thank the conference organizers for their invitation and a stimulating and enjoyable conference.

References

1. CDF Collaboration, F. Abe *et al.*, *Phys. Rev. Lett.* **74**, 2626 (1995); DØ Collaboration, S. Abachi *et al.*, *Phys. Rev. Lett.* **74**, 2632 (1995).
2. R. Bonciani, S. Catani, M.L. Mangano, and P. Nason, *Nucl. Phys.* B **529**, 424 (1998); N. Kidonakis and R. Vogt, *Phys. Rev.* D **68**, 114014 (2003); M. Cacciari *et al.*, *JHEP* **404**, 68 (2004).
3. N. Cabbibo, *Phys. Rev. Lett.* **10**, 531 (1963); M. Kobayashi and T. Maskawa, *Prog. Theor. Phys.* **49**, 652 (1973).
4. S. Eidelman *et al.*, *Phys. Lett.* B **592**, 1 (2004).
5. J.H. Kühn, *Acta Phys. Polon.* B **12**, 347 (1981); J.H. Kühn, *Act. Phys. Austr.* Suppl. XXIV, 203 (1982); I.Y. Bigi *et al.*, *Phys. Lett.* B **181**, 157 (1986).
6. CDF Collaboration, T. Affolder *et al.*, *Phys. Rev.* D **64**, 032002 (2001); DØ Collaboration, V.M. Abazov *et al.*, *Phys. Rev.* D **67**, 012004 (2003).
7. N. Kidonakis, *Phys. Rev.* D **64**, 014009 (2001); N. Kidonakis, E. Laenen, S. Moch, and R. Vogt, *Phys. Rev.* D **64**, 114001 (2001).
8. CDF Collaboration, D. Acosta *et al.*, *Phys. Rev. Lett.* **93**, 142001 (2004).
9. DØ Collaboration, V.M. Abazov *et al.*, *hep-ex/0505082*.
10. CDF Collaboration, D. Acosta *et al.*, *Phys. Rev.* D **71**, 052003 (2005).
11. DØ Collaboration, V.M. Abazov *et al.*, *hep-ex/0504058*.
12. CDF Collaboration, D. Acosta *et al.*, *Phys. Rev.* D **72**, 032002 (2005).
13. DØ Collaboration, V.M. Abazov *et al.*, *hep-ex/0504043*.
14. CDF Collaboration, DØ Collaboration and Tevatron Electroweak Working Group, *hep-ex/0404010*.
15. DØ Collaboration, V.M. Abazov *et al.*, *Nature* **429**, 638 (2004).
16. U. Baur, A. Juste, L.H. Orr, and D. Rainwater, *Phys. Rev.* D **71**, 054013 (2005).
17. B.W. Harris *et al.*, *Phys. Rev.* D **66**, 054024 (2002); Z. Sullivan, *Phys. Rev.* D **70**, 114012 (2004).
18. T. Tait and C.-P. Yuan, *Phys. Rev.* D **63**, 014018 (2001).
19. CDF Collaboration, D. Acosta *et al.*, *Phys. Rev.* D **71**, 012005 (2005).
20. DØ Collaboration, V.M. Abazov *et al.*, *Phys. Lett.* B **622**, 265 (2005).
21. DØ Collaboration, V.M. Abazov *et al.*, *Phys. Rev.* D **72**, 011104(R) (2005).
22. CDF Collaboration, D. Acosta *et al.*, *Phys. Rev.* D **71**, 031101 (2005); DØ Collaboration, V.M. Abazov *et al.*, *Phys. Lett.* B **617**, 1 (2005).
23. CDF Collaboration, T. Affolder *et al.*, *Phys. Rev. Lett.* **86**, 3233 (2001).
24. CDF Collaboration, D. Acosta *et al.*, *Phys. Rev. Lett.* **95**, 102003 (2005).
25. CDF Collaboration, F. Abe *et al.*, *Phys. Rev. Lett.* **80**, 2525 (1998).
26. Only the most sensitive result quoted: L3 Collaboration, P. Achard *et al.*, *Phys. Lett.* B **549**, 290 (2002).
27. ZEUS Collaboration, S. Chekanov *et al.*, *Phys. Lett.* B **559**, 153 (2003).
28. H1 Collaboration, A. Aktas *et al.*, *Eur. Phys. J.* C **33**, 9 (2004).
29. CDF Collaboration, T. Affolder *et al.*, *Phys. Rev. Lett.* **85**, 2062 (2000); DØ Collaboration, V.M. Abazov *et al.*, *Phys. Rev. Lett.* **92**, 221801 (2004).
30. R.M. Harris, C.T. Hill, and S.J. Parke, *hep-ph/9911288* (1999).
31. CDF Collaboration, F. Abe *et al.*, *Phys. Rev. Lett.* **80**, 2779 (1998); DØ Collaboration, S. Abachi *et al.*, *Phys. Rev. Lett.* **79**, 1203 (1997).
32. R.M. Barnett and L.J. Hall, *Phys. Rev. Lett.* **77**, 3506 (1996).
33. CDF Collaboration, D. Acosta *et al.*, *Phys. Rev. Lett.* **95**, 022001 (2005).

DISCUSSION

Peter Schleper (University of Hamburg):
High order corrections are large for top quark production. How are the corresponding uncertainties taken into account in the kinematic reconstruction of the top mass?

Aurelio Juste: High order QCD corrections are indeed large, especially regarding the normalization of the total cross section, which does not enter the top mass analysis. They also have sizable effects in more differential distributions, which can affect the measurement of the top mass. The most direct effect would arise from hard gluon radiation, i.e. the production of extra jets in the event. While the plan for the experimental analyses is to start using the available NLO event generators (e.g. MC@NLO), they are still based on LO MCs (PYTHIA, HERWIG, ALPGEN, etc). Systematic uncertainties due to gluon radiation are being evaluated e.g. by varying PYTHIA parameters related to ISR within the allowed region from $Z + jets$ data, or by comparing $t\bar{t}$ vs $t\bar{t} + jets$ LO calculations using ALPGEN. Gluon radiation, as well as other similar theory-related systematic uncertainties (fragmentation functions, parton distribution functions, etc) are expected to be limiting components on the top mass measurement by the end of Run II, expected to achieve uncertainties ≤ 1.5 GeV. By then, we are hoping we will be making use of large available control samples to further constrain these systematics.

Tord Ekelöf (Uppsala University):
The reanalysis of Run I in DØ, a few years ago, produced a higher mass for the top quark, ~ 178 GeV. Now the combined analysis of Run II doesn't confirm that rise from earlier measurements. These variations are within the errors anyhow, but is there any explanation?

Aurelio Juste: The new combination of CDF Run II and DØ Run I measurements has a $\chi^2/dof = 3.6/3$, which corresponds to a probability of 47%, indicating that all measurements are in good agreement with each other. So far we don't have any indication that there is a problem with the reanalysis of Run I in DØ, which has a pull with respect to the average of only $+1.44$.

ELECTROWEAK MEASUREMENTS

CRISTINEL DIACONU

Centre de Physique des Particules de Marseille, IN2P3-CNRS, case 902, 163 Avenue de Luminy, Marseille, France

and Deutsches Elektronen Synchrotron, DESY, Notkestr. 85, 22607 Hamburg, Germany
E-mail: diaconu@cppm.in2p3.fr

The measurements of electroweak sector of the Standard Model are presented, including most recent results from LEP, Tevatron and HERA colliders. The robustness of the Standard Model is illustrated with the precision measurements, the electroweak fits and the comparisons to the results obtained from low energy experiments. The status of the measurements of the W boson properties and rare production processes involving weak bosons at colliders is examined, together with the measurements of the electroweak parameters in ep collisions.

1 Introduction

The Standard Model (SM) of the elementary particles has proven its robustness in the past decades due to extensive tests with increasing precision. In the present paper, the status of electroweak measurements in summer 2005 are presented. First, the precision measurements from LEP and SLD colliders will be summarised[a] and the confrontation with the low energy experiments will be reviewed. Then production of weak bosons at LEP, Tevatron and HERA will be presented. The constraints obtained from an electroweak fit over DIS data will be described together with the latest measurements from polarized electron data at HERA. Finally, prospects for electroweak measurements at future colliders will be briefly reviewed.

2 The experimental facilities

The LEP collider stopped operation in 2000, after providing an integrated luminosity of more than 0.8 fb^{-1} accumulated by each of the four experiments (ALEPH, DELPHI, L3 and OPAL). The first stage (LEP I) running at Z peak was continued with a second pe-

riod (LEPII) with centre-of-mass energies up to 209 GeV, beyond the W pair production threshold. The physics at the Z peak was greatly enforced due to the polarised e^+e^- collisions programme at the SLC. The SLD detector recorded a data sample corresponding to an integrated luminosity of 14 pb^{-1}, with a luminosity weighted electron beam polarisation of 74%.

The Tevatron $p\bar{p}$ collider completed a first period (Run I) in 1996. After an upgrade, including the improvement of the two detectors CDF and D0, a second high luminosity period started in 2002. In summer 2005 the delivered integrated luminosity reached 1 fb^{-1}. The completion of the second part of the programme (Run II) is forseen in 2009 with a goal of 4 to 8 fb^{-1}.

The unique $e^{\pm}p$ collider HERA is equiped with two detectors in collider mode H1 and ZEUS. After a first period with unpolarised collisions (HERA I, 1993–2000), the collider provides in the new stage (HERA II) both electron and positron–proton collisions with $e^{\pm}$ beam polarisation of typically 40%. The HERA programme will end in 2007 with a delivered luminosity around 700 pb^{-1}.

The situation of the high energy colliders in the last 15 years was therefore a favourable one, with all three combinations of colliding beams e^+e^-, $p\bar{p}$ and ep. The most pre-

---

[a]The new averages of the W boson and top quark masses, made public[1] in july 2005 after the conference, are included in this paper together with the corresponding results of the electroweak fit.

cise testing of the weak interactions is done at e^+e^- colliders (LEP and SLC), where Z bosons are produced in the s–channel in a clean environment with sufficient luminosities. The hadronic collider (Tevatron) produces large samples of weak bosons and enables complementary studies at higher energies, including the measurement of the top quark properties. In ep collisions, the exchange of space–like electroweak bosons in the t channel leads to new experimental tests of the Standard Model. The high energy experiments are complemented by low energy measurements that test the electroweak theory with high precision far below the weak boson masses.

3 The precision measurements, the electroweak fits and comparisons with low energy data

3.1 The precision measurements from high energy experiments

The Standard Model is tested using a set of precison mesurements at e^+e^- colliders close to the Z peak. Those measurements, which were finalised recently[2], include data from the LEP experiments and the SLD detector at the SLC.

The two fermion production is measured using the flavour tag of the final state (leptons ℓ and the b and the c quarks). More than 1000 measurements are used to extract a few observables that have simple relations to the fundamental parameters of the SM. The observables set include: the cross sections and its dependence on the $\sqrt{s}$ (line shape given by Z mass m_Z, width Γ_Z and the hadronic pole cross section σ^0_{had}) and on final state flavour (partial widths $R_{\ell,b,c}$), the forward-backward asymmetries (A_{FB}), the left-right asymmetries (A_{LR}, measured for polarised beams or using the measured polarisation of the final state tau leptons). Moreover, the measured asymmetries are used in order to extract the asymmetry parameters that are

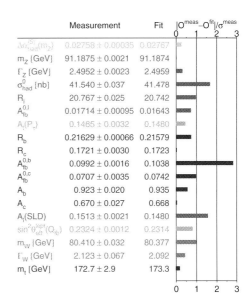

	Measurement	Fit	$|O^{meas}-O^{fit}|/\sigma^{meas}$
$\Delta\alpha^{(5)}_{had}(m_Z)$	0.02758 ± 0.00035	0.02767	
m_Z [GeV]	91.1875 ± 0.0021	91.1874	
Γ_Z [GeV]	2.4952 ± 0.0023	2.4959	
σ^0_{had} [nb]	41.540 ± 0.037	41.478	
R_l	20.767 ± 0.025	20.742	
$A^{0,l}_{fb}$	0.01714 ± 0.00095	0.01643	
$A_l(P_\tau)$	0.1465 ± 0.0032	0.1480	
R_b	0.21629 ± 0.00066	0.21579	
R_c	0.1721 ± 0.0030	0.1723	
$A^{0,b}_{fb}$	0.0992 ± 0.0016	0.1038	
$A^{0,c}_{fb}$	0.0707 ± 0.0035	0.0742	
A_b	0.923 ± 0.020	0.935	
A_c	0.670 ± 0.027	0.668	
$A_l(SLD)$	0.1513 ± 0.0021	0.1480	
$\sin^2\theta^{lept}_{eff}(Q_{fb})$	0.2324 ± 0.0012	0.2314	
m_W [GeV]	80.410 ± 0.032	80.377	
Γ_W [GeV]	2.123 ± 0.067	2.092	
m_t [GeV]	172.7 ± 2.9	173.3	

Figure 1. The observables of the SM electroweak test compared with the values predicted from the fit. The pulls are also graphically shown and display a consistent picture of the Standard Model. The largest deviation is found for A^b_{FB}, slightly below 3σ.

directly related to the ratio of axial and vector couplings $A_f = 2\frac{g^f_V/g^f_A}{1+(g^f_V/g^f_A)^2}$. In the SM, $g^f_V/g^f_A = 1 - 4|Q_f|\sin^2\theta^f_{\text{eff}}$, where Q_f is the fermion charge and $\sin^2\theta_{\text{eff}}$ the effective weak mixing angle, defined as the weak mixing angle including the radiative corrections.

In order to relate the measurements to the fundamental constants of the SM, a simple parameter set is chosen. This includes the fine structure constant $\alpha(0)$, the strong coupling α_s, the mass of the Z boson m_Z and the Fermi constant G_F (related in practice to the W boson mass). All fermion masses are neglected except the top quark mass m_{top}. The Higgs boson mass m_{Higgs} plays a special role, due to its contribution to the radiative corrections. The observables are corrected for experimental effects and compared to the predictions from the Standard Model $O(\alpha, \alpha_s, m_Z, G_F, m_{\text{top}}, m_{\text{Higgs}})$. The QCD and electroweak radiative corrections are needed to match the experimental accuracy. The observables are therefore calculated with a precision beyond two loops. The fermion couplings $g^f_{A,V}$ that enter most of the observables depend via the radiative correction logarithmically on m_{Higgs} and quadratically on m_{top}. This dependence allows the indirect determination of m_{Higgs} and m_{top}.

The running with energy of the electromagnetic coupling has to be taken into account in the radiative corrections. The running can be calculated analytically with high precision for the photon vacuum polarisation induced by the leptons and by the top quark. In constrast, the contribution due to the five light quark flavours $\Delta\alpha^{(5)}_{\text{had}}$ is non-perturbative and has to be deduced from the measured $e^+e^- \rightarrow$ hadrons cross section via the dispersion relations. The determination of $\Delta\alpha^{(5)}_{\text{had}}$ has been recently updated[3] by including the new data from the ρ resonance measured by CMD-2[4] and KLOE[5]. Despite a precision improvement by more than a factor of two in the ρ region, the impact on $\Delta\alpha^{(5)}_{\text{had}}$ precision is modest. QCD based assumptions may lead to an improved accuracy of the $\Delta\alpha^{(5)}_{\text{had}}$ extraction[6]. More precision measurements of hadron production in e^+e^- collisions at low energy (in preparation) can bring significant improvements for the consistency checks of the SM. In addition, the hadronic vacuum polarisation estimates based on this data are also of high interest for the $(g-2)_\mu$ measurement, for which a 2.7σ discrepancy between the observation and the theory persists[6,7].

The list of the measured observables, using latest input from the LEP and SLC experiments[1] is shown in figure 1. The fit of the observables in the SM framework is taken as a consistency check of the SM. The pulls of the observables plus the $\Delta\alpha^{(5)}_{\text{had}}$ are also shown in figure 1. The picture dis-

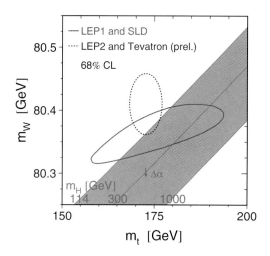

Figure 2. The comparison of the direct measurements and indirect determinations of the top quark and W boson masses. The band indicates the SM constraint from the G_F precise measurement for a range of Higgs masses. The confidence domain from the LEP1 and SLD data is compared with the direct measurement from LEP2 and Tevatron.

plays both the tremendous precision achieved by the electroweak tests and also the very good consistency of the SM. The most significant deviation, close to 3σ, is given by the forward-backward asymmetry of the b-quarks, A^b_{FB}. For this combined measurement, the individual values from various experiments and using different methods show very consistent results.

Subtracting the visible partial widths $\Gamma_{\ell,b,c}$ deduced form the individual cross section from the total Z width measured form the line shape, an invisible width can be deduced. Assuming that the invisible width is due to neutrinos with the same couplings as predicted by the SM, the number of neutrino flavours is determined to be $N_\nu = 2.9840 \pm 0.0082$, in agreement with the SM expectation of three fermion generations.

An important ingredient for the SM consistency check is the direct measurement of the top quark mass. Together with the direct

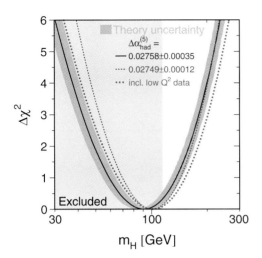

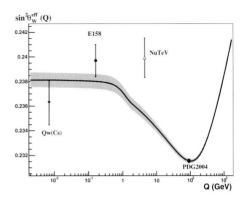

Figure 4. The measurements of effective mixing angle at various energies (Q) compared with the theoretical prediction.

Figure 3. The χ^2 of the SM fit, including the measured top quark mass, as a function of the Higgs boson mass. The results using two different estimations of $\Delta\alpha_{had}^{(5)}$ are shown together with the χ^2 of the fit including the low energy data, described in the section 3.2. The theoretical error is indicated as a band.

determination of the W boson mass, to be discussed later, it consitutes a powerful test of the SM consistency. The new techniques and data samples from Run II improved the measurement of the top quark mass at Tevatron. The new average[8], including recent measurement by the CDF and D0 collaborations is $m_{top} = 172.7 \pm 2.9$ GeV, a measurement which displays a dramatic improvement with respect to the previous error of 4.3 GeV (before the summer 2005). The comparison of the measured and fitted top and W masses is shown in figure 2. The direct and indirect determinations are in agreement and favour low m_{Higgs}. Due to close connections between the electroweak correction involving the top quark and Higgs boson, the precison of the top mass measurement is crucial for the indirect constraints on the Higgs mass. The new fit of the Higgs boson mass from the electroweak model is shown in figure 3. The new value of the fitted Higgs mass is $M_H = 91^{+45}_{-32}$ GeV with an upper limit within the SM

formalism $M_H < 186$ GeV at 95% CL. When the direct lower limit $M_{Higgs} > 114$ GeV is taken into account, the upper bound is found to be $M_H < 219$ GeV at 95% CL.

3.2 The electroweak precision measurements at lower energies

Measurements of parity violation in the highly forbidden 6S-7S transition in Cs offer a way to test with high precision the SM[11]. Due to its specific configuration with one valence electron above compact electronic shells, the theoretical calculation of the transition amplitude achieves 0.5% precision which allows the extraction with low ambiguities of the value for the nuclear weak charge[12]. The weak charge of the nucleus depends linearly on the weak charges on the u and d quarks contained by the nucleons and interacting with the valence electron via γ/Z in the t-channel. The weak charge Q_w can therefore be written as a function of the effective weak mixing angle that can be measured in this way at very low (atomic-like) energies.

The weak interactions can be tested at low energies via parity violating reactions. Using the end of beam at the SLC, polarized electrons are scattered off unpolarised atomic electrons (E158 expriment). The polarised Moller scattering $e^-e^- \to e^-e^-$ offers the opportunity to extract $\sin\theta_{\text{eff}}$ from cross section helicity asymmetry. This observable is related to the weak mixing angle that can be inferred with high precision[10] at an energy of 160 MeV.

A classical way to access the weak sector at any energy is to measure neutrino induced processes. In the case of neutrino-nucleon scattering studied by the NuTeV experiment[9], the ratio of neutral current to charged current cross sections $R_\nu = \sigma_{\text{CC}}/\sigma_{\text{NC}}$ is sensitive to the effective weak mixing angle, but subject to many systematic uncertainties related to the nucleon structure. Using both neutrino and anti-neutrino beams, the experimental results are combined using the Pachos–Wolfenstein method $R_- = (\sigma_{\text{NC}}^\nu - \sigma_{\text{NC}}^{\bar\nu})/(\sigma_{\text{CC}}^\nu - \sigma_{\text{CC}}^{\bar\nu})$, for which large cancellations of systematical errors are expected. This ratio accesses the effective weak mixing angle and is also sensitive to the neutrino and quark weak couplings. When SM couplings are assumed, the mixing angle measured by NuTeV is different from the SM prediction at 3.2σ level. Missing pieces in either the theoretical prediction or in the theory error associated to the measurement are still under investigation[14].

The measurement of the effective weak mixing angle at high and low energy can be used to test the electroweak running[13]. The result is shown in figure 4. Good agreement is found with the theoretical prediction, except for the NuTeV measurement discussed above. Precise measurements at energies beyond M_Z, as expected at the next e^+e^- collider will test the predicted increase of $\sin^2\theta_W^{\text{eff}}$ with energy.

3.3 The direct measurement of the running of α

The running of the electromagnetic coupling has been observed by the OPAL experiment using low angle Bhabba scattering $e^+e^- \rightarrow e^+e^-$ [15]. The scattered elec-

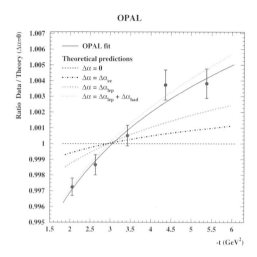

Figure 5. $|t|$ spectrum normalized to the theoretical prediction for a fixed coupling ($\Delta\alpha = 0$).

trons and positrons are detected close to the beampipe by two finely segmented calorimeters that allow the measurement of the scattering angles. The transfered momentum t is therefore reconstructed and the variation of the cross section as a function of t can be measured. The cross section is directly proportional to the square of the electromagnetic coupling and inversely proportional to t^2. The electromagnetic coupling is expected to run with the collision scale, given by t. The t spectrum normalised to the theoretical prediction for a fixed coupling is shown in figure 5. The difference of the measured event rates in t bins and the theoretical prediction for no α running shows a clear dependence on t. This evidence at 5σ level is compatible with the interpretation of $\alpha(t)$ running. When the pure electromagnetic running is taken into account in the theory, the remaining difference can be attributed to the hadronic component running that is in this way directly measured at 3σ level.

4 The weak bosons production and properties

4.1 The production of W and Z bosons

The weak boson production mechanisms at LEP2 and Tevatron provide a test of the SM. In addition, the weak boson samples can be used to study their decay properties and further constrain the Standard Model.

W pair production at LEP has been studied as a function of the centre-of-mass energy. The production cross-section variation with energy, flattens off at values around 15 pb, as expected from the SM, including the triple boson coupling ZWW. This behaviour is therefore directly related to the gauge structure of the Standard Model and constitutes an evidence for the non-abelian internal symmetry of the electroweak sector.

The W and Z bosons can be singly produced in $p\bar{p}$ collisions at Tevatron via the Drell-Yan process $q\bar{q} \rightarrow W$. The production crosss section is sensitive to the parton distribution functions. From this point of view, the production mechanism is also a convenient test ground for QCD, since the radiative corrections apply only to colliding partons and decouple from the produced bosons. The W and Z production cross sections measured in $p\bar{p}$ collisions at Tevatron are measured using the leptonic decay channels in e, μ or τ. The results[16] obtained from Run I ($\sqrt{(s)} = 1.8$ TeV) and Run II ($\sqrt{(s)} = 1.96$ TeV) are shown in figures 6 and show a good agreement with the NNLO calculation[17].

4.2 The W mass, width and branching ratios

The W harvest is also used to study the W properties like the mass, the branching ratios and the width. The latest world average between the LEP and the Tevatron Run I measurements yields a $M_W = 80.410 \pm 0.032$ GeV. The direct measurement agrees with the indirect determination from the

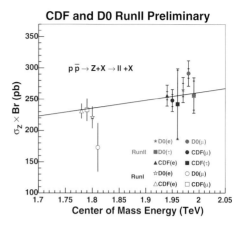

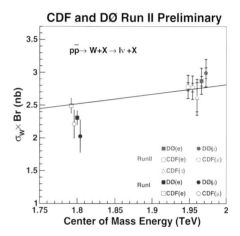

Figure 6. Measurements of the W and Z boson production cross section as a function of $\sqrt{s}$ at Tevatron. Data from CDF and D0 experiments obtained from various channels are compared with the theoretical prediction based on a NNLO QCD calculation.

LEP and SLD electroweak fit, including the M_{top} constraint $M_W = 80.364 \pm 0.021$ GeV. The average include a recent final result published by the OPAL Collaboration[18], for which a careful evaluation of the main systematical errors related to the colour reconnection and Bose–Einstein correlations together with an increased data sample allowed an improvement of the sytematical error from 70 to 56 MeV. This final precision of one LEP experiment is already better than the one of the combined result from Tevatron Run I and UA2 measurements[19] $M_W^{\text{runI+UA2}} =$

80.456 ± 0.059 GeV.

W pair production at LEP is a favourable configuration to measure W branching fractions. The decay to electron channel and to muon channel are found to be in very good agreement. However, the tau decay branching fraction is measured consistently by the four LEP experiments higher than the averaged electron and muon channel. This effect at 2.9σ is for the time being one of the largest deviations in the SM precision tests. Future measurements of $W \to \tau$ branching ratio are expected from Tevatron.

The W width can be measured directly from the invariant mass spectrum at LEP, where high precision can be achieved via a kinematic fit based on the energy-momentum conservation. At Tevatron, where only the leptonic channel is measurable, the transverse mass spectrum is sensitive to the W width in the tail at high mass. Finally, an indirect determination can be achieved exploiting the ratio of the W and Z cross section and using the precisely measured Z parameters and the theoretical prediction of the cross section ratios, for which most of the QCD uncertainties cancel out.

The present average of direct determination from LEP and Tevatron Run I is $\Gamma_W^{\mathrm{direct}} = 2.123 \pm 0.067$, while the indirect determination from Run I data is 2.141 ± 0.057. A recent direct determination[20] from Run II data by D0 still display large errors 2.011 ± 0.136 GeV, while an indirect determination using the cross section ratio measured by CDF already improves the Run I value 2.079 ± 0.041 GeV. The value obtained from the LEP1 and SLD electroweak fit is extremely precise 2.091 ± 0.002 GeV and in agreement with the direct and indirect determinations from LEP and Tevatron.

4.3 The A_{FB} from e^+e^- production at Tevatron.

The measurement of lepton pair production at Tevatron, produced via the Drell–Yan process $q\bar{q} \to \ell^+\ell^-$ provides complex information about both the proton structure and the electroweak effects in new energy domain. In particular, electron pair production can be used to measure the forward–backward asymmetry as a function of the pair mass. The result obtained by the D0 collaboration[21] is shown in figure 7. The characteristic change of sign is observed around the Z mass, similar to the much more precise measurement from LEP. From the measured asymmetry, the effective weak mixing angle is extracted by CDF[22] $\sin^2\theta_W^{\mathrm{eff}} = 0.2238 \pm 0.0050$, in good agreement with the value measured at LEP from the forward–backward asymmetry 0.2324 ± 0.0012. At large invariant masses, the deviation from the SM prediction may indicate the production of a heavier neutral boson Z', in case it has similar couplings to fermions as in the SM.

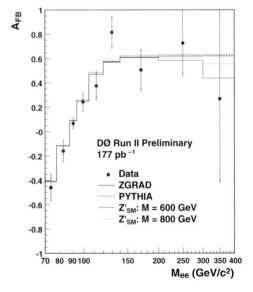

Figure 7. The A_{FB} as a function of the e^+e^- invariant mass measured at Tevatron.

Table 1. Summary of the results of searches for events with isolated leptons, missing transverse momentum and large hadronic transverse momentum p_T^X at HERA. The number of observed events is compared to the SM prediction. The $W^{\pm}$ component is given in parentheses in percent. The statistical and systematic uncertainties added in quadrature are also indicated.

obs./exp.(W)		Electron	Muon	Tau$^{H1:105\ pb^{-1}}$
H1	Full sample	25 / 20.4 $\pm$ 2.9 (68%)	9 / 5.4 $\pm$ 1.1 (82%)	5 / 5.8 $\pm$ 1.4 (15%)
211 pb^{-1}	$p_T^X > 25 GeV$	11 / 3.2 $\pm$ 0.6 (77%)	6 / 3.2 $\pm$ 0.5 (81%)	0 / 0.5 $\pm$ 0.1 (49%)
ZEUS	Full sample	24 / 20.6 $^{+1.7}_{-4.6}$ (17%)	12 / 11.9 $^{+0.6}_{-0.7}$ (16%)	3 / 0.40 $^{+0.12}_{-0.13}$ (49%)
130 pb^{-1}	$p_T^X > 25 GeV$	2 / 2.90 $^{+0.59}_{-0.32}$ (45%)	5 / 2.75 $^{+0.21}_{-0.21}$ (50%)	2 / 0.20 $^{+0.05}_{-0.05}$ (49%)
106 pb$^{-1}_{new}$	$p_T^X > 25 GeV$	1 / 1.5 $\pm$ 0.2 (78%)		

4.4 Rare W and Z production processes

At LEP2, in contrast to W pair production, single boson production (W or Z) is a rare process with cross sections below 1 pb. The final state contains four fermions, with only one fermion pair consistent with the boson mass. The comparison to the Standard Model provides a test in a low density phase space region, where new phenomena can occur. The cross section is typically 0.6-0.9 pb for single W production and 0.5–0.6 pb for single Z production at $\sqrt{s} = 182 - 209$ GeV.

At Tevatron, where weak bosons are massively singly produced, boson pair occur with a much lower rate. The associated $W\gamma$ or $Z\gamma$ production processes, with the weak bosons decaying into leptons, have cross sections close to 20 pb and 5 pb respectively[23]. The pair production of weak bosons is a particularly interesting process due to the spectacular final state and because it constitutes the main background for the search of the Higgs boson for $m_{Higgs} > 150$ GeV. While WW production has been measured[24,25], the search for WZ and ZZ production[26,27] have not been successful with the present luminosity and upper limits around $13-15$ pb at 95% CL have been calculated, for a total SM prediction of 5 pb.

The single W can also be produced in $e^{\pm}p$ collisions at HERA, with a cross section around 1 pb. The main production mechanism involves a fluctuation of photon emitted

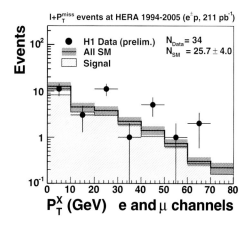

Figure 8. The transverse momentum of the hadronic system in events with isolated electrons or muons and missing P_T measured by the H1 experiment at HERA.

by the electron into a hadronic state, followed by the collision with the proton which leads to a qq' fusion into a W boson. In case of leptonic decay of the W, the final state consist of an high transverse momentum isolated lepton, missing transverse energy and possibly a low P_T hadronic system X. The H1 collaboration reported[28,29] the observation of such events and measured the cross section as a function of the hadrons transverse momentum (P_T^X). While a good agreement is observed at low P_T^X, a few spectacular candidates are observed at large P_T^X. The events continue to be observed at HERA II[30] by the H1 Collaboration which has analysed a total sample corresponding to an integrated luminosity of 211 pb^{-1}. The distribution of the transverse momentum of the hadronic system is shown in figure 8, where the excess of observed events at large P_T^X is visible.

The ZEUS collaboration has investigated their data with a different analysis strategy, with less purity for the SM W signal, the full HERA I data set and observes some events at large P_T^X, but no prominent excess above the SM prediction[31]. Recent modified analysis of the electron channel only, performed using a similar amount of data (but combining partial HERA I and II data sets) also do not support the H1 observation[32]. ZEUS collaboration observes events with tau leptons, missing transverse momentum and large hadronic transverse momentum, while no such event is observed by H1. The results are summarized in table 1. More incoming data will help to clarify this issue, which is at present one of the most intriguing results from HERA.

5 The measurement of the electroweak effects at HERA

5.1 Combined QCD/Electroweak fit of DIS data

The deep inelastic collisions at HERA are classically used to extract the proton structure information[33,34]. More than 600 measurement points of the charged and neutral current double differential cross section $d\sigma^{\mathrm{CC,NC}}/dxdQ^2$, where x is the proton momentum fraction carried by the struck quark and Q^2 is the boson virtuality, have been used together with other (fixed target) measurements to extract the parton distribution functions. Due to the high ep centre–of–mass energy (320 GeV), the proton is investigated down to scales of 10^{-18}m. The point–like nature of quarks is tested in the electroweak regime, where the proton is "flashed" with weak bosons. This experimental configuration allows to separate quark flavours within the proton and to improve the precision with

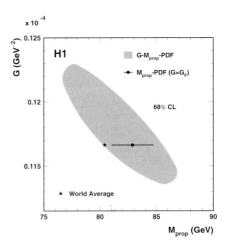

Figure 9. The allowed region at 65% CL in the plane $(G_F, M_W^{\mathrm{prop}})$ obtained from the combined electroweak–QCD fit of the DIS data.

which the parton distribution functions are extracted. Conversely, the electroweak sector can be investigated using the knowledge of the proton structure.

Recently, a consistent approach has been adopted by the H1 Collaboration[35], performing a combined QCD–electroweak fit. The strategy is to leave free in the fit the EW parameters together with the parameterisation of the parton distribution functions.

An interesting result is related to the so–called propagator mass M_W^{prop}, that enters a model independent parameterisation of the CC cross section:

$$\frac{d^2\sigma_{\mathrm{CC}}^{\pm}}{dxdQ^2} = \frac{G_F^2}{2\pi x}\left(\frac{M_W^2}{M_W^2 + Q^2}\right)^2 \tilde{\Phi}_{CC},$$

where G_F is the Fermi constant and $\tilde{\Phi}_{CC}$ is the reduced cross section that encapsulates the proton structure in terms of parton distribution functions. If the Fermi constant G_F and the propagator mass are left free in the fit, an allowed region in the $(G_F, M_W^{\mathrm{prop}})$ plane can be measured. The result is shown in figure 9. By fixing G_F to the very precise experimental measurement, the propagator mass can be extracted

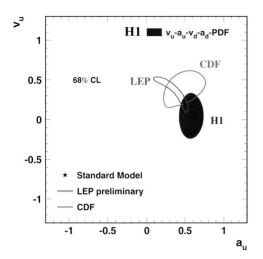

Figure 10. Axial and vector couplings of the u–quark measured from the combined electroweak–QCD fit at HERA and compared with measurements from LEP and Tevatron.

and amounts in this analysis to $M_W^{\text{prop}} = 82.87 \pm 1.82(\text{exp.})^{+0.30}_{-0.16}(\text{model})$ GeV, in agreement with the direct measurements.

If the framework of SM model is assumed, the W mass can be considered as a parameter constrained by the SM relations and entering both the cross section and the higher order correction. In this fitting scheme, where M_W depends on the top and Higgs masses, the obtained value from DIS is $M_W = 80.709 \pm 0.205(\text{exp})^{+0.048}_{-0.029}(\text{mod}) \pm 0.025(\text{top}) \pm 0.033(\text{th}) - 0.084(\text{Higgs})$ GeV, in good agreement with other indirect determinations and with the world average. The fit value can be converted into an indirect $\sin\theta_W$ determination using the relation $\sin^2\theta_W = 1 - \frac{M_W^2}{M_Z^2}$, assumed in the on mass shell scheme. The result $\sin^2\theta_W = 0.2151 \pm 0.0040_{exp.}{}^{+0.0019}_{-0.0011}|_{th}$, obtained for the first time in from $e^{\pm}p$ collisions, is in good agreement with the value of 0.2228 ± 0.0003 obtained from the measurements in e^+e^- collisions at LEP and SLC.

Due to the t-channel electron-quark scattering via Z bosons, the DIS cross sections at high Q^2 are sensitive to light quark ax-

ial (a_q) and vector (v_q) coupling to the Z. This dependence includes linear terms with significant weight in the cross section which allow to determin not only the value but also the sign of the couplings. In contrast, the measurements at the Z resonance (LEP1 and SLD) only access av or $a^2 + v^2$ combinations. Therefore there is an ambiguity between axial and vector couplings and only the relative sign can be determined. In addition, since the flavour separation for light quarks cannot be achieved experimentally, flavour universality assumptions have to be made. The Tevatron measurement[22] of the Drell-Yan process allows to access the couplings at an energy beyond the Z mass resonance, where linear contributions are significant. The measurements of the u–quark couplings obtained at HERA, LEP and Tevatron are shown in figure 10. The data to be collected at Tevatron and HERA as well as the use of polarized $e^{\pm}$ beams at HERA open interesting oportunities for the light quarks couplings measurements in the near future.

5.2 $e^{\pm}$ collision with polarised lepton beam

The polarisation of the electron beam at HERA II allows a test of the parity non-conservation effects typical for the electroweak sector. The most prominent effect is predicted in the CC process, for which the cross section depends linearly on the $e^{\pm}$– beam polarisation: $\sigma^{e^{\pm}p}(P) = (1 \pm P)\sigma^{e^{\pm}p}_{P=0}$. The results[36,37] obtained for the first time in $e^{\pm}p$ collisions are shown in figure 11. The expected linear dependence is confirmed and constitute supporting evidence for the V-A structure of charged currents in the Standard Model, a property already verified more than 25 years ago, by measuring the polarisation of positive muons produced from ν_{μ}–Fe scattering by the CHARM experiment[38].

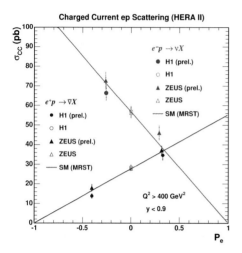

Figure 11. The dependence of the total CC cross section of the $e^{\pm}$-beam polarization at HERA.

6 Outlook

The present experimental activity towards electroweak measurements continues to provide increasing endurance tests for the Standard Model. The LEP analyses are final in many aspects and the results still play a key role in the present understanding of the electroweak symmetry breaking mechanism. The incoming data from Tevatron has good chance to take over in many aspects, especially concerning the weak boson properties, but also to extend the area of the measurements beyond LEP energies. At HERA, a consistent approach of the electroweak and QCD processes will certainly bring valuable information in the near future. The low energy measurements provide not only a cross check but also a solid testing ground for the electroweak sector, for which surprises are not excluded.

The future colliders will test the electroweak sector with high precision[39]. At the LHC, the electroweak physics will mainly profit from the huge increase in the weak boson production cross-section. In the foreseen experimental condition the precision on the W mass measurement should approach 15 MeV while the top quark mass will be measured at 1 GeV level. The next $e^{+}e^{-}$ linear collider will improve the precision on M_W to below 10 MeV while the top mass will be measured to 100 MeV. Similarly to the present situation, the precise measurements of the electroweak sector will allow to set indirect limits on the new physics, that might well be beyond the direct reach of the future colliders.

Acknowledgments

I would like to thank the following people for kind assistance during the preparation of this contribution: Max Klein, Martin Grunewald, Pippa Wells, Dmitri Denisov, Jan Timmermans, Bolek Pietrzyk, Emmanuelle Perez, Matthew Wing, Richard Hawkings, David Waters, Chris Hays and Dave South.

References

1. The LEP Collaborations, the LEP Electroweak Working Group, the SLD Electroweak, Heavy Flavour Groups, hep-ex/0412015, update for summer 2005 at http://lepewwg.web.cern.ch/LEPEWWG
2. The LEP Collaborations, the SLD Collaboration, the LEP Electroweak Working Group, the SLD Electroweak, Heavy Flavour Groups, hep-ex/0509008
3. H. Burkhardt and B. Pietrzyk, LAPP-EXP 2005-02
4. R. R. Akhmetshin *et al.* [CMD-2 Collaboration], *Phys. Lett. B* **578** 285 (2004), [hep-ex/0308008].
5. A. Aloisio *et al.* [KLOE Collaboration], *Phys. Lett. B* **606** 12 (2005), [hep-ex/0407048].
6. J. F. de Troconiz and F. J. Yndurain, *Phys. Rev. D* **71**, 073008 (2005), [hep-ph/0402285].
7. G. W. Bennett *et al.* [Muon g-2 Collaboration], *Phys. Rev. Lett.* **92**, 161802 (2004), [hep-ex/0401008].

8. [The CDF and D0 Collaborations], "Combination of CDF and D0 results on the top-quark mass", hep-ex/0507091.

9. G. P. Zeller *et al.* [NuTeV Collaboration], *Phys. Rev. Lett.* **88** 091802 (2002); [Erratum-ibid. 90 (2003) 239902] [hep-ex/0110059].

10. P. L. Anthony *et al.* [SLAC E158 Collaboration], hep-ex/0504049.

11. C.S. Wood *et al.*, *Science* **275** 1759 (1997).

12. J. S. M. Ginges and V. V. Flambaum, *Phys. Rept.* **397**, 63 (2004), [physics/0309054].

13. A. Czarnecki and W. J. Marciano, *Int. J. Mod. Phys. A* **15** 2365 (2000), [hep-ph/0003049].

14. J. T. Londergan, *AIP Conf. Proc.* **747**, 205 (2005).

15. G. Abbiendi *et al.* [OPAL Collaboration], hep-ex/0505072.

16. D. Acosta *et al.* [CDF II Collaboration], *Phys. Rev. Lett.* **94**, 091803 (2005), [hep-ex/0406078].

17. R. Hamberg, W. L. van Neerven and T. Matsuura, *Nucl. Phys. B* **359**, 343 (1991); [Erratum-ibid. B **644**, 403 (2002)].

18. G. Abbiendi [OPAL Collaboration], hep-ex/0508060.

19. V. M. Abazov *et al.* [CDF Collaboration], *Phys. Rev. D* **70**, 092008 (2004), [hep-ex/0311039].

20. Oliver Stelzer-Chilton, for the CDF and D0 Collaborations, hep-ex/0506016.

21. D0 Collaboration, Note 4757-CONF7, http://www-d0.fnal.gov/Run2Physics/WWW/results/ew.htm

22. D. Acosta *et al.* [CDF Collaboration], *Phys. Rev. D* **71**, 052002 (2005), [hep-ex/0411059].

23. D. Acosta *et al.* [CDF Collaboration], *Phys. Rev. Lett.* **94**, 041803 (2005), [hep-ex/0410008].

24. D. Acosta *et al.* [CDF Collaboration], *Phys. Rev. Lett.* **94**, 211801 (2005), [hep-ex/0501050].

25. V. M. Abazov *et al.* [D0 Collaboration], *Phys. Rev. Lett.* **94**, 151801 (2005), [hep-ex/0410066].

26. V. M. Abazov *et al.* [D0 Collaboration], *Phys. Rev. Lett.* **95**, 141802 (2005), [hep-ex/0504019].

27. D. Acosta *et al.* [CDF Collaboration], *Phys. Rev. D* **71**, 091105 (2005), [hep-ex/0501021].

28. C. Adloff *et al.* [H1 Collaboration], *Eur. Phys. J. C* **5**, 575 (1998), [hep-ex/9806009].

29. V. Andreev *et al.* [H1 Collaboration], *Phys. Lett. B* **561**, 241 (2003), [hep-ex/0301030].

30. V. Andreev *et al.* [H1 Collaboration], EPS2005 abstract 637, http://www-h1.desy.de/h1/www/publications/htmlsplit/H1prelim-05-164.long.html.

31. S. Chekanov *et al.*, [ZEUS Collaboration] *Phys. Lett.* B**559** 153 (2003).

32. S. Chekanov *et al.* [ZEUS Collaboration], abstract 257, http://www-zeus.desy.de/physics/phch/conf/lp05_eps05/

33. C. Adloff *et al.* [H1 Collaboration], *Eur. Phys. J. C* **30**, 1 (2003), [hep-ex/0304003].

34. S. Chekanov *et al.* [ZEUS Collaboration], *Phys. Rev. D* **67**, 012007 (2003), [hep-ex/0208023].

35. A. Aktas *et al.* [H1 Collaboration], hep-ex/0507080.

36. V. Andreev *et al.* [H1 Collaboration], abstract 388, http://www-h1.desy.de/h1/www/publications/htmlsplit/H1prelim-05-042.long.html.

37. S. Chekanov *et al.* [ZEUS Collaboration], abstract 255, http://www-zeus.desy.de/physics/phch/conf/lp05_eps05/

38. M. Jonker *et al.*, *Phys. Lett. B* **86**, 229 (1979).

39. G.Weiglein *et al.* [The LHC/LC Study Group], hep-ph/0410364

DISCUSSION

Stephen Olsen (Uni. of Hawaii):

The pi-beta experiment at PSI report a large discrepancy with SM predictions for radiative π_{e2} decay. Is there some reason why we should not be concerned about this?

Cristinel Diaconu: In this measurement[b] three phase space cases are investigated. A discrepancy between the measured branching ratio and the theoretical prediction is observed in one phase space region, while two other measurement cases with the same apparatus found a good agreement. This situation indicates that more theoretical and experimental work is needed before a definitive conclusion can be drawn.

Lee Roberts (Boston University):

As you mentioned our muon (g-2) experiment differs from the Standard Model value by 2.7 standard deviations. We have proposed an upgrade experiment to improve the accuracy from 5 parts in 10^7 to 2 parts in 10^7. We have received scientific approval but no funding yet.

Harvey Newman (Caltech):

Is there a prospect of a new W mass measurement from Run II that will significantly affect the world average on M_W?

Cristinel Diaconu: The measurement of the W mass at Tevatron is in progress[20]. First estimation of the precision obtained with the present data sample is of 76 MeV. The incoming data should allow to considerably improve this precision and challenge the existing world average.

---

[b]E. Frlez *et al.* *Phys. Rev. Lett.* **93**, 181804-1-4 (2004)

SEARCH FOR HIGGS AND NEW PHENOMENA AT COLLIDERS

STEPHAN LAMMEL

Fermi National Accelerator Laboratory, Batavia, Illinois 60510, USA
E-mail: lammel@fnal.gov

The present status of searches for the Higgs boson(s) and new phenomena is reviewed. The focus is on analyses and results from the current runs of the HERA and Tevatron experiments. The LEP experiments have released their final combined MSSM Higgs results for this conference. Also included are results from sensitivity studies of the LHC experiments and lepton flavour violating searches from the B factories, KEKB and PEP-II.

1 Introduction

A scalar Higgs particle[1] has been postulated over 30 years ago as the mechanism of electroweak symmetry breaking in the Standard Model (SM) of particle physics. This spontaneous breaking introduces a huge hierarchy between the electroweak and Planck scales that is unsatisfying. Extensions to the SM have been proposed over the years to avoid unnatural fine-tuning. Supersymmetry[2] (SUSY) is one such attractive extensions. Depending on its internal structure and SUSY breaking mechanism, a variety of new phenomena are expected to be observed. Rare signatures, as in high-mass tails or from SM suppressed processes, are good places for generic beyond–the–Standard Model searches.

The Large Electron Positron (LEP) collider at CERN completed operation about four years ago. It ran at a center–of–mass energy of up to 209 GeV and delivered about 1 fb^{-1} of data to the four experiments, ALEPH, DELPHI, L3, and OPAL. The data are analysed. Extensive searches for Higgs and new phenomena have come up negative. For many new particles coupling to the Z boson LEP still holds the most stringent limits.

Two machines, the Hadron Electron Ring Accelerator (HERA) and the Tevatron, are currently running at the energie frontier with ever increasing luminosities. HERA at DESY collides electrons or positrons with protons at a center–of–mass energy of 319 GeV. The HERA upgrade increased the luminosity by a factor of 4.7. So far the machine has delivered over 180 pb^{-1} of electron–proton (about half) and positron–proton data to the two experiments, H1 and ZEUS. The experiments are particularly sensitive to new particles coupling to electron/positron and up/down quarks. HERA II can also deliver polarized lepton beams.

The Fermilab Tevatron collides proton and antiprotons at a center–of–mass energy of 1.96 TeV. Luminosity upgrades are continuing. So far the machine has delivered over 1 fb^{-1} of data to the two experiments, CDF and DØ. The improved detectors, higher cener–of–mass energy, and ten fold increase in luminosity enable the experiments not only to significantly extend previous searches but provide them with a substantial discovery potential.

The Large Hadron Collider (LHC) at CERN and the International Linear Collider (ILC) are machines under construction and in the planning phase. The LHC will collide protons with protons at a center–of–mass energy of 14 TeV. First collisions are expected in 2007. The two experiments, AT-LAS and CMS, have made detailed studies of their reach to new physics. LHC is expected to boost our sensitivity to new physics by an order of magnitude in energy/mass. The ILC will collide electrons and positrons with a center–of–mass energy of several hundred

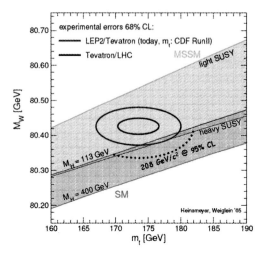

Figure 1. One sigma contours of the current W and top mass measurements compared to SM and MSSM Higgs masses. Plot from Heinemeyer[4] updated for new CDF top mass measurment.

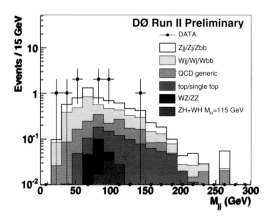

Figure 2. Dijet mass spectrum of the DØ ZH analysis after two b-tags.

GeV. It will be the next generation machine for precision measurments, like LEP was.

1.1 Precision Electroweak and Top Measurments

Precision electroweak measurements allow us to check the SM for consistency or derive the mass of the unknown Higgs particle. For Higgs prediction, the W boson mass and top quark mass are key ingredients. With the new preliminary CDF Run II top mass measurment[3], the world average is pulled down to $m_t = 174.3 \pm 3.4 \, \text{GeV}/c^2$. Figure 1 shows the 1σ and 95% confidence level (CL) contours of the W and top mass with overlaid Higgs mass. Current measurements put the SM Higgs below $208 \, \text{GeV}/c^2$ at 95% CL.

However, the top mass is an even more important ingredient for the Higgs in Minimal Supersymmetric extensions of the Standard Model (MSSM). The MSSM exclusion at low $\tan(\beta)$ derived from the SM Higgs limit of LEP depends very sensitively on the mass of the top quark and vanishes when the top mass is large.

2 Standard Model Higgs

The current lower limit on the Higgs mass of $114.4 \, \text{GeV}/c^2$ at 95% CL comes from the LEP experiments[5]. They did a fantastic job of pushing the Higgs mass limit well above the Z pole where it would be hard for proton–antiproton experiments to detect. The Tevatron is the current place for Higgs searches with an expected sensitivity to about $130 \, \text{GeV}/c^2$. Here the main Higgs production mechanism is via gluon-gluon fusion. Associated production with a W or Z has a factor five lower cross-section. For low Higgs masses, below $135 \, \text{GeV}/c^2$, the $b\bar{b}$ decay mode is dominant. With a leptonic W or Z decay we get signatures of zero, one, or two charged leptons, an imbalance of energy in the transverse plane, missing E_T (in case of zero or one charged lepton), and two b-jets. For heavier Higgs the WW^* decay dominates and then Higgs production via gluon fusion yields a viable signature.

The WH analyses of CDF and DØ were performed early on and results are updated regularly with increased luminosity[6]. The DØ experiment has also completed a search in the ZH channel where the Z decays into neutrinos[7]. The analysis compares the missing E_T ($\not{E}_T$) as calculated from all energy in

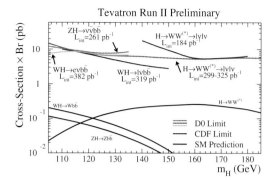

Figure 3. Current CDF and DØ Run II Higgs cross-section times branching ratio limits from the WH, ZH, and WW* channel.

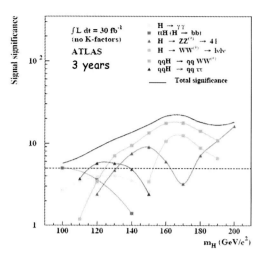

Figure 4. SM Higgs signal significance of the ATLAS experiment for the different search channels.

the detector with the calculation from just clustered energy and the jet energy vector sum with the track momentum vector sum to reduce instrumental background which comes mainly from jet mismeasurments. The main background in the analysis comes from Z plus multijet production and W plus $b\bar{b}$ production with W decay into $\tau\nu$ decay. Figure 2 shows the dijet mass spectrum after two b-tags are required. No excess of events over background expectation is observed in this search nor in any other Higgs analysis of CDF and DØ. The cross-section times branching ratio limit of this analysis is shown in Fig. 3 together with the limits from the other Tevatron SM Higgs searches.

The sensitivity of CDF and DØ is currently between 3 and 10 pb while a SM Higgs is at about 0.2 pb. The difference between the current and the final Run II Higgs sensitivity projection[8] is understood. In addition to the luminosity accumulation, improvements in lepton and b-tagging acceptance, the dijet mass resolution, and analysis techniques will bring the sensitivity of the experiments to the projections made before Run II.

At LHC the Higgs production cross-section is huge. Even a decay mode with small branching ratio, like Higgs into a photon pair, yields a sizable event number. The two experiments each have an electromag-

netic calorimeter with very precise energy resolution to be able to observe a diphoton mass bump from Higgs[9] on top of the huge diphoton continuum. For LHC vector boson fusion, however, will be the most important production for Higgs. Both ATLAS and CMS can observe a SM Higgs up to several hundred GeV/c^2 after a few years of running, Fig. 4. For LHC the observation of a Higgs boson would be just the initial step. The two experiments can measure the ratio of couplings and decay widths to an uncertainty of 20 and 30%.

3 MSSM Higgs

Current and next generation experiments cover a SM Higgs well. The Higgs sector, however, can be richer than a single doublet. Supersymmetry extends the symmetry concept, that has been so successfull in particle physics, to the spin sector. It provides a consistent framework for gauge unification and solves the hierarchy problem of the SM. No SUSY particles have been observed so far. Several SUSY breaking scenarios are under consideration which determine the SUSY structure. The MSSM is the general mini-

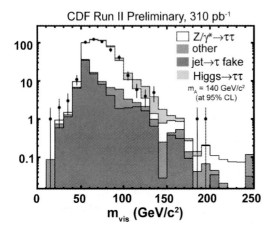

Figure 5. Distribution of the "visible" Higgs mass in the CDF ditau analysis.

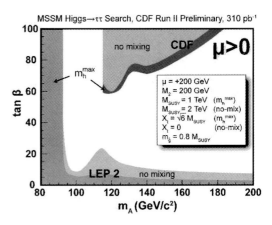

Figure 6. Excluded regions in the mass of A versus $\tan(\beta)$ plane for the m_h^{max} and no mixing scenario for the Higgsino mass mamareter $\mu > 0$.

mal supersymmetric extension of the SM. It has two Higgs doublets yielding five physical Higgs particles: h, H, A, H^+, and H^-. At tree level the Higgs sector is described by two parameters, the pseudoscalar Higgs mass, m_A, and ratio of the two Higgs vacuum expectation values, $\tan(\beta)$. The MSSM, although the minimal extension, has a lot of free parameters. One normally uses models constrained based on SUSY breaking scenarios and GUT scale relations or special benchmarking models.

At the Tevatron the Higgses of the MSSM are of particular interest. The Yukawa coupling to down-type fermions is enhanced, boosting the cross-section by a factor of $\tan(\beta)^2$. For large $\tan(\beta)$ the pseudoscalar Higgs and either h or H are expected to be almost mass degenerate. The branching ratio into $b\bar{b}$ is at around 90% independent of mass. Decays into tau pairs account for close to 10%.

Two neutral MSSM Higgs searches are performed at CDF and DØ. The first is based on Higgs plus $b\bar{b}$ production: $b\bar{b}A \rightarrow b\bar{b}b\bar{b}$. It yields a striking four b-jet signature. The second search is based on the tau decay mode: $A \rightarrow \tau^+\tau^-$.

Tau leptons are not as easily identified as electrons or muons. The CDF analysis[10] is based on one leptonic tau decay and one hadronic tau decay. Jets from hadronic tau decays are very narrow, pencil like, compared to quark/gluon jets. CDF uses a double cone algorithm to identify hadronic tau decays. An efficiency of 46% is achieved with a misidentification rate between 1.5% to 0.1% per jet depending on the jet energy. For the Higgs search the experiment uses a data sample selected by an electron or muon plus track trigger to achieve high efficiency. Figure 5 shows the visible mass of the ditau system, calculated from the momentum vector of the lepton, hadronic tau, and $\not{E}_T$. The main background comes from Z and Drell-Yan ditau production. No excess of events is observed in the first $310\,\mathrm{pb}^{-1}$ of Run II data. A binned likelihood fit in the visible mass is used to set limits on the mass of A versus $\tan(\beta)$, Fig 6.

The DØ analysis[11] for the four b-jet channel requires three b-tag jets in the event. The first jet has to have $E_T > 35\,\mathrm{GeV}$ while the third can be as low as $15\,\mathrm{GeV}$. To estimate the background from light quark and gluon jets, the probability of mis-tagging a jet is measured on the three jet sample before b-tagging, subtracting any true heavy

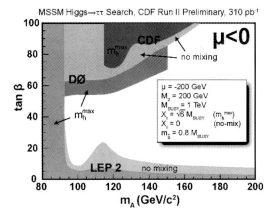

Figure 7. Excluded regions in the mass of A versus $\tan(\beta)$ plane for the m_h^{max} and no mixing scenario for the Higgsino mass marameter $\mu < 0$.

flavour contribution. Those mistag functions are then applied to the untagged jets in the double b-tag sample to get the shape of the multijet background to the triple b-tag sample. DØ determines the overall background normalization by fitting the dijet mass outside the hypothesized signal region in the triple b-tag sample. Figure 7 shows the mass versus $\tan(\beta)$ limit obtained by this analysis. For $\mu > 0$ the sensitivity of the four b-jet channel is very low due to the lower cross-section and lower branching ratio into $b\bar{b}$, while for the tau channel cross-section reduction and branching ratio enhancement compensate.

The final combined MSSM Higgs mass limits from the four LEP experiments have been released[12]. There are no signals of Higgsstrahlung or pair production. Sensitivity is evaluated in several benchmark models. A top mass of $179\,\mathrm{GeV}/c^2$ is assumed for all limits. Figure 8 shows the excluded mass of A versus $\tan(\beta)$ for the classic no-stop mixing benchmark model with $\mathrm{M_{SUSY}} = 1000\,\mathrm{GeV}/c^2$, $\mathrm{M_2} = 200\,\mathrm{GeV}/c^2$, $\mu = -200\,\mathrm{GeV}/c^2$, $\mathrm{m_{gluino}} = 800\,\mathrm{GeV}/c^2$, $A = 0 + \mu \cdot \cot(\beta)$. The excluded area is reduced in the case of stop mixing and is quite sensitive to the top mass. The LEP

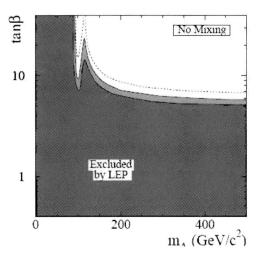

Figure 8. Excluded regions in the mass of A versus $\tan(\beta)$ plane for the no mixing scenario. Dark areas are excluded at over 99% CL, light areas at 95% CL.

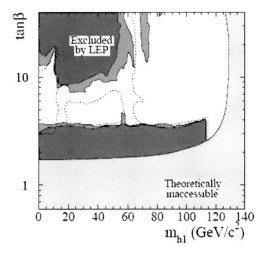

Figure 9. Excluded regions in the mass of h_1 versus $\tan(\beta)$ plane for the CP-violating scenario. Dark areas are excluded at over 99% CL, light areas at 95% CL.

34

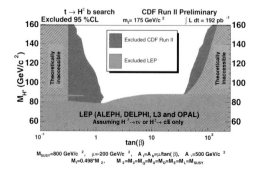

Figure 10. Excluded regions in the $\tan(\beta)$ versus mass of charged Higgs plane for a tevatron benchmarking scenario. The dark area is excluded at 95% CL, the dark line shows the expected limit with 1σ band.

Table 1. Observed events and expected number of background events in the six channels of the DØ chargino–neutralino analysis.

Channel	Expected	Observed
e e t	0.21 ± 0.12	0
e μ t	0.31 ± 0.13	0
$\mu\mu$ t	1.75 ± 0.57	2
$\mu^{\pm} \mu^{\pm}$	0.64 ± 0.38	1
e τ_{h} t	0.58 ± 0.14	0
$\mu \tau_{\mathrm{h}}$ t	0.36 ± 0.13	1
Total	3.85 ± 0.75	4

experiments also considered the case of CP-violation in the Higgs sector. Such a scenario appeals in explaining the cosmic matter–antimatter asymmetry. Experimentally such a scenario is much more challenging as the lightest Higgs can decouple from the Z. Figure 9 shows the LEP results. An inconsistency in the prediction from CPH and Feyn-Higgs for the $h_2 \rightarrow h_1 h_1$ branching ratio causes the hole at $\tan(\beta) \sim 6$ to open up[13].

There are also two charged Higgs particles in the MSSM. CDF uses its top cross-section measurements from the various decay channels to search for top decays into charged Higgs plus b-quark[14]. Such a decay would change the expected number of events differently in the dilepton, lepton plus single b-tag, lepton plus double b-tag, and lepton plus tau channel, especially for small and large $\tan(\beta)$ values. Figure 10 shows the excluded mass as function of $\tan(\beta)$ in one of the benchmark models studied.

4 Supersymmetry

From the LEP experiments[15] we know that the chargino has to be heavier than $103.5 \, \mathrm{GeV}/c^2$. At the Tevatron the cross-section for chargino–neutralino production is rather small. However, in R_{P} con-

serving minimal supergravity inspired SUSY (mSUGRA) one can get a very distinct signature. In case of leptonic chargino and neutralino decay, the event will contain only three charged leptons and missing E_{T} from the escaping neutrinos and the lightest SUSY particles (LSP). The challenge in the analysis is the charged lepton acceptance times efficiency since it enters with third power. For $\tan(\beta)$ values above 8 to 10, tau decays become significant and tau identification thus very important.

The DØ analysis[16] searches in six separate channels and combines the results. In all the channels known dilepton resonances are removed and a combined cut on the $\not{E}_{\mathrm{T}}$ and p_{T} of the third lepton used to suppress background from mainly misidentified leptons and diboson production. Table 1 shows the expected background and observed number of events in each of the six channels. In the $320 \, \mathrm{pb}^{-1}$ of data analysed, no excess is observed. DØ continues to set cross-section times branching ratio into three lepton limits. The analysis also improves the LEP chargino mass limit to $116 \, \mathrm{GeV}/c^2$ in case of light sleptons, i.e. small m_0.

The production cross-section of coloured SUSY particles is much larger than that of chargino–neutralino. The squarks of the first two generations are assumed to be degenerate in mass. Stop and sbottom quarks could be significantly lighter due to the large top

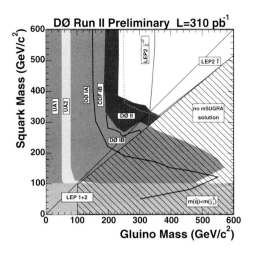

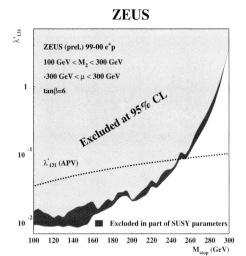

Figure 11. Excluded region in the gluino versus squark mass plane of the DØ missing E_T plus multi-jet analysis.

Figure 12. Excluded stop mass as function of λ'_{131} coupling of the ZEUS analysis.

Yukawa coupling. CDF and DØ have dedicated analyses for those[17]. The analyses assume direct decay of the third generation squark into LSP: $\tilde{b} \rightarrow b\tilde{\chi}_1^0$ or $\tilde{t} \rightarrow c\tilde{\chi}_1^0$. Both direct production of $\tilde{t}\bar{\tilde{t}}$ and $\tilde{b}\bar{\tilde{b}}$ and indirect production through, for instance, $\tilde{g} \rightarrow \tilde{b}\bar{b}$ are researched.

For the gluinos and squarks of the first two generation the signature depends on the mass hierarchy. If the squarks are lighter, squark production is dominant and squarks will decay via $\tilde{q} \rightarrow q\tilde{\chi}^0$ or $\tilde{q} \rightarrow q'\tilde{\chi}^\pm$ with $\tilde{\chi}^\pm \rightarrow q\bar{q}\tilde{\chi}_1^0$. In case the gluino is the lighter one, gluino pair production is dominant with $\tilde{g} \rightarrow q\bar{q}\tilde{\chi}^0$ or $\tilde{g} \rightarrow q\bar{q}'\tilde{\chi}^\pm$. With the jets from the chargino decay generally being softer, this yields a 2, 3, or 4 jet signature together with missing E_T. The DØ analysis[18] makes a preselection, vetoing jet back-to-back topologies, events with leptons, and events where the missing E_T is close in azimuthal angle to a jet. The main SM backgrounds are from W/Z plus multijet and QCD multijet production. After the preselection dedicated analyses are made for each jet multiplicity. In all three cases the observed events are explained by the background estimate. Fig-

ure 11 shows the new excluded region in gluino versus squark mass.

R_P conserving SUSY yields a natural dark matter candidate. However, R-parity conservation is really put into the models *ad hoc* and nature may not conserve it. In case of R_P violation ($\not{R}_P$), different signatures arise. In the case of a non-vanishing λ' coupling electrons and u/d-quarks can couple, ideal for HERA. In electron–proton mode, λ'_{11k} couplings are accessible while in positron–proton mode, λ'_{1j1} couplings would produce resonant $\tilde{u}_L$. Both H1 and ZEUS searched for a large variety of $\not{R}_P$ signatures. The most striking signature is "wrong" sign electrons, i.e. events with energetic electrons while in positron–proton mode and events with positrons, jets, and no missing E_T while in electron–proton mode. No signals of $\not{R}_P$ SUSY has been found. Figures 12 and 13 show the stop and sbottom mass limits as function of the coupling constant from ZEUS and H1. Squarks with R_P violating couplings of electroweak strength are excluded up to $275\,\mathrm{GeV}/c^2$.

In case SUSY is broken via gauge interactions (GMSB) the gravition acquires

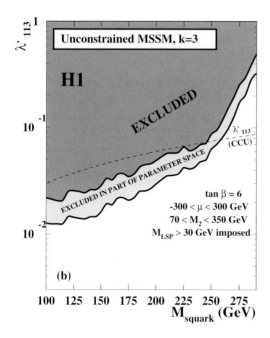

Figure 13. Excluded sbottom mass as function of λ'_{113} coupling of the H1 analysis.

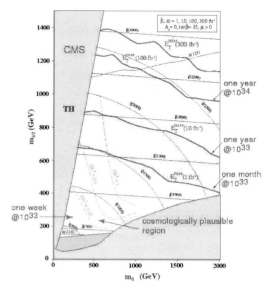

Figure 14. SUSY sensitivity of CMS in the m_0 versus $m_{1/2}$ plane for different integrated luminosity.

a small mass and becomes the LSP. The next-to-lightest SUSY particle will decay into a photon plus gravitino for the distinct GMSB photon signature. CDF and DØ have both searched in the diphoton plus $\not{E}_T$ channel[20]. The experiments use chargino–neutralino production as reference model for the search. The two experiment have combined their results from the first $250\,\mathrm{pb}^{-1}$ of data and exclude charginos in GMSB models below $209\,\mathrm{GeV}/c^2$ at 95% CL.

In the case of low energy supersymmetry, LHC will be a great machine[21]. It will provide a definite answer to the question and with a small luminosity of only a few month probe SUSY scales of over a TeV. But LHC can do more and measure sparticle masses, for instance, for the second lightest neutralino from the dilepton spectrum endpoint or even the gluino mass from the top–bottom endpoint.

The ILC[22], however, will be required for precision mass and coupling measurements.

5 Isolated Lepton and Missing Energy

In Run I of HERA H1 observed an excess of events with isolated lepton $p_T > 10\,\mathrm{GeV}$ and missing $E_T > 12\,\mathrm{GeV}$ beyond what one would expect from W production[23]. The excess was pronounced at large $p_T^X > 25\,\mathrm{GeV}$ and did not fit well any new physics model. Both H1 and ZEUS have searched for an isolated lepton plus $\not{E}_T$ signature in the new Run II data[24]. H1 has used the identical selection in the analysis of the new data, separately for positron–proton and electron–proton data. The muon channel shows no more the excess seen in Run I while an event excess remains in the electron channel. H1 has also analysed the tau data from Run I which show no excess either. ZEUS did not observe an event excess in Run I. With $40\,\mathrm{pb}^{-1}$ of Run II data analysed ZEUS finds also no excess in the electron channel. Table 2 shows the current results of all the searches.

Table 2. Expected and observed number of isolated electron, muon, and tau events of H1 and ZEUS.

	electron		muon		tau new	
	all	p_T^X>25GeV	all	p_T^X>25GeV	all	p_T^X>25GeV
H1 HERA I 118 pb^{-1}	11 11.54±1.50	5 1.76±0.30	8 2.94±0.50	6 1.68±0.30	5 5.8±1.36	0 0.53±0.10
H1 e$^+$p new 53 pb^{-1}	9 4.75±0.76	5 0.84±0.19	1 1.33±0.19	0 0.85±0.13		
H1 e$^-$p new 39 pb^{-1}	5 4.09±0.61	1 0.62±0.11	0 1.10±0.17	0 0.67±0.11		
ZEUS HERA I 130 pb^{-1}	24 20.6$^{+1.7}_{-4.6}$	2 2.90$^{+0.59}_{-0.32}$	12 11.9$^{+0.6}_{-0.7}$	5 2.75±0.21	3 0.40$^{+0.12}_{-0.13}$	2 0.20±0.05
ZEUS e$^+$p new 40 pb^{-1}	0 0.46±0.10	0 0.58$^{+0.08}_{-0.09}$				

6 High Mass Searches

High-mass searches were one of the first results presented from Run II of the Tevatron. New gauge bosons and other high mass resonances yield energetic objects when they decay. Searches based on energetic leptons, photons, and missing E_T give access to a large variety of new physics. For instance, events with an energetic electron and positron are sensitive to Z', large extra dimensions, Randall-Sundrum gravitons, R_P sneutrinos, and technicolor particles, ρ and ω. The analyses[25] of CDF and DØ are constantly refined, on one side to cover signatures in a generic way, by for instance calculating sensitivity based on the spin, or to incorporate new models and interpretations, like expressing Z' sensitivity based on d-xu, or B-xL couplings, on the other side to include additional event kinematics like $\cos\theta^*$ in the analysis to enhance sensitivity to new physics. About $450\,\mathrm{pb}^{-1}$ of Run II data are analysed for high mass objects. No excess or deviation are observed so far.

7 Indirect Searches

With no signals of new physics in any of the direct searches, we can search for signs of new physics where new particles are in virtual states. Processes that are rare in the SM provide an excellent place to search for

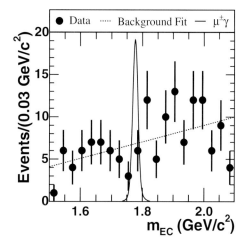

Figure 15. The energy constraint muon photon mass spectrum of the BaBar tau decay into muon plus photon analysis.

signs of new physics.

Tau decays into a muon and a photon are tiny in the SM with a branching ratio around 10^{-40} but allowed if one includes neutrino mixing. The decay violates lepton flavour which occures naturally in SUSY grand unified theories. Both Belle and BaBar[26] have recorded over 20 million ditau events. BaBar uses one tau as tag and then the other as probe. A neural network is used to discriminate signal from background. The main background comes from dimuon production and ditau production with tau decays into a muon plus neutrinos and a photon from initial or

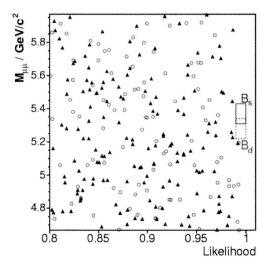

Figure 16. Distribution of events in the likelihood versus dimuon mass plane of the CDF B $\to \mu^+\mu^-$ analysis.

final state radiation. Figure 15 shows the energy constraint muon photon mass with a curve of how a potential signal would look. Observation agrees with the background expectation and BaBar sets a 90% CL limit on the branching ratio of tau into a muon plus a photon at $6.8 * 10^{-8}$.

Another interesting channel is the B_s into $\mu^+\mu^-$ decay. The flavor changing neutral current (FCNC) decay is heavily suppressed in the SM. In the MSSM, however, the branching ratio is enhanced, proportional to $\tan(\beta)^6$. CDF has a long tradition of searching for B $\to \mu^+\mu^-$. The analysis is normalized to the observed $B^+ \to J/\psi K^+$ decays to become independent of the b production cross-section. A likelihood function is used to separate dimuons that originate from a decay of a particle with lifetime from prompt dimuons. CDF observes no events in the B_d and B_s window, Fig. 16. The combined CDF/DØ analyses set a 95% CL branching ratio limit of $1.2 * 10^{-7}$ for B_s and $3.1 * 10^{-8}$ for B_d. This excludes first regions in SUSY parameter space at high $\tan(\beta)$.

8   Summary and Outlook

Scientists have explored nature to smaller and smaller scales over the years. In the last 50 years particle physics has made tremendous progress, revealing and exploring the next smaller layer of particles. We have developed a self-consistent, although incomplete, model that describes our current knowledge. Nature still surprises us, like with the observation of neutrino oscillation and the accelerating expansion of the universe. Our current understanding strongly suggests new physics to be close to the electroweak scale. However, no significant evidence of new physics has been observed so far. The current experiments search extensively in a large variety of signatures for deviations from the Standard Model. Some of the most interesting and promising search channels were presented in this review. Both HERA and the Tevatron are running well with record luminosities and the experiments are keeping up analysing the data. The hope is on the current experiments to unveil the next layer or the next symmetry of nature. A new generation of experiments is only a few years away and should answer our question about new electroweak scale physics. The transfer of expertise and experience to those new experiments has started.

References

1. P.W. Higgs, *Phys. Lett.* **12**, 132 (1964);
   F. Englert and R. Brout, *Phys. Rev. Lett.* **13**, 321 (1964);
   G.S. Guralnik, C.R. Hagen, and T.W. Kibble, *Phys. Rev. Lett.* **13** 585 (1964).
2. K.A. Olive, (ed.), S. Rudaz, (ed.), M.A. Shifman, (ed.) *Nucl. Phys. Proc. Suppl.* **101** 1 (2001).
3. A. Juste, these proceedings.
4. S. Heinemeyer and G. Weiglein, *Proceedings of the LCWS2000*, hep-ph/0012364 (2000).
5. G. Abbiendi *et al.* *Phys. Lett.* B **565**, 61 (2003).
6. For the most updated results please use the Higgs and New Particle Searches web site of

the experiments at:
http://www-cdf.fnal.gov/physics/exotic
/exotic.html and http://www-d0.fnal.go
v/Run2Physics/higgs/pubresults.html

7. DØ Collab., DØnote 4774-CONF (2005).
8. Physics at Run II, Fermilab-PUB-00/349 (2000);
CDF Collab., CDF note 6353 (2003)
9. ATLAS Collab., CERN/LHCC 99-15, ATLAS-TDR-15 (1999);
CMS Collab., CERN/LHCC 97-33, CMS-TDR-4 (1997);
F. Gianotti, these proceedings.
10. A. Abulencia et al. hep-ex/0508051 (2005).
11. V.M. Abazov et al. hep-ex/0504018 (2005).
12. LEP Higgs Working Group, LHWG-Note 2005-01 (2005), paper 249 submitted to this conference.
13. M. Carena, J.R. Ellis, A. Pilaftsis, and C.E. Wagner *Phys. Lett.* B **495** 155 (2000);
S. Heinemeyer, W. Hollik, and G. Weiglein *Comp. Phys. Comm.* **124** 76 (2000);
M. Frank, S. Heinemeyer, W. Hollik, and G. Weiglein, hep-ph/0212037 (2002).
14. CDF Collab., CDF note 7712 (2005).
15. LEP SUSY Working Group, LEPSUSYWG/01-07.1 (2002).
16. V.M. Abazov et al. hep-ex/0504032 (2005);
DØ Collab., DØnote 4740-CONF (2005).
17. CDF Collab., CDF note 7136 (2004);
CDF Collab., CDF note 7457 (2005);
DØ Collab., DØnote 4832-CONF (2005).
18. DØ Collab., DØnote 4737-CONF (2005).
19. A. Aktas et al. *Eur. Phys. J.* C **36** 425 (2004);
ZEUS Collab., ZEUS-prel-05-002, paper 258 submitted to this conference.
20. D. Acosta et al. *Phys. Rev.* D **71** 031104(R) (2005);
V.M. Abazov et al. *Phys. Rev. Lett.* **94**, 041801 (2005);
V. Buscher et al. hep-ex/0504004 (2005).
21. ATLAS Collab., CERN/LHCC 99-15 (1999);
M. Dittmar, CMS CR-1999/007 (1999);
A. Tricomi, CMS CR-2004/019 (2004);
22. J.L. Feng and M.M. Nojiri, hep-ph/0210390 (2002).
23. C. Adloff et al. *Eur. Phys. J.* C **5** 575 (1998);
V. Andreev et al. *Phys. Lett.* B **561** 241 (2003).
24. H1 Collab., paper 421 submitted to this conference;
H1 Collab., H1prelim-04-061, paper 417 submitted to this conference;
J. Breitweg et al. *Phys. Lett.* B **471**, 411 (2000);
S. Chekanov et al. *Phys. Lett.* B **583** 41 (2004);
ZEUS Collab., paper 257 submitted to this conference.
25. CDF Collab., CDF note 7711 (2005);
CDF Collab., CDF note 7286 (2004);
CDF Collab., CDF note 7167 (2004);
CDF Collab., CDF note 7098 (2004);
CDF Collab., CDF note 7459 (2005);
DØ Collab., DØnote 4336-CONF (2004);
DØ Collab., DØnote 4922-CONF (2005);
DØ Collab., DØnote 4552-CONF (2004);
DØ Collab., DØnote 4893-CONF (2005);
DØ Collab., DØnote 4829-CONF (2005).
26. BaBar Collab., hep-ex/0502032, paper 217 submitted to this conference;
BaBar Collab., hep-ex/0409036 (2004);
Belle Collab., hep-ex/0508044 (2005).
27. CDF Collab., hep-ex/0508036 (2005);
DØ Collab., D0-Note 4733-Conf (2005).

DISCUSSION

Inti Lehmann (Uppsala University):
What is the long term plans for the Tevatron after LHC comes into operation?

Stephan Lammel: As long as the Tevatron
is at the energy frontier it would not be
wise to switch off the machine. Once
LHC has taken over one has to see if any
measurments remain that can be done
better at the Tevatron or if it is better
to focus on LHC and may be prepare at
Fermilab for a new experiment/machine.

Dieter Zeppenfeld (Uni. Karlsruhe):
With about 300 pb^{-1} or 5-10% of the ultimate data sample the Tevatron Higgs
cross section bounds are more than an
order of magnitude above SM expectations. This indicates that the present
sensitivity is well below what was predicted. Is this correct? And what are
the main problems?

Stephan Lammel: At the moment there is
more than an order of magnitude difference between experimental limits and
theoretical expectation. The important
ingredients for Higgs search at the Tevatron are lepton and b-tagging efficiency.
Both experiments are still improving on
their detection efficiencies. The $b\bar{b}$ dijet mass resolution is also extremly important. Right now the experiments
are back to their Run I mass resolution of 15% (from 17% at the start of
Run II). The goal is to use b-jet specific corrections and global event variables which should bring the resolution
to about 10%.

ELECTROWEAK SYMMETRY BREAKING CIRCA 2005

SALLY DAWSON

Physics Department, Brookhaven National Laboratory, Upton, NY, 11973, USA
E-mail: dawson@bnl.gov

Recent progress in both the experimental and theoretical explorations of electroweak symmetry breaking is surveyed.

1 Introduction

Particle physicists have a Standard Model of electroweak interactions which describes a large number of measurements extraordinarily well at energies on the few hundred GeV scale. In fact, we have become extremely blasé about tables such as that of Fig. 1,[1] which shows an impressive agreement between experiment and theory. Virtual probes, using the sensitivity of rare decays to high scale physics, are also in good agreement with the predictions of the Standard Model. This agreement, however, assumes the existence of a light, scalar Higgs boson, without which the theory is incomplete. There has thus been an intense experimental effort at the Tevatron aimed at discovering either the Standard Model Higgs boson or one of the Higgs bosons associated with the minimal supersymmetric model (MSSM).

In the Standard Model, using G_F, α, and M_Z as inputs, along with the fermion masses, the W mass is a predicted quantity. The comparison between the prediction and the measured value can not only be used to check the consistency of the theory, but also to infer limits on possible extentions of the Standard Model. The relationship between M_W and M_t is shown in Fig. 2. The curve labelled "old" does not include the new values (as of Summer, 2005), for the W mass and width from LEP-2 and the new mass of the top quark from the Tevatron. (These new values are reflected in Fig. 1.)

The measurements of Fig. 1 can be used to extract limits on the mass of a Standard Model Higgs boson. The limit on the Higgs boson mass depends quadratically on the top quark mass and logarithmically on the Higgs boson mass, making the limit exquisitely sensitive to the top quark mass. The limit is also quite sensitive to which pieces of data are included in the analysis. The fit of Fig. 2 includes only the high energy data and so does not include results from NuTeV or atomic parity violation.

The precision electroweak measurements of Fig. 1 give a 95% confidence level upper limit on the value of the Higgs boson mass of,[1]

$$M_H < 186 \text{ GeV}. \tag{1}$$

If the LEP-2 direct search limit of $M_H > 114$ GeV is included, the limit increases to

$$M_H < 219 \text{ GeV}. \tag{2}$$

Both CDF and D0 have presented experimental limits on the production rate for a Standard Model Higgs boson, which are shown in Fig. 3.[2] For most channels, the limits are still several orders of magnitude away from the predicted cross sections in the Standard Model. With an integrated luminosity of 4 fb^{-1} (8 fb^{-1}), the 95% exclusion limit will increase to $M_H > 130$ GeV ($M_H > 135$ GeV). A much more optimistic viewpoint is to note that with 4 fb^{-1} there is a 35% chance that the Tevatron will find 3σ evidence for a Higgs boson with a mass up to $M_H = 130$ GeV.

Despite the impressive agreement between the precision electroweak data and the theoretical predictions of the Standard

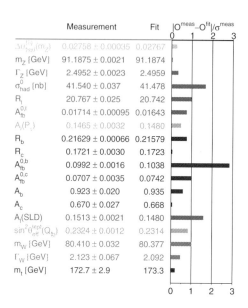

Measurement	Fit	$\|O^{meas}-O^{fit}\|/\sigma^{meas}$
$\Delta\alpha_{had}^{(5)}(m_Z)$	0.02758 ± 0.00035	0.02767
m_Z [GeV]	91.1875 ± 0.0021	91.1874
Γ_Z [GeV]	2.4952 ± 0.0023	2.4959
σ_{had}^0 [nb]	41.540 ± 0.037	41.478
R_l	20.767 ± 0.025	20.742
$A_{fb}^{0,l}$	0.01714 ± 0.00095	0.01643
$A_l(P_\tau)$	0.1465 ± 0.0032	0.1480
R_b	0.21629 ± 0.00066	0.21579
R_c	0.1721 ± 0.0030	0.1723
$A_{fb}^{0,b}$	0.0992 ± 0.0016	0.1038
$A_{fb}^{0,c}$	0.0707 ± 0.0035	0.0742
A_b	0.923 ± 0.020	0.935
A_c	0.670 ± 0.027	0.668
A_l(SLD)	0.1513 ± 0.0021	0.1480
$\sin^2\theta_{eff}^{lept}(Q_{fb})$	0.2324 ± 0.0012	0.2314
m_W [GeV]	80.410 ± 0.032	80.377
Γ_W [GeV]	2.123 ± 0.067	2.092
m_t [GeV]	172.7 ± 2.9	173.3

Figure 1. Precision electroweak measurements and the best theoretical fit to the Standard Model as of September, 2005. Also shown is the deviation of the fit for each measurement from the value predicted using the parameters of the central value of the fit.[1]

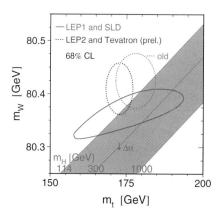

Figure 2. The relationship between M_W and M_t in the Standard Model. The curve labelled "old" does not include the Summer, 2005 updates on the W boson mass and width from LEP-2 and the new top quark mass from the Tevatron.[1]

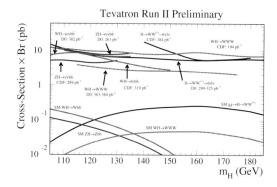

Figure 3. CDF and D0 limits on the production cross section times the branching ratios for various Higgs boson production channels as a function of the Higgs boson mass, along with the Standard Model expectations for each channel.[2]

Model with a light Higgs boson, theorists have been busy inventing new models where mechanisms other than a light Higgs boson are responsible for the electroweak symmetry breaking. We begin in Section 2 by reviewing the theoretical arguments for the existence of a Higgs boson and continue in Section 3 to discuss the reasons why a light Higgs boson is unattractive to many theorists. In the following sections, we review a sampling of models of electroweak symmetry breaking.

2 Who needs a Higgs Boson?

The Standard Model requires a Higgs boson for consistency with precision electroweak data, as is clear from Fig. 2. The Standard Model Higgs boson also serves two additional critical functions.

The first is to generate gauge invariant masses for the fermions. Since left- (ψ_L) and right- (ψ_R) handed fermions transform differently under the chiral $SU(2)_L \times U(1)_Y$ gauge groups, a mass term of the form

$$L_{mass} \sim m_f \left(\overline{\psi}_L \psi_R + \overline{\psi}_R \psi_L \right) \qquad (3)$$

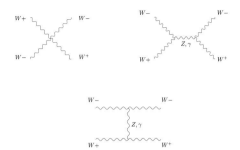

Figure 4. Feynman diagrams contributing to the process $W^+W^- \to W^+W^-$ with the Higgs boson removed from the theory.

Table 1. Representative limits (90 % c.l.) on the scale of new dimension-6 operators corresponding to $L = \mathcal{O}_i/\Lambda^2$[4].

	Operator, $\mathcal{O}_i$	Λ_{min} (TeV)
LEP	$H^\dagger \tau H W^a_{\mu\nu} B^{\mu\nu}$	10
LEP-2	$\bar{e}\gamma_\mu e \bar{l}\gamma^\mu l$	5
Flavor	$H^\dagger \bar{d}_R \sigma_{\mu\nu} q_L F^{\mu\nu}$	9

is forbidden by the gauge symmetry. A Higgs doublet, Φ, with a vacuum expectation value, v, generates a mass term of the required form,

$$L_{mass} \sim \frac{m_f}{v}\left(\overline{\psi}_L \Phi \psi_R + \overline{\psi}_R \Phi^\dagger \psi_L\right). \quad (4)$$

The second important role of the Standard Model Higgs boson is to unitarize the gauge boson scattering amplitudes. The $J = 0$ partial wave amplitude for the process $W^+W^- \to W^+W^-$ (Fig. 4) grows with energy when the Higgs boson is not included in the amplitude and violates partial wave unitarity at an energy around $E \sim 1.6$ TeV.[3] The Higgs boson has just the right couplings to the gauge bosons to restore partial wave unitarity as long as the Higgs boson mass is less than around $M_H < 800$ GeV. With a Higgs boson satisfying this limit, the Standard Model preserves unitarity at high energies and is weakly interacting.

3 Problems in Paradise

The Standard Model is theoretically unsatisfactory, however, because when loop corrections are included, the Higgs boson mass contains a quadratic dependence on physics at some unknown higher energy scale, Λ. When the one-loop corrections to the Higgs boson

mass, δM_H^2, are computed we find,

$$\delta M_H^2 = \frac{G_F \Lambda^2}{4\sqrt{2}\pi^2}\left(6M_W^2 + 3M_Z^2 + M_H^2 - 12M_t^2\right)$$
$$\sim -\left(\frac{\Lambda}{.7 \text{ TeV}}\, 200 \text{ GeV}\right)^2. \quad (5)$$

In order to have a light Higgs boson as required by the precision electroweak measurements, the scale Λ must be near 1 TeV. The quantum corrections thus suggest that there must be some new physics lurking at the TeV scale.

We therefore need new physics at the 1 TeV scale to get a light Higgs boson. However, much of the possible new physics at this scale is already excluded experimentally. A model independent analysis which looked at various dimension-6 operators found that typically new physics cannot occur below a scale $\Lambda > 5$ TeV. A representative sampling of limits on possible dimension-6 operators is shown in Table 1 and a more complete list can be found in Ref.[3]. This tension between needing a low scale Λ for new physics in order to get a light Higgs boson and the experimental exclusion of much possible new physics at the TeV scale has been dubbed the "little hierarchy problem". However, a global fit to 21 flavor- and CP- conserving operators found that there are certain directions in parameter space where the limit on Λ can be lowered considerably[5] (even to below 1 TeV) raising the possibility that in specific models the "little hierarchy problem" may not be a problem at all.

In recent years, there have been a variety of creative new models constructed which attempt to find a mechanism to lower the scale Λ, while at the same time not violating the existing experimental limits. Supersymmetric models are the trusty standard for addressing this problem and we discuss progress and variations on the minimal supersymmetric model in the next section. In the following sections, we discuss attempts to address electroweak symmetry breaking with Little Higgs models[7,8] and with Higgsless models.[6] There are many other novel models for electroweak symmetry breaking–fat Higgs models,[9] strong electroweak symmetry breaking[10] (and many more!) –which will not be addressed here due to space limitations.

4 Supersymmetry

The classic model of new physics at the TeV scale is supersymmetry, where a cancellation between the contributions of the Standard Model particles and the new partner particles of a supersymmetric model keeps the Higgs boson mass at the TeV scale. This cancellation occurs as long as the supersymmetric partner particles have masses on the order of the weak scale. For example, the top quark contribution to Eq. 5 becomes,[13]

$$\delta M_H^2 \sim G_F \Lambda^2 \left(M_t^2 - \tilde{m}_{t1,t2}^2 \right), \quad (6)$$

where $\tilde{m}_{t1,t2}$ are the masses of the scalar partners of the top quark.

The simplest version of a supersymmetric model, the MSSM, has many positive aspects:

- The MSSM predicts gauge coupling unification at the GUT scale.

- The MSSM contains a dark matter candidate, the LSP (Lightest Supersymmetric Particle).

- The MSSM predicts a light Higgs boson, $M_H < 140$ GeV.

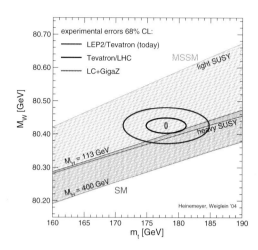

Figure 5. Fit to precision electroweak data in the MSSM. The curve labelled heavy SUSY assumes the supersymmetric parameters are set at 2 TeV.[11]

- The MSSM agrees with precision electroweak measurements.[11]

The fit to the electroweak precision data can be performed in the context of the MSSM and is shown in Fig. 5 for supersymmetric partner masses below 2 TeV. The MSSM with supersymmetric partner particles in the 1-2 TeV region is actually a slightly better statistical fit to the data than the Standard Model.[14]

There are also many negative things about the supersymmetric model, the most obvious of which is: *Where is it?*

In the MSSM, the lightest Higgs boson mass has a theoretical upper bound,

$$M_H^2 < M_Z^2 \cos^2 2\beta + \frac{3 G_F M_t^4}{\sqrt{2}\pi^2 \sin^2 \beta} \log\left(\frac{\tilde{m}_{t1}\tilde{m}_{t2}}{M_t^2} \right), \quad (7)$$

where $\tan \beta$ is the ratio of the neutral Higgs boson vacuum expectation values. Requiring that the Higgs boson mass satisfy the LEP direct search limit, $M_H > 114$ GeV, implies that the stop squarks must be relatively heavy,[12]

$$\tilde{m}_{t1}\tilde{m}_{t2} > (950 \text{ GeV})^2. \quad (8)$$

However, the supersymmetric partner particles in the MSSM are naturally on the order

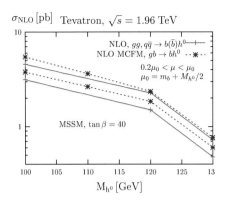

Figure 6. Total next-to-leading order cross section in the MSSM for bH production at the Tevatron. The bands show the renormalization/factorization dependence. The solid (red) curves correspond to the four-flavor number scheme with no b partons, and the dotted (blue) curves are the prediction from the five-flavor number scheme with b partons in the initial state.[15]

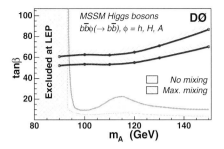

Figure 7. 95% c.l. upper limit from the D0 experiment at the Tevatron on $\tan\beta$ in the MSSM from $gg \to b\bar{b}\phi$, where ϕ is any of the three neutral Higgs bosons of the MSSM.[17]

of the weak scale, so there is a tension between the desire for them to be light (to fill their required role in cancelling the quadratic contributions to the Higg boson mass as in Eq. 6) and the limit of Eq. 8.

The couplings of the Higgs boson to the bottom quark are enhanced in the MSSM for large values of $\tan\beta$ and the dominant production mechanism becomes $gg \to b\bar{b}H$, where 0, 1, or 2 b quarks are tagged.[15,16] Fig. 6 shows the total next-to-leading order cross section for bH production at the Tevatron as a function of the mass of the lightest Higgs boson of the MSSM for $\tan\beta = 40$.[15] D0 has a new limit on this process, which is shown in Fig. 7.[17]

Many variants of the MSSM have been constructed. One of the simplest is the NMSSM (next-to-minimal- supersymmetric model) which is obtained by adding a Higgs singlet superfield $\hat{S}$ to the MSSM.[18,19] The superpotential in the NMSSM is,

$$W = W_{MSSM} + \lambda \hat{H}_1 \hat{H}_2 \hat{S} + \frac{\kappa}{3} \hat{S}^3, \qquad (9)$$

where $\hat{H}_1$ and $\hat{H}_2$ are the Higgs doublet superfields of the MSSM, and $\hat{S}$ is the Higgs singlet superfield. When the scalar component of the singlet, S, gets a vacuum expectation value, the term $\lambda \hat{H}_1 \hat{H}_2 \langle S \rangle$ in the superpotential naturally generates the $\mu \hat{H}_1 \hat{H}_2$ term of the MSSM superpotential and it is straightforward to understand why $\mu \sim M_Z$. This is the major motivation for constructing the NMSSM.

In the NMSSM model, the bound on the lightest Higgs boson mass becomes,

$$M_H^2 < M_Z^2 \cos^2 2\beta + v^2 \lambda^2 \sin^2 2\beta$$
$$+1\text{-loop corrections}, \qquad (10)$$

and the lightest Higgs boson can be significantly heavier than in the MSSM. If we further assume that the couplings remain perturbative to the GUT scale, the theoretical upper bound on the lightest Higgs boson mass becomes $M_H < 150$ GeV.[20]

The phenomenology in the NMSSM is significantly different than in the MSSM. There are three neutral Higgs bosons and two pseudoscalar Higgs bosons. A typical scenario for the masses is shown in Fig. 8. New decays such as the Higgs pseudoscalar into two scalar Higgs bosons are possible and changes the LHC Higgs search strategies. In addition, the lightest Higgs boson can have a large CP-odd component and so can evade the LEP bound on M_H.[18,19]

The minimal version of the MSSM conserves CP, but CP violation in the Higgs

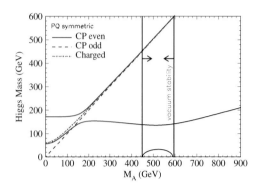

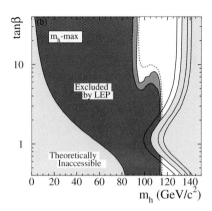

Figure 8. Typical mass scenario for the Higgs bosons in the NMSSM with $\tan\beta = 3$. The region between the vertical lines (denoted by arrows) is the region allowed by vacuum stability.[19]

Figure 9. Excluded region in the M_{H_1}-$\tan\beta$ plane in the CP conserving version of the MSSM. The light (dark) green is the 95 % (97 %cl) excluded region in the M_H(max) benchmark scenario. The solid lines from left to right vary the top quark mass: $M_t =$ 169.3, 174.3, 179.3 and 183 GeV.[21]

sector can easily be accommodated in the MSSM. Non-zero phases in the scalar trilinear couplings can generate large CP violating effects from radiative corrections, especially those involving the third generation. If there is CP violation in the Higgs sector of the MSSM, then the three neutral Higgs mass eigenstates, H_1, H_2, and H_3, are mixtures of the CP-even and CP-odd Higgs states.[21] The production and decay properties of the Higgs bosons can be very different from those of the Higgs bosons in the CP conserving version of the MSSM since the CP-odd components of the Higgs mass eigenstates do not couple to the Z boson.

Experimental searches for the Higgs boson in a version of the MSSM with CP violation in the Higgs sector have been performed by the LEP collaborations[22] using the benchmark parameters of the CPX model.[21] For large values of M_{H_2}, H_1 is almost completely CP-even and the exclusion limit for the lightest Higgs boson mass is similar to the CP conserving limit. If $M_{H_2} > 130$ GeV, then $M_{H_1} > 113$ GeV. For lighter M_{H_2}, the H_1 has a large mixture of the CP-odd component and the result is that there are unex-

cluded regions in the $M_{H_1} - \tan\beta$ parameter space and the excluded region disappears completely for $4 < \tan\beta < 10$. At 95% c.l., $\tan\beta < 3.5$ and $M_{H_1} < 114$ GeV and also $\tan\beta > 2.6$ are excluded in the CPX scenario.[a]

It is interesting to compare the excluded regions in the $M_{H_1} - \tan\beta$ plane for the CP conserving and CP nonconserving versions of the MSSM, as shown in Figs. 9 and 10. We observe that the shape of the excluded region is significantly different in the two cases. As noted in Ref.[22], the limit is extremely sensitive to small variations in the top quark mass.

5 Little Higgs Models

Little Higgs models[7,8] are an attempt to address the hierarchy problem by cancelling the quadratic contributions to the Higgs boson mass in the Standard Model with the contributions resulting from the addition of new particles which are assumed to exist at a scale around 1-3 TeV. The cancellation of the quadratic contributions occurs between

---

[a]These limits assume $M_t = 179.3$ GeV[22].

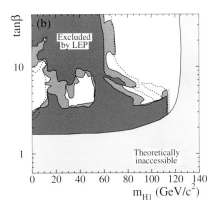

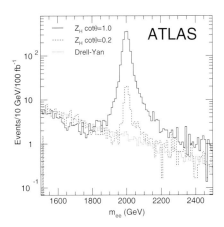

Figure 10. Excluded region in the $M_{H_1}-\tan\beta$ plane in the CPX CP violating version of the MSSM. The curves are as in Fig. 9.[21]

Figure 11. ATLAS simulation with 300 fb^{-1} of data of the e^+e^- invariant mass distribution in a Little Higgs model resulting from the decay $Z_H \rightarrow ZH$ for $Z_H = 2$ TeV. The lower dotted histogram is the background.[26]

states with the same spin statistics. Thus contributions to Eq. 5 from the Standard Model W, Z, and photon are cancelled by the contributions from new heavy gauge bosons, W_H, Z_H and A_H, with Standard Model quantum numbers, while Standard Model contributions from the top quark are cancelled by those from a heavy charge 2/3 top-like quark, and those from the Higgs doublet by contributions from a scalar triplet. A clear prediction of the Little Higgs models is the existence of these new particles. Decays such as $Z_H \rightarrow ZH$ should be particularly distinctive[8] as demonstrated in Fig. 11.[26]

The basic idea of the Little Higgs models is that a continuous global symmetry is broken spontaneously and the Higgs boson is the Goldstone boson of the broken symmetry. There are many variants of this idea, with the simplest being a model with a global SU(5) symmetry broken to a global SO(5) symmetry by the vacuum expectation value of a non-linear sigma field $\Sigma = \exp(2i\Pi/f)$. The Goldstone bosons contain both a Higgs doublet and a Higgs triplet and reside in the field Π. The parameter f sets the scale of the symmetry breaking, which occurs at a scale $\Lambda \sim 4\pi f \sim 10$ TeV where the theory becomes strongly interacting. The quadratic contribu-

tions to the Higgs boson mass of the Standard Model are cancelled by the new states at a scale $gf \sim 1-3$ TeV. Furthermore, the gauge symmetries are arranged in such a manner that the Higgs boson gets a mass only at two-loops, $M_H \sim g^2 f/(4\pi)$, and so the Higgs boson is naturally light, as required by the precision electroweak data.

The mixing of the Standard Model gauge bosons with the heavy gauge bosons of Little Higgs models typically gives strong constraints on the scale $f > 1-4$ TeV.[23] It is possible to evade many of these limits by introducing a symmetry (T parity) which requires that the new particles be produced in pairs.[24,25] This allows the scale f to be as low as 500GeV. The lightest particle with T-odd parity is stable and is a viable dark matter candidate for M_H between around 200 and 400 GeV and the scale f in the $1-2$ TeV region, as seen in Fig. 12.

Little Higgs models allow the lightest neutral Higgs boson to be quite heavy, as is demonstrated in Fig. 13.[27] The relaxation of the strong upper bound on the Higgs mass of the Standard Model is a generic feature of

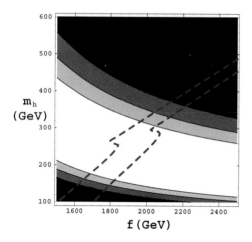

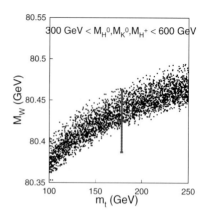

Figure 12. Excluded regions at 95%, 99% and 99.9% confidence level (from lightest to darkest) in the little Higgs model with T-Parity. In the band between the two dashed lines the lightest T-Parity odd particle is a consistent dark matter candidate and contributes to a relic density within 2σ of the WMAP data.[25]

Figure 13. Predictions for the W mass as a function of the top quark mass in a theory with a Higgs triplet. The masses of the three scalars in the theory, H^0, K^0, and $H^{\pm}$, are varied between 300 and 600 GeV. The red point is the experimental data point with the 1σ errors.[27]

models with Higgs triplets.

6 Higgsless Models

Finally, we consider a class of models in which the Higgs boson is completely removed from the theory. These models face a number of basic challenges:

- How to break the electroweak symmetry?

- How to restore unitarity without a Higgs boson?

- How to generate gauge boson and fermion masses?

- How to ensure

$$\rho = \frac{M_W^2}{M_Z^2 \cos^2 \theta_W} = 1? \qquad (11)$$

Models with extra dimensions offer the possibility of removing the Higgs boson from the theory and generating the electroweak symmetry breaking from boundary conditions on the branes of the extra dimensions.[6] Before even constructing such a Higgsless

model, it is obvious that models of this class will have problems with the electroweak precision data. As can be seen from Fig. 14, as the Higgs boson gets increasingly massive, the predictions of the Standard Model get further and further away from the data. A heavy Higgs boson gives too large a value of S and too small a value of T. This figure gives a hint as to what the solution must eventually be: The Higgsless models must have a large and positive contribution to T and must not have any additional contributions to S.[28]

The Higgsless models all contain a tower of Kaluza Klein (KK) particles, V_n, with the quantum numbers of the Standard Model gauge bosons. The lightest particles in the KK tower are the Standard Model W, Z, and γ. These Kaluza Klein particles contribute to the elastic scattering amplitudes for gauge bosons. In general, the elastic scattering amplitudes have the form, (where E is the scattering energy):

$$A = A_4 \frac{E^4}{M_W^4} + A_2 \frac{E^2}{M_W^2} + A_0 + ... \qquad (12)$$

In the Standard Model, A_4 vanishes by gauge invariance and A_2 vanishes because of the

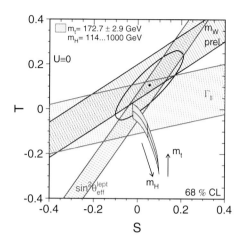

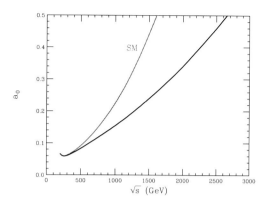

Figure 15. $J = 0$ partial wave for elastic gauge boson scattering in the Standard Model with the Higgs boson removed (red) and with the inclusion of a single Kaluza Klein excitation with $M = 500$ GeV (blue) in a deconstructed Higgsless model.[30]

Figure 14. Limits on S and T from precision electroweak measurements, as of September, 2005. The Standard Model reference values (which give $S = T = U = 0$) are $M_t = 175$ GeV and $M_H = 150$ GeV.[1]

cancellation between the gauge boson and Higgs boson contributions. In the Higgsless models, the contributions to A_4 and A_2 cancel if,

$$g^2_{nnnn} = \Sigma_k g^2_{nnk}$$
$$4g^2_{nnnn} = 3\Sigma_k g^2_{nnk} \frac{M^2_k}{M^2_n}, \quad (13)$$

where g_{nnk} is the cubic coupling between V_n, V_n, and V_k, g_{nnnn} is the quartic self coupling of V_n, and M_k is the mass of the kth KK particle.

The amazing fact is that the 5-dimensional Higgsless models satisfy these sum rules exactly due to 5-dimensional gauge invariance. Similarly, 4-dimensional deconstructed versions of the Higgsless models[33] satisfy these sum rules to an accuracy of a few percent. The Kaluza Klein particles play the same role as the Higgs boson does in the Standard Model and unitarize the scattering amplitudes. Of course, the lightest Kaluza Klein mode needs to be light enough for the cancellation to occur before the amplitude is already large, which restricts the masses of

the Kaluza Klein particles to be less than $1 - 2$ TeV.[29,30]

Fig. 15 shows the growth of the $J = 0$ partial wave in the Standard Model with the Higgs boson removed and in a Higgsless model with a single Kaluza Klein particle with mass $M = 500$ GeV included. The inclusion of the Kaluza Klein contributions pushes the scale of unitarity violation from $E \sim 1.6$ TeV in the Standard Model with no Higgs boson to around $E \sim 2.6$ TeV in the Higgsless models.

The Kaluza Klein particles contribute to the electroweak precision measurements. In general, the corrections are too large for KK particles with masses on the TeV scale.[31] Considerable progress in addressing this problem has been made in the last year with the realization that the contributions of the Kaluza Klein particles to the precision electroweak observables depend on where the fermions are located in the extra dimensions. In the Randall-Sundrum model, S is positive if the fermions are located on the Planck brane and negative if they are located on the TeV brane. The trick is to find an intermediate point where there is a weak coupling between the KK modes and the fermions.[31,32] It appears to be possible to

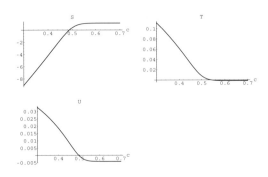

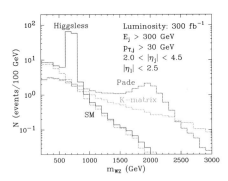

Figure 17. The number of events per 100 GeV bin in the 2-jet plus 3 lepton plus ν channel at the LHC, coming from the subprocess $WZ \to WZ$ in a Higgsless model.[34]

Figure 16. Oblique parameters, S, T, and U in a Higgsless model as a function of the fermion localization parameter, c. If the fermions are localized on the TeV brane, $c << \frac{1}{2}$, while fermions localized on the Planck brane have $c >> \frac{1}{2}$. A flat fermion wavefunction corresponds to $c = \frac{1}{2}$.[32]

Fig. 17.[34] The lightest KK resonance should be clearly observable above the background.

construct models which are consistent with the electroweak precision measurements by having the fermion wavefunction be located between the branes.[32]

Fig. 16 shows the oblique parameters as a function of the variable c, which characterizes the location of the fermion wavefunction. If the fermions are localized on the TeV brane, $c << \frac{1}{2}$, while fermions localized on the Planck brane have $c >> \frac{1}{2}$. A flat fermion wavefunction corresponds to $c = \frac{1}{2}$. For $c \sim 1/2$ it is possible to satisfy the bounds from precision electroweak data. Fermions with a flat wavefunction are weakly coupled to the Kaluza Klein particles and so such Kaluza Klein particles would have escaped the direct searches for heavy resonances at the Tevatron.

The next challenge for Higgsless models is to generate the large mass splitting between the top and the bottom quarks.[35]

Weakly coupled Kaluza Klein particles are a generic feature of Higgsless models and can be searched for in a model independent fashion. These KK particles appear as massive W-, Z-, and γ- like resonances in vector boson fusion and they will appear as narrow resonances in the WZ channel as shown in

7 Conclusions

The mechanism of electroweak symmetry breaking could be far more complicated than a simple Higgs boson. Almost all models, however, have distinctive signatures which should be observed at the LHC. Soon, with data from the LHC, we should have some indication what mechanism nature has chosen! A complete understanding of the unknown physics awaiting us at the TeV scale will probably require a future linear collider.[36]

Acknowledgments

This research supported by Contract No. DE-AC02-76CH1-886 with the U.S. Department of Energy. I thank my collaborators, M.C. Chen, C. Jackson, T. Krupovnickas, L. Reina, and D. Wackeroth for countless discussions.

References

1. LEP Electroweak Working Group, http://lepewwg.web.cern.ch/LEPEWWG/.
2. Preliminary result posted at http://www-cdf.fnal.gov/physics/exotic.html.

3. B. Lee, C. Quigg, and H. Thacker, *Phys. Rev.* D **16**, 1519 (1977); D. Dicus and V. Mathur, *Phys. Rev. D* **7**, 3111 (1973)

4. G. Giudice, *Int. J. Mod. Phys.* A **19**, 835 (2004), hep-ph/0311344; R. Barbieri, A. Pomerol, and A. Strumia, *Nucl. Phys.* B **703**, 127 (2004), hep-ph/0405040.

5. Z. Han and W. Skiba, *Phys. Rev.* D **71**, 075009, hep-ph/0412166.

6. C. Csaki, C. Grojean, H. Murayama, L. Pilo, and J. Terning, *Phys. Rev. D* **69**, 055006 (2004), hep-ph/0305237; C. Csaki, C. Grojean, L. Pilo, and J. Terning, *Phys. Rev. Lett.* **92**, 101802 (2004), hep-ph/0308038; R. Chivukula, E. Simmons, H.-J. He, M. Kurachi, and M. Tanabashi, *Phys. Rev D* **70**, 075008 (2004), hep-ph/0406077. For a review and references to the original literature see G. Cacciapaglia, C. Csaki, C. Grojean,J. Terning, eConf C040802:FRT004, 2004, C. Csaki, hep-ph/0412339.

7. N. Arkani-Hamed, A.Cohen, E. Katz, and A. Nelson, *JHEP* **0207**, 034 (2002), hep-ph/0206021; N. Arkani-Hamed, A.Cohen, E. Katz, A. Nelson, T. Gregoire, and J. Wacker, *JHEP* **0208**, 021 (2002), hep-ph/0206020; D. Kaplan and M. Schmaltz, *JHEP* **0310**, 039 (2003), hep-ph/0302049; W. Skiba and J. Terning, *Phys. Rev. D* **68**, 075001 (2003), hep-ph/0305302.

8. For a review and references to the original literature, see T. Han, H. Logan, and L.-T. Wang, hep-ph/0506313.

9. R. Harnik, G. Kribs, D. Larson, and H. Murayama, *Phys. Rev.* D **70**, 015002 (2004), hep-ph/0311349.

10. C. Hill and E. Simmons, *Phys. Rept* **381**, 235 (2003), hep-ph/0203079.

11. S. Heinemeyer, W. Hollik, and G. Weiglein, hep-ph/0412214.

12. R. Dermisek and J. Gunion, *Phys. Rev. Lett.***95**, 041801 (2005), hep-ph/0502105.

13. M.Drees, hep-ph/9611409 and S. Dawson, TASI97, hep-ph/9712464, and references therein.

14. W. deBoer and C. Sander, *Phys. Lett.* B **585**, 276 (2004), hep-ph/0307049.

15. S. Dawson, C.Jackson, L. Reina, and D. Wackeroth, hep-ph/0508293.

16. S. Dawson, C.Jackson, L. Reina, and D. Wackeroth, *Phys. Rev. Lett.* **94**, 031802 (2005), hep-ph/0408077; J. Campbell *et. al.*, hep-ph/0405302; S. Dittmaier, M. Kramer, and M. Spira, *Phys. Rev.* D **70**, 074010 (2004); F. Maltoni, Z. Sullivan, and S. Willenbrock, *Phys. Rev.* D **67**, 093005 (2003), hep-ph/0301033.

17. D0 Collaboration, *Phys. Rev. Lett.* **95**, 151801 (2005).

18. U. Ellwanger, J. Gunion, and C. Hugonie, *JHEP* **0507**, 041 (2005), hep-ph/0503203; D. Miller, R. Nevzorov, and P. Zerwas, *Nucl. Phys.* B **681**, 3 (2003), hep-ph/0304049.

19. U. Ellwanger, J. Gunion, C. Hugonie, and S. Moretti, hep-ph/0401228.

20. G. Kane, C. Kolda, and J. Wells, *Phys. Rev. Lett.* **70**, 2686 (1993), hep-ph/9210242; M. Quiros and J. Espinosa, hep-ph/9809269.

21. M. Carena, J. Ellis, A. Pilaftsis, C. Wagner, *Phys. Lett.* B **495**, 155 (2000); *Nucl. Phys.* B **586**, 92 (2000).

22. ALEPH, DELPHI, L3, and OPAL Collaborations, Search for Neutral Higgs Bosons at LEP, LHWG-Note-2005-01.

23. J. Hewett, F. Petriello, and T. Rizzo, *JHEP***0310**, 062 (2003), hep-ph/0211218; C. Csaki, J. Hubisz, P. Meade, and J. Terning, *Phys. Rev.* D **68**, 035009 (2003), hep-ph/0303236; R. Casalbuoni, A. Deandrea, and M. Oertel, *JHEP* **0402**, 032 (2004), hep-ph/0311038; W. Kilian and J. Reuter, *Phys. Rev.* D **70**, 015004(2004), hep-ph/0311095; M. Chen and S. Dawson, *Phys. Rev.* D **70**,015003 (2004), hep-ph/0311032; T. Gregoire, D. Smith and G. Wacker, *Phys. Rev.* D **69**,115008 (2004),hep-ph/0305275.

24. H. Cheng and I. Low, *JHEP* **0408**, 061 (2004), hep-ph/0405243; I. Low, *JHEP* **0410**, 067 (2004), hep-ph/0409025; J. Hubisz and P. Meade, *Phys. Rev.* D **71**, 035016 (2005), hep-ph/0411264.

25. J. Hubisz, P. Meade, A. Noble, and M. Perelstein, hep-ph/0506042.

26. G. Azuelos *et.al.*, *Eur. Phys. Jour.* C **39S2**, 13 (2005), hep-ph/0402037.

27. T. Blank and W. Hollik, *Nucl. Phys.* B **514**, 113 (1998), hep-ph/9703392; M. Chen, S. Dawson, and T. Krupovnickas, hep-ph/0504286.

28. R. Chivukula, C. Hoelbling, and N. Evans, *Phys. Rev. Lett* **85**, 511 (2000), hep-ph/0002022; M. Peskin and J. Wells, *Phys. Rev.* D **64**, 093003 (2001), hep-ph/0101342.

29. H. Davoudiasl, J. Hewett, B. Lillie, and T. Rizzo, *JHEP* **0405**, 015 (2004), hep-

ph/0403300; *Phys. Rev.* D **70**, 015006 (2004); M. Papucci, hep-ph/0408058; R. Casalbuoni, S. DeCurtis, and D. Dominici, *Phys. Rev.* D **70**, 055010 (2004), hep-ph/0405188.

30. R. Foadi, S. Gopalakrishna, and C. Schmidt, *JHEP* **403**, 042 (2004), hep-ph/0312324.

31. K. Agashe, A. Delgado, M. May, R. Sundrum, *JHEP* **0308**, 050 (2003), hep-ph/0308036; J. Hewett, B. Lillie, and T. Rizzo, *JHEP* **0410**, 014 (2004), hep-ph/0407059; R. Foadi, S. Gopalakrishna, and C. Schmidt, *Phys. Lett.* B **606**, 157 (2005), hep-ph/0409266.

32. C. Cacciapaglia, C. Csaki, C. Grojean, J. Terning, *Phys. Rev.* D **71**, 035015 (2005), hep-ph/0409126.

33. R. Chivukula, H.-J. He, M. Kurachi, E. Simmons, and M. Tanabachi, *Phys. Rev.* D **72**, 015008 (2005), hep-ph/0504114; R. Chivukula, H.-J. He, M. Kurachi, E. Simmons, and M. Tanabachi, *Phys. Rev.* D **70**, 075008 (2004), hep-ph/0406077; R. Chivukula, H.-J. He, M. Kurachi, E. Simmons, and M. Tanabachi, *Phys. Lett.* B *603*, 210 (2004), hep-ph/0408262.

34. A. Birkedal, K. Matchev, and M. Perelstein, *Phys Rev. Lett.* **94**, 191803 (2005).

35. G. Cacciapaglia, C. Csaki, C. Grojean, M. Reece, and J. Terning, hep-ph/0505001; R. Foadi and C. Schmidt, hep-ph/0509071; H. Davoudiasl, B. Lillie, and T. Rizzo, hep-ph/0508279; R. Chivukula, H.-J. He, M. Kurachi, E. Simmons, and M. Tanabachi, hep-ph/0508147.

36. S. Dawson and M. Oreglia, *Ann. Rev. Nucl. Part. Sci.* **54**, 269 (2004), hep-ph/0403015.

DISCUSSION

Daniel Kaplan (Illinois Inst. of Techn.): How does the new state possibly seen in the HyperCP experiment at Fermilab fit into SUSY models? It has a mass of 214.3 MeV and decays into $\mu^+\mu^-$.

Sally Dawson: This state is very difficult to understand in terms of SUSY models.

Anna Lipniacka (University of Bergen):

Is gauge coupling unification natural in Large Extra Dimension models?

Sally Dawson: No. These theories typically violate unitarity and become strongly interacting at a scale between 1 and 10 TeV.

Ignatios Antoniadis (CERN):

What is the prize to pay in models that solve the little hierarchy problem, such as the little Higgs models, in particular on the number of parameters and the unification of gauge couplings?

Sally Dawson: Obviously, there is a large increase in the number of parameters and gauge unification is forfeited.

Luca Silvestrini (Munich and Rome): Maybe one should comment about the statement that you made that new physics has to have a scale Λ greater than 5 TeV. Of course this is a conventional scale that is only valid if the coupling in front of the operator is one, which is generally not true in any weakly interacting theory and generally not true if new physics enters through loops. So I do not want that anybody in the audience really believes that new physics must be at a scale larger than 5 TeV. It can easily be around the electroweak scale as we know very well.

Sally Dawson: Absolutely true. The limits depend on the couplings to the operators, which in turn depend on the model.

PROBING THE HIERARCHY PROBLEM WITH THE LHC

FABIOLA GIANOTTI

CERN, PH Department, 1211 Genève 23, Switzerland
E-mail: fabiola.gianotti@cern.ch

The hierarchy problem is one of the main motivations for building the CERN Large Hadron Collider (LHC). The prospects for the ATLAS and CMS experiments to understand this problem are discussed, with emphasis on early studies with the first LHC data. The longer-term potential for constraining the parameters of the underlying new theory is also addressed.

1 Introduction

The hierarchy problem[1] is the huge gap, 17 orders of magnitude, between two fundamental scales of physics: the electroweak scale and the gravity scale M_{Planck}. One of the consequences is that, if no new physics exists between these two scales, so that the Standard Model (SM) is valid all the way up to M_{Planck}, then the Higgs mass (m_H) diverges, unless it is unnaturally fine-tuned. New physics to stabilize m_H is already needed at the TeV scale, a well-known problem that has motivated the construction of the LHC and has been addressed by several theories beyond the SM. In Supersymmetry[2], new particles with masses at the TeV scale cancel divergent loop corrections to the Higgs mass. In theories with extra-dimensions[3], the fundamental scale of gravity is lowered from M_{Planck} to the electroweak scale, thereby removing the gap. In so-called "little Higgs" models[4], the SM is embedded in a larger symmetry group broken in such a way to provide the exact amount of new physics at the TeV scale to stabilize m_H. In Technicolour[5], the source of the problem, *i.e.* the Higgs boson being a fundamental scalar, is removed since the Higgs is a fermion condensate. New particles and new (strong) interactions are predicted at the TeV scale. Finally, in the so-called "split Supersymmetry"[6], the fine-tuning of m_H (and of the cosmological constant) is accepted on the basis of the anthropic principle: we live in a fine-tuned universe, out of all *a priori* possible landscapes, because only in this way can stable galaxy structures, atoms, *etc.* form. Supersymmetry sits at a high energy scale, since it is not required to stabilize the Higgs mass, but part of the spectrum (the gauginos) can be at the TeV scale for convenience (*i.e.*, to provide a dark matter candidate and gauge-coupling unification).

Many studies have been performed and presented over the past years about the LHC potential for exploring the above (and other) scenarios and for discovering the Higgs boson, a key element of the hierarchy problem. The results will not be repeated here. Instead, two specific questions are addressed in the following sections. Firstly, what are the prospects for elucidating at least part of the hierarchy problem at the beginning of the LHC operation, in two years from now (Secs. 2 and 3) ? Secondly, assuming that a signal from new physics will be observed, how well can the experiments discriminate among different scenarios and constrain the underlying theory (Sec. 4) ?

2 Early physics goals and measurements

Table 1 shows the data samples expected to be recorded by each of the two experiments, ATLAS or CMS, for some representative physics processes and for an integrated luminosity of 1 fb^{-1}. The latter corresponds to about six months of data taking at an

Table 1. For some physics processes, and for each experiment (ATLAS or CMS), the expected numbers of events on tape for an integrated luminosity of 1 fb^{-1}.

Channel	Events for 1 fb^{-1}
$W \rightarrow \mu\nu$	$\sim 7 \times 10^6$
$Z \rightarrow \mu\mu$	$\sim 1 \times 10^6$
$t\bar{t} \rightarrow \mu + X$	$\sim 1 \times 10^5$
$\tilde{g}\tilde{g}$, m($\tilde{g}$)=1 TeV	$10^2 - 10^3$

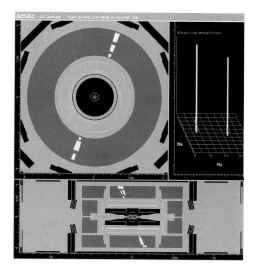

Figure 1. Display of a cosmic muon event recorded by the ATLAS Tilecal hadron calorimeter in the underground cavern (in the garage position). Transverse (top left) and longitudinal (bottom) views of the ATLAS detector are shown, as well as a map of the calorimeter towers (top right). The muon signal (in white-yellow) corresponds to about 2.5 GeV per calorimeter tower.

instantaneous luminosity of 10^{32} cm^{-2} s^{-1} (this is 1% of the LHC design luminosity) with an efficiency of 50%. It can be seen that already in the first months of operation huge event samples should be produced by known SM processes, larger than those collected by the LEP and Tevatron colliders over their whole life. We note that this will be true in many cases also for very modest integrated luminosities, of the order of 100 pb^{-1}.

The most urgent tasks to undertake with the first data will be to understand and calibrate the detectors *in situ* using well-known final states, such as $Z \rightarrow \ell\ell$ and $t\bar{t}$ events, and to perform extensive measurements of the SM processes, *e.g.* $W, Z, t\bar{t}$ production. The latter are important on their own, but also as potential backgrounds to new physics. This phase, that will take a lot of time given the complexity of the experiments and of the LHC environment but will be crucial to prepare a solid road to discoveries, is illustrated in the next sections with two examples.

2.1 Understanding the detector performance

Construction, integration and installation of the ATLAS and CMS experiments is progressing well[7]. Figure 1 shows one of the first cosmic muons recorded in June 2005 by the ATLAS Tilecal hadron calorimeter in the underground cavern.

The strategy to prepare the detectors to explore the hierarchy problem is illustrated here with a specific example: the ATLAS electromagnetic calorimeter (ECAL), which is a lead-liquid argon sampling calorimeter with accordion shape[8]. Construction is completed, and the barrel detector has been installed in the underground cavern.

One crucial performance requirement for the LHC electromagnetic calorimeters is to provide a mass resolution of about 1% in the hundred GeV range, needed *e.g.* to observe a possible $H \rightarrow \gamma\gamma$ signal as a narrow peak on top of a huge $\gamma\gamma$ irreducible background. This in turn demands a response uniformity, *i.e.* a total constant term of the energy resolution, of $\leq 0.7\%$ over the full calorimeter coverage ($|\eta| < 2.5$). This performance is difficult to obtain, especially in the initial phases of data taking, but can hopefully be achieved in four steps.

The first step is the construction quality of the detector mechanics. A response uniformity of 0.7% requires that the thickness of

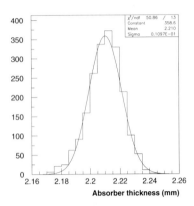

Figure 2. Distribution of the thicknesses of the 2048 absorber plates (3 m long and 0.5 m wide) for the ATLAS barrel ECAL, as obtained from ultrasound measurements. The mean value of the distribution is 2.2 mm and the r.m.s. is 11 μm.

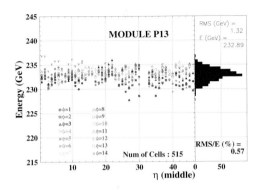

Figure 3. Energy response of one module of the ATLAS barrel ECAL, as a function of rapidity, as measured from a position scan over about 500 calorimeter cells with test-beam electrons. The various symbols indicate different ϕ rows.

the calorimeter absorber plates be uniform to $\sim$0.5%, i.e. 10 μm. This goal has been achieved, as shown in Fig. 2.

As a second step, test-beam measurements of some calorimeter modules have been performed, in order to verify the construction uniformity and to prepare correction factors to the detector response. Figure 3 shows the results of a position scan of one module (of size $\Delta\eta \times \Delta\phi = 1.4 \times 0.4$) made with high-energy test-beam electrons. For all tested modules, the response uniformity was found to be about 1.5% before correction, i.e. at the end of the construction process, and better than 0.7% after calibration with test-beam data.

As a third step, the calorimeter calibration can be checked in the underground cavern with physics-like signals by using cosmic muons (the signal-to-noise ratio is larger than seven for muons in the ATLAS ECAL). A few million events can be collected in three months of cosmics runs in the first half of 2007, during the machine cool-down and commissioning. This data sample is large enough to check the response uniformity of

the barrel calorimeter as a function of rapidity with a precision of $\sim$0.5%.

Finally, as soon as the first LHC data will become available, $Z \rightarrow ee$ events, which will be produced at the rate of $\sim$0.1 Hz at $L = 10^{32}$ cm^{-2} s^{-1}, will be used to correct long-range response non-uniformities from module to module, possible temperature effects, the impact of the upstream material, etc. Simulation studies indicate that 10^5 $Z \rightarrow ee$ events (i.e. an integrated luminosity of $\sim$100 pb^{-1}) should be sufficient to achieve the goal overall constant term of 0.7%, thanks to the knowledge and experience gained with the three previous steps. Therefore, after a few weeks of data taking the ATLAS ECAL should be fairly well calibrated.

However, let's consider a very pessimistic scenario, as an academic exercise. That is, ignoring the results and expectations discussed above, let's assume the raw (non-)uniformity of the calorimeter modules at the end of the construction phase, with no corrections applied (neither based on test-beam data, nor using $Z \rightarrow ee$ events). In this case the ECAL constant term would be 2% instead of 0.7%, and the significance of a possible $H \rightarrow \gamma\gamma$ signal would be reduced by about 30%. A factor 1.7 more integrated luminosity would therefore be needed to achieve the same sen-

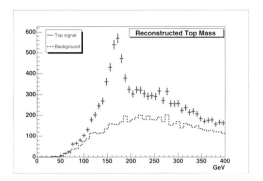

Figure 4. Three-jet invariant mass distribution for events selected as described in the text, as obtained from a simulation of the ATLAS detector. The crosses show the expected signal from $t\bar{t}$ events plus the background, and the dashed line shows the $W+4$-jet background alone (ALPGEN Monte Carlo[9]). The number of events corresponds to an integrated luminosity of 150 pb^{-1}.

sitivity.

2.2 First measurements of SM physics

Top-quark pair production, a process relevant to the hierarchy problem since it is a potential background to many new physics channels, is discussed here as an example of an initial measurement.

Figure 4 shows the expected signal from the gold-plated $t\bar{t} \rightarrow bjj\ b\ell\nu$ semileptonic channel, as obtained from a simulation of the ATLAS detector. The event sample corresponds to an integrated luminosity of only 150 pb^{-1}. A very simple analysis was used to select these events, requiring an isolated electron or muon with transverse momentum $p_T > 20$ GeV, missing transverse energy larger than 20 GeV, and four and only four jets with $p_T > 40$ GeV. The invariant mass of the three jets with the highest p_T of the three-jet system was then plotted. No kinematic fit was made, and no b-tagging of some of the jets was required, assuming conservatively that the b-tagging performance of the ATLAS tracker would not have been well understood yet. Figure 4 demonstrates that, even in these pessimistic conditions, a clear

top signal should be observed above the background after a few weeks of data taking (a data sample of 30 pb^{-1} would be sufficient). In turn, this signal can be used to understand the detector performance. For instance, if the top mass were to be wrong by several GeV, this would indicate a problem with the jet energy scale. Also, top events are an excellent sample to study the b-tagging performance of ATLAS and CMS. We note that, unlike at the LHC, at the Tevatron today the number of recorded $t\bar{t}$ events is not sufficient for detector calibration purposes.

3 Early discoveries

Only after both, detector performance and Standard Model physics, will have been well understood can the LHC experiments hope to extract convincing signals of new physics from their data. Three examples relevant to the hierarchy problem are discussed below, ranked by increasing difficulty for discovery in the first year(s) of operation: an easy case, namely an extra-dimension graviton decaying into an e^+e^- pair; an intermediate case, Supersymmetry; and a difficult case, a light SM Higgs boson.

3.1 Extra-dimension gravitons

A narrow resonance of mass about 1 TeV decaying into e^+e^- pairs, such as the gravitons (G) predicted by Randall-Sundrum extra-dimension theories[3], is probably the easiest object to discover at the LHC. Indeed, if the couplings to SM particles are reasonable, e.g. branching ratios for decays into electron pairs at the percent level, large enough signals are expected with integrated luminosities of 1 fb^{-1} or less for masses up to $\sim$1 TeV. In addition, the signal should provide a narrow mass peak on top of a much smaller and smooth Drell-Yan background, as shown in Fig. 5, and not just an excess of events in the tails of a distribution.

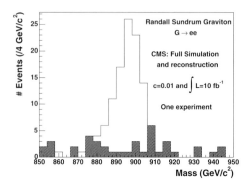

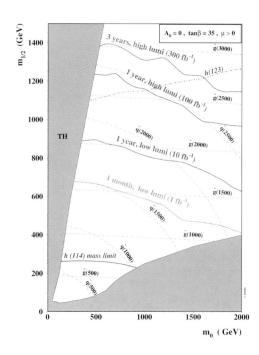

Figure 5. Expected signal from Randall-Sundrum gravitons $G \to e^+e^-$ (white histogram) on top of the Drell-Yan background (shaded histogram) in the CMS experiment[10]. The distributions correspond to a single experiment and to an integrated luminosity of 10 fb^{-1}.

Understanding the nature of the observed resonance, *e.g.* if it is a graviton or a heavy gauge boson Z', will require more time. It has been shown[11] that, with an integrated luminosity of 100 fb^{-1}, the angular distribution of the electron pairs in the final state can provide discrimination between a spin-1 Z' and a spin-2 graviton.

3.2 Supersymmetry

If Supersymmetry (SUSY) has something to do with the hierarchy problem, it must be at the TeV scale, and therefore could be found quickly at the LHC. This is because of the huge production cross-sections for squark and gluino masses as large as ~1 TeV (see Table 1), and the clear signatures expected from the cascade decays of these particles in most models. Therefore, by looking for final states containing for instance several high-p_T jets and large missing transverse energy, the LHC experiments should be able to discover squarks and gluinos up to masses of ~1.5 TeV in only one month of data taking at $L = 10^{33}$ cm^{-2} s^{-1} (Fig. 6), provided that the detectors and the main backgrounds are

Figure 6. The CMS discovery potential[12] for squarks and gluinos in minimal Supergravity (mSUGRA) models[13], parametrized in terms of the universal scalar mass m_0 and universal gaugino mass $m_{1/2}$, as a function of the integrated luminosity. Squark and gluino mass isolines are shown as dot-dashed lines (masses are given in GeV).

well understood (which will take more time than for the case discussed in Sec. 3.1).

Probing the underlying theory after discovery requires detailed studies and measurements of exclusive SUSY channels, as discussed in Sec. 4.2.

3.3 Standard Model Higgs boson

The possibility of discovering a SM Higgs boson at the LHC during the first year(s) of operation depends very much on the Higgs mass, as shown in Fig. 7. The most difficult case is the low-mass region close to the LEP limit ($m_H > 114.4$ GeV). As an example, the expected sensitivity for a Higgs mass of 115 GeV and for the first good (i.e. collected with well-understood detectors) 10 fb^{-1} is

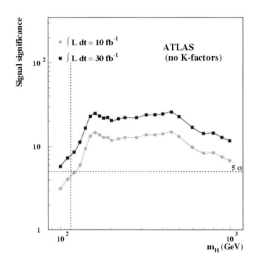

ATLAS
(no K-factors)

Figure 7. The expected signal significance for a SM Higgs boson in ATLAS[14,15] as a function of mass, for integrated luminosities of 10 fb^{-1} (dots) and 30 fb^{-1} (squares). The vertical line shows the mass lower limit from searches at LEP. The horizontal line indicates the minimum significance (5σ) needed for discovery.

Table 2. For $m_H = 115$ GeV and an integrated luminosity of 10 fb^{-1}, the expected numbers of signal (S) and background (B) events after all cuts and the expected signal significance ($S/\sqrt{B}$) in ATLAS for the three most sensitive channels.

	$H \to \gamma\gamma$	$t\bar{t}H \to$ $\to t\bar{t}b\bar{b}$	qqH with $H \to \tau\tau \to$ $\to \ell + X$
S	130	15	~ 10
B	4300	45	~ 10
$S/\sqrt{B}$	2.0	2.2	~ 2.7

summarized in Table 2. The total signal significance of about 4σ per experiment is more or less equally shared among three channels: $H \to \gamma\gamma$, $t\bar{t}H$ production with $H \to b\bar{b}$, and Higgs production in vector-boson fusion followed by $H \to \tau\tau$. It will not be easy to extract an indisputable signal with only 10 fb^{-1}, because the significances of the individual channels are small, and because an excellent knowledge of the large backgrounds and close-to-optimal detector performances are required, as discussed below. Therefore the contribution of both experiments, and the observation of possibly all three channels, will be crucial for as-fast-as-possible a discovery.

The channels listed in Table 2 are complementary. They are characterized by different Higgs production mechanisms and decay modes, and therefore by different backgrounds and different detector requirements. Good uniformity of the electromagnetic calorimeters is crucial for the $H \to \gamma\gamma$ channel, as already mentioned. Powerful b-tagging is the key performance issue for the $t\bar{t}H$ channel, since there are four b-jets in the final state which all need to be tagged in order to reduce the background. Efficient and precise jet reconstruction over ten rapidity units ($|\eta| < 5$) is needed for the $H \to \tau\tau$ channel, since tagging the two forward jets accompanying the Higgs boson and vetoing additional jet activity in the central region of the detector are necessary tools to defeat the background. Finally, all three channels demand relatively low trigger thresholds (at the level of 20-30 GeV on the lepton or photon p_T), and a control of the backgrounds to a few percent. These requirements are challenging especially in the first year(s) of operation.

On the other hand, if the Higgs boson is heavier than 180 GeV early discovery should be easier thanks to the gold-plated $H \to 4\ell$ channel. As shown in Fig. 8, the expected number of events is small for integrated luminosities of a few fb^{-1}, but these events are very pure, since the background is essentially negligible, and should cluster in a narrow mass peak.

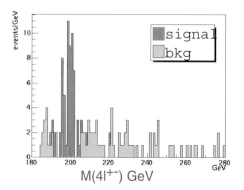

Figure 8. The expected $H \to ZZ \to 4\ell$ signal ($\ell = e$ or μ) in CMS[16] for a Higgs mass of 200 GeV (dark-shaded histogram) on top of the background (light-shaded histogram), for an integrated luminosity of 10 fb^{-1}.

4 Constraining the underlying theory

With more time and more data, signals from new physics will hopefully become more numerous and convincing. Many scenarios related to the hierarchy problem have been studied by the LHC Collaborations. The results demonstrate that the ATLAS and CMS detectors are sensitive to a large number of different topologies and final states, which indicates robustness and potential ability to cope with surprises. Furthermore, it has also been shown that the direct discovery potential, and therefore the sensitivity to the hierarchy problem, extends up to particle masses of 5-6 TeV.

However, discovery is not enough. In order to constrain the underlying fundamental theory, and hopefully understand the origin of the scale hierarchy, precise measurements are needed. Although the expected accuracy is in general not competitive with that achievable at a Linear Collider[17], the LHC data should nevertheless provide a huge amount of information about the observed new physics. A few examples are discussed below.

4.1 Higgs sector

ATLAS and CMS can measure the mass of a SM Higgs boson with the ultimate experimental accuracy of $\sim 0.1\%$ up to $m_H \simeq 500$ GeV[14].

Ratios of Higgs couplings to fermions and bosons should be measured with typical precisions of 20%[18]. The Higgs self-coupling λ, an important parameter providing direct access to the Higgs potential in the SM Lagrangian, is not measurable at the LHC, but may be constrained to about 20% at an upgraded LHC operating at a luminosity of 10^{35} cm^{-2} s^{-1}.

Finally, the scalar nature of the Higgs boson and its positive CP-state ($J^P = 0^+$) can be distinguished unambiguously from other spin-CP hypotheses ($J^P = 1^+$, 1^-, 0^-) for $m_H > 180$ GeV. Indeed, in this mass range the almost background-free $H \to 4\ell$ channel can be used to infer the Higgs spin-CP information from the angular distributions of the decay products in the final state[19]. In contrast, hopes are very modest at lower masses, where the large backgrounds dilute the signal angular distribution.

The above and other measurements should provide very useful insight into the electroweak symmetry breaking mechanism.

4.2 Supersymmetry

Mass measurements of the observed SUSY particles (sparticles) are essential to constrain the underlying theory. If Nature has chosen R-parity conserving scenarios, that are best motivated because of dark matter arguments (see below), sparticle mass peaks cannot be reconstructed directly, since each sparticle produces, at the end of the decay chain, the lightest neutralino (χ_1^0) that is stable and escapes detection. However, mass combinations of the visible SM particles produced in the various steps of the long SUSY decay chains (an example is given in Fig. 9) can be formed, and from the shapes of their

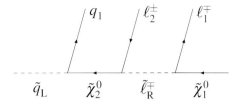

Figure 9. An example of squark decay chain (from Ref. 20).

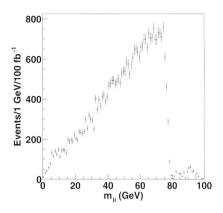

Figure 10. Expected di-lepton invariant mass distribution (after background subtraction) for point SPS1a of the mSUGRA parameter space and an integrated luminosity of 100 fb^{-1}, as obtained from a simulation of the ATLAS detector (from Ref. 20).

spectra, in particular from kinematic edges and end-points, constraints can be deduced on the unknown masses of the involved sparticles.

As an illustration, for the $\tilde{q}_L$ decay chain depicted in Fig. 9, which is quite typical over a broad region of the parameter space, Fig. 10 shows the reconstructed mass of the lepton pair obtained from an ATLAS simulation of a given point in the mSUGRA parameter space, the so-called SPS1a point[20]. In this case, the sparticle masses are about 540 GeV, 177 GeV, 143 GeV, 96 GeV for $\tilde{q}_L$, χ_2^0, $\tilde{\ell}_R$, χ_1^0 respectively. The distribution exhibits a sharp end-point, due to the two-body decay kinematics, whose position depends on the masses of the involved sparticles (χ_2^0, $\tilde{\ell}_R$, χ_1^0). Similarly, the invariant mass of the two leptons and one jet is characterized by an end-point that depends on the $\tilde{q}_L$, χ_2^0, χ_1^0 masses. These edges can be measured with experimental precisions between a few permil and a few percent in many cases, thereby providing a set of constraints on the involved sparticle masses. Therefore, by combining all experimental measurements, and solving the system of kinematic equations, it should be possible to derive the masses of several sparticles. This procedure is "model-independent", because it is based on a kinematic method and does not rely on any *a priori* assumption about the underlying theory.

Further progress can be made if some preliminary observations point toward a specific class of models. Indeed, by fitting the model(s) to the ensemble of experimental measurements (masses, as described above, and others), it should be possible to extract some of the fundamental parameters of the underlying theory, as well as other non-observable quantities. The most interesting ones are the mass of the escaping neutralino χ_1^0, a candidate for the universe cold dark matter; the expected dark matter density, assuming it is made only of relic neutralinos; and the neutralino-nucleon scattering cross-section. These results could then be compared to cosmological predictions, to astrophysics measurements[21], and to the findings of experiments performing direct searches for dark matter. This method has been demonstrated so far in the framework of very minimal models, like mSUGRA, where the expected precisions on masses[14] and dark matter parameters[22] are between a few percent and 20-30%, and is now being extended to more general models.

It would be a spectacular achievement for particle physics if we could demonstrate, at the LHC and with more precision later on at a Linear Collider[17], that the neutralinos produced in our laboratories are indeed the constituent of the universe dark matter because their features, as determined at accelerators, are consistent with cosmological predictions, astrophysical measurements and direct dark matter searches.

4.3 General strategy

From the examples presented above, it is possible to deduce some lessons about the general strategy to understand the underlying theory. An approach by steps will be followed:

1. Discovery phase. Inclusive analyses will be used in this phase (*e.g.* searches for events with jets and large missing transverse energy, searches for events with leptons), since the goal is to be as little model-dependent as possible, and therefore as much sensitive as possible to any scenario from new physics.

2. First characterization of the new theory. The combination of several inclusive signatures and distributions may offer strong discrimination power among classes of models. These general observations will include events with leptons (same-sign and opposite-sign), with taus, with b-tagged jets, with top-quarks, with large missing transverse energy, with exotic signatures (such as long-lived massive particles), *etc.*

3. Interpretation phase. The selection of increasingly more exclusive topologies and channels is required at this stage, in order to perform detailed precise measurements such as those described in Sec. 4.2. These exclusive samples may contain di-lepton edges, $h \to b\bar{b}$ peaks coming from the cascade decays of heavier particles, events with supersymmet-

ric Higgs bosons decaying as $A/H \to \mu\mu$, $\tau\tau$, events with $t\bar{t}$ pairs, *etc.*

It should be possible, at each of the above steps, to narrow the *a priori* huge variety of possible scenarios by excluding some of them, and to get guidance on how to continue. It will not be easy to pin down the correct framework. However, the LHC data will likely provide much more information than we can predict today, and the consistency with other data (*e.g.* from astrophysics observations, from precision experiments measuring rare decays) should also help to draw the global picture.

The ultimate goal will be to reconstruct the fundamental theory at high energy, and therefore understand the hierarchy of scales. For instance, if Nature is supersymmetric at the TeV scale, by measuring the masses of squarks, gluinos, charginos and neutralinos at the electroweak scale, and evolving them to the GUT scale using Renormalization Group Equations, it could be possible to test unification hypotheses at GUT energies. This goal can be best achieved by measuring the features of strongly-interacting sparticles at the LHC and those of electroweak sparticles at a Linear Collider[23].

5 Conclusions

In about two years from now the LHC will start operation, and particle physics will enter a new epoch, hopefully the most glorious and fruitful of its history. Indeed, the hierarchy problem indicates that the Standard Model is incomplete already at the TeV scale, and that new physics should be expected there.

The LHC will be able to explore this scale in detail, with a direct discovery potential up to particle masses of ∼5-6 TeV. Hence, if new physics is there the LHC will find it, and will also provide definitive answers about the SM Higgs mechanism, Supersymmetry, and several other TeV-scale predictions that have resisted experimental verification for decades [a]. More importantly, perhaps, the LHC will tell us which are the right questions to ask and how to continue.

References

1. S. Dawson, these Proceedings.

2. For a phenomenological review see for instance: P. Fayet and S. Ferrara, *Phys. Rep.* C **32**, 249 (1977); H. P. Nilles, *Phys. Rep.* C **110**, 1 (1984).

3. N. Arkani-Hamed, S. Dimopoulos and G. Dvali, *Phys. Lett.* B **429**, 263 (1998); L. Randall and R. Sundrum, *Phys. Rev. Lett.* **83**, 3370 (1999).

4. N. Arkani-Hamed, A.G. Cohen and H. Georgi, *JHEP* **0207**, 020 (2002); N. Arkani-Hamed, A.G. Cohen and H. Georgi, *Phys. Lett.* B **513**, 232 (2001).

5. S. Weinberg, *Phys. Rev.* D **19**, 1277 (1979); L. Susskind, *Phys. Rev.* D **20**, 2619 (1979).

6. N. Arkani-Hamed, S. Dimopoulos, G.F. Giudice and A. Romanino, *Nucl. Phys.* B **709**, 3 (2005).

7. L. Rolandi, these Proceedings.

8. ATLAS Collaboration, *Liquid Argon Calorimeter Technical Design Report*, CERN/LHCC/96-41.

9. M.L. Mangano, eConf **C030614**, 15 (2003).

10. C. Collard and M.-C. Lemaire, *Eur. Phys. J.* C **40**, Num 5, 15 (2005).

11. B.C. Allanach, K. Odagiri, A.M. Parker and B.R. Webber, *JHEP* **0009**, 004 (2000).

12. CMS Collaboration, *Technical Proposal*, CERN/LHCC/94-38.

13. A.H. Chamseddine, R. Arnowitt and P. Nath, *Phys. Rev. Lett.* **49**, 970 (1982).

14. ATLAS Collaboration, *Detector and Physics Performance Technical Design Report*, CERN/LHCC/99-15.

15. S. Asai *et al.*, *Eur. Phys. J.* C **32**, 19 (2004).

16. M. Sani, CMS Note 2003/046.

17. K. Moenig, these Proceedings.

18. M. Duehrssen, ATLAS Note ATL-PHYS-2003-030.

19. C.P. Buszello, I. Fleck, P. Marquard and J.J. van der Bij, *Eur. Phys. J.* C **32**, 209 (2004).

20. G. Weiglein *et al.*, hep-ph/0410364.

21. C. L. Bennet *et al.*, *Astrophys. J. Suppl.* **148** 1, (2003); D. N. Spergel *et al.*, *Astrophys. J. Suppl.* **148**, 175 (2003); H. V. Peiris *et al.*, *Astrophys. J. Suppl.* **148**, 213 (2003).

22. G. Polesello and D. Tovey, *JHEP* **0405**, 071 (2004).

23. B.C. Allanach *et al.*, hep-ph/0403133.

---

[a] We note that the future of particle physics and the planning for new facilities (Linear Colliders, underground dark matter experiments, *etc.*) would benefit a lot from a quick (*i.e.* by the end of this decade) determination of the scale of new physics.

DISCUSSION

Edmond Berger (Argonne):

The measurement of the inclusive hadronic jet cross section as a function of transverse energy should offer an excellent opportunity to observe evidence of a new physics scale, based on the deviations from the Standard Model QCD predictions. Scaling from the Tevatron results, I would guess that LHC should be able to access scales $E_\perp \simeq 4$ TeV after one year of operation. Many new physics scenarios will produce deviations from Standard Model QCD predictions. The measurement should be easy as long as the detectors have adequate energy resolution.

Fabiola Gianotti: Indeed, observables related to QCD jets are potentially sensitive to a huge number of scenarios beyond the Standard Model (*e.g.* Compositeness) and, as you mentioned, hundreds of events with jet $E_\perp \simeq 4$ TeV are expected with (very modest) integrated luminosities of order 100 pb^{-1}. However, extracting an indisputable signal of new physics from jet cross-section measurements will not be easy and will take a lot of time. This is because jet cross-sections are affected by large theoretical and instrumental uncertainties (calorimeter response, parton distribution functions, *etc.*). In particular, it will take time to understand the calorimeter response to jets as a function of energy up to the TeV scale. Possible uncorrected response non-linearities can produce deviations of the jet $E_\perp$ spectra from the QCD expectation, and therefore fake a signal of new physics. This is why I stressed that narrow resonances decaying into lepton pairs are more reliable for early discoveries.

Francesco Forti (INFN-Pisa):

Many new physics processes do not have clear and clean signatures, and you will have to detect them on top of the Standard Model processes. Is our understanding and simulation of the Standard Model advanced enough and detailed enough?

Fabiola Gianotti: The Standard Model backgrounds will be determined at the LHC by using a combination of data and Monte Carlo simulations (data alone or Monte Carlo alone are not enough, we need both). Recently, a lot of progress has been made from the theoretical and phenomenological side. The three main processes, the so-called "candles", W, Z and $t\bar{t}$ production, are theoretically known to 5% or better. Furthermore, matrix-element generators (*e.g.* ALPGEN, MC@NLO) for dominant background processes like W+jets, $t\bar{t}$+jets have been developed. They have also been interfaced to parton-shower Monte Carlo using methods avoiding jet double-counting.

Tord Ekelöf (Uppsala University):

Just as a comment to the previous question by Ed Berger on the knowledge of QCD as opposed to looking for leptons and peaks. I think that the real way is the combination of the two, and you really need to go in great detail with both. You have the early evaluation of the jet-multiplicity — you have to have a very good knowledge of that, and then combine it with lepton signals. So I do not see a choice there — it is really a combination.

Fabiola Gianotti: The experiments will study everything, all possible final states and topologies.

Ignatios Antoniadis (CERN):

What is the upper limit for a LSP neutralino that can be measured in LHC.

Will LHC be able to explore the whole region of a LSP dark matter?

Fabiola Gianotti: LSP neutralinos (χ_1^0) are produced at the end of the cascade decays of squarks and gluinos. Hence their observability depends on the observability of the strongly-interacting sparticles. Squarks and gluinos can be detected at the LHC up to masses of 2.5-3 TeV, which corresponds to neutralino masses of up to about 400 GeV. There are regions of the (mSUGRA) parameter space with neutralino physics consistent with dark matter where squark and gluino masses are heavier than 2.5-3 TeV (an example is given by the so-called "rapid-annihilation funnels"). These regions are not accessible at the LHC.

THE LHC MACHINE AND EXPERIMENTS

GIGI ROLANDI

European Laboratory for Particle Physics (CERN), 1211 Geneva 23, Switzerland
E-mail: Gigi.Rolandi@cern.ch

This paper is a very short report on the status of the Large Hadron Collider(LHC) and its four experiments at the end of August 2005

This paper is a very short report on the status of the Large Hadron Collider(LHC)[1] and its four experiments at the end of August 2005. A copy of the slides[2] and a video[3] of the presentation can be found on the web site of the conference.

The first superconducting magnet for the LHC was lowered into the accelerator tunnel at 2.00 p.m. on Monday, 7th March. This is the first of the 1232 dipole magnets for the future collider, which measures 27 km in circumference and is scheduled to be commissioned in 2007. The magnets production proceeds very well and is on schedule, to date more than 800 magnets have been delivered and about 100 have been already installed in the tunnel following the installation of the cryogenic pipes. The quality of the magnets is very good. The first full LHC cell (120 m long) comprising 6 dipoles + 4 quadrupoles has been successful tests at nominal current (12 kA) since 2002.

The installation of the LHC in the tunnel is on the critical path for the first collisions. Critical items are the cryogenic services lines (QRL). The QRL runs along side the magnets in the tunnel and supplies the liquid helium necessary to get the super conducting magnets down to 1.9 K. The QRL had problems in the past and a recovery plan was implemented successfully one year ago. To date QRL components have been delivered for 4 of the 8 LHC sectors and almost two sectors have been already equipped. The first QRL subsector has been cooled few weeks ago: there were some small problems during a pressure test before the cool down attempt but these were resolved quickly and thereafter things went well.

The LHC schedule[4] foresees a parallel installation of pairs of sectors. The last sector will be completed in June 2007. The first pair of sectors (sector 7 and sector 8) will be completed in May 2006 and cooled down for a first test with beam. The test will involve injection of beam from the SPS down TI8 into LHC at the injection point right of point 8. The beam would then pass though IP8 (LHCb) and then through sector 8-7 to a temporary beam dump located after the Q6 quadrupole just right of the warm insertion of point 7. Around 2 weeks beam time is foreseen for the test. The scope of this test is to check that the ensemble of installed equipment works as foreseen, that there are no problems with ongoing installation, and to pre-commission essential acquisition and correction procedures.

LHC will be cooled down in summer 2007. After the machine check-out the commissioning with beam will start and will be followed by a pilot run in fall 2007. The startup of the machine is foreseen in four stages approaching gradually the final number of bunches (2808), the final bunch intensity (1.15 10^{11}) and the final squeeze (0.5 m). In the first stage LHC will run with 43 x 43 bunches (moving to 156 x 156) with moderate intensities. First collisions unsqueezed, followed by partial squeeze. The maximal luminosity will gradually approach 10^{32} cm^{-2} sec^{-1}. In the second stage LHC will move to 75 ns bunch crossing with the aim of moving to intensities around $3 \cdot 10^{10}$ particles per bunch. The maximal luminosity will gradually approach $10^{33} \ cm^{-2} \ sec^{-1}$. In the third stage LHC will move to 25 ns bunch crossing and the luminosity will approach again $10^{33} \ cm^{-2} \ sec^{-1}$ with fewer interactions per bunch crossing. The fourth stage will be done after the installation of the final collimator system and of the complete dump system. This will allow reaching the design luminosity of $10^{34} \ cm^{-2} \ sec^{-1}$.

The four LHC experiments LHCb, Alice, CMS and ATLAS are installing their apparatus and preparing for the commissioning phase.

LHCb[5] have already installed the magnet, the electromagnetic calorimeter, the hadron calorimeter and the iron for the muon filters. Good achievement has been made for the construction of many subsystems and the construction of the muon chambers is proceeding with tight planning. They will be able to fully exploit LHC since day one, also because their design luminosity is lower than ATLAS and CMS.

The Alice[6] magnet (former L3 magnet) is ready for the installation of the experiment. The very large Time Projection Chamber is very much advanced as many other subsystems. An almost complete initial detector will be ready for the first pp collisions and for the first Heavy Ions run. The detector will be completed in 2008 with the installation of the remaining 50% of the Transition Radiation Detector and the remaining 3 out of 8 elements of the photon calorimeter.

The construction of the CMS[7] magnet and its coil has been completed in the surface building while the infrastructures are being prepared in the experimental cavern. The hadron calorimeter is completed and the installation of the muon chambers in the magnet joke is proceeding. The magnet will be powered for the first time near the end of the year. After the test of the magnet in the surface building the heavy lowering will start: 15 heavy lifts of about 1 week duration each, the heaviest piece (central wheel + solenoid) is about 2000 tons. The CMS sub-detectors are being commissioned with cosmic rays for addressing system issues. An important integration milestone is the slice test during the test of the magnet beginning of 2006. Test with cosmic rays will continue in the pit after installation and re-cabling. CMS foresee to install the ECAL endcaps and the pixel vertex detector (even though ready) after the pilot run in the 2006/2007 shutdown. The procurement of the Ecal barrel crystals and the integration of the Tracker proceed with very tight planning. The objectives for the pilot run are: to verify data coherence, sub-system synchronization; to inter-calibrate ECAL barrel crystals to 2%; to cross check and complete source calibration for HCAL channels to 2%; to align the tracker strip detector significantly below the 100 μm level and to align the muon chambers at the 100 μm level.

ATLAS[8] are progressing in the installation in the experimental cavern. They have recently completed the installation of the eight huge coils of the barrel muon system. The construction and installation of the muon chambers is proceeding on schedule. The electromagnetic calorimeter is completed. A full vertical slice of ATLAS was tested in the beam in fall 2004: for the first time, all ATLAS sub-detectors were integrated and run together with common DAQ, "final" electronics and DCS. ATLAS are using cosmic rays for initial physics alignment and calibration of the detector, debugging of sub-systems, mapping dead channels etc. These activities will continue during the machine commissioning. The objectives for the pilot run are : to prepare the trigger and the detector; to tune the trigger menus; to begin to measure reconstruction efficiencies, fake rates, energy scales, resolutions; to begin to understand backgrounds to discovery channels. ATLAS is on track for collisions in summer 2007 and physics still in 2007.

References

1. http://lhc.web.cern.ch/lhc/
2. http://www.uu.se/LP2005/
   LP2005 /programme/presentationer/
   E2_rolandi_Uppsala_Gigi.ppt
3. http://lp2005.tsl.uu.se/∼lp2005/
   LP2005/webcast/real/Rolandi/web/index.htm
4. http://sylvainw.home.cern.ch/
   sylvainw/planning-follow-up/Schedule.pdf
5. http://lhcb-new.web.cern.ch/
   LHCb-new/
6. http://aliceinfo.cern.ch/
7. http://cmsinfo.cern.ch/
   Welcome.html/
8. http://atlas.ch/

EXPERIMENTAL SIGNATURES OF STRINGS AND BRANES

IGNATIOS ANTONIADIS

Department of Physics, CERN - Theory Division, 1211 Geneva 23, Switzerland
E-mail: ignatios.antoniadis@cern.ch

(on leave from CPHT (UMR CNRS 7644) Ecole Polytechnique, F-91128 Palaiseau)

Type I string theory provides a D-brane world description of our universe and leads to two new scenaria for physics beyond the Standard Model: low string scale and split supersymmetry. Lowering the string scale in the TeV region provides a theoretical framework for solving the mass hierarchy problem and unifying all interactions. The apparent weakness of gravity can then be accounted by the existence of large internal dimensions, in the submillimeter region, and transverse to a braneworld where we must be confined. I review the main properties of this scenario and its implications for observations at both particle colliders, and in non-accelerator gravity experiments. I also present a concrete realization of split supersymmetry which guarantees gauge coupling unification at the conventional scale $M_{\rm GUT} \simeq 2 \times 10^{16}$ GeV.

1 Introduction

During the last few decades, physics beyond the Standard Model (SM) was guided from the problem of mass hierarchy. This can be formulated as the question of why gravity appears to us so weak compared to the other three known fundamental interactions corresponding to the electromagnetic, weak and strong nuclear forces. Indeed, gravitational interactions are suppressed by a very high energy scale, the Planck mass $M_P \sim 10^{19}$ GeV, associated to a length $l_P \sim 10^{-35}$ m, where they are expected to become important. In a quantum theory, the hierarchy implies a severe fine tuning of the fundamental parameters in more than 30 decimal places in order to keep the masses of elementary particles at their observed values. The reason is that quantum radiative corrections to all masses generated by the Higgs vacuum expectation value (VEV) are proportional to the ultraviolet cutoff which in the presence of gravity is fixed by the Planck mass. As a result, all masses are "attracted" to become about 10^{16} times heavier than their observed values.

Besides compositeness, there are three main theories that have been proposed and studied extensively during the last years, corresponding to different approaches of dealing with the mass hierarchy problem. (1) Low energy supersymmetry with all superparticle masses in the TeV region. Indeed, in the limit of exact supersymmetry, quadratically divergent corrections to the Higgs self-energy are exactly canceled, while in the softly broken case, they are cutoff by the supersymmetry breaking mass splittings. (2) TeV scale strings, in which quadratic divergences are cutoff by the string scale and low energy supersymmetry is not needed. (3) Split supersymmetry, where scalar masses are heavy while fermions (gauginos and higgsinos) are light. Thus, gauge coupling unification and dark matter candidate are preserved but the mass hierarchy should be stabilized by a different way and the low energy world appears to be fine-tuned. All these ideas are experimentally testable at high-energy particle colliders and in particular at LHC. Below, I discuss their implementation in string theory.

The appropriate and most convenient framework for low energy supersymmetry and grand unification is the perturbative heterotic string. Indeed, in this theory, gravity and gauge interactions have the same origin, as massless modes of the closed heterotic string, and they are unified at the string scale M_s. As a result, the Planck mass M_P is predicted to be proportional to M_s:

$$M_P = M_s/g\,, \qquad (1)$$

where g is the gauge coupling. In the simplest constructions all gauge couplings are the same at the string scale, given by the four-dimensional (4d) string coupling, and thus no grand unified group is needed for unification. In our conventions $\alpha_{\rm GUT} = g^2 \simeq 0.04$, leading to a discrepancy between the string and grand unification scale $M_{\rm GUT}$ by almost two orders of magnitude. Explaining this gap introduces in general new parameters or a new scale, and the predictive power is essentially lost. This is the main defect of this framework, which remains though an open and interesting possibility.

The other two ideas have both as natural framework of realization type I string theory with D-branes. Unlike in the heterotic string, gauge and gravitational interactions have now different origin. The latter are described again by closed strings, while the former emerge as excitations of open strings with endpoints confined on D-branes[1]. This leads to a braneworld description of our universe, which should be localized on a hypersurface, i.e. a membrane extended in p spatial dimensions, called p-brane (see Fig. 1). Closed strings propagate in all nine dimensions of string theory: in those extended along the p-brane, called parallel, as well as in the transverse ones. On the contrary, open strings are attached on the p-brane. Obviously, our p-brane world must have at least the three known dimensions of space. But it may contain more: the extra $d_\parallel = p-3$ parallel dimensions must have a finite size, in order to be unobservable at present energies, and can be as large as TeV$^{-1} \sim 10^{-18}$ m^2. On the other hand, transverse dimensions interact with us only gravitationally and experimental bounds are much weaker: their size should be less than about 0.1 mm [3]. In the following, I review the main properties and experimental signatures of low string scale[4,5]

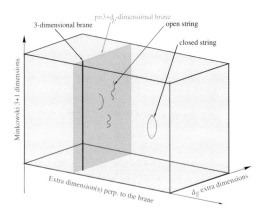

Figure 1. In the type I string framework, our Universe contains, besides the three known spatial dimensions (denoted by a single blue line), some extra dimensions ($d_\parallel = p-3$) parallel to our world p-brane (green plane) where endpoints of open strings are confined, as well as some transverse dimensions (yellow space) where only gravity described by closed strings can propagate.

and split supersymmetry[6,7] proposals.

2 Low string scale

2.1 Framework

In type I theory, the different origin of gauge and gravitational interactions implies that the relation between the Planck and string scales is not linear as (1) of the heterotic string. The requirement that string theory should be weakly coupled, constrain the size of all parallel dimensions to be of order of the string length, while transverse dimensions remain unrestricted. Assuming an isotropic transverse space of $n = 9 - p$ compact dimensions of common radius $R_\perp$, one finds:

$$M_P^2 = \frac{1}{g^4} M_s^{2+n} R_\perp^n\,, \qquad g_s \simeq g^2\,. \qquad (2)$$

where g_s is the string coupling. It follows that the type I string scale can be chosen hierarchically smaller than the Planck mass[8,4] at the expense of introducing extra large transverse dimensions felt only by gravity, while keeping the string coupling small[4]. The

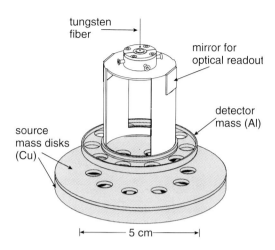

Figure 2. Torsion pendulum that tested Newton's law at 130 nm. Several sources of background noise were eliminated using appropriate devices.

weakness of 4d gravity compared to gauge interactions (ratio M_W/M_P) is then attributed to the largeness of the transverse space $R_\perp$ compared to the string length $l_s = M_s^{-1}$.

An important property of these models is that gravity becomes effectively $(4+n)$-dimensional with a strength comparable to those of gauge interactions at the string scale. The first relation of Eq. (2) can be understood as a consequence of the $(4+n)$-dimensional Gauss law for gravity, with

$$M_*^{(4+n)} = M_s^{2+n}/g^4 \qquad (3)$$

the effective scale of gravity in $4+n$ dimensions. Taking $M_s \simeq 1$ TeV, one finds a size for the extra dimensions $R_\perp$ varying from 10^8 km, .1 mm, down to a Fermi for $n = 1, 2$, or 6 large dimensions, respectively. This shows that while $n = 1$ is excluded, $n \geq 2$ is allowed by present experimental bounds on gravitational forces[3,9]. Thus, in these models, gravity appears to us very weak at macroscopic scales because its intensity is spread in the "hidden" extra dimensions. At distances shorter than $R_\perp$, it should deviate from Newton's law, which may be possible to explore in laboratory experiments (see Fig. 2).

The main experimental implications of

TeV scale strings in particle accelerators are of three types, in correspondence with the three different sectors that are generally present: (i) new compactified parallel dimensions, (ii) new extra large transverse dimensions and low scale quantum gravity, and (iii) genuine string and quantum gravity effects. On the other hand, there exist interesting implications in non accelerator table-top experiments due to the exchange of gravitons or other possible states living in the bulk.

2.2 World-brane extra dimensions

In this case $RM_s \gtrsim 1$, and the associated compactification scale $R_\parallel^{-1}$ would be the first scale of new physics that should be found increasing the beam energy[2,10]. There are several reasons for the existence of such dimensions. It is a logical possibility, since out of the six extra dimensions of string theory only two are needed for lowering the string scale, and thus the effective p-brane of our world has in general $d_\parallel \equiv p-3 \leq 4$. Moreover, they can be used to address several physical problems in braneworld models, such as obtaining different SM gauge couplings, explaining fermion mass hierarchies due to different localization points of quarks and leptons in the extra dimensions, providing calculable mechanisms of supersymmetry breaking, etc.

The main consequence is the existence of Kaluza-Klein (KK) excitations for all SM particles that propagate along the extra parallel dimensions. Their masses are given by:

$$M_m^2 = M_0^2 + \frac{m^2}{R_\parallel^2} \quad ; \quad m = 0, \pm 1, \pm 2, \ldots \ (4)$$

where we used $d_\parallel = 1$, and M_0 is the higher dimensional mass. The zero-mode $m = 0$ is identified with the 4d state, while the higher modes have the same quantum numbers with the lowest one, except for their mass given in (4). There are two types of experimental signatures of such dimensions[10,11,12]: (i) virtual exchange of KK excitations, leading to deviations in cross-sections compared to the

SM prediction, that can be used to extract bounds on the compactification scale; (ii) direct production of KK modes.

On general grounds, there can be two different kinds of models with qualitatively different signatures depending on the localization properties of matter fermion fields. If the latter are localized in 3d brane intersections, they do not have excitations and KK momentum is not conserved because of the breaking of translation invariance in the extra dimension(s). KK modes of gauge bosons are then singly produced giving rise to generally strong bounds on the compactification scale and new resonances that can be observed in experiments. Otherwise, they can be produced only in pairs due to the KK momentum conservation, making the bounds weaker but the resonances difficult to observe.

When the internal momentum is conserved, the interaction vertex involving KK modes has the same 4d tree-level gauge coupling. On the other hand, their couplings to localized matter have an exponential form factor suppressing the interactions of heavy modes. This form factor can be viewed as the fact that the branes intersection has a finite thickness. For instance, the coupling of the KK excitations of gauge fields $A^\mu(x,y) = \sum_m A^\mu_m \exp i\frac{my}{R_\parallel}$ to the charge density $j_\mu(x)$ of massless localized fermions is described by the effective action[13]:

$$\int d^4x \sum_m e^{-\ln 16 \frac{m^2 l_s^2}{2R_\parallel^2}} j_\mu(x) A^\mu_m(x). \quad (5)$$

After Fourier transform in position space, it becomes:

$$\int d^4x\, dy \frac{1}{(2\pi \ln 16)^2} e^{-\frac{y^2 M_s^2}{2\ln 16}} j_\mu(x) A^\mu(x,y), \quad (6)$$

from which we see that localized fermions form a Gaussian distribution of charge with a width $\sigma = \sqrt{\ln 16}\, l_s \sim 1.66\, l_s$.

To simplify the analysis, let us consider first the case $d_\parallel = 1$ where some of the gauge fields arise from an effective 4-brane, while

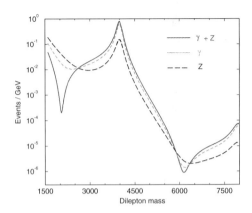

Figure 3. Production of the first KK modes of the photon and of the Z boson at LHC, decaying to electron-positron pairs. The number of expected events is plotted as a function of the energy of the pair in GeV. From highest to lowest: excitation of $\gamma + Z$, γ and Z.

fermions are localized states on brane intersections. Since the corresponding gauge couplings are reduced by the size of the large dimension $R_\parallel M_s$ compared to the others, one can account for the ratio of the weak to strong interactions strengths if the $SU(2)$ brane extends along the extra dimension, while $SU(3)$ does not. As a result, there are 3 distinct cases to study[12], denoted by (t,l,l), (t,l,t) and (t,t,l), where the three positions in the brackets correspond to the three SM gauge group factors $SU(3) \times SU(2) \times U(1)$ and those with l (longitudinal) feel the extra dimension, while those with t (transverse) do not.

In the (t,l,l) case, there are KK excitations of $SU(2) \times U(1)$ gauge bosons: $W^{(m)}_\pm$, $\gamma^{(m)}$ and $Z^{(m)}$. Performing a χ^2 fit of the electroweak observables, one finds that if the Higgs is a bulk state (l), $R_\parallel^{-1} \gtrsim 3.5$ TeV [14]. This implies that LHC can produce at most the first KK mode. Different choices for localization of matter and Higgs fields lead to bounds, lying in the range $1 - 5$ TeV [14].

In addition to virtual effects, KK excitations can be produced on-shell at LHC as new resonances[11] (see Fig. 3). There are two different channels, neutral Drell–Yan processes $pp \to l^+l^-X$ and the charged channel $l^\pm\nu$, corresponding to the production of the KK modes $\gamma^{(1)}, Z^{(1)}$ and $W_\pm^{(1)}$, respectively. The discovery limits are about 6 TeV, while the exclusion bounds 15 TeV. An interesting observation in the case of $\gamma^{(1)} + Z^{(1)}$ is that interferences can lead to a "dip" just before the resonance. There are some ways to distinguish the corresponding signals from other possible origin of new physics, such as models with new gauge bosons. In fact, in the (t,l,l) and (t,l,t) cases, one expects two resonances located practically at the same mass value. This property is not shared by most of other new gauge boson models. Moreover, the heights and widths of the resonances are directly related to those of SM gauge bosons in the corresponding channels.

In the (t,l,t) case, only the $SU(2)$ factor feels the extra dimension and the limits set by the KK states of $W^\pm$ remain the same. On the other hand, in the (t,t,l) case where only $U(1)_Y$ feels the extra dimension, the limits are weaker and the exclusion bound is around 8 TeV. In addition to these simple possibilities, brane constructions lead often to cases where part of $U(1)_Y$ is t and part is l. If $SU(2)$ is l the limits come again from $W^\pm$, while if it is t then it will be difficult to distinguish this case from a generic extra $U(1)'$. A good statistics would be needed to see the deviation in the tail of the resonance as being due to effects additional to those of a generic $U(1)'$ resonance. Finally, in the case of two or more parallel dimensions, the sum in the exchange of the KK modes diverges in the limit $R_\parallel M_s >> 1$ and needs to be regularized using the form factor (5). Cross-sections become bigger yielding stronger bounds, while resonances are closer implying that more of them could be reached by LHC.

On the other hand, if all SM particles propagate in the extra dimension (called universal)[a], KK modes can only be produced in

pairs and the lower bound on the compactification scale becomes weaker, of order of 300-500 GeV. Moreover, no resonances can be observed at LHC, so that this scenario appears very similar to low energy supersymmetry. In fact, KK parity can even play the role of R-parity, implying that the lightest KK mode is stable and can be a dark matter candidate in analogy to the LSP[15].

2.3 Extra large transverse dimensions

The main experimental signal is gravitational radiation in the bulk from any physical process on the world-brane. In fact, the very existence of branes breaks translation invariance in the transverse dimensions and gravitons can be emitted from the brane into the bulk. During a collision of center of mass energy $\sqrt{s}$, there are $\sim (\sqrt{s}R_\perp)^n$ KK excitations of gravitons with tiny masses, that can be emitted. Each of these states looks from the 4d point of view as a massive, quasi-stable, extremely weakly coupled (s/M_P^2 suppressed) particle that escapes from the detector. The total effect is a missing-energy cross-section roughly of order:

$$\frac{(\sqrt{s}R_\perp)^n}{M_P^2} \sim \frac{1}{s}(\frac{\sqrt{s}}{M_s})^{n+2}. \qquad (7)$$

Explicit computation of these effects leads to the bounds given in Table 1. However, larger radii are allowed if one relaxes the assumption of isotropy, by taking for instance two large dimensions with different radii.

Fig. 4 shows the cross-section for graviton emission in the bulk, corresponding to the process $pp \to jet+graviton$ at LHC, together with the SM background[16]. For a given value of M_s, the cross-section for graviton emission decreases with the number of large transverse dimensions, in contrast to the case of parallel dimensions. The reason is that gravity becomes weaker if there are more dimensions

---

[a]Although interesting, this scenario seems difficult

to be realized, since 4d chirality requires non-trivial action of orbifold twists with localized chiral states at the fixed points.

Table 1. Limits on $R_\perp$ in mm.

Experiment	$n = 2$	$n = 4$	$n = 6$
Collider bounds			
LEP 2	5×10^{-1}	2×10^{-8}	7×10^{-11}
Tevatron	5×10^{-1}	10^{-8}	4×10^{-11}
LHC	4×10^{-3}	6×10^{-10}	3×10^{-12}
NLC	10^{-2}	10^{-9}	6×10^{-12}
Present non-collider bounds			
SN1987A	3×10^{-4}	10^{-8}	6×10^{-10}
COMPTEL	5×10^{-5}	-	-

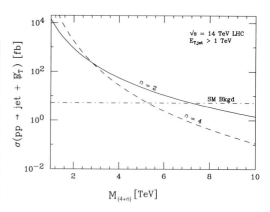

Figure 4. Missing energy due to graviton emission at LHC, as a function of the higher-dimensional gravity scale M_*, produced together with a hadronic jet. The expected cross-section is shown for $n = 2$ and $n = 4$ extra dimensions, together with the SM background.

because there is more space for the gravitational field to escape. There is a particular energy and angular distribution of the produced gravitons that arise from the distribution in mass of KK states of spin-2. This can be contrasted to other sources of missing energy and might be a smoking gun for the extra dimensional nature of such a signal.

In Table 1, there are also included astrophysical and cosmological bounds. Astrophysical bounds[17,18] arise from the requirement that the radiation of gravitons should not carry on too much of the gravitational binding energy released during core collapse of supernovae. In fact, the measurements of Kamiokande and IMB for SN1987A suggest that the main channel is neutrino fluxes. The best cosmological bound[19] is obtained from requiring that decay of bulk gravitons to photons do not generate a spike in the energy spectrum of the photon background measured by the COMPTEL instrument. Bulk gravitons are expected to be produced just before nucleosynthesis due to thermal radiation from the brane. The limits assume that the temperature was at most 1 MeV as nucleosynthesis begins, and become stronger if temperature is increased.

2.4 String effects

At low energies, the interaction of light (string) states is described by an effective field theory. Their exchange generates in par-

ticular four-fermion operators that can be used to extract independent bounds on the string scale. In analogy with the bounds on longitudinal extra dimensions, there are two cases depending on the localization properties of matter fermions. If they come from open strings with both ends on the same stack of branes, exchange of massive open string modes gives rise to dimension eight effective operators, involving four fermions and two space-time derivatives[20,13]. The corresponding bounds on the string scale are then around 500 GeV. On the other hand, if matter fermions are localized on non-trivial brane intersections, one obtains dimension six four-fermion operators and the bounds become stronger: $M_s \gtrsim 2 - 3$ TeV [13,5]. At energies higher than the string scale, new spectacular phenomena are expected to occur, related to string physics and quantum gravity effects, such as possible micro-black hole production[21]. Particle accelerators would then become the best tools for studying quantum gravity and string theory.

2.5 Sub-millimeter forces

Besides the spectacular predictions in accelerators, there are also modifications of gravitation in the sub-millimeter range, which can be tested in "table-top" experiments that measure gravity at short distances. There are three categories of such predictions:

(i) Deviations from the Newton's law $1/r^2$ behavior to $1/r^{2+n}$, which can be observable for $n = 2$ large transverse dimensions of sub-millimeter size. This case is particularly attractive on theoretical grounds because of the logarithmic sensitivity of SM couplings on the size of transverse space[22], that allows to determine the hierarchy[23].

(ii) New scalar forces in the sub-millimeter range, related to the mechanism of supersymmetry breaking, and mediated by light scalar fields φ with masses[24,4]:

$$m_\varphi \simeq \frac{m_{susy}^2}{M_P} \simeq 10^{-4} - 10^{-6} \text{ eV} , \quad (8)$$

for a supersymmetry breaking scale $m_{susy} \simeq 1 - 10$ TeV. They correspond to Compton wavelengths of 1 mm to 10 μm. m_{susy} can be either $1/R_\parallel$ if supersymmetry is broken by compactification[24], or the string scale if it is broken "maximally" on our world-brane[4]. A universal attractive scalar force is mediated by the radion modulus $\varphi \equiv M_P \ln R$, with R the radius of the longitudinal or transverse dimension(s). In the former case, the result (8) follows from the behavior of the vacuum energy density $\Lambda \sim 1/R_\parallel^4$ for large $R_\parallel$ (up to logarithmic corrections). In the latter, supersymmetry is broken primarily on the brane, and thus its transmission to the bulk is gravitationally suppressed, leading to (8). For $n = 2$, there may be an enhancement factor of the radion mass by $\ln R_\perp M_s \simeq 30$ decreasing its wavelength by an order of magnitude[23].

The coupling of the radius modulus to matter relative to gravity can be easily computed and is given by:

$$\sqrt{\alpha_\varphi} = \frac{1}{M}\frac{\partial M}{\partial \varphi} ; \quad \alpha_\varphi = \begin{cases} \frac{\partial \ln \Lambda_{QCD}}{\partial \ln R} \simeq \frac{1}{3} & \text{for } R_\parallel \\ \\ \frac{2n}{n+2} = 1 - 1.5 & \text{for } R_\perp \end{cases}$$

$$(9)$$

where M denotes a generic physical mass. In the longitudinal case, the coupling arises dominantly through the radius dependence of the QCD gauge coupling[24], while in the case of transverse dimension, it can be deduced from the rescaling of the metric which changes the string to the Einstein frame and depends slightly on the bulk dimensionality ($\alpha = 1 - 1.5$ for $n = 2 - 6$) [23]. Such a force can be tested in microgravity experiments and should be contrasted with the change of Newton's law due the presence of extra dimensions that is observable only for $n = 2$ [3,9]. The resulting bounds from an analysis of the radion effects are[3]:

$$M_* \gtrsim 3 - 4.5 \text{ TeV} \quad \text{for} \quad n = 2 - 6 . \quad (10)$$

In principle there can be other light moduli which couple with even larger strengths. For example the dilaton, whose VEV determines the string coupling, if it does not acquire large mass from some dynamical supersymmetric mechanism, can lead to a force of strength 2000 times bigger than gravity[25].

(iii) Non universal repulsive forces much stronger than gravity, mediated by possible abelian gauge fields in the bulk[17,26]. Such fields acquire tiny masses of the order of M_s^2/M_P, as in (8), due to brane localized anomalies[26]. Although their gauge coupling is infinitesimally small, $g_A \sim M_s/M_P \simeq 10^{-16}$, it is still bigger that the gravitational coupling E/M_P for typical energies $E \sim 1$ GeV, and the strength of the new force would be $10^6 - 10^8$ stronger than gravity. This is an interesting region which will be soon explored in micro-gravity experiments (see Fig. 5). Note that in this case supernova constraints impose that there should be at least four large extra dimensions in the bulk[17].

In Fig. 5 we depict the actual information from previous, present and upcoming experiments[23]. The solid lines indicate the present limits from the experiments indicated. The excluded regions lie above these solid lines. Measuring gravitational strength forces at short distances is challenging. The dashed thick lines give the expected sensitiv-

76

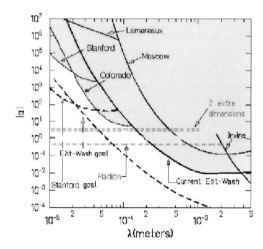

Figure 5. Present limits on non-Newtonian forces at short distances (yellow regions), as a function of their range λ and their strength relative to gravity α. The limits are compared to new forces mediated by the graviton in the case of two large extra dimensions, and by the radion.

ity of the various experiments, which will improve the actual limits by roughly two orders of magnitude, while the horizontal dashed lines correspond to the theoretical predictions for the graviton in the case $n = 2$ and for the radion in the transverse case. These limits are compared to those obtained from particle accelerator experiments in Table 1.

2.6 Brane non-linear supersymmetry

When the closed string sector is supersymmetric, supersymmetry on a generic brane configuration is non-linearly realized even if the spectrum is not supersymmetric and brane fields have no superpartners. The reason is that the gravitino must couple to a conserved current locally, implying the existence of a goldstino on the brane world-volume. The goldstino is exactly massless in the infinite (transverse) volume limit and is expected to acquire a small mass suppressed by the volume, of order (8). In the standard realization, its coupling to matter is given via the energy

momentum tensor[27], while in general there are more terms invariant under non-linear supersymmetry that have been classified, up to dimension eight[28,29].

An explicit computation was performed for a generic intersection of two brane stacks, leading to three irreducible couplings, besides the standard one[29]: two of dimension six involving the goldstino, a matter fermion and a scalar or gauge field, and one four-fermion operator of dimension eight. Their strength is set by the goldstino decay constant κ, up to model-independent numerical coefficients which are independent of the brane angles. Obviously, at low energies the dominant operators are those of dimension six. In the minimal case of (non-supersymmetric) SM, only one of these two operators may exist, that couples the goldstino χ with the Higgs H and a lepton doublet L:

$$\mathcal{L}_\chi^{int} = 2\kappa (D_\mu H)(L D^\mu \chi) + h.c., \qquad (11)$$

where the goldstino decay constant is given by the total brane tension

$$\frac{1}{2\,\kappa^2} = N_1\,T_1 + N_2\,T_2\,; \quad T_i = \frac{M_s^4}{4\pi^2 g_i^2}\,, \quad (12)$$

with N_i the number of branes in each stack. It is important to notice that the effective interaction (11) conserves the total lepton number L, as long as we assign to the goldstino a total lepton number $L(\chi) = -1$ [30]. To simplify the analysis, we will consider the simplest case where (11) exists only for the first generation and L is the electron doublet[30].

The effective interaction (11) gives rise mainly to the decays $W^\pm \to e^\pm \chi$ and $Z, H \to \nu\chi$. It turns out that the invisible Z width gives the strongest limit on κ which can be translated to a bound on the string scale $M_s \gtrsim 500$ GeV, comparable to other collider bounds. This allows for the striking possibility of a Higgs boson decaying dominantly, or at least with a sizable branching ratio, via such an invisible mode, for a wide range of the parameter space (M_s, m_H), as seen in Fig. 6.

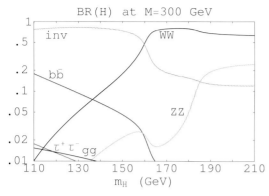

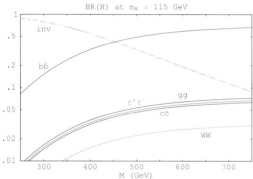

Figure 6. Higgs branching rations, as functions either of the Higgs mass m_H for a fixed value of the string scale $M_s \simeq 2M = 600$ GeV, or of $M \simeq M_s/2$ for $m_H = 115$ GeV.

2.7   Electroweak symmetry breaking

Non-supersymmetric TeV strings offer also a framework to realize gauge symmetry breaking radiatively. Indeed, from the effective field theory point of view, one expects quadratically divergent one-loop contributions to the masses of scalar fields. The divergences are cut off by M_s and if the corrections are negative, they can induce electroweak symmetry breaking and explain the mild hierarchy between the weak and a string scale at a few TeV, in terms of a loop factor[31]. More precisely, in the minimal case of one Higgs doublet H, the scalar potential is:

$$V = \lambda(H^\dagger H)^2 + \mu^2(H^\dagger H), \qquad (13)$$

where λ arises at tree-level. Moreover, in any model where the Higgs field comes from an open string with both ends fixed on the same brane stack, it is given by an appropriate truncation of a supersymmetric theory. Within the minimal spectrum of the SM, $\lambda = (g_2^2 + g'^2)/8$, with g_2 and g' the $SU(2)$ and $U(1)_Y$ gauge couplings. On the other hand, μ^2 is generated at one loop:

$$\mu^2 = -\varepsilon^2\, g^2\, M_s^2\,, \qquad (14)$$

where ϵ is a loop factor that can be estimated from a toy model computation and varies in the region $\epsilon \sim 10^{-1} - 10^{-3}$.

The potential (13) has the usual minimum, given by the VEV of the neutral component of the Higgs doublet $v = \sqrt{-\mu^2/\lambda}$. Using the relation of v with the Z gauge boson mass, $M_Z^2 = (g_2^2 + g'^2)v^2/4$, and the expression of the quartic coupling λ, one obtains for the Higgs mass a prediction which is the MSSM value for $\tan\beta \to \infty$ and $m_A \to \infty$: $m_H = M_Z$. The tree level Higgs mass is known to receive important radiative corrections from the top-quark sector and rises to values around 120 GeV. Furthermore, from (14), one can compute M_s in terms of the Higgs mass $m_H^2 = -2\mu^2$:

$$M_s = \frac{m_H}{\sqrt{2}\, g\varepsilon}\,, \qquad (15)$$

yielding naturally values in the TeV range.

3   Split supersymmetry

3.1   Motivations

Recent precision tests of the SM, implying the absence of any deviation to a great accuracy, suggest that any new physics at a TeV needs to be fine-tuned at the per-cent level. Thus, either the underlying theory beyond the SM is very special, or our notion of naturalness should be reconsidered. The latter is also motivated from the recent evidence of a tiny cosmological constant creating another more severe hierarchy problem. This raises the possibility that the same mechanism may solve both problems and casts some doubts on all previous proposals on gauge hierarchy.

On the other hand, the necessity of a Dark Matter (DM) candidate and the fact that LEP data favor the unification of SM gauge couplings are smoking guns for the presence of new physics at high energies. Supersymmetry is then a nice candidate offering both properties and arising naturally in string theory. It was then proposed to consider that supersymmetry might be broken at high energies without solving the gauge hierarchy problem. More precisely, making squarks and sleptons heavy does not spoil unification and the existence of a DM candidate while at the same time it gets rid of all unwanted features of the supersymmetric SM related to its complicated scalar sector. On the other hand, experimental hints to the existence of supersymmetry persist since there are still gauginos and higgsinos at the electroweak scale. This is the split supersymmetry framework[6]. Its main experimental signals are long lived gluinos that may give rise to displaced vertices or decays outside the detector, and several relations among Yukawa couplings involving Higgs, higgsinos and electroweak gauginos, valid at the supersymmetry breaking scale $m_{susy} \sim m_0$.

3.2 String realization

Split supersymmetry has a natural realization in type I string theory with magnetized D9-branes, or equivalently with branes at angles[7]. Indeed, internal magnetic fields can be turned on around any non-contractible 2-cycle of the internal compactification manifold. The Gauss law for the magnetic flux implies that the field H is quantized in terms of the area of the corresponding 2-cycle A:

$$ H = \frac{m}{nA} , \qquad (16) $$

where the integers m, n correspond to the respective magnetic and electric charges; m is the quantized flux and n is the wrapping number of the brane around the 2-cycle.

For simplicity, consider the case where the internal manifold is a product of three factorized tori $\prod_{I=1}^{3} T_{(I)}^2$. Then, in the weak field limit $|H| < M_s^2$, the mass shifts of charged states are given by:

$$ \delta M^2 = \sum_I (2k_I + 1)|qH_I| + 2qH_I\Sigma_I , \quad (17) $$

where q is the charge and Σ_I the projection of the internal helicity along the I-th plane. For a ten-dimensional (10d) spinor, its eigenvalues are $\Sigma_I = \pm 1/2$, while for a 10d vector $\Sigma_I = \pm 1$ in one of the planes $I = I_0$ and zero in the other two ($I \neq I_0$). Thus, charged higher dimensional scalars become massive, massless fermions lead to chiral 4d zero modes if all $H_I \neq 0$, while the lightest scalars coming from 10d vectors have masses

$$ M_0^2 = \sum_{I \neq I_0} |qH_I| - |qH_{I_0}| . \qquad (18) $$

All of them ($I_0 = 1, 2, 3$) can be made positive, avoiding the Nielsen-Olesen instability, if all $H_I \neq 0$. For arbitrary magnetic fields, supersymmetry is spontaneously broken and described by effective D-terms in the 4d theory[32]. However, if a scalar mass vanishes, some supersymmetry remains unbroken[33,34].

We now turn on several magnetic fields H_I^a in different Cartan generators $U(1)_a$, so that the gauge group is a product of unitary factors $\prod_a U(N_a)$ with $U(N_a) = SU(N_a) \times U(1)_a$. In an appropriate T-dual representation, it amounts to consider several stacks of D6-branes intersecting in the three internal tori at angles. An open string with one end on the a-th stack has charge ± 1 under the $U(1)_a$, and is neutral with respect to all others. It is now clear that this framework leads to models with a tree-level spectrum realizing split supersymmetry. Embedding the SM in an appropriate configuration of D-brane stacks, one obtains massless gauginos, since they are neutral under all magnetized $U(1)$'s, while all scalar superpartners of quarks and leptons correspond to charged open strings stretched among various stacks and become massive. On the other hand, the condition to obtain a (tree-level) massless Higgs in the spectrum implies that supersymmetry remains unbroken in the Higgs sector, leading to a pair of massless higgsinos, as required by anomaly cancellation.

It turns out that equality of the two non-abelian couplings is a consequence of the correct SM spectrum for weak magnetic fields, while the value for the weak angle $\sin^2 \theta_W = 3/8$ is easily obtained even in simple constructions. Indeed, a general study of SM embedding in three brane stacks reveals a simple model realizing the conditions for unification[7]. In general, split supersymmetry offers new possibilities for realistic string model building, that were previously unavailable because they were mainly restricted in the context of low scale strings. In this scenario, the string and compactification scales are of order $M_{\mathrm{GUT}} \simeq 2 \times 10^{16}$ GeV. Moreover, light masses in the TeV region can be generated for gauginos and higgsinos by higher dimensional operators[35], yielding $m_{1/2} \sim m_0^4/M_s^3$, for scalar masses m_0 of order 10^{13} GeV.

Acknowledgments

This work was supported in part by the European Commission under the RTN contract MRTN-CT-2004-503369, and in part by the INTAS contract 03-51-6346.

References

1. C. Angelantonj and A. Sagnotti, *Phys. Rept.* **371**, 1 (2002) [Erratum-ibid. **376**, 339 (2003)] [arXiv:hep-th/0204089].

2. I. Antoniadis, *Phys. Lett.* B **246**, 377 (1990).

3. C. D. Hoyle, D. J. Kapner, B. R. Heckel, E. G. Adelberger, J. H. Gundlach, U. Schmidt and H. E. Swanson, *Phys. Rev.* D **70**, 042004 (2004).

4. N. Arkani-Hamed, S. Dimopoulos and G. R. Dvali, *Phys. Lett.* B **429**, 263 (1998) [arXiv:hep-ph/9803315]; I. Antoniadis, N. Arkani-Hamed, S. Dimopoulos and G. R. Dvali, *Phys. Lett.* B **436**, 257 (1998) [arXiv:hep-ph/9804398].

5. For a review see e.g. I. Antoniadis, *Prepared for NATO Advanced Study Institute and EC Summer School on Progress in String, Field and Particle Theory, Cargese, Corsica, France (2002)*; and references therein.

6. N. Arkani-Hamed and S. Dimopoulos, arXiv:hep-th/0405159; G. F. Giudice and A. Romanino, *Nucl. Phys.* B **699**, 65 (2004) [Erratum-ibid. B **706**, 65 (2005)] [arXiv:hep-ph/0406088].

7. I. Antoniadis and S. Dimopoulos, *Nucl. Phys.* B **715**, 120 (2005).

8. J. D. Lykken, *Phys. Rev.* D **54**, 3693 (1996) [arXiv:hep-th/9603133].

9. J.C. Long and J.C. Price, *Comptes Rendus Physique* **4**, 337 (2003); R. S. Decca, D. Lopez, H. B. Chan, E. Fischbach, D. E. Krause and C. R. Jamell, *Phys. Rev. Lett.* **94**, 240401 (2005); S.J. Smullin, A. A. Geraci, D. M. Weld, J. Chiaverini, S. Holmes and A. Kapitulnik, arXiv:hep-ph/0508204; H. Abele, S. Haeßler and A. Westphal, in 271th WE-Heraeus-Seminar, Bad Honnef (2002).

10. I. Antoniadis and K. Benakli, *Phys. Lett.* B **326**, 69 (1994).

11. I. Antoniadis, K. Benakli and M. Quirós, *Phys. Lett.* B **331**, 313 (1994) and *Phys. Lett.* B **460**, 176 (1999); P. Nath, Y. Yamada and M. Yamaguchi, *Phys. Lett.* B **466**, 100 (1999) T.G. Rizzo and J.D. Wells, *Phys. Rev.* D **61**, 016007 (2000); T.G. Rizzo, *Phys. Rev.* D **61**, 055005 (2000); A. De Rujula, A. Donini, M.B. Gavela and S. Rigolin, *Phys. Lett.* B **482**, 195 (2000);

12. E. Accomando, I. Antoniadis and K. Benakli, *Nucl. Phys.* B **579**, 3 (2000).

13. I. Antoniadis, K. Benakli and A. Laugier, *JHEP* **0105**, 044 (2001).

14. P. Nath and M. Yamaguchi, *Phys. Rev.* D **60**, 116004 (1999); *Phys. Rev.* D **60**, 116006 (1999); M. Masip and A. Pomarol, *Phys. Rev.* D **60**, 096005 (1999); W.J. Marciano, *Phys. Rev.* D **60**, 093006 (1999); A. Strumia, *Phys. Lett.* B **466**, 107 (1999); R. Casalbuoni, S. De Curtis, D. Dominici and R. Gatto, *Phys. Lett.* B **462**, 48 (1999); C.D. Carone, *Phys. Rev.* D **61**, 015008 (2000); A. Delgado, A. Pomarol and M. Quirós, *JHEP* **1**, 30 (2000).

15. G. Servant and T. M. P. Tait, *Nucl. Phys.*

B **650**, 391 (2003).

16. G.F. Giudice, R. Rattazzi and J.D. Wells, *Nucl. Phys.* B **544**, 3 (1999); E.A. Mirabelli, M. Perelstein and M.E. Peskin, *Phys. Rev. Lett.* **82**, 2236 (1999); T. Han, J.D. Lykken and R. Zhang, *Phys. Rev.* D **59**, 105006 (1999); K. Cheung and W.-Y. Keung, *Phys. Rev.* D **60**, 112003 (1999); C. Balázs *et al.*, *Phys. Rev. Lett.* **83**, 2112 (1999); L3 Collaboration (M. Acciarri *et al.*), *Phys. Lett.* B **464**, 135 (1999) and **470**, 281 (1999): J.L. Hewett, *Phys. Rev. Lett.* **82**, 4765 (1999).

17. N. Arkani-Hamed, S. Dimopoulos and G. Dvali, *Phys. Rev.* D **59**, 086004 (1999).

18. S. Cullen and M. Perelstein, *Phys. Rev. Lett.* **83**, 268 (1999); V. Barger, T. Han, C. Kao and R.J. Zhang, *Phys. Lett.* B **461**, 34 (1999).

19. K. Benakli and S. Davidson, *Phys. Rev.* D **60**, 025004 (1999); L.J. Hall and D. Smith, *Phys. Rev.* D **60**, 085008 (1999).

20. S. Cullen, M. Perelstein and M.E. Peskin, *Phys. Rev.* D **62**, 055012 (2000); D. Bourilkov, *Phys. Rev.* D **62**, 076005 (2000); L3 Collaboration (M. Acciarri *et al.*), *Phys. Lett.* B **489**, 81 (2000).

21. S. B. Giddings and S. Thomas, *Phys. Rev.* D **65**, 056010 (2002); S. Dimopoulos and G. Landsberg, *Phys. Rev. Lett.* **87**, 161602 (2001).

22. I. Antoniadis, C. Bachas, *Phys. Lett.* B **450**, 83 (1999).

23. I. Antoniadis, K. Benakli, A. Laugier and T. Maillard, *Nucl. Phys.* B **662**, 40 (2003) [arXiv:hep-ph/0211409].

24. I. Antoniadis, S. Dimopoulos and G. Dvali, *Nucl. Phys.* B **516**, 70 (1998); S. Ferrara, C. Kounnas and F. Zwirner, *Nucl. Phys.* B **429**, 589 (1994).

25. T.R. Taylor and G. Veneziano, *Phys. Lett.* B **213**, 450 (1988).

26. I. Antoniadis, E. Kiritsis and J. Rizos, *Nucl. Phys.* B **637**, 92 (2002).

27. D. V. Volkov and V. P. Akulov, *JETP Lett.* **16**, 438 (1972) and *Phys. Lett.* B **46**, 109 (1973).

28. A. Brignole, F. Feruglio and F. Zwirner, *JHEP* **9711**, 001 (1997); T. E. Clark, T. Lee, S. T. Love and G. Wu, *Phys. Rev.* D **57**, 5912 (1998); M. A. Luty and E. Ponton, *Phys. Rev.* D **57**, 4167 (1998); I. Antoniadis, K. Benakli and A. Laugier, *Nucl. Phys.* B **631**, 3 (2002).

29. I. Antoniadis and M. Tuckmantel, *Nucl. Phys.* B **697**, 3 (2004).

30. I. Antoniadis, M. Tuckmantel and F. Zwirner, *Nucl. Phys.* B **707**, 215 (2005) [arXiv:hep-ph/0410165].

31. I. Antoniadis, K. Benakli and M. Quirós, *Nucl. Phys.* B **583**, 35 (2000).

32. C. Bachas, arXiv:hep-th/9503030.

33. C. Angelantonj, I. Antoniadis, E. Dudas and A. Sagnotti, *Phys. Lett.* B **489**, 223 (2000) [arXiv:hep-th/0007090].

34. M. Berkooz, M.R. Douglas and R.G. Leigh, *Nucl. Phys.* B **480**, 265 (1996).

35. I. Antoniadis, K. S. Narain and T. R. Taylor, arXiv:hep-th/0507244.

DISCUSSION

Jonathan Rosner (University of Chicago):
I would like you to comment on the relative merits of electron-positron and hadron colliders for exploring the scenarios you describe.

Ignatios Antoniadis: Hadron colliders, such as LHC, are good for discovery of which general theory is realized beyond the Standard Model. For instance, if there are missing energy events and from which source (supersymmetry or higher dimensional graviton), or a production of a new resonance that could correspond to some Z' or to a KK excitation. Once we know the general framework, electron-positron colliders are ideal to explore in detail the physics.

Bennie Ward (Baylor University):
In the scenario in which the Kaluza-Klein momentum is not conserved so that we produce a single resonance, what is its natural width?

Ignatios Antoniadis: The typical width is given by the gauge coupling times its mass. Thus, a KK resonance is quite narrow with a width roughly a tenth of its mass.

Luca Silvestrini (Munich and Rome):
I was a little bit confused when you discussed your gauge-Higgs unification model with Benakli and Quiros. If I remember correctly, there was a problem there not only with the Higgs mass but also with the downtype quarks being massless and not having any flavor mixing between fermions. Now you show the toy model computation and you say that the Higgs mass can be pushed up to 120 GeV, but what happens to the fermion masses?

Ignatios Antoniadis: The model I presented is not the one with gauge-Higgs unification and therefore does not have these problems. On the other hand, the advantage of the model you mention is that the loop factor correction is calculable within the effective field theory leading to an additional prediction for the compactification scale. On the contrary, in the model I presented, the loop correction depends on the details of the string construction. and thus only the Higgs mass is predictable and not the string scale.

FLAVOUR PHYSICS

Previous page:

The Museum Gustavianum (above). It holds collections related to the history of the University, like the first thermometer built by Celsius. It houses the famous anatomical theater from 1663.

The Old University building (below), the premises of the conference. Inaugurated in 1887, it housed the lecture rooms and professors' offices.

CP VIOLATION IN *B* MESONS

KAZUO ABE

KEK, Tsukuba, Ibaraki 305-0801 Japan
E-mail: kazuo.abe@kek.jp

Experimental status of *CP* violation in *B* mesons is summarized and the measurements are compared with the Standard Model expectations.

1 Introduction

The origin of *CP* violation in the Standard Model is the presence of complex phase in the CKM quark-mixing matrix. In this picture, all *CP* violating phenomena must be described in terms of three angles, ϕ_1, ϕ_2, and ϕ_3 of the unitarity triangle (Fig. 1).

$$\phi_1 = \beta = arg\left(-\frac{V_{cb}^*V_{cd}}{V_{tb}^*V_{td}}\right)$$
$$\phi_2 = \alpha = arg\left(-\frac{V_{tb}^*V_{td}}{V_{ub}^*V_{ud}}\right) \quad (1)$$
$$\phi_3 = \gamma = arg\left(-\frac{V_{ub}^*V_{ud}}{V_{cb}^*V_{cd}}\right)$$

The angle ϕ_1, which is the phase of V_{td} and appears in B^0-$\bar{B}^0$ mixing, induces mixing-assisted *CP* violation. This category of *CP* violation was observed by Babar and Belle in 2001 in the $B^0 \to J/\psi K^0$ decay. The Standard model allows *CP* violation in B^0-$\bar{B}^0$ mixing itself. This has not been seen yet. The angle ϕ_3, which is the phase of V_{ub}, causes direct *CP* violation through interference of $b \to u$ transition diagram and other diagrams. This was seen by Belle in 2003 in the $B^0 \to \pi^+\pi^-$ decay, but not supported by BaBar. In 2004, BaBar and Belle saw an evidence of direct *CP* violation in the $B^0 \to K^+\pi^-$ decay.

There is another category of *CP* violation where both ϕ_1 and ϕ_3 are involved. Mixing assisted *CP* violation for decays containing V_{ub} contribution belongs to this category, such as $B^0 \to \pi^+\pi^-$ and $B^0 \to D^{*-}\pi^+$. An evidence for this category of *CP* violation was seen by Belle in 2003 in the $B^0 \to \pi^+\pi^-$,

but not supported by BaBar. For some cases, this category leads to ϕ_2 measurements. The ϕ_2 related subjects will be covered by F. Forti's talk.

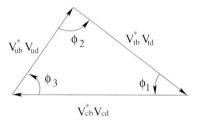

Figure 1. CKM unitarity triangle.

CP is violated if a process $B \to f$ is different from corresponding anti-particle process $\bar{B} \to \bar{f}$. The most straightforward method for detecting *CP* violation is to perform a simple counting experiment as used in the $B^0 \to K^+\pi^-$ case

$$\frac{N_{\bar{B}} - N_B}{N_{\bar{B}} + N_B}. \quad (2)$$

If *CP* is conserved, not only the decay rates must be the same in $B \to f$ and $\bar{B} \to \bar{f}$, but all aspect of the two processes must be the same. Therefore we can examine the time-dependent decay rates

$$\frac{N_{\bar{B}}(\Delta t) - N_B(\Delta t)}{N_{\bar{B}}(\Delta t) + N_B(\Delta t)}. \quad (3)$$

This method was used for the first observations of *CP* violation in $B^0 \to J/\psi K^0$ and subsequent $\sin 2\phi_1$ measurements. *CP* violation can also be detected as a difference of Dalitz distributions of any final state particle that subsequently decays into three-body state

$$\left|M_{\bar{B}}(m_+^2, m_-^2)\right|^2 \quad \text{vs} \quad \left|M_B(m_+^2, m_-^2)\right|^2 . \quad (4)$$

This method is used to extract ϕ_3 from $B^+ \to D^0 K^+$ by examining $D^0 \to K_S^0 \pi^+ \pi^-$ Dalitz distributions. We can go even further and examine time-dependence of the Dalitz distributions

$$\left|M_{\bar{B}}(m_+^2, m_-^2)(\Delta t)\right|^2 \quad \text{vs} \quad \left|M_B(m_+^2, m_-^2)(\Delta t)\right|^2 \quad (5)$$

which is used for resolving four-fold ambiguity in ϕ_1.

2 Asymmetric Energy e^+e^- Collision at $\Upsilon(4S)$

Studies of time-dependent CP violations require a large data sample of moving B mesons and measurements of proper decay time for each detected B meson. PEP-II and KEKB are the asymmetric energy e^+e^- colliders operating at $\Upsilon(4S)$ (Table 1) and provide such data sample to BaBar and Belle experiments, respectively. Figure 2 schematically shows how this scheme works. Since the B flight-

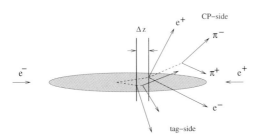

Figure 2. Asymmetric energy e^+e^- collision at $\Upsilon(4S)$.

length in x-y is only $\sim 30\mu$m as compared with $\sim 200\mu$m in z-direction, we can approximate as $\Delta z = z_{cp} - z_{\text{tag}}$, where z_{cp} and z_{tag} are the z decay vertexes of the B for which we try to measure CP violation and accompanying B, respectively.

It should be noted that the interaction point is much larger than Δz so that we must use the z_{tag} as a reference for the proper

decay time. The $\Upsilon(4S) \to B\bar{B}$ decay is a strong interaction process and the charge-conjugation must be conserved. This requires a relation $\psi(t) = |B_1^0 > |\bar{B}_2^0 > -|\bar{B}_1^0 > |B_2^0 >$ to hold so that one is B^0 and other is $\bar{B}^0$ at any time. Therefore the tag-side B also provides flavor information of the CP-side B at $\Delta t = 0$.

Table 1. Characteristics of the two B factories.

Parameters	BaBar	Belle
e^+e^- energies (GeV)	3.1×9	3.5×8.5
$\gamma\beta$	0.56	0.425
IP size x (μm)	120	80
y (μm)	5	2
z (mm)	8.5	3.4
Typical Δz (μm)	260	200
σ_z (CP-side)(μm)	50	75
σ_z (tag-side)(μm)	125	140

Figure 3 shows the integrated luminosity versus years of operation for the two B factories. The $1fb^{-1}$ luminosity corresponds roughly to one million $B\bar{B}$ events. For KEKB, it corresponds to a data sample accumulated in one day running.

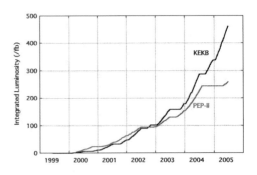

Figure 3. Integrated luminosity versus operating years for the two B factories.

3 $\sin 2\phi_1$ from $b \to c\bar{c}s$ decays

Interference between $B^0 \to f$ and $B^0 \to \bar{B}^0 \to f$ leads to time-dependent asymmetry.

$$\frac{\Gamma(\bar{B}^0(\Delta t) \to f) - \Gamma(B^0(\Delta t) \to f)}{\Gamma(\bar{B}^0(\Delta t) \to f) + \Gamma(B^0(\Delta t) \to f)}$$
$$= S_f \sin(\Delta m_d \Delta t) - C_f \cos(\Delta m_d \Delta t) \quad (6)$$

where

$$S_f = \frac{2 Im\lambda}{1 + |\lambda|^2}, \quad C_f = \frac{1 - |\lambda|^2}{1 + |\lambda|^2} \quad (7)$$

with $\lambda = \frac{q}{p} \frac{A(\bar{B}^0 \to f)}{A(B^0 \to f)}$.

The $J/\psi K_S^0$ and other $b \to c\bar{c}s$ decays ($J/\psi K_L^0$, $\psi(2S)K_S^0$, $\chi_{c1}K_S^0$, $\eta_c K_S^0$, and $J/\psi K^{*0}(K_S^0\pi^0)$) provide theoretically clean measurements of $\sin 2\phi_1$. They are dominated by only one diagram and $f = f_{cp}$ (Fig. 4), and we obtain $S_f = \sin 2\phi_1$ for $J/\psi K_S^0$. Extending to other $b \to c\bar{c}s$ decays, we obtain $\sin 2\phi_1 = -\eta_f \times S_f$, where η_f is the CP value of the final state f. The Standard Model corrections to this relation are believed to be very small and an oder of $\mathcal{O}(10^{-4})$ [1].

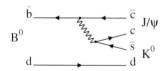

Figure 4. Diagram for $B^0 \to J/\psi K_S^0$.

The Δt distributions and their raw asymmetry of B^0-tag and $\bar{B}^0$-tag samples from BaBar's $227 \times 10^6 B\bar{B}$ data are shown in Figure 5 [2]. Recently, Belle updated their results using only $J/\psi K^0$ mode from $386 \times 10^6 B\bar{B}$ data [3] (Fig. 6) and obtained $\sin 2\phi_1 = 0.652 \pm 0.039 \pm 0.020$ and $C = -0.010 \pm 0.026 \pm 0.036$. Using these values and the BaBar results based on $227 \times 10^6 B\bar{B}$ data, new BaBar-Belle averages are [4]

$$\sin 2\phi_1 = 0.685 \pm 0.032$$
$$C = 0.016 \pm 0.046 \quad (8)$$

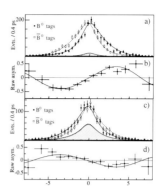

Figure 5. The Δt distributions for BaBar, $\eta_f = -1$ events (above) and $\eta_f = +1$ (below).

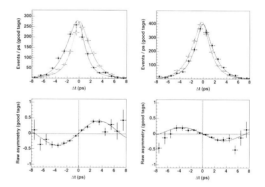

Figure 6. The Δt distributions and raw asymmetries for Belle, $J/\psi K_S^0$ ($\eta_f = -1$) events (left) and $J/\psi K_L^0$ ($\eta_f = +1$) events (right).

3.1 Four-fold ambiguity of ϕ_1

We still have four-fold ambiguity for the value of ϕ_1, $23°$, $(23 + 180)°$, $67°$, and $(67 + 180)°$. One approach for resolving this ambiguity is to measure $\cos 2\phi_1$, using time-dependent angular analysis of $B^0 \to J/\psi K^{*0}(K_S^0\pi^0)$. BaBar performed this analysis [5] using $88 \times 10^6 B\bar{B}$ data sample and obtained $\cos 2\phi_1 = +2.72^{+0.50}_{-0.79} \pm 0.27$ (fix $\sin 2\phi_1 = 0.731$). Based on this result, they concluded that the $23°$ $(+180°)$ solution is prefered at 86% CL. Belle performed a similar analysis [6] using $275 \times 10^6 B\bar{B}$ data and obtained $\cos 2\phi_1 = +0.87 \pm 0.74 \pm 0.12$ (fix $\sin 2\phi_1 = 0.726$). Error is too large for Belle to resolve the four-fold ambiguity.

3.2 Direct ϕ_1 measurement in $B^0 \to D\pi^0$

Time evolution of neutral B meson initially created as B^0 is expressed by

$$|B^0(\Delta t) >= e^{-|\Delta t|/2\tau_{B^0}} \times$$
$$\left[\cos(\frac{\Delta m \Delta t}{2})|B^0 > -i\,\frac{q}{p}\sin(\frac{\Delta m \Delta t}{2})|\bar{B}^0 > \right]$$
$$(9)$$

where $q/p = e^{-2i\phi_1}$. Ignoring CKM-suppressed decay, B^0 primarily decays into $\bar{D}^0\pi^0$ and $\bar{B}^0$ primarily decays into $D^0\pi^0$. If we reconstruct neutral D mesons with the $K_S^0\pi^+\pi^-$ decay mode, D^0 and $\bar{D}^0$ are indistinguishable (call these states $\tilde{D}^0$) and the two decay amplitudes can interfere. We make use of the fact that $\tilde{D}^0 \to K_S\pi^+\pi^-$ is dominated by quasi-two body amplitudes and described by $m_+ = m_{K_S\pi^+}$ and $m_- = m_{K_S\pi^-}$. If $\bar{D}^0 \to K_S\pi^+\pi^-$ is described by $f(m_+^2, m_-^2)$, $D^0 \to K_S\pi^+\pi^-$ must be described by $f(m_-^2, m_+^2)$.

The time dependent $\tilde{D}^0$ Dalitz distributions would then be

$$|\cos(\frac{\Delta m \Delta t}{2})f(m_\pm^2, m_\mp^2) - i\,e^{\mp 2i\phi_1} \times$$
$$\sin(\frac{\Delta m \Delta t}{2})\eta_{h^0}(-1)^l f(m_\mp^2, m_\pm^2)|^2 \quad (10)$$

for the B^0 sample (upper sign) and the $\bar{B}^0$ sample (lower sign). Here we include $D\eta$ and $D\omega$ modes and η_{h^0} for CP values of h^0 and l for the orbital angular momentum of final state must be included. The time-dependent Dalitz distributions will look different for B^0 and $\bar{B}^0$ because of $2\phi_1$.

The $\bar{D}^0 \to K_S^0\pi^+\pi^-$ decay amplitude $f(m_+^2, m_-^2)$ is determined using the $\bar{D}^0$ sample collected in the continuum data ($e^+e^- \to q\bar{q}$) and specifying the D flavor using the charge information of slow pions in the $D^{*-} \to \bar{D}^0(K_S^0\pi^+\pi^-)\pi_{slow}^-$. The $\bar{D}^0 \to K_S\pi^+\pi^-$ amplitude is expressed as

$$f(m_+^2, m_-^2) = \sum_{j=1}^N a_j e^{i\alpha_j} A_j(m_+^2, m_-^2) + be^{i\beta}$$
$$(11)$$

where 18 resonance amplitudes ($K_S\sigma$, $K_S\rho^0$, $K_S\omega$, $K_S f$, $K^*\pi$, and higher mass ρ, f, and K^*) and one non-resonant amplitude are included, and their relative fractions and phases are determined from unbinned maximum likelihood fit. Figure 7 shows the Dalitz distribution obtained from Belle's 253fb^{-1} data sample, and the projections onto m_+^2, m_-^2, and $m_{\pi\pi}^2$ axes, together with the fit results.

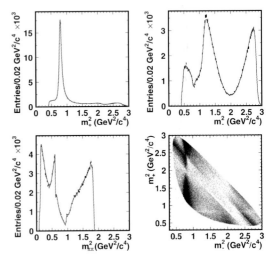

Figure 7. The $\bar{D}^0 \to K_S^0\pi^+\pi^-$ Dalitz distribution and its projections onto m_+^2, m_-^2, and $M_{\pi\pi}^2$ axes. Curves are fit results.

Belle performed a time-dependent Dalitz analysis [7] using the $B^0 \to \tilde{D}^0(\tilde{D}^{*0})h^0$ data samles, where $h^0 = \pi^0, \eta, \omega$ and $\tilde{D}^{*0} \to \tilde{D}^0\pi^0$. Using the $386 \times 10^6 B\bar{B}$ data, a total of 309 ± 31 signal events are obtained with 63% purity: $D\pi^-$ (157 ± 24), $D\omega$ (67 ± 10), $D\eta$ (58 ± 13), $D^{*}\pi^0$ and $D^{*}\eta$ combined (27 ± 11). Figure 8 shows time-integrated Dalitz distribution of the B^0-tag sample. The $K^*(890)$ signal is present in both $M_{K_S^0\pi^+}^2$ and $M_{K_S^0\pi^-}^2$ projections since tha data sample is a mixture of B^0 and $\bar{B}^0$ when integrated in time. However, we can see more $K^{*+}(890)$ than $K^{*-}(890)$ as we expect that more B^0 are present even after integrated in time and they decay as $B^0 \to \bar{D}^0 \to K^{*+}(890)$. Figure 9 shows the raw asymmetry for a region in

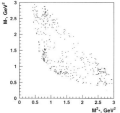
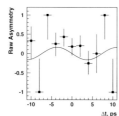

Figure 8. Time-integrated Dalitz distribution for B^0-tag sample.

Figure 9. Δt dependence of raw asymmetry for the ρ^0 region.

order of $A\lambda^2$, has an identical CKM element as $b \to c\bar{c}s$ and leads to $\sin 2\phi_1$. The second term is suppressed compared to the first term since its CKM factor is on the order of $A\lambda^4(\rho - i\eta)$.

For other $b \to s$ penguin modes, there are additional small contributions. For $B^0 \to \eta' K^0$, $f^0 K^0$, a contribution from $b \to u$ tree diagram of the order of $\mathcal{O}(A\lambda^4(\rho - i\eta))$ can be present. The $B^0 \to \pi^0 K^0$ and ωK^0 modes can also have contribution from $b \to u$ tree diagram. In addition, these modes contain $b \to s d\bar{d}$ instead of $b \to s\bar{s}s$, and this can cause a different behaviour. Considering these effects, the Standard Model corrections of up to $\mathcal{O}(\lambda^2) \sim 5\%$ can be possible in the extraction of $\sin 2\phi_1$ from the S_f. Magnitude of the corrections can differ in different modes [8]. A larger deviation exceeding these corrections will be an indication of new physics in penguin loops.

Table 2 summarizes the Standard Model expectations of S_f and C_f for the $b \to s$ penguin decays. The fraction of CP even component in the $B^0 \to K^+ K^- K_S^0$ events is measured by BaBar and Belle, and is about 90%.

the Dalitz distributions where ρ^0 are concentrated, $(N_\rho(\bar{B}^0) - (N_\rho(B^0))/(N_\rho(\bar{B}^0) + (N_\rho(B^0)))$. It behaves as the $B^0 \to J/\psi K_L^0$ sample. This is expected because $K_S \rho^0 \pi^0$ is a CP eigenstate with $\eta_f = +1$. Result of fit gives

$$\phi_1 = (16 \pm 21 \pm 12)° \qquad (12)$$

corresponding to 95% CL region of $-30° < \phi_1 < 62°$. This result exclude $\phi_1 = 67°$ solution at 95% CL.

4 $\sin 2\phi_1$ from $b \to s$ Penguin Decays

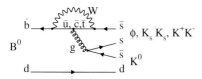

Figure 10. Diagram for $b \to s$ penguin decays

The $B^0 \to \phi K^0$ decay is dominated by penguin diagram (Fig. 10) and the Standard Model amplitude is expressed by

$$P \sim V_{ub}^* V_{us} P^u + V_{cb}^* V_{cs} P^c + V_{tb}^* V_{ts} P^t$$
$$\sim V_{cb}^* V_{cs}(P^c - P^t) + V_{ub}^* V_{us}(P^u - P^t).$$
$$(13)$$

Here one of the CKM unitarity relation, $V_{cb}^* V_{cs} + V_{ub}^* V_{us} + V_{tb}^* V_{ts} = 0$ is used to eliminate $V_{tb}^* V_{ts}$. The first term, which is on the

The B decay candidates are reconstructed using two kinematical variables, $M_{\mathrm{bc}} = \sqrt{E_{\mathrm{beam}}^2 - (\sum_i \vec{p}_i)^2}$ (called M_{ES} at BaBar) and $\Delta E = \sum_i E_i - E_{\mathrm{beam}}$. The $(\sum_i E_i)^2$ is replaced by E_{beam}^2 to improve the resolution. The E_i and $\vec{p}_i$ are the energy and momentum vector for the i-th daughter particle of the B candidate. The signal events should concentrate at $M_{\mathrm{bc}} = 5.28$ GeV/c^2 and $\Delta E = 0$ GeV regions (Figs. 11 and 12). For K_L^0, two-body decay kinematics must be assumed for computing the momentum since the K_L^0 energy is too low to be be measured. Belle uses a likelihood ratio and calculated K_L^0 momentum in the cms selecting the signal candidates for the K_L^0 modes. The Δt dependences are shown in Figure 13 for BaBar [9] $(230 \times 10^6 B\bar{B})$ in terms of raw asymmetries of B^0-tag and $\bar{B}^0$-tag events, and in Figure 14 for Belle [3] $(386 \times 10^6 B\bar{B})$ in terms of Δt

Table 2. Standard Model expectations of S_f and C_f for $b \to s$ penguin decays.

Final State	η_{cp}	S_f	C_f	Corrections
ϕK_S^0	-1	$\sin 2\phi_1$	0	u-quark penguin
ϕK_L^0	+1	$-\sin 2\phi_1$	0	
$K_S^0 K_S^0 K_S^0$	+1	$-\sin 2\phi_1$	0	
$\eta' K_S^0$	-1	$\sin 2\phi_1$	0	u-quark penguin, $b \to u$ tree
$\eta' K_L^0$	+1	$-\sin 2\phi_1$	0	
$f_0(980) K_S^0$	+1	$-\sin 2\phi_1$	0	
$K^+ K^- K_S^0$	mixture	$-(f_+ - f_-)\sin 2\phi_1$	0	
$K^+ K^- K_L^0$	mixture	$-(f'_+ - f'_-)\sin 2\phi_1$	0	
$\pi^0 K_S^0$	-1	$\sin 2\phi_1$	0	$b \to s d\bar{d}$ different from $b \to s s\bar{s}$?
ωK_S^0	-1	$\sin 2\phi_1$	0	$b \to u$ tree

distributions and raw asymmetry. Effective $\sin 2\phi_1$, "$\sin 2\phi_1''$, that are extracted from the Δt distributions of ϕK^0 are $+0.50 \pm 0.25^{+0.07}_{-0.04}$ from BaBar, and $+0.44 \pm 0.27 \pm 0.05$ from Belle.

The "$\sin 2\phi_1''$ values for $K^+ K^- K_S^0$ are $+0.55 \pm 0.22 \pm 0.04 \pm 0.11(CP)$ for BaBar [9] and $+0.60 \pm 0.18 \pm 0.04^{+0.19}_{-0.12}$ (CP) for Belle [3], respectively, where the last errors are due to uncertainty of CP even component in the $B^0 \to K^+ K^- K_S^0$ sample. BaBar also measured "$\sin 2\phi_1''$ = $+0.09 \pm 0.33^{+0.13}_{-0.14} \pm 0.10(CP)$ for the $K^+ K^- K_L^0$ mode [10].

The $B^0 \to K_S^0 K_S^0 K_S^0$ mode is a clean $b \to s$ penguin although the Δt measurement was thought be difficult. BaBar and Belle have now results for this mode [11]. The BaBar result (Fig. 15) is "$\sin 2\phi_1'' = 0.63^{+0.32}_{-0.28} \pm 0.04$, and corresponding Belle result is $0.58 \pm 0.36 \pm 0.08$.

The $\eta' K_S^0$ mode has the highest statistical power among the $b \to s$ penguin modes. Belle has extended the analysis by adding the $\eta' K_L^0$ mode (Fig. 16). They obtained a combined result of $\eta' K_S^0$ and $\eta' K_L^0$ "$\sin 2\phi_1''$ = $0.62 \pm 0.12 \pm 0.04$ [3], whereas BaBar result from only $\eta' K_S^0$ is $0.30 \pm 0.14 \pm 0.02$ [12].

Figure 17 summarizes the "$\sin 2\phi_1''$ results for all $b \to s$ penguin decays [4]. Plotted here are individual results from BaBar and Belle, BaBar and Belle averages for each

mode, and most recent $\sin 2\phi_1$ result as a reference. All "$\sin 2\phi_1''$ except the one from $\eta' K^0$ are within $\sim 1\sigma$ from the Standard Model value of $\sin 2\phi_1 = 0.69 \pm 0.03$. Somewhat larger deviations of $\Delta S \equiv$ "$\sin 2\phi_1''$ − $\sin 2\phi_1$ from zero and larger descrepancies between the two experiments, that was reported earlier [13] seem to be settling down. On the other hand $\Delta S \equiv$ "$\sin 2\phi_1''$ − $\sin 2\phi_1$ is all negative in exception of the $\eta' K^0$ mode.

5 $\sin(2\phi_1 + \phi_3)$ from $B^0 \to D^{(*)-}\pi^+$

Interference of the amplitudes for Cabibbo-favored diagram and B^0-$\bar{B}^0$ mixing followed by doubly-Cabibbo-suppressed diagram (Fig. 18) results in an mixing-assisted CP violation in the $B^0 \to D^{(*)-}\pi^+$ decay. The S and C terms in the Δt distributions are summarized in Table 3. Here r and δ are

Table 3. Expressions of S and C terms for $B^0 \to D^{(*)-}\pi^+$ decays.

Final State	S	C
$D^{*+}\pi^-$	$-2r^* \sin(2\phi_1 + \phi_3 + \delta^*)$	$+1$
$D^{*-}\pi^+$	$-2r^* \sin(2\phi_1 + \phi_3 - \delta^*)$	-1
$D^+\pi^-$	$+2r \sin(2\phi_1 + \phi_3 + \delta)$	$+1$
$D^-\pi^+$	$+2r \sin(2\phi_1 + \phi_3 - \delta)$	-1

the amplitude ratio and strong phase differ-

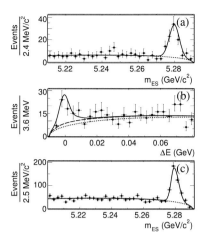

Figure 11. Distributions of $M_{\rm ES}$ for ϕK_S^0 (a), ΔE for ϕK_L^0 (b), and $M_{\rm ES}$ for $K^+K^-K_S^0$ (c) from BaBar's $230 \times 10^6 B\bar{B}$ data.

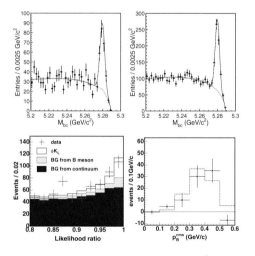

Figure 12. Distributions of $M_{\rm bc}$ for ϕK_S^0 (upper-left), $M_{\rm bc}$ for $K^+K^-K_S^0$ (upper-right), likelihood ratio and $P_B^{\rm cms}$ for ϕK_L^0 (lower) from Belle's $386 \times 10^6 B\bar{B}$ data.

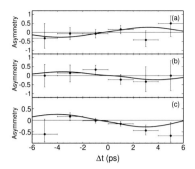

Figure 13. Raw asymmetries for the ϕK_S^0 (a), ϕK_L^0 (b), $K^+K^-K_S^0$ (c) samples in the BaBar analysis. Curves are the result of fit.

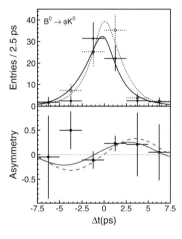

Figure 14. The Δt distribution and raw asymmetry for the ϕK^0 sample in the Belle analysis. For the Δ distributions, dotted and solid points and curves are for B^0-tag and $\bar{B}^0$-tag events and fit results. For the raw asymmetry plot, the fit result is given by the solid line and the Standard Model expectation is indicated by the dotted line.

ence of DCSD and CFD. We expect $r \sim 0.02$ so that $C = \pm(1 - r^2)/(1 + r^2) \simeq \pm 1$.

The Δt distributions for the $D^{*\mp}\pi^\pm$ samples selected by a partial reconstruction technique in BaBar's $232 \times 10^6 B\bar{B}$ data [14], and for the $D^{*\mp}\pi^\pm$ samples selected by full reconstruction technique in Belle's $152 \times 10^6 B\bar{B}$ data [15] are shown in Figures 19 and 20.

Significant differences in $\Delta t > 0$ and $\Delta t < 0$ in any of these distribution would be an indication of CPV. We need a lot more data before we begin to see definitive effects.

6 CP Violation in $B\bar{B}$ Mixing

The Standard Model allows CP violation in the $B\bar{B}$ mixing itself in analogy to ϵ_K in K^0 System [16]. It lead to $\rightarrow |q/p| \neq 1$, where p and q are the coefficients that relate the mass and flavor eigenstates of the neutral B mesons, as a result of M_{12} and Γ_{12} having different phases. In the Standard Model, we

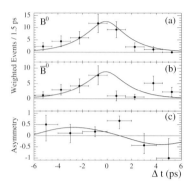

Figure 15. The Δt and raw asymmetry distributions for BaBar's $B^0 \to K_S^0 K_S^0 K_S^0$ candidates.

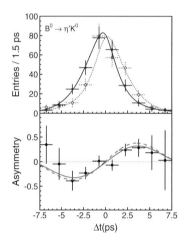

Figure 16. The Δt and raw asymmetry distributions for Belle's combined sample of $\eta' K_S^0$ and $\eta' K_L^0$. The sample is divided into B^0-tagged (dotted line) and $\bar{B}^0$-tagged (solid line). Dotted curve in the asymmetry distribution is the Standard Model expectation.

$\sin(2\beta^{\mathrm{eff}})/\sin(2\phi_1^{\mathrm{eff}})$ HFAG LP 2005 PRELIMINARY

$b \to c\bar{c}s$	Average	0.69 ± 0.03
ϕK^0	BaBar	$0.50 \pm 0.25 ^{+0.07}_{-0.04}$
	Belle	$0.44 \pm 0.27 \pm 0.05$
	Average	0.47 ± 0.19
$\eta' K^0$	BaBar	$0.30 \pm 0.14 \pm 0.02$
	Belle	$0.62 \pm 0.12 \pm 0.04$
	Average	0.48 ± 0.09
$f_0 K_S$	BaBar	$0.95 ^{+0.23}_{-0.32} \pm 0.10$
	Belle	$0.47 \pm 0.36 \pm 0.08$
	Average	0.75 ± 0.24
$\pi^0 K_S$	BaBar	$0.35 ^{+0.30}_{-0.33} \pm 0.04$
	Belle	$0.22 \pm 0.47 \pm 0.08$
	Average	0.31 ± 0.26
ωK_S	BaBar	$0.50 ^{+0.34}_{-0.38} \pm 0.02$
	Belle	$0.95 \pm 0.53 ^{+0.12}_{-0.15}$
	Average	0.63 ± 0.30
$K^+ K^- K^0$	BaBar	$0.41 \pm 0.18 \pm 0.07 \pm 0.11$
	Belle	$0.60 \pm 0.18 \pm 0.04 ^{+0.19}_{-0.12}$
	Average	$0.51 \pm 0.14 ^{+0.11}_{-0.08}$
$K_S K_S K_S$	BaBar	$0.63 ^{+0.28}_{-0.32} \pm 0.04$
	Belle	$0.58 \pm 0.36 \pm 0.08$
	Average	0.61 ± 0.23

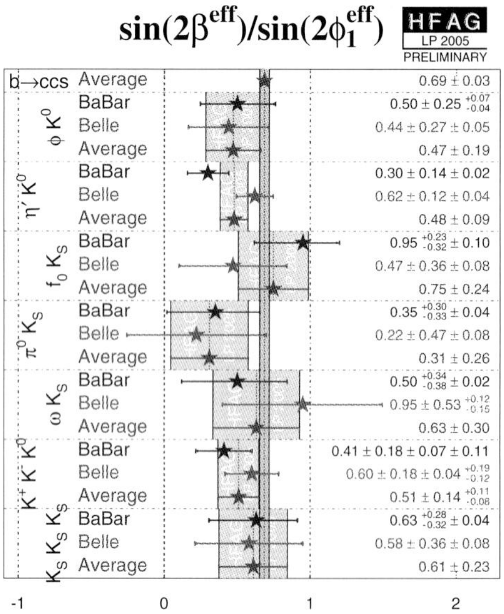

Figure 17. Summary of effective $\sin 2\phi_1$ values for the $b \to s$ penguin decays. Most recent $\sin 2\phi_1$ value is also plotted as a reference.

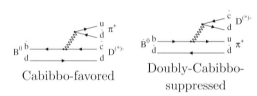

Cabibbo-favored   Doubly-Cabibbo-suppressed

Figure 18. Cabibbo-favored and doubly-Cabibbo-suppressed diagrams in the $B \to D^{(*)}\pi$ decay.

expect the observable effect to be

$$1 - |\frac{q}{p}|^2 \simeq Im\left(\frac{\Gamma_{12}}{M_{12}}\right) \sim \mathcal{O}(10^{-3}). \quad (14)$$

Observation of significantly larger effect will certainly be very exciting.

BaBar and Belle have measured $|q/p|$ using charge asymmetry in the same-sign dilepton events from $\Upsilon(4S)$ decays. BaBar obtained $0.998 \pm 0.006 \pm 0.007$ using $23 \times 10^6 B\bar{B}$ data while Belle obtained $1.0005 \pm 0.0040 \pm 0.0035$ using $85 \times 10^6 B\bar{B}$ data [17]. One can translate these values to ϵ_B in analogy to $|\epsilon_K| = (2.284 \pm 0.014) \times 10^{-3}$ in the K^0 system. The Belle result corresponds to $\frac{Re(\epsilon_B)}{1+|\epsilon_B|^2} = (-0.3 \pm 2.0 \pm 1.7) \times 10^{-3}$. There is no sign of CP violation yet. This is the only missing category of CPV for B mesons in the Standard Model. The measurements are already limitted by the systematic error with less than $100 fb^{-1}$ data. We need another factor of ~ 10 improvement in sensitivity.

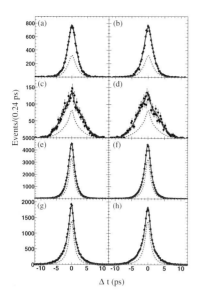

Figure 19. The Δt distributions for partially reconstructed $D^{*\mp}\pi^{\pm}$ events in BaBar analysis. (a) - (d) are lepton-tag, and (e) - (h) are kaon-tag. (a) $\bar{B}^0 \to D^{*+}\pi^-$, (b) $B^0 \to D^{*-}\pi^+$ (c) $\bar{B}^0 \to D^{*-}\pi^+$, (d) $B^0 \to D^{*+}\pi^-$. (e) - (h) are in the same order.

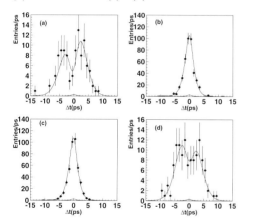

Figure 20. The Δt distributions for fully reconstructed $D^{*\mp}\pi^{\pm}$ events in Belle analysis. (a) $\bar{B}^0 \to D^{*+}\pi^-$, (b) $B^0 \to D^{*-}\pi^+$ (c) $\bar{B}^0 \to D^{*-}\pi^+$, (d) $B^0 \to D^{*+}\pi^-$.

7 Direct CP Violation in $B^0 \to K^+\pi^-$

In the Standard Model, direct CP violation can occur when the decay amplitude contains two or more diagrams. The amplitude for $B \to f$ is then given as

$$A_f = |a_1| + |a_2|e^{i(\delta + \phi_w)} \qquad (15)$$

where δ and ϕ_w are the strong and weak phases of the second amplitude with respect to the first amplitude. For $\bar{B} \to \bar{f}$, ϕ_w changes sign, but δ remains same. The decay rate asymmetry is then expressed as

$$
\begin{aligned}
A_{CP} &= \frac{\Gamma(\bar{B} \to \bar{f}) - \Gamma(B \to f)}{\Gamma(\bar{B} \to \bar{f}) + \Gamma(B \to f)} \\
&= \frac{-2|a_1 a_2|\sin\delta\sin\phi_w}{|a_1|^2 + |a_2|^2 + 2|a_1 a_2|\cos\delta\cos\phi_w},
\end{aligned}
\qquad (16)
$$

where A_{CP} can be non-zero if ϕ_w and δ are simultaneously non-zero.

Figure 21. Two diagrams contributing to $B^0 \to K^+\pi^-$, $\pi^+\pi^-$.

The $B^0 \to K^+\pi^-$ and $B^0 \to \pi^+\pi^-$ have been thought as likely places to see the direct CP violation because they are contributed by $b \to u$ tree diagram and gluon-penguin diagram (Fig. 21).

Clear evidence of direct CP violation in $B^0 \to K^+\pi^-$ was seen by BaBar and Belle last year [18]: $A_{CP}(K^+\pi^-) = -0.133 \pm 0.030 \pm 0.009$ from BaBar ($227 \times 10^6 B\bar{B}$), and $-0.101 \pm 0.025 \pm 0.005$ from Belle ($275 \times 10^6 B\bar{B}$). This year Belle updated their measurement using $386 \times 10^6 B\bar{B}$ data and obtained $-0.113 \pm 0.022 \pm 0.008$ [19].

Belle reported an evidence of another direct CP violation in $B^0 \to \pi^+\pi^-$ in 2004 based on the measurement of the C parameter, which is equal to $-A_{CP}$, in the Δt distributions using $152 \times 10^6 B\bar{B}$ data [20], but it was not supported by BaBar [21]. This year both experiments updated their results. Using $275 \times 10^6 B\bar{B}$ data, Belle obtained $C_{\pi\pi} = -0.56 \pm 0.12 \pm 0.06$, which is consistent with their previous result and shows a

94

significant deviation from zero [22]. However, BaBar result using $227 \times 10^6 B\bar{B}$ data gives $-0.09 \pm 0.15 \pm 0.04$, which is consistent with no direct CP violation [23].

Table 4 summarizes the A_{CP} measurements for the charmless two body decays and contributing diagrams (Fig. 22). Extraction of ϕ_3 from A_{CP} may be difficult due to hadronic effects. We hope to learn about them from the measurements.

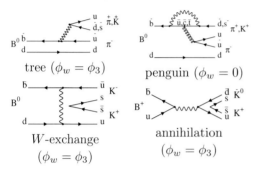

Figure 22. Diagrams contributing to charmless two body decays.

8 Extraction of ϕ_3 from Direct CP Violation in $B^+ \to DK^+$

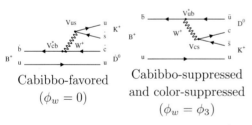

Figure 23. Two diagrams contributing to $B^+ \to DK^+$.

Two diagrams contribute to the $B^+ \to \tilde{D}^0 K^+$ decay (Fig. 23). If we reconstruct D^0 and $\bar{D}^0$ with common decay modes, the two processes are indistinguishable and an interfere can occur, which can then cause differences in the yields or decay patterns of B^+ and B^- sample. Several methods have been pursued using different common decay modes such as i) CP eigenstates (GLW method [24]),

ii) suppressed $K\pi$ charge combinations (ADS method [25]), and iii) $K_S^0\pi^+\pi^-$ decays (GSZB method [26]). The GLW modes have been established but no significant CP violation has been seen. The ADS modes have not been seen [27]. For the GSZB, we are getting very close to seeing an evidence of CP violation.

Neutral D from $B^+ \to \tilde{D}^0 K^+$ is mostly $\bar{D}^0$, but contains $\sim 10\%$ D^0. Reconstructing $\tilde{D}^0$ in the $K_S^0\pi^+\pi^-$ decay, the Dalitz distributions for $K_S^0\pi^+\pi^-$ should be give by

$$\left| f(m_\pm^2, m_\mp^2) + re^{i(\pm\phi_3+\delta)} f(m_\mp^2, m_\pm^2) \right|^2 \quad (17)$$

for the B^+ sample (upper sign) and the B^- sample (lower sign). Here r and δ are the ratio and strong phase difference of the CSD and CFD amplitudes. The two patters would look different if $r \neq 0$ and $\phi_3 \neq 0$.

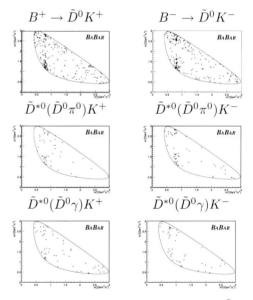

Figure 24. Dalitz distributions for the $\tilde{D} \to K_S\pi^+\pi^-$ decays in the $B^\pm \to \tilde{D}K^\pm$ decays from BaBar analysis.

Dalitz distributions from BaBar's $227 \times 10^6 B\bar{B}$ data and Belle's $275 \times 10^6 B\bar{B}$ data are shown in Figures 24 and 25. Fit results in terms of 1σ and 2σ allowed region in $x_\pm = Re(r_\pm e^{i(\pm\phi_3+\delta)})$ and $y_\pm = Im(r_\pm e^{i(\pm\phi_3+\delta)})$ are shown in Figures 26 and 27 for each decay mode. Extracted values for ϕ_3 are $(70 \pm$

Table 4. Summary of A_{CP} results for charmless two body decays.

Decay Mode	BaBar	Belle	SM diagrams
$K^+\pi^-$	$-0.133 \pm 0.030 \pm 0.009$	$-0.113 \pm 0.021 \pm 0.008$	tree, penguin
$K^+\pi^0$	$+0.06 \pm 0.06 \pm 0.01$	$+0.04 \pm 0.04 \pm 0.02$	tree, penguin
$K_S^0\pi^+$	$-0.09 \pm 0.05 \pm 0.01$	$+0.05 \pm 0.05 \pm 0.01$	penguin
$K_S^0\pi^0$	$-0.06 \pm 0.18 \pm 0.03$	$-0.11 \pm 0.18 \pm 0.09$	penguin
$\pi^+\pi^-$	$+0.09 \pm 0.15 \pm 0.04$	$+0.52 \pm 0.14$	tree, penguin
$\pi^+\pi^0$	$-0.01 \pm 0.10 \pm 0.02$	$+0.02 \pm 0.08 \pm 0.01$	tree
$\pi^0\pi^0$	$+0.12 \pm 0.56 \pm 0.06$	$0.44^{+0.53}_{-0.52} \pm 0.17$	tree, penguin
K^+K^-	signal not seen	signal not seen	W-exchange
K^+K^0	seen	seen	penguin, annihilation
$K^0\bar{K}^0$	seen	seen	penguin

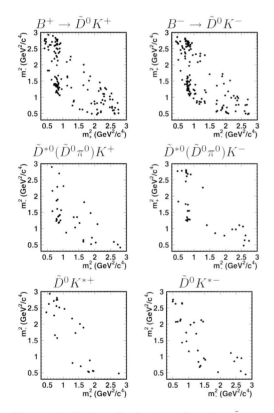

Figure 25. Dalitz distributions for the $\tilde{D} \to K_S\pi^+\pi^-$ decays in the $B^\pm \to \tilde{D}K^\pm$ decays from Belle analysis.

$31^{+12+14}_{-10-11})^\circ$ from BaBar and $(68^{+14}_{-15}\pm13\pm11)^\circ$ from Belle, where the last errors come from uncertainty of the $\bar{D}^0 \to K_S^0\pi^+\pi^-$ decay model [28]. These results correspond to 2σ al-

lowed intervals are 12° - 137° and 22° - 113°, respectively. Significance of direct CPV is about 2.4σ in both measurements.

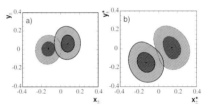

Figure 26. Allowed regions in $(x_\pm, y_\pm)$ in BaBar analysis. Right circles correspond to the B^- sample in $(x_\pm, y_\pm)$, and the B^+ sample in $(x_\pm^*, y_\pm^*)$.

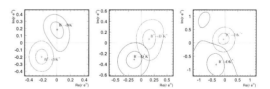

Figure 27. Allowed regions in $(x_\pm, y_\pm)$ in Belle analysis.

9 Summary

Updated result for $\sin 2\phi_1$ is $\sin 2\phi_1 = 0.685 \pm 0.032$ as compared to the previous value of 0.725 ± 0.037. Deviations of effective $\sin 2\phi_1$ measured in $b \to s$ penguin decays from the Standard Model, $\Delta S \equiv'' \sin 2\phi_1'' - \sin 2\phi_1$, are getting smaller. Updated results for ΔS

in most modes are within about 1σ of zero. However, for most modes ΔS are negative rather than scattering around zero. Whether the results of ΔS measurements can be fully explained in terms of the Standard Model corrections or they require new physics to explain remains as one of the most important issue in our field. Only significant increase of data sample will lead us to a definitive conclusion. We are beginning to see the first useful measurement of ϕ_3, where Dalitz analysis played an important role. In fact this old method has been used not only for the ϕ_3, but also to resolve quadratic ambiguity in ϕ_1 and a masurement of ϕ_2. CP violation in $B\bar{B}$ mixing is not seen yet. This is the only missing category of Standard Model CP violation in B mesons.

References

1. H. Boos, J. Reuter, T. Mannel, *Phys. Rev. D* **70**, 036006 (2004).
2. BABAR Collaboration, B. Aubert *et al,Phys. Rev. Lett.* **94**, 161803 (2005).
3. Belle Collaboration, K. Abe *et al.*, hep-ex/0507037.
4. Heavy Flavor Averaging Group, http://www.slac.stanford.edu/xorg/hfag/.
5. BABAR Collaboration, B. Aubert,*Phys. Rev. D* **71**, 032005 (2005).
6. Belle Collaboration, R. Itoh *et al,Phys. Rev. Lett.* **95**, 091601 (2005).
7. Belle Collaboration, K. Abe *et al.*, hep-ex/0507065.
8. Yu. Grossman, Z. Ligeti, Y. Nir, and H. Quinn,*Phys. Rev. D* **68**, 015004 (2003).
9. BaBar Collaboration, B. Aubert *et al,Phys. Rev. D* **71**, 091102(R) (2005).
10. BaBar Collaboration, B. Aubert *et al,* hep-ex/0507016.
11. BABAR Collaboration, B. Aubert *et al,Phys. Rev. Lett.* **95** , 011801 (2005); Belle Collaboration, K. Sumisawa *et al,Phys. Rev. Lett.* **95**, 061801 (2005).
12. BABAR Collaboration, B. Aubert *et al,Phys. Rev. Lett.* **94**, 191802 (2005).
13. Z. Ligeti, hep-ph/0408267.
14. BABAR Collaboration, B. Aubert *et al,* hep-ex/0504035.
15. Belle Collaboration, T.R. Sarangi *et al,*
16. Particle Data Group, S. Eidelman *et al,Phys. Lett. B* **592** (2004).
17. BABAR Collaboration, B. Aubert *et al,Phys. Rev. Lett.* **88**, 231801 (2002); Belle Collaboration, E. Nakano *et al,* hep-ex/0505017.
18. BABAR Collaboration, B. Aubert *et al,Phys. Rev. Lett.* **93**, 131801 (2004); Belle Collaboration, Y. Chao *et al,Phys. Rev. Lett.* **93**, 191802 (2004).
19. Belle Collaboration, K. Abe *et al,* hep-ex/0507045.
20. Belle Collaboration, K. Abe *et al,Phys. Rev. Lett.* **93**, 021601 (2004).
21. BABAR Collaboration, B. Aubert *et al,Phys. Rev. Lett.* **89**, 281802 (2002).
22. Belle Collaboration, K. Abe *et al,Phys. Rev. Lett.* **95**, 101801 (2005).
23. BABAR Collaboration, B. Aubert *et all,* hep-ex/0501071.
24. M. Gronau and D. London,*Phys. Lett. B* **252**, 483 (1991); M. Gronau and D. Wyler,*Phys. Lett. B* **265**, 172 (1991).
25. D. Atwood, I. Dunietz, and A. Soni,*Phys. Rev. Lett.* **78**, 3257 (1997); Phys. Rev. D **63**, 036005 (2001).
26. A. Giri, Yu. Grossman, A. Soffer, and J. Zupan,*Phys. Rev. D* **68**, 054018 (2003); A. Bondar, Proceedings of BINP Special Analysis Meeting (2002) (unpublished).
27. BABAR Collaboration, B. Aubert *et al,* hep-ex/0508001; Belle Collaboration, K. Abe *et al,* hep-ex/0508048.
28. BABAR Collaboration, B. Aubert *et al,* hep-ex/0504039; Belle Collaboration, K. Abe *et al,* hep-ex/0411049.

Phys. Rev. Lett. **93**, 031802 (2004).

DISCUSSION

Sheldon Stone (Syracuse University):
The last line of your conclusions CPV in $B\overline{B}$ mixing is not seen yet. The only missing category of Standard Model CPV in B mesons must refer to B_s mesons where no measurements yet been made and where new physics may yet be present and possibly large.

Kazuo Abe: It is true that we dont know anything about CPV in B_s mesons and they have to be explored in detail as B_d system. However, what I emphasized here is that we observe all categories of CPV which was laid out in the Standard Model except for that in $B\overline{B}$ mixing itself. These findings strongly supports the Standard Model explanation for origin of CPV.

Jonathan Rosner (University of Chicago):
(1) The direct CP asymmetry in $B_0 \rightarrow K^+\pi^-$ should be correlated with a large direct CP asymmetry in $B_0 \rightarrow \pi^+\pi^-$, favoring the average between the Belle and Babar values.

(2) The inequality between $A_{CP}(K^+\pi^-)$ and $A_{CP}(K^+\pi^0)$ is not a problem if one notes the important contribution of color-suppressed amplitude to the latter process. This amplitude implies a non-zero A_{CP} of $K^0\pi^0$ of approximate magnitude 0.15 and sign depending on whether one is speaking of C or A. [See Gronan and Rosner, Phys. Rev. D71 (2005), and Iain Stewart's talk this conference]

Kazuo Abe: I fully agree with your comments. We are aware of your works on a relation between $A_{CP}(\pi^+\pi^-)$ and $A_{CP}(K^+\pi^-)$, and an explanation of $A_{CP}(K^+\pi^-) \neq A_{CP}(K^+\pi^0)$. Further measurements will provide A_{CP} and branching fractions for many different modes. We like to hear theoretical predictions on their relations, hopefully before the measurements.

Jonathan Dorfan (SLAC):
What value of $sin2\beta$ for the new ΨK_s data? Your first $150fb^{-1}$ gave a measurement of 0.72, the new data must have $sin2\beta < 0.60$?? Is the new data all with the 4 layer vertex detector?

Kazuo Abe: (I did not have these numbers with me during the presentation , and they were added later.) The Belle $sin2\beta$ from $\Psi K_s(\pi^+\pi^-)$ in the old vertex detector data $(140fb^{-1})$ was 0.67 ± 0.08. All modes combined result from this data set was $0.728 \pm 0.056 \pm 0.023$. With new data (old and new vertex detector data combined), $\sin 2\phi_1$ from $J/\psi K_S$ and $J/\psi K_L$ separate samples are 0.668 ± 0.047 and 0.619 ± 0.069, respectively. The $\sin 2\phi_1$ from $J/\psi K^0$ subsample was on the lower side and seems to stay that way. However, we think this is just a statistical fluctuation.

CKM PARAMETERS AND RARE B DECAYS

FRANCESCO FORTI

INFN-Pisa, L.go Pontecorvo, 3, 56127 Pisa, Italy
E-mail: Francesco.Forti@pi.infn.it

Measurements of the angles and sides of the unitarity triangle and of the rates of rare B meson decays are crucial for the precise determination of Standard Model parameters and are sensitive to the presence of new physics particles in the loop diagrams. In this paper the recent measurements performed in this area by *BABAR* and Belle will be presented. The direct measurement of the angle α is for the first time as precise as the indirect determination. The precision of the $|V_{ub}|$ determination has improved significantly with respect to previous measurement. New limits on $B \to \tau\nu$ decays are presented, as well as updated measurements on $b \to s$ radiative transitions and a new observation of $b \to d\gamma$ transition made by Belle.

1 Introduction

In the Standard Model (SM), the interaction between up-type and down-type quarks is described by a unitarity matrix called the Cabibbo-Kobayashi-Maskawa matrix (in brief CKM).[1,2] This matrix can be parametrized with 3 real angles and one complex phase, which gives rise to CP violation. A widely used parametrization of the matrix[3,4] uses the four parameters $A, \lambda, \overline{\rho}, \overline{\eta}$, with $\overline{\eta}$ controlling the CP violation in this framework. The unitarity of the CKM matrix imposes 9 complex relations amongst the matrix elements, one of which is given by

$$V_{ub}^* V_{ud} + V_{cb}^* V_{cd} + V_{tb}^* V_{td} = 0,$$

where $V_{qq'}$ is the matrix element relating the quark q and q'. This relation can be represented as a triangle (called the unitarity trangle) in the complex $\overline{\rho}, \overline{\eta}$ plane, as shown in Fig. 1. B-meson decays are sensitive probes to measure both the angles and sides of the unitarity triangle and can unveil physics beyond the SM. In fact, most B decay amplitudes receive contributions from diagrams containing loops, where the presence of new particles can be detected through effects on the branching ratios, asymmetries, or spectra. Another possible route to detecting new physics is the high precision measurement of the unitarity triangle parameters to uncover

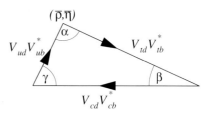

Figure 1. Unitarity triangle.

any inconsistency among them or between different determinations of the same parameter.

After having clearly established CP-violation in the B sector, the *BABAR* and Belle experiments are now pursuing an extended program of precision measurements of the unitarity triangle parameters and of rare B decays, taking advantage of the very large data sample collected at the B-Factories. The recent results of this measurement program are reported at this conference in two papers. The measurement of $\sin 2\beta$ and the direct CP violation measurements are presented by Kazuo Abe. In this paper, after introducing the *BABAR* and Belle experiments in Sec. 2, I will cover the α measurements in Sec. 3 and the $|V_{ub}|$ and $|V_{cb}|$ measurements in Sec. 4. The rest of the paper will be devoted to rare decays: $B \to \tau\nu$ in Sec. 5, $b \to s\gamma$ in Sec. 6, and $b \to d\gamma$ in Sec. 7. I will finally give some concluding remarks in Sec. 8.

2 The B-Factory experiments and datasets

The data used in the analyses presented in this paper have been collected with the *BABAR* detector at the PEP-II machine at SLAC and with the Belle detector at the KEKB machine in the KEK laboratory between 1999 and 2005. Both detectors, whose detailed description can be found elsewhere,[5,6] have been designed and optimized to study time-dependent CP-violation in B decays at the $\Upsilon(4S)$ resonance. Their major components are: a vertexing and tracking system based on silicon and gas detectors; a particle identification system; an electromagnetic calorimeter based on CsI(Tl) crystals operating within a 1.5 T magnetic field; an iron flux return located outside of the coil, instrumented to detect K_L^0 and identify muons. The $\Upsilon(4S)$ resonance decays most of the time in a pair of B-mesons, either B^+B^- or $B^0\overline{B}^0$, which acquire a boost thanks to the asymmetry of the beam energies: 9 GeV e^- on 3.1 GeV e^+ for PEP-II and 8 GeV e^- on 3.5 GeV e^+ for KEKB. Because of this boost, the decay vertices of the two mesons are separated, thus allowing their individual determination and the measurement of time-dependent CP asymmetries. In these analyses, the signal B is reconstructed in a CP-eigenstate (such as $B \to \pi\pi$) while the other B (the tagging B) is reconstructed in a decay mode that allows the determination of its flavor at the time of decay, such as exclusive hadronic or semileptonic modes, or inclusive modes with a lepton or a kaon, whose sign carries the information of the B flavor.

The data samples used in the measurements presented in this paper vary for the two experiments. Most measurements are based on $232 \times 10^6 B\overline{B}$ pairs for *BABAR* and $275 \times 10^6 B\overline{B}$ for Belle, but there are several results obtained with smaller statistics, while Belle performed the $b \to d\gamma$ analysis with $385 \times 10^6 B\overline{B}$.

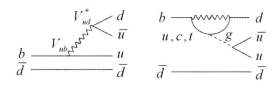

Figure 2. The tree (left) and penguin (right) diagrams contributing to charmless B decays $B^0 \to \pi^+\pi^-$, $B^0 \to \rho^+\pi^-$ and $B^0 \to \rho\rho$.

3 Determination of the angle $\alpha(\Phi_2)$

The angle α is the relative phase of the V_{ub} and V_{td} CKM matrix elements and can be measured in the charmless B decays $B \to \pi\pi$, $B \to \rho\pi$ and $B \to \rho\rho$ which arise from tree-level $b \to u(\overline{u}d)$ transitions (Fig. 2,left). A complication to this approach is the presence of loop level penguin diagrams leading to the same final states (Fig. 2, right), which introduce different CKM matrix elements. While in the absence of penguin contribution, the measurement of time dependent CP asymmetries in neutral B charmless decays would directly yield the angle α, the interference between tree and penguin diagrams obscures the simple relationship between CP observables and the angle α and requires the development of specific techniques to disentangle the penguin contribution.

Time-dependent CP asymmetries arise from the intereference of two possible paths reaching the same final state: $B \to f$ and $B \to \overline{B} \to f$, and can be expressed in terms of the complex parameter $\lambda_f = \eta_f \frac{p}{q} \frac{\overline{A}}{A}$, where $A = |\langle f|T|B^0\rangle|$, $\overline{A} = |\langle f|T|\overline{B}^0\rangle|$, η_f is the CP eigenvalue of the final state and q, p are the parameters describing how B^0 and $\overline{B}^0$ mix to form the mass eigenstates. The time dependent CP asymmetry follows

$$A_{CP}(\Delta t) = S_f \sin(\Delta m \Delta t) + C_f \cos(\Delta m \Delta t),$$

where $S_f = 2\frac{\Im(\lambda_f)}{1+|\lambda_f|^2}$ measures the CP violation arising from the interference of the decays with and without mixing, and $C_f = \frac{1-|\lambda_f|^2}{1+|\lambda_f|^2}$ measures the direct CP violation

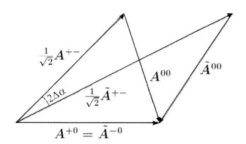

Figure 3. Isospin triangles for the charmless B decays $B^0 \to \pi^+\pi^-$, $B^0 \to \rho\rho$.

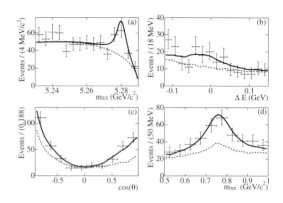

Figure 4. The distributions for the highest purity tagged events in the *BABAR* $B \to \rho\rho$ analysis for the variables m_{ES} (a), ΔE (b), cosine of the ρ helicity angle (c), and $m_{\pi^\pm\pi^0}$ (d). The dotted lines are the sum of backgrounds and the solid lines are the full PDF.

in the decay. For the tree diagram in (Fig. 2,left)

$$\lambda_f = \eta_f \frac{V_{tb}^* V_{td} V_{ub} V_{ud}^*}{V_{tb} V_{td}^* V_{ub}^* V_{ud}} = \eta_f e^{2i\alpha},$$

with $C_f = 0$ and $S_f = \sin(2\alpha)$. In the presence of the penguin diagram, the expression becomes:

$$\lambda_f = \eta_f e^{2i\alpha} \frac{T + Pe^{+i\gamma}e^{i\delta}}{T + Pe^{-i\gamma}e^{i\delta}}$$

where T and P are the tree and penguin amplitudes, and δ is the strong phase. The effect of penguin diagram interference is the possibility of direct CP violation ($C_f \propto \sin\delta$) and a shift $\Delta\alpha$ in the measurement of the angle α: $S_f = \sqrt{1 - C_f^2} \sin(2\alpha_{\text{eff}})$ with $\Delta\alpha = \alpha_{\text{eff}} - \alpha$.

Isospin relations amongst rates of the various $B \to \pi\pi$ and $B \to \rho\rho$ decays can be used[7] to extract the shift $\Delta\alpha$. The isospin analysis involves the separate measurement of B^0 and $\bar{B}^0$ decay rates into h^+h^- (h indicates either π or ρ) and h^0h^0, as well as the measurement of the rate of the charged B decay $B^{+(-)} \to h^{+(-)}h^0$. Constructing a B^0 and a $\bar{B}^0$ triangle from the 6 amplitudes (Fig. 3) one can extract $\Delta\alpha$ from the mismatch of the two triangles.

It has also been shown[8] that, in alternative to full isospin analysis, one can use the branching fractions for $B \to h^0h^0$ and $B \to h^+h^0$ averaged over meson and antimeson to impose an upper bound on $\Delta\alpha$:

$$\sin^2\Delta\alpha < \frac{\overline{\mathcal{B}}(B^0 \to h^0h^0)}{\overline{\mathcal{B}}(B^\pm \to h^\pm h^0)}$$

Other relations have also been developed,[9,10] but with the current level of accuracy of the measurements none improves significantly over the above limit. The constraints on α derived from a full isospin analysis in the $\pi\pi$ channel[11,12] are very weak, as shown in Fig. 5 explained later in the text, mainly due to the fact that the branching ratio $\mathcal{B}(B^0 \to \pi^0\pi^0) = (1.45 \pm 0.29) \times 10^{-6}$ (averaged by HFAG[13] on the basis of the *BABAR*[14] and Belle[15] measurements) is too large to be effective in setting the above limit, but is also too small for the full isospin analysis.

The $\rho\rho$ channel has three polarization amplitudes, which introduce dilution in the measurement because they have different CP eigenvalues, and has been considered in the past as less promising than $\pi\pi$. Both *BABAR* and Belle have recently performed full analyses of this decay.[16,17] The charged ρ is reconstructed through the decay $\rho^\pm \to \pi^\pm\pi^0$, and the events are selected through a kinematical signal identification based on the beam-energy substituted mass (also known as beam constrained mass) $m_{bc} \equiv m_{ES} = \sqrt{E_{\text{beam}}^{*2} - p_B^{*2}}$ and the energy difference between the reconstructed B and the beam

$\Delta E = E_B^* - E_{\text{beam}}^*$. All quantities are computed in the CM frame. The distribution of these variables for the signal and the background is shown in Fig. 4

It is found that the fraction of longitudinal polarization (f_L) in the $\rho\rho$ final state is almost 100%, and that therefore there is no dilution effect in the measurement of α. In addition, the $\rho^+\rho^-$ and $\rho^+\rho^0$ branching fractions are a factor of 5 larger than the corresponding ones in the $\pi\pi$ decays, but at the same time the $\rho^0\rho^0$ is not yet observed, with a relatively small limit on $\Delta\alpha$. The results are summarized in Table 1

Using the *BABAR* limit on $\mathcal{B}(B^0 \to \rho^0\rho^0)$ and the average between the two experiments for the other quantities one arrives at a relatively stringent limit on $\Delta\alpha$ ($\Delta\alpha < 11°$) and at the determination $\alpha[\rho\rho] = (96 \pm 13)°$.

The isospin analysis has an intrinsic two-fold ambiguity that can be removed with a full time-dependent Dalitz plot analysis of the $B \to \rho\pi$ decay.[18] Results on this analysis have been presented the ICHEP04 conference.[19,20]

The results of the three analysis are summarized in Fig. 5, where a combined fit[21] is also shown. The result of this combined fit is $\alpha = (99^{+12}_{-9})°$. The result from the indirect measurement of α obtained by fitting all the other CKM triangle measurements, $\alpha[\text{CKM}] = (96^{+11}_{-12})°$ is shown for comparison on the same plot. This is the first time that the direct measurement of α has a better precision than the its indirect determination from the CKM triangle fit.

4 Measurement of $|V_{ub}|$ and $|V_{cb}|$

The magnitude of the CKM matrix elements V_{ub} and V_{cb} can be extracted from the semileptonic decay rate of B mesons. At the parton level the decay rates for $b \to u\ell\nu$ and $b \to c\ell\nu$ can be calculated accurately; they are proportional to $|V_{ub}|^2$ and $|V_{cb}|^2$, respectively, and depend on the quark masses,

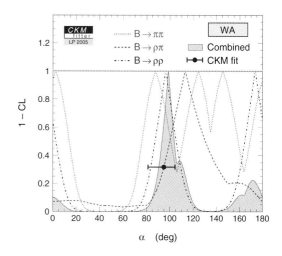

Figure 5. Alpha determination from the charmless B decays $B^0 \to \pi^+\pi^-$, $B^0 \to \rho\rho$ and $B^0 \to \rho^+\pi^-$. The dotted lines represent the results of the three individual analyses. The green (shaded) area is the result of the combined fit. The CKM triangle fit independent determination of alpha, which is not included in the fit, is shown by the blue point.

m_b, m_u, and m_c. To relate measurements of the semileptonic decay rate to $|V_{ub}|$ and $|V_{cb}|$, the parton-level calculations have to be corrected for effects of strong interactions, thus introducing significant theoretical uncertainties for both exclusive and inclusive analyses.

For of exclusive decays, the effect is parametrized by form factors(FF), such as in the simple case of the $B \to \pi\ell\nu$ decay, neglecting the π mass:

$$\frac{d\Gamma(B^0 \to \pi^-\ell^+\nu)}{dq^2} = \frac{G_F^2 |V_{ub}|^2}{24\pi^3} |f_+(q^2)|^2 p_\pi^3,$$

where G_F is the Fermi constant, q^2 is the invariant-mass squared of the lepton-neutrino system and p_π is the pion momentum in the B frame. The FF $f_+(q^2)$ can be calculated with a variety of approaches based on quark model,[22] Light Cone Sum Rules,[23] and lattice QCD.[24,25] In inclusive decays, the main difficulty is to relate the partial rate obtained by the experimental event selection process to the matrix elements. This is a particularly serious issue for $|V_{ub}|$, where

102

Table 1. Summary of measurements for the $B \to \rho\rho$ decays

Quantity	BABAR	Belle
f_L	$0.978 \pm 0.014^{+0.021}_{-0.029}$	$0.951^{+0.033+0.029}_{-0.039-0.031}$
$S_{\rho\rho,L}$	$-0.33 \pm 0.24^{+0.08}_{-0.14}$	$0.09 \pm 0.42 \pm 0.08$
$C_{\rho\rho,L}$	$-0.03 \pm 0.18 \pm 0.09$	$0.00 \pm 0.30^{+0.09}_{-0.10}$
$\mathcal{B}(B^0 \to \rho^+\rho^-)\ [10^{-6}]$	$30 \pm 4 \pm 5$	$24.4 \pm 2.2^{+3.8}_{-4.1}$
$\mathcal{B}(B^\pm \to \rho^+\rho^0)\ [10^{-6}]$	$22.5^{+5.7}_{-5.4} \pm 5.8$	$31.7 \pm 7.1^{+3.8}_{-6.7}$
$\mathcal{B}(B^0 \to \rho^0\rho^0)\ [10^{-6}]$	< 1.1	-

only a small fraction of the total rate can be determined experimentally because of the severe background rejection cuts. Heavy-Quark Expansions (HQEs)[26] have become a useful tool for calculating perturbative and non-perturbative QCD corrections and for estimating their uncertainties. These expansions contain parameters such as the b quark mass and the average Fermi momentum of the b quark inside the B meson. These parameters must be determined experimentally, for instance from the photon energy spectrum in $B \to X_s\gamma$ decays and the spectrum of the hadronic mass in $B \to X_c\ell\nu$ decays.

For the determination of $|V_{cb}|$, a global analysis of inclusive B decays has been performed,[27] leading to a very precise measurement:

$$|V_{cb}|_{\text{incl.}} = (41.4 \pm 0.6_{\text{exp}} \pm 0.1_{\text{th}}) \times 10^{-3}.$$

The measurement obtained from the world average of $\mathcal{B}(B \to D^*\ell\nu)$,[28]

$$|V_{cb}|_{D^*\ell\nu} = (41.3 \pm 1.0_{\text{exp}} \pm 1.8_{\text{th}}) \times 10^{-3},$$

is fully compatible, although less precise.

These accurate measurements demonstrate the rapid experimental and theoretical advancements in these area.

4.1 $b \to u\ell\nu$ inclusive decays.

Several methods have been used to isolate inclusive $b \to u\ell\nu$ decays from the much more frequent $b \to c\ell\nu$ decays.

In the lepton endpoint method[29,30] one uses the fact that, due to the mass difference between c and u quarks, the lepton spectrum in the $b \to u$ transition extends to slightly higher energies than in the $b \to c$ decays. The lepton momentum window is typically $1.9 < p_{\text{lept}} < 2.6\,\text{GeV}/c$, and a selection is applied on the basis of event shape variables and missing momentum. The background remains in any case significant with typically $S/B \approx 1/14$.

One can refine the selection by using a q^2-dependent electron energy cut (the $E_e - q^2$ method)[31] where the neutrino momentum is estimated from the event missing momentum and q^2 is calculated from $q^2 = (p_e + p_\nu)^2$. For each E_e and q^2 one can calculate the maximum kinematically allowed hadronic mass square s_h^{max} and veto $b \to c\ell\nu$ decays by requiring $s_h^{\text{max}} < 3.5\,\text{GeV}^2 \approx m_D^2$. This technique significantly improves the S/B ratio to about 1/2. Figure 6 shows the electron energy and s_h^{max} spectra, along with signal and sideband regions.

Reconstructing the other B in the event in an exclusive channel allows the direct reconstruction of the hadronic system (called

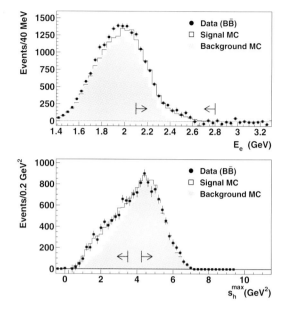

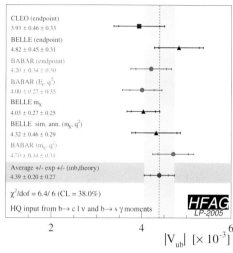

Figure 7. Summary and average of inclusive $|V_{ub}|$ determinations using HQE parameters extracted from $B \to X_s\gamma$ and $B \to X_c\ell\nu$ moments.

Figure 6. The electron energy, E_e, and $s_h^{\max}$ spectra in the $\Upsilon(4S)$ frame for continuum-subtracted data and simulated $B\bar{B}$ events satisfying all the selection criteria except for the variable shown. The arrows denote the signal and sideband regions.

X) produced in $b \to u\ell\nu$ decays by assigning all the remaining particles to it. $BABAR$ uses the mass of the hadronic system to perform a 2-dimensional fit for the partial branching fraction in the area $\{M_X < 1.7\,\text{GeV}/c^2, q^2 > 8\,\text{GeV}^2\}$,[32] while Belle also introduces the variable $P_+ \equiv E_X - |\mathbf{p}_X|$, where E_X and $\mathbf{p}_X$ are the energy and 3-momentum of the hadronic system, analyzing data in three kinematical regions $M_X < 1.7\,\text{GeV}/c^2$, $\{M_X < 1.7\,\text{GeV}/c^2, q^2 > 8\,\text{GeV}^2\}$, and $P_+ < 0.66\,\text{GeV}/c$.[33]

The extraction of $|V_{ub}|$ from these partial branching fractions involves the determination of HQE parameters, which can be done following a variety of schemes and using different physical processes.[34] This extraction is the object of a very active discussion with the goal of improving the precision of the measurement. A summary of $|V_{ub}|$ inclusive determinations based on HQE parameters de-

rived from the moments of the photon energy spectrum in $B \to X_s\gamma$ decays and from the hadronic-mass and lepton-energy moments in $B \to X_c\ell\nu$ decays is shown[13] in Fig. 7:

$$|V_{ub}|_{\text{incl.}} = (4.39 \pm 0.20_{\text{exp}} \pm 0.27_{\text{th}}) \times 10^{-3}.$$

An alternative determination, using HQE parameters[35] obtained fitting the Belle $B \to X_s\gamma$ photon energy spectrum, yields:[29]

$$|V_{ub}|_{\text{incl.}} = (5.08 \pm 0.47_{\text{exp}} \pm 0.48_{\text{th}}) \times 10^{-3}.$$

4.2 $B \to \pi\ell\nu, \rho\ell\nu$ decays.

Various methods have been devised to isolate exclusive $B \to \pi\ell\nu, \rho\ell\nu$ decays from the large backgrounds from $b \to c\ell\nu$ and continuum events. Estimating the neutrino momentum from the missing momentum in the event allows the usage of the mass of the B candidate m_{ES} as a discriminating variable. In addition, one can analyze the data in bins of q^2 (three bins for CLEO[36] and five bins for $BABAR$[37]) and measure the q^2 dependance of the form factor, thus discriminating among theoretical models.

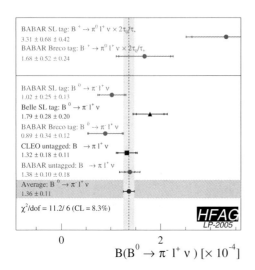

Figure 8. Summary and average of exclusive $B \to \pi\ell\nu$ branching fractions.

Tagging the other B in the event is another powerful method to reduce backgrounds. As in the case of inclusive decays one can reconstruct the other B in an exclusive hadronic channel[38] (BReco tag) which allows the reconstruction of the hadronic system on the signal side. Alternatively, one can tag the other B through semileptonic decays, and use the kinematics of 2 back-to-back semileptonic decays to reduce the background.[39,40,41]

A summary of exclusive $B \to \pi\ell\nu$ branching fractions measurements is shown in Fig. 8. The extraction of $|V_{ub}|$ from these branching fractions requires a theoretical calculation of the form factor, which depends on the q^2 range used. Reducing the q^2 range usually improves the error on the form factor calculation while the experimental error increases because of the loss of statistics. For $q^2 < 15\,\mathrm{GeV}^2$ Light Cone Sum Rules[23] provide the most accurate calculation, whereas lattice calculation are limited to $q^2 > 15\,\mathrm{GeV}^2$ due to the restriction to π energies smaller than the inverse lattice spacing. Using the FNAL04 lattice calculations[25]

for $q^2 > 16\,\mathrm{GeV}^2$ one obtains

$$|V_{ub}|_{\mathrm{excl.}} = (3.75 \pm 0.27^{+0.64}_{-0.42}) \times 10^{-3}.$$

It should be noted that the inclusive and exclusive determinations of $|V_{ub}|$ are experimentally and theoretically independent. The previously reported hints of discrepancy[42] between the two measurements are now reduced in size and the results are compatible. Theory errors have been progressively reduced and have broken the 10% limit for the inclusive measurement.

5 $B \to \tau\nu$ decay

In the SM, the purely leptonic decay $B^+ \to \ell^+\nu$ (charge conjugate modes are implied) proceeds via the annihilation of the $\bar{b}$ and u quark into a virtual W boson. Its amplitude is proportional to the product of $|V_{ub}|$ and the B meson decay constant f_B, with a predicted branching fraction given by:[43]

$$\mathcal{B}(B^+ \to \ell^+\nu) =$$
$$\frac{G_F^2 m_B}{8\pi} m_\ell^2 \left(1 - \frac{m_\ell^2}{m_B^2}\right)^2 f_B^2 |V_{ub}|^2 \tau_B,$$

where G_F is the Fermi coupling constants, m_ℓ and m_B are the lepton and B meson masses, and τ_B is the B^+ meson lifetime. The dependance on the lepton mass arises from helicity conservation, which suppresses the electron and muon channels. The branching ratio in the τ channel is predicted in the SM to be roughly 10^{-4}, but physics beyond the SM, such as supersymmetry or two-Higgs-doublets models could significantly modify the process. Observation of $B \to \tau\nu$ would allow a direct determination of f_B, which is currently estimated with a 15% theoretical uncertainty[44] using lattice QCD calculations. Besides, the ratio of $\mathcal{B}(B^+ \to \tau^+\nu)$ to ΔM_{B_d}, the mass difference between heavy and light B_d mesons, can be used to determine the ratio of $|V_{ub}|^2/|V_{td}|^2$, constraining an area in the $\bar{\rho}, \bar{\eta}$ plane with small theoretical uncertainties.[45] Conversely, from the

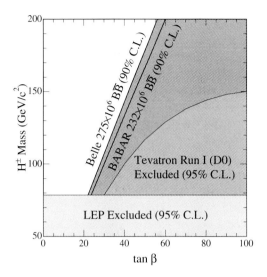

Figure 9. Exclusion are in the $[m_H, \tan\beta]$ plane obtained from the upper limit on $.\mathcal{B}(B^+ \to \tau^+\nu)$.

global CKM fit one can derive[21] the constraint $\mathcal{B}(B^+ \to \tau^+\nu) = (8.1^{+1.7}_{-1.3}) \times 10^{-5}$.

Due to the presence of at least two neutrinos in the final state, the $B^+ \to \tau^+\nu$ decay lacks the kinematical constraints that are usually exploited in B decay searches to reject both continuum and $B\bar{B}$ backgrounds. The strategy adopted is to exclusively identify the other B in the event through a semileptonic or hadronic decay, and assign all the remaining tracks to the signal B. The τ lepton is then searched in one or three prongs decays, with a maximum of one π^0. After applying kinematical cuts and requiring a large missing mass in the event, the most powerful variable for separating signal and background is remaining energy (E_{ECL}) non associated with either B. Applying a cut $E_{ECL} < 0.3\,\text{GeV}$, Belle[46] finds no significant eccess of events over the expected backgrounds, that ranges between 3 and 12 events depending on the τ decay mode, and sets an upper limit $\mathcal{B}(B^+ \to \tau^+\nu) < 1.8 \times 10^{-4}$ @ 90%C.L.. *BABAR* finds a slightly higher upper limit.

This result can be interpreted in the con-text of extensions to the SM. In the two-Higgs doublet model the decay can occur via a charged Higgs particle, and the $\mathcal{B}(B^+ \to \tau^+\nu)$ upper limited can be translated in a constraint in the $[m_H, \tan\beta]$ plane, as seen in Fig. 9 where m_H is the mass of the Higgs particle and $\tan\beta$ is the ratio of the vacuum expectation values of the two Higgs doublets.[47]

6 $b \to s$ radiative decays

Radiative decays involving the $b \to s$ flavour-changing neutral current transition occur in the SM via one-loop penguin diagrams containing an up-type quark (u, c, t) and a W boson. Example of these decays are: $B \to X_s\gamma, K^*\gamma, K_S^0\pi^0\gamma, K\pi\pi\gamma, K^{(*)}\ell^+\ell^-, K\nu\nu, \cdots$.

New physics particles replacing the SM ones in the penguin loop, e.g. a charged Higgs boson or squarks, can affect both the total rate of these processes and the decay properties, such as photon polarization, direct CP violation, and forward-backward asymmetry in $B \to K^{(*)}\ell^+\ell^-$.

6.1 $B \to X_s\gamma$ decays

Within the SM, the inclusive $B \to X_s\gamma$ rate is predicted by next-to-leading order (NLO) calculations[48] to be $\mathcal{B}(B \to X_s\gamma) = (3.57 \pm 0.30) \times 10^{-4}$ for $E_\gamma > 1.6\,\text{GeV}$. The photon energy spectrum provides access to the distribution function of the b quark inside the B meson,[49] whose knowledge is crucial for the extraction of $|V_{ub}|$ from inclusive semileptonic $B \to X_u\ell\nu$ decays, as discussed in Sec. 4. The heavy quark parameters m_b and μ_π^2, which describe the effective the b-quark mass and the kinetic energy inside the B meson, can be determined from the photon energy spectrum, either by fitting the spectrum directly or by fitting the spectrum moments.[50,51]

The branching fraction and the photon energy spectrum can be measured with two methods, originally introduced by CLEO:[52] in the fully inclusive method the photon en-

Table 2. Summary of partial branching fraction measurements for the $B \rightarrow X_s\gamma$ process. As explained in the text, Belle uses a photon energy cut $E_\gamma > 1.8\,\text{GeV}$, while *BABAR* uses $E_\gamma > 1.9\,\text{GeV}$. The errors are statistical, systematical, and model dependent.

Experiment	$\mathcal{B}(B \rightarrow X_s\gamma)[10^{-4}]$
Belle, incl.[53]	$3.55 \pm 0.32^{+0.30+0.11}_{-0.31-0.07}$
BABAR, incl.[54]	$3.67 \pm 0.29 \pm 0.34 \pm 0.29$
BABAR, excl.[55]	$3.27 \pm 0.18^{+0.55+0.04}_{-0.40-0.09}$

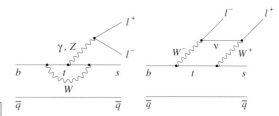

Figure 10. Feynmann diagrams decribing the $B \rightarrow s\ell^+\ell^-$ decay.

ergy spectrum is measured without reconstructing the X_s system, and backgrounds are suppressed using event shape variables and high-momentum lepton tagging of the other B; the semi-inclusive method uses a sum of exclusive final states where possible X_s systems are combined with the photon, and kinematic constraints are used to suppress backgrounds. The semi-inclusive method suffers from uncertainties on the fragmentation of the X_s system and on the assumptions made as to the fraction of unmeasured final states. On the other hand the fully-inclusive method has much larger residual backgrounds that must be carefully subtracted using off-resonance data. Table 2 summarizes the $\mathcal{B}(B \rightarrow X_s\gamma)$ measurements. Belle uses a photon energy cut $E_\gamma > 1.8\,\text{GeV}$, while *BABAR* uses $E_\gamma > 1.9\,\text{GeV}$. The results are fully consistent with the SM expectations.

6.2 Photon polarization

In the SM, the photon from the $b \rightarrow s\gamma$ ($\bar{b} \rightarrow \bar{s}\gamma$) decays has an almost complete left-handed (right-handed) polarization. This pattern was generally assumed to be valid up to a $O(m_s/m_b)$ correction,[56] but it has been recently shown[57] that the corrections can be significantly larger. A different polarization pattern would be a marker of new physics, and can be explored in different ways. In one method[56] photon helicity is probed in mixing-induced CP asymmetries, exploiting the fact that left-handed and right-handed photons cannot interfere, thus suppressing time-dependent CP asymmetries in decays such as $B \rightarrow K^{*0}\gamma$. In another method[58] one uses the kaon resonances decays $B \rightarrow K_{res}\gamma \rightarrow K\pi\pi\gamma$ to measure the up-down asymmetry of the photon direction relative to the $K\pi\pi$ decay plane. Experimentally, many $B \rightarrow K\pi\pi\gamma$ decay channels have been observed,[59,60] with branching fractions varying in the range $(1.8 - 4.3) \times 10^{-5}$, although with a statistics still insuffient for the helicity analysis. Both decays $B \rightarrow K^{*0}\gamma$ and $B \rightarrow K_S\pi^0\gamma$ have been observed and their time-dependent CP asymmetry measured.[61,62] With the present statistics all the results are consistent with zero.

6.3 $B \rightarrow K^{(*)}\ell^+\ell^-$ decays

As shown in Fig. 10, $b \rightarrow s\ell^+\ell^-$ decays proceed in the SM both via a radiative penguin diagram with a photon or a Z, and via a W-mediated box diagram. The magnitude of the photon penguin amplitude is known from the $b \rightarrow s\gamma$ rate measurement, while the Z penguin and W box amplitudes provide new information on FCNC processes. The predicted total branching fraction is[63] $\mathcal{B}(b \rightarrow s\ell^+\ell^-) = (4.2 \pm 0.7) \times 10^{-6}$, in agreement with measurements.[64,65]

The $B \rightarrow K^{(*)}\ell^+\ell^-$ exclusive decays are predicted to have branching fractions of 0.4×10^{-6} for $B \rightarrow K\ell^+\ell^-$ and about 1.2×10^{-6}

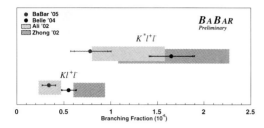

Figure 11. Experimental measurements (points) and theoretical predictions for $B \to K^{(*)}\ell^+\ell^-$ branching fractions. Red (upper) points are the *BABAR*[70] result, while blue (lower) points are the Belle[71] result. The width of the boxes indicates the estimated precision of the predictions.[63,73]

for $B \to K^*\ell^+\ell^-$, with a theoretical uncertainty of about 30% mainly due the lack of precision in predicting how often the s quark will result in a single $K^{(*)}$ meson in the final state. Since the electroweak couplings to electron and muon are identical, the ratio $R_K = \mathcal{B}(B \to K\mu^+\mu^-)/\mathcal{B}(B \to Ke^+e^-)$ is expected to be unity, while in $B \to K^*\ell^+\ell^-$ decays a phase space contribution from a pole in the photon penguin amplitude at $q^2 = m_{\ell^+\ell^-}^2 \simeq 0$ enhances the lighter lepton pair, with a prediction of $R_{K^*} = \mathcal{B}(B \to K^*\mu^+\mu^-)/\mathcal{B}(B \to K^*e^+e^-) = 0.752$. Neglecting the pole region ($q^2 < 0.1\,\mathrm{GeV}^2$) for $B \to K^*e^+e^-$, both ratios R_K and R_{K^*} are predicted to be very close to unity. However, an enhancement of order 10% is expected in the presence of a supersymmetric neutral Higgs boson with large $\tan\beta$.[66] New physics at the electroweak scale could also enhance direct CP asymmetries, defined as $A_{CP} = \frac{\Gamma(\overline{B}\to K^{(*)}\ell^+\ell^-)-\Gamma(B\to K^{(*)}\ell^+\ell^-)}{\Gamma(\overline{B}\to K^{(*)}\ell^+\ell^-)+\Gamma(B\to K^{(*)}\ell^+\ell^-)}$, to values of order one,[67] while the SM expectations[68] are much less than 1%. Finally, the q^2-dependance of the lepton forward backward asymmetry is sensitive to some new physics effects, such as a change of sign[69] of the Wilson coefficient C_7 of the Operator Product Expansion, that would not show up in other channels.

Experimentally, the $B \to K^{(*)}\ell^+\ell^-$ decays are identified through kinematical constraints following a positive K identification. Care must be taken to reject dilepton pairs with a mass consistent with the J/ψ and the $\psi(2S)$, which are produced abundantly in B decays. Both processes are well established,[70,71] and the branching fractions are compared to theoretical calculations in Fig. 11 CP asymmetries measurements are consistente with zero with an error of 0.25. Belle also reports the first measurement of the lepton forward-backward asymmetry[71] and of the ratio of Wilson coefficients[72], although the statistical power is not yet sufficient to identify new physics effects.

6.4   Other radiative decays

Several other exclusive B radiative decay modes have been looked at, searching for deviation from SM expectations. No signal has been found yet, but some of the limits (given below at 90% C.L.) are getting close to the SM values. $B \to D^{*0}\gamma$ proceeds via a W-exchange diagram and the branching fraction is expected to be around 10^{-6} in the SM. The measured limit[74] is $\mathcal{B}(\overline{B}^0 \to D^{*0}\gamma) < 2.5 \times 10^{-5}$. For $B \to \phi\gamma$, which proceeds through a penguin annihilation diagram[75] the SM expectations are around 10^{-12}, while the experimental limit is $\mathcal{B}(B^0 \to \phi\gamma) < 8.5 \times 10^{-7}$. The double radiative decay $B \to \gamma\gamma$ has a clean experimental signature and is expected to be around 3×10^{-8} in the SM. The measurements[76] limit its rate at $\mathcal{B}(B^0 \to \gamma\gamma) < 5.4 \times 10^{-7}$

7   Observation of $b \to d$ radiative decays

The $b \to d\gamma$ process is suppressed with respect to $b \to s\gamma$ by a factor $|V_{td}/V_{ts}|^2 \simeq 0.04$. Due to the large background from continuum events, only exclusive modes such as $B^- \to \rho^-\gamma$, $\overline{B}^0 \to \rho^0\gamma$, $\overline{B}^0 \to \omega\gamma$ (charge conjugate

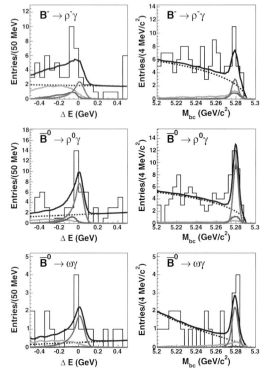

Figure 12. Projection of the fit results to M_{bc} and ΔE for the individual $b \to d\gamma$ modes. Lines represent the signal (magenta), continuum (blu-dashed), $B \to K^*\gamma$ (red), other B decay background components (green), and the total fit result (blue-solid).

modes are implied), have been searched so far. Measurement of these exclusive branching fractions, which are predicted to be in the range $(0.9 - 2.7) \times 10^{-6}$ in the SM,[77] gives a precise determination of $|V_{td}/V_{ts}|$ and provides sensitivity to physics beyond the SM. Belle reports the first observation[78] of these decays, reconstructing the ρ and ω with final states with at most one π^0. Background rejection is obtained through the use of event shape variables, vertex separtion, and by tagging the other B in the event. All the variables are used in an unbinned maximum likelyhood fit where the $B \to (\rho, \omega)\gamma$ and $B \to K^*\gamma$ yields are simultaneuosly determined.

Figure 12 shows the projection of the likelihood fit onto the M_{bc} and ΔE axes for the individual modes. A clear peak is always

visible. The individual branching ratios are determined as follows:

$$\mathcal{B}(B^- \to \rho^-\gamma) = (0.55^{+0.43+0.12}_{-0.37-0.11}) \times 10^{-6},$$
$$\mathcal{B}(\overline{B}^0 \to \rho^0\gamma) = (1.17^{+0.35+0.09}_{-0.31-0.08}) \times 10^{-6},$$
$$\mathcal{B}(\overline{B}^0 \to \omega\gamma) = (0.58^{+0.35+0.07}_{-0.27-0.11}) \times 10^{-6},$$

where the first error is statistical and the second error is systematical. The significance figures of the three measurements are $1.5\sigma, 5.1\sigma$, and 2.6σ, respectively. A simultaneous fit is also perfomerd using the isospin relation:

$$\mathcal{B}(B \to \rho/\omega\gamma) \equiv \mathcal{B}(B^- \to \rho^-\gamma) =$$
$$2\frac{\tau_{B^+}}{\tau_{B^0}}\mathcal{B}(\overline{B}^0 \to \rho^0\gamma) = 2\frac{\tau_{B^+}}{\tau_{B^0}}\mathcal{B}(\overline{B}^0 \to \omega\gamma)$$

where $\tau_{B^+}/\tau_{B^0} = 1.076 \pm 0.008$ is the ratio of charged B lifetime to the neutral B lifetime, yielding

$$\mathcal{B}(B \to \rho/\omega\gamma) = 1.34^{+0.345+0.14}_{-0.31-0.10} \quad (5.5\sigma)$$

It should be noted that the individual fit results (especially $\mathcal{B}(B \to \rho^0\gamma)$) are in marginal agreement with the isospin relation above or with the previous limits. More statistics will hopefully clarify the issue. The simultaneous determination of $\mathcal{B}(B \to K^*\gamma)$ allows the determination of $|V_{td}/V_{ts}|$:[79]

$$|V_{td}/V_{ts}| = 0.200^{+0.026+0.038}_{-0.025-0.029},$$

where the errors are respectively from experiment and theory. This value is in agreement with global fit to the unitarity triangle,[21] but the $b \to d\gamma$ observation provides an independent constraint on the unitarity triangle which will become more and more effective as statistics increase.

8 Summary and conclusions

The accuracy of the analyses performed by the *BABAR* and Belle experiments has been steadily improving, and the precision measurement of CKM parameters is now a reality. The direct $\alpha(\Phi_2)$ determination $\alpha = (99^{+12}_{-9})°$ is for the first time more precise than its indirect determination from the CKM triangle

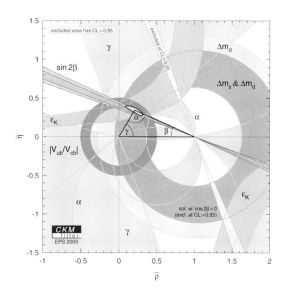

Figure 13. Allowed region in the $\bar{\rho}, \bar{\eta}$ plane once all the constraints are included.

fit. $|V_{cb}|$ is known at the 1.5% level, while $|V_{ub}|_{\text{incl.}} = (4.39 \pm 0.20_{\text{exp}} \pm 0.27_{\text{th}}) \times 10^{-3}$. is determined at the 8% level, and is the object of intense activity to further reduce the error.

Rare decays are very powerful tools for testing the consistency of the SM and are sensitive to new physics particles in the loop. They also allow the investigation of the inner structure of the B meson, thus reducing theory uncertainties in many measurements. $b \to d\gamma$ penguin transitions have been observed at the 5.5σ level, $\mathcal{B}(B \to \rho/\omega\gamma) = 1.34^{+0.345+0.14}_{-0.31-0.10}$, starting to provide new constraints on the unitarity triangle.

Figure 13 shows the allowed region in the $\bar{\rho}, \bar{\eta}$ plane after all the constraints have been applied. The figure represents the experimental situation after the summer conferences 2005. As more data will be necessary to disentangle all the effects and identify the signals of new physics, each experiment is set to reach a data sample of about 1 ab^{-1} within a few years. Larger samples will require significant machine and detector upgrades which are being actively studied by the community.

Acknowledgments

I wish to thank my colleagues in Babar and Belle who are responsible for most of the experimental results presented here. In particular I am grateful to Vera Luth, Christofer Hearty, Jeff Richman, Marcello Giorgi, David MacFarlane, Giancarlo Piredda, Riccardo Faccini, Bob Kowalewski, Jeffrey Berryhill, Iain Stewart, and Yoshi Sakai, for their comments and help in preparing the talk and this document. A special thank to Maurizio Pierini and Andreas Hoecker for producing the UTFit and CKMFitter plots only hours after the results have been available. This work has been in part supported by a grant from the Italian National Institute for Nuclear Physics (INFN).

References

1. N. Cabibbo, *Phys. Rev. Lett.* **10** 531 (1963).
2. M. Kobayashi and T. Maskawa, *Prog. Theor. Phys.* **49** 652 (1973).
3. L. Wolfenstein, *Phys. Rev. Lett.* **51** 1945 (1983).
4. G. Buchalla, A. J. Buras and M. E. Lautenbacher, *Rev. Mod. Phys.* **68** 1125 (1996), [arXiv:hep-ph/9512380].
5. B. Aubert *et al.* [BABAR Coll.], *Nucl. Instrum. Meth. A* **479** 1 (2002), [arXiv:hep-ex/0105044].
6. A. Abashian *et al.* [BELLE Coll.], *Nucl. Instrum. Meth. A* **479** 117 (2002).
7. M. Gronau and D. London, Decays," *Phys. Rev. Lett.* **65** 3381 (1990). PRLTA,65,3381;
8. Y. Grossman and H. R. Quinn, *Phys. Rev. D* **58** 017504 (1998), [arXiv:hep-ph/9712306].
9. J. Charles, *Phys. Rev. D* **59** 054007 (1999), [arXiv:hep-ph/9806468].
10. M. Gronau, D. London, N. Sinha and

R. Sinha, *Phys. Lett. B* **514** 315 (2001), [arXiv:hep-ph/0105308].

11. B. Aubert *et al.* [BaBar Coll.], *Phys. Rev. Lett.* **95** 151803 (2005), [arXiv:hep-ex/0501071].

12. K. Abe *et al.* [BELLE Coll.], arXiv:hep-ex/0502035.

13. Heavy Flavor Averaging Group, www.slac.stanford.edu/xorg/hfag/

14. B. Aubert *et al.* [BABAR Coll.], *Phys. Rev. Lett.* **94** 181802 (2005), [arXiv:hep-ex/0412037].

15. K. Abe *et al.* [BELLE Coll.], *Phys. Rev. Lett.* **94** 181803 (2005), [arXiv:hep-ex/0408101].

16. B. Aubert *et al.* [BABAR Coll.], *Phys. Rev. Lett.* **95** 041805 (2005), [arXiv:hep-ex/0503049].

17. K. Abe *et al.* [BELLE Coll.], arXiv:hep-ex/0507039.

18. A. E. Snyder and H. R. Quinn, *Phys. Rev. D* **48** 2139 (1993).

19. C. C. Wang *et al.* [BELLE Coll.], *Phys. Rev. Lett.* **94** 121801 (2005), [arXiv:hep-ex/0408003].

20. B. Aubert *et al.* [BABAR Coll.], arXiv:hep-ex/0408099.

21. J. Charles *et al.* [CKMfitter Group], *Eur. Phys. J. C* **41** 1 (2005), [arXiv:hep-ph/0406184]. updated plots at www.slac.stanford.edu/xorg/ckmfitter

22. D. Scora, N. Isgur, *Phys. Rev. D* **52** 2783 (1995).

23. P. Ball and R. Zwicky, *Phys. Rev. D* **71** 014015 (2005); P. Ball and R. Zwicky, *Phys. Rev. D* **71** 014029 (2005).

24. J. Shigemitsu *et al.*, hep-lat/0408019, Contribution to Lattice 2004, FNAL, June 21–26, 2004.

25. M. Okamoto *et al.*, hep-lat/0409116, Contribution to Lattice 2004, FNAL, June 21–26, 2004.

26. M. Voloshin and M. Shifman, *Sov. J. Nucl. Phys.* **41**, 120 (1985); J. Chay, H. Georgi, and B. Grinstein, *Phys. Lett.* **B247**, 399 (1990); I. I. Bigi, and

N. Uraltsev, *Phys. Lett.* **B280**, 271 (1992).

27. C. W. Bauer, Z. Ligeti, M. Luke, A. V. Manohar and M. Trott, *Phys. Rev. D* **70** 094017 (2004), [arXiv:hep-ph/0408002].

28. H. F. A. Group(HFAG), arXiv:hep-ex/0505100.

29. A. Limosani *et al.* [BELLE Coll.], *Phys. Lett. B* **621** 28 (2005), [arXiv:hep-ex/0504046].

30. B. Aubert *et al.* [BaBar Coll.], arXiv:hep-ex/0408075.

31. B. Aubert *et al.* [BABAR Coll.], *Phys. Rev. Lett.* **95** 111801 (2005), [arXiv:hep-ex/0506036].

32. B. Aubert *et al.* [BABAR Coll.], arXiv:hep-ex/0507017.

33. I. Bizjak *et al.* [BELLE Coll.], arXiv:hep-ex/0505088.

34. B. O. Lange, M. Neubert and G. Paz, arXiv:hep-ph/0504071.

35. I. Bizjak, A. Limosani and T. Nozaki, arXiv:hep-ex/0506057.

36. S. B. Athar *et al.* [CLEO Coll.], *Phys. Rev. D* **68** 072003 (2003), [arXiv:hep-ex/0304019].

37. B. Aubert *et al.* [BABAR Coll.], arXiv:hep-ex/0507003.

38. B. Aubert *et al.* [BABAR Coll.], arXiv:hep-ex/0408068.

39. B. Aubert *et al.* [BABAR Coll.], arXiv:hep-ex/0506064.

40. B. Aubert *et al.* [BABAR Coll.], arXiv:hep-ex/0506065.

41. K. Abe *et al.* [BELLE Coll.], arXiv:hep-ex/0408145.

42. Summer 2004 averages of the HFAG, www.slac.stanford.edu/xorg/hfag/

43. P. F. . Harrison and H. R. . Quinn [BABAR Collaboration], SLAC-R-0504

44. S. M. Ryan, *Nucl. Phys. Proc. Suppl.* **106** 86 (2002), [arXiv:hep-lat/0111010].

45. T. E. Browder *et al.* [CLEO Collaboration], *Phys. Rev. Lett.* **86**, 2950 (2001), [arXiv:hep-ex/0007057].

46. K. Abe *et al.* [Belle Collaboration], arXiv:hep-ex/0507034.

47. W. S. Hou, *Phys. Rev. D* **48** 2342 (1993).

48. P. Gambino and M. Misiak, *Nucl. Phys. B* **611** 338 (2001), [arXiv:hep-ph/0104034].
A. J. Buras, A. Czarnecki, M. Misiak and J. Urban, *Nucl. Phys. B* **631** 219 (2002), [arXiv:hep-ph/0203135].
T. Hurth, E. Lunghi and W. Porod, *Nucl. Phys. B* **704** 56 (2005), [arXiv:hep-ph/0312260].

49. M. Neubert, *Phys. Rev. D* **49** 4623 (1994), [arXiv:hep-ph/9312311].

50. M. Neubert, *Phys. Lett. B* **612** 13 (2005), [arXiv:hep-ph/0412241].

51. D. Benson, I. I. Bigi and N. Uraltsev, *Nucl. Phys. B* **710** 371 (2005). [arXiv:hep-ph/0410080].

52. M. S. Alam *et al.* [CLEO Collaboration], *Phys. Rev. Lett.* **74** 2885 (1995).

53. P. Koppenburg *et al.* [Belle Collaboration], *Phys. Rev. Lett.* **93** 061803 (2004), [arXiv:hep-ex/0403004].

54. B. Aubert *et al.* [BaBar Collaboration], arXiv:hep-ex/0507001.

55. B. Aubert *et al.* [BABAR Collaboration], arXiv:hep-ex/0508004.

56. D. Atwood, M. Gronau and A. Soni, *Phys. Rev. Lett.* **79** 185 (1997), [arXiv:hep-ph/9704272].

57. B. Grinstein, Y. Grossman, Z. Ligeti and D. Pirjol, *Phys. Rev. D* **71** 011504 (2005), [arXiv:hep-ph/0412019].

58. M. Gronau, Y. Grossman, D. Pirjol and A. Ryd, *Phys. Rev. Lett.* **88** 051802 (2002), [arXiv:hep-ph/0107254].

59. B. Aubert *et al.* [BABAR Collaboration], arXiv:hep-ex/0507031.

60. H. Yang *et al.*, *Phys. Rev. Lett.* **94** 111802 (2005), [arXiv:hep-ex/0412039].

61. Y. Ushiroda *et al.*, *Phys. Rev. Lett.* **94** 231601 (2005), [arXiv:hep-ex/0503008].

62. B. Aubert *et al.* [BaBar Collaboration], arXiv:hep-ex/0507038.

63. A. Ali, E. Lunghi, C. Greub and G. Hiller, *Phys. Rev. D* **66** 034002 (2002), [arXiv:hep-ph/0112300].

64. K. Abe *et al.* [Belle Collaboration], arXiv:hep-ex/0408119.

65. B. Aubert *et al.* [BABAR Collaboration], *Phys. Rev. Lett.* **93** 081802 (2004), [arXiv:hep-ex/0404006].

66. G. Hiller and F. Kruger, *Phys. Rev. D* **69** 074020 (2004), [arXiv:hep-ph/0310219].

67. F. Kruger and E. Lunghi, *Phys. Rev. D* **63** 014013 (2001), [arXiv:hep-ph/0008210].

68. F. Kruger, L. M. Sehgal, N. Sinha and R. Sinha, *Phys. Rev. D* **61** 114028 (2000), [Erratum-ibid. D **63** 019901 (2001)], [arXiv:hep-ph/9907386].

69. J. L. Hewett and J. D. Wells, *Phys. Rev. D* **55** (1997) 5549, [arXiv:hep-ph/9610323].

70. B. Aubert *et al.* [BaBar Collaboration], arXiv:hep-ex/0507005.

71. K. Abe *et al.* [Belle Collaboration], arXiv:hep-ex/0410006.

72. K. Abe *et al.* [The Belle Collaboration], arXiv:hep-ex/0508009.

73. M. Zhong, Y. L. Wu and W. Y. Wang, *Int. J. Mod. Phys. A* **18** 1959 (2003), [arXiv:hep-ph/0206013].

74. B. Aubert *et al.* [BABAR Collaboration], arXiv:hep-ex/0506070.

75. B. Aubert *et al.* [BABAR Collaboration], arXiv:hep-ex/0501038.

76. K. Abe *et al.* [Belle Collaboration], arXiv:hep-ex/0507036.

77. A. Ali and E. Lunghi, *Eur. Phys. J. C* **26** 195 (2002), [arXiv:hep-ph/0206242].

78. K. Abe *et al.*, arXiv:hep-ex/0506079.

79. A. Ali, E. Lunghi and A. Y. Parkhomenko, *Phys. Lett. B* **595** 323 (2004), [arXiv:hep-ph/0405075].

NEW K DECAY RESULTS

EDWARD BLUCHER

Department of Physics, University of Chicago. 5640 S. Ellis Ave., Chicago, IL 60637. USA
E-mail: e-blucher@uchicago.edu

NO CONTRIBUTION RECEIVED

CHARM DECAYS WITHIN THE STANDARD MODEL AND BEYOND

MARINA ARTUSO

Department of Physics, Syracuse University, Syracuse NY 13244, USA
E-mail: artuso@phy.syr.edu

The charm quark has unique properties that make it a very important probe of many facets of the Standard Model. New experimental information on charm decays is becoming available from dedicated experiments at charm factories, and through charm physics programs at the b-factories and hadron machines. In parallel, theorists are working on matrix element calculations based on unquenched lattice QCD, that can be validated by experimental measurements and affect our ultimate knowledge of the quark mixing parameters. Recent predictions are compared with corresponding experimental data and good agreement is found. Charm decays can also provide unique new physics signatures; the status of present searches is reviewed. Finally, charm data relevant for improving beauty decay measurements are presented.

1 Introduction

The charm quark has played a unique role in particle physics for more than three decades. Its discovery by itself was an important validation of the Standard Model, as its mass and most of its relevant properties were predicted before any experimental signature for charm was available. Since then, much has been learned about the properties of charmed hadronic systems.

Experiments operating at the $\psi(3770)$ resonance, near threshold for $D\bar{D}$ production, such as MARK III at SPEAR, performed the initial exploration of charm phenomenology.[1] Later, higher energy machines, either fixed target experiments operating at hadron machines or higher energy e^+e^- colliders, entered this arena, with much bigger data samples. In recent years, we have seen a renewed interest in studying open charm in e^+e^- colliders with a center-of-mass energy close to $D\bar{D}$ threshold. The CLEO-c experiment[2] at CESR, has collected a sample of 281 pb^{-1} at the $\psi(3770)$ center-of-mass energy. This experiment is poised to accumulate a total integrated luminosity of the order of 1 fb^{-1} at the $\psi(3770)$ and a similar size sample at an energy optimal to study D_S decays. The BES-II experiment, at BEPC, has published results based on 33 pb^{-1} accumulated around the $\psi(3770)$. It has an ongoing upgrade program both for the detector (BESIII) and the machine (BEPCII), designed as a charm factory with 10^{33}cm^{-2}s^{-1} peak luminosity.[3]

Several features distinguish charm Its mass ($\mathcal{O}(1.5)$ GeV) makes it an ideal laboratory to probe QCD in the non-perturbative domain. In particular, a comparative study of charm and beauty decays may lead to more precise theoretical predictions for key quantities necessary for accurate determination of important Standard Model parameters. On the other hand, once full QCD calculations have demonstrated control over hadronic uncertainties, charm data can be used to probe the Yukawa sector of the Standard model. Finally, charm decays provide a unique window on new physics affecting the u-type quark dynamics. For example, it is the only u-type quark that can have flavor oscillations. Moreover, some specific new physics models predict enhancements on CP violation phases in D decays, beyond the 10^{-3} level generally predicted within the Standard Model.[4]

The charge-changing transitions involving quarks feature a complex pattern, that is summarized by a 3×3 unitary matrix, the Cabibbo-Kobayashi-Maskawa (CKM) matrix:

$$V_{CKM} = \begin{pmatrix} V_{ud} & V_{us} & V_{ub} \\ V_{cd} & V_{cs} & V_{cb} \\ V_{td} & V_{ts} & V_{tb} \end{pmatrix}. \qquad (1)$$

These 9 complex couplings are described by 4 independent parameters. In the Wolfenstein approximation,[5] the CKM matrix is expressed in terms of the four parameters λ, A, ρ, and η, and is expanded in powers of λ:

$$\begin{pmatrix} 1 - \lambda^2/2 & \lambda & A\lambda^3(\rho - i\eta) \\ -\lambda & 1 - \lambda^2/2 & A\lambda^2 \\ A\lambda^3(1 - \rho - i\eta) & -A\lambda^2 & 1 \end{pmatrix}. \qquad (2)$$

The parameters λ, A, ρ and η are fundamental constants of nature, just as basic as G, Newton's constant, or α_{EM}.

B meson semileptonic decays (determining $|V_{ub}|/|V_{cb}|$) and neutral B flavor oscillations provide crucial constraints to determine the CKM parameters ρ and η. In both cases, hadronic matrix elements need to be evaluated to extract these parameters from the experimental data. Due to the relatively small masses of the b and c quarks, strong interactions effects are of a non-perturbative nature. Lattice QCD calculations seem the ideal approach to tackle this problem. However, a realistic simulation of quark vacuum polarization has eluded theorists for several decades, thus limiting lattice QCD results to the so-called "quenched approximation." A new unquenched approach, based on a Symanzik-improved staggered-quark formalism,[6] bears the promise of precise predictions on some key observables.[7] The main ingredients of the new approach are: improved staggered quarks representing sea and valence quarks, chiral perturbation theory for staggered quarks and heavy quark effective theory (HQET) for the heavy quarks.[7] This formalism is expected to deliver predictions soon on some "golden" physical quantities with errors of a few %. They are matrix elements that involve one hadron in the initial state and one or no stable hadrons

in the final state, and they require that the chiral perturbation theory is "well-behaved" for the specific mode under consideration. Several processes relevant for the study of quark mixing fall in this category. Important examples include the leptonic decay constants $f_{B_{(s)}}$ and $f_{D_{(s)}}$ and semileptonic decay form factors. Checks on theory predictions for key "golden quantities" are under way[2,3] and may validate the theory inputs for the corresponding quantities in beauty decays.

2 The decay constant f_{D^+}.

CKM unitarity tests include constraints from $B^0_{(s)}\bar{B}^0_{(s)}$ oscillations. The theoretical inputs, are $\sqrt{\hat{B}_d}f_{B_d}$, $\sqrt{\hat{B}_s}f_{B_s}$, or $\xi \equiv \sqrt{\hat{B}_s}f_{B_s}/\sqrt{\hat{B}_d}f_{B_d}$, where $\hat{B}_i$ represents the relevant "bag parameter", the correction for the vacuum insertion approximation, and f_{B_i} represents the corresponding decay constant. It is thus important to validate the theoretical uncertainties, and a proposed strategy is to use the corresponding observables in D decays for this purpose. The decay $D^+ \rightarrow \ell^+\nu$ proceeds by the c and $\bar{d}$ quarks annihilating into a virtual W^+, with a decay width[8] given by:

$$\Gamma(D^+ \rightarrow \ell^+\nu) = \frac{G_F^2}{8\pi}f_{D^+}^2 m_\ell^2 M_{D^+} \qquad (3)$$

$$\left(1 - \frac{m_\ell^2}{M_{D^+}^2}\right)^2 |V_{cd}|^2 \quad,$$

where M_{D^+} is the D^+ mass, m_ℓ is the mass of the final state lepton, $|V_{cd}|$ is a CKM matrix element that we assume to be equal to $|V_{us}|$, and G_F is the Fermi coupling constant. Due to helicity suppression, the rate goes as m_ℓ^2; consequently the electron mode $D^+ \rightarrow e^+\nu$ has a very small rate in the Standard Model. The relative widths are $2.65 : 1 : 2.3 \times 10^{-5}$ for the $\tau^+\nu$, $\mu^+\nu$ and $e^+\nu$ final states, respectively.

CLEO-c was the first experiment to have a statistically significant $D^+ \rightarrow \mu\nu$ signal,[9] and has now published an improved measurement of f_{D^+}.[10] They use a tagging technique similar to that developed by the MARK III collaboration,[1] where one D meson is reconstructed in a low background hadronic channel and the remaining tracks and showers are used to study a specific decay mode. The relatively high single tag yield makes this technique extremely useful.[a] They reconstruct the D^- meson in one of six different decay modes and search for $D^+ \to \mu\nu$ in the rest of the event. The existence of the neutrino is inferred by requiring the missing mass squared (MM^2) to be consistent with zero. Here:

$$MM^2 = (E_{beam} - E_{\mu^+})^2 - (-\vec{p}_{D^-} - \vec{p}_{\mu^+})^2, \tag{4}$$

where $\vec{p}_{D^-}$ is the three-momentum of the fully reconstructed D^-. Events with additional charged tracks originating from the event vertex or unmatched energy clusters in the calorimeters with energy greater than 0.250 GeV are vetoed. These cuts are very effective in reducing backgrounds. Efficiencies are mostly determined using data, while backgrounds are evaluated either with large Monte Carlo samples or with data. Fig. 1 shows the measured MM^2, with a 50 event peak in the interval [-0.050 GeV^2,+0.050 GeV^2], approximately $\pm 2\sigma$ wide. The background is evaluated as $2.81 \pm 0.30 \pm 0.27$ events. This implies:

$$\mathcal{B}(D^+ \to \mu^+\nu_\mu) = (4.40 \pm 0.66^{+0.09}_{-0.12}) \times 10^{-4}. \tag{5}$$

The decay constant f_{D^+} is derived from Eq. 3 using $\tau_{D^+} = 1.040 \pm 0.007$ ps,[11] and $|V_{cd}| = 0.2238 \pm 0.0029$,[12] yielding:

$$f_{D^+} = (222.6 \pm 16.7^{+2.8}_{-3.4}) \text{ MeV}. \tag{6}$$

The same tag sample is used to search for $D^+ \to e^+\nu_e$. No signal is found, corresponding to a 90% cl upper limit $\mathcal{B}(D^+ \to e^+\nu_e) < 2.4 \times 10^{-5}$. These measurements are much more precise than previous observations or limits.[13] The very small systematic

---

[a]Throughout this paper charge conjugate particles are implied unless specifically noted.

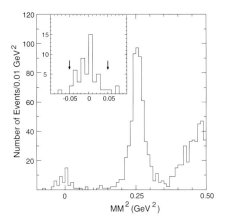

Figure 1. CLEO-c MM^2 using D^- tags and one opposite charged track with no extra energetic clusters.[10] The insert shows the signal region for $D^+ \to \mu\nu_\mu$ enlarged; the defined signal region is shown between the two arrows.

error is achieved through very careful background and efficiency studies, involving large Monte Carlo and data samples.

Fig. 2 summarizes the present experimental data[10,13] and the various theoretical predictions for the decay constant.[14−21] The latest lattice QCD result, performed by the Fermilab lattice, MILC and HPQCD collaborations, working together,[22] is the first to include three quark flavors fully unquenched and was published shortly before the CLEO-c updated result. It is consistent with the CLEO-c result with a 37% confidence level.

3 Semileptonic decays

The study of D meson semileptonic decays is another important area of investigation. In principle, charm meson semileptonic decays provide the simplest way to determine the magnitude of quark mixing parameters: the charm sector allows direct access to $|V_{cs}|$ and $|V_{cd}|$. Semileptonic decay rates are related to $|V_{cx}|^2$ via matrix elements that describe strong interactions effects. Traditionally, these hadronic matrix elements have been described in terms of form factors cast as a function of the Lorentz invariant q^2, the

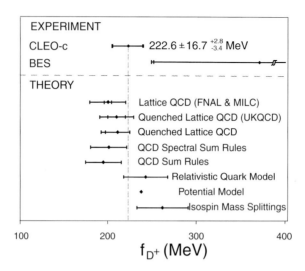

Figure 2. Summary of theoretical predictions and experimental data for f_{D^+}.

invariant mass of the electron-ν pair. Experimental determinations of these form factors are performed through the study of the differential decay width $d\Gamma/dq^2$.

3.1 Goals in semileptonic decays

If we assume that V_{cs} and V_{cd} are known, experiments can determine the form factor shape as well as their normalization. Form factors have been evaluated at specific q^2 points in a variety of phenomenological models,[23] where the shape is typically assumed. More recently, lattice QCD calculations[24] have predicted both the normalization and shape of the form factors in $D \rightarrow K\ell\nu$ and $D \rightarrow \pi\ell\nu$. Note, that we can form ratios between leptonic and exclusive semileptonic branching fractions that can provide direct theory checks without any CKM input.

On the other hand, if we use validated theoretical results as inputs, we can derive direct measurements for V_{cs} and V_{cd}; the most accurate determinations of these parameters presently require some additional input information, such as unitarity. Thus we could extend the unitarity checks of the CKM matrix beyond the first row.

The study of charm semileptonic decays may contribute to a precise determination of the CKM parameter $|V_{ub}|$. A variety of theoretical approaches have been proposed to use constraints provided by charm decays to reduce the model dependence in the extraction of $|V_{ub}|$ from exclusive charmless B semileptonic decays. In particular, if HQET is applicable both to the c and b quarks, there is a SU(2) flavor symmetry that relates the form factors in D and B semileptonic decays.[25] For example, a flavor symmetry relates the form factors in $D \rightarrow \pi\ell\nu$ are related to the ones in $B \rightarrow \pi\ell\bar{\nu}$, at the same $E \equiv \mathbf{v} \cdot \mathbf{p}_\pi$, where $\mathbf{v}$ is the heavy meson 4-velocity and $\mathbf{p}_\pi$ is the π 4-momentum. The original method has been further refined;[26] the large statistics needed to implement these methods may be available in the near future.

3.2 Semileptonic branching fractions: the data

BES-II[27] and CLEO-c[28] have recently presented data on exclusive semileptonic branching fractions. BES-II results are based on 33 pb^{-1}; CLEO-c's results are based on the first 57 pb^{-1} data set. Both experiments use tagged samples and select a specific final state

Table 1. Summary of recent absolute branching fraction measurements of exclusive D^+ and D^0 semileptonic decays.

Decay mode	B(%) [CLEO-c][28]	B(%) [BES][27]	B(%) [average][ℵ]
$D^0 \to K^- e^+ \nu_e$	$3.44 \pm 0.10 \pm 0.10$	$3.82 \pm 0.40 \pm 0.27$	3.54 ± 0.11
$D^0 \to \pi^- e^+ \nu_e$	$0.262 \pm 0.025 \pm 0.008$	$0.33 \pm 0.13 \pm 0.03$	0.285 ± 0.018
$D^0 \to K^{\star -} e^+ \nu_e$	$2.16 \pm 0.15 \pm 0.08$		2.14 ± 0.16
$D^0 \to \rho^- e^+ \nu_e$	$0.194 \pm 0.039 \pm 0.013$		$0.194 \pm 0.039 \pm 0.013$
$D^+ \to \bar{K}^0 e^+ \nu_e$	$8.71 \pm 0.38 \pm 0.37$		8.31 ± 0.44
$D^+ \to \pi^0 e^+ \nu_e$	$0.44 \pm 0.06 \pm 0.0.03$		0.43 ± 0.06
$D^+ \to \bar{K}^{\star 0} e^+ \nu_e$	$5.56 \pm 0.27 \pm 0.23$		5.61 ± 0.32
$D^+ \to \rho^0 e^+ \nu_e$	$0.21 \pm 0.04 \pm 0.01$		0.22 ± 0.04
$D^+ \to \omega e^+ \nu_e$	$0.16^{+0.07}_{-0.06} \pm 0.01$		$0.16^{+0.07}_{-0.06}$

[ℵ]The averages reported here include all the branching fractions reported in the PDG 2004 for $D \to X e^+ \nu_e$ and the CLEO-c and BES-II data. Indirect measurements are normalized with respect to the hadronic[58] and average semileptonic branching ratios included in this report.

through the kinematic variable:

$$U \equiv E_{miss} - |c\vec{p}_{miss}|, \qquad (7)$$

where E_{miss} represents the missing energy and $\vec{p}$ represents the missing momentum of the D meson decaying semileptonically. For signal events, U is expected to be 0, while other semileptonic decays peak in different regions. Fig. 3 shows the U distribution for 5 exclusive D^+ decay modes reported by CLEO-c, which demonstrate that U resolution is excellent, thus allowing a full separation between Cabibbo suppressed and Cabibbo favored modes. Table 1 summarizes the recent measurements from CLEO-c and BES-II, as well world averages obtained from the results presented in this paper and the previous measurements of $\mathcal{B}(D \to X_i e^+ \nu_e)$ reported in the PDG 2004.[11]

CLEO-c uses the two tagging modes with lowest background ($\bar{D}^0 \to K^+ \pi^-$ and $D^- \to K^+ \pi^- \pi^-$) to measure the inclusive D^0 and D^+ semileptonic branching fractions.[29] Table 2 summarizes the measured semileptonic branching fractions, and it also includes the sum of the branching fractions for D decay into all the known exclusive modes. The CLEO-c data have been used in this compari-

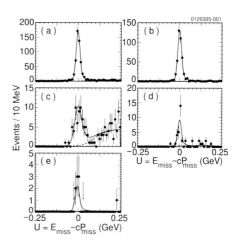

Figure 3. Fits (solid lines) to the U distributions in CLEO-c[28] data (dots with error bars) for the five D^+ semileptonic modes: (a) $D^+ \to \bar{K}^0 e^+ \nu_e$, (b)$D^+ \to \bar{K}^{\star 0} e^+ \nu_e$, (c) $D^+ \to \pi^0 e^+ \nu_e$, (d)$D^+ \to \rho^0 e^+ \nu_e$, (e)$D^+ \to \omega e^+ \nu_e$. The arrows in (e) show the signal region. The background (in dashed lines) is visible only in (c) and (d).

son, as they dominate the present world average: the exclusive modes are consistent with saturating the inclusive semileptonic branching fraction at a 41% confidence level in the case of the D^+ and 18% confidence level in the case of the D^0.

The preliminary inclusive branching fractions can be translated into inclusive semileptonic widths $\Gamma^{sl}_{D^+}$ and $\Gamma^{sl}_{D^0}$, using the known D lifetimes,[11]. These widths are expected to be equal, modulo isospin violations, and indeed the measured ratio $\Gamma^{sl}_{D^+}/\Gamma^{sl}_{D^0} = 1.01 \pm 0.03 \pm 0.03$: thus isospin violations are limited to be below $\sim 4\%$.

Table 2. Comparison between exclusive[28] and preliminary inclusive[29] results from CLEO-c.

Mode	$\mathcal{B}$ (%)
$(D^0 \to X\ell\nu_e)$	$6.45 \pm 0.17 \pm 0.15$
$\Sigma_i \mathcal{B}((D^0 \to X_i\ell\nu_e)$	$6.1 \pm 0.2 \pm 0.2$
$(D^+ \to X\ell\nu_e)$	$16.19 \pm 0.20 \pm 0.36$
$\Sigma_i \mathcal{B}((D^+ \to X_i\ell\nu_e)$	$15.1 \pm 0.50 \pm 0.50$

3.3 Form factors for $D \to K(\pi)\ell\nu$

Recently, non-quenched lattice QCD calculations for $D \to K\ell\bar{\nu}$ and $D \to \pi\ell\nu$ have been reported.[24] The chiral extrapolation is performed at fixed $E = \vec{v} \cdot \vec{p}_P$, where E is the energy of the light meson in the center-of-mass D frame, $\vec{v}$ is the unit 4-velocity of the D meson, and $\vec{p}_P$ is the 4-momentum of the light hadron P (K or π). The results are presented in terms of a parametrization originally proposed by Becirevic and Kaidalov (BK):[30]

$$f_+(q^2) = \frac{F}{(1 - \tilde{q}^2)(1 - \alpha\tilde{q}^2)}, \qquad (8)$$

$$f_0(q^2) = \frac{F}{1 - \tilde{q}^2/\beta},$$

where q^2 is the 4-momentum of the electron-ν pair, $\tilde{q}^2 = q^2/m^2_{D^*_x}$, and $F = f_+(0)$, α and β are fit parameters. This formalism models the effects of higher mass resonances other than the dominant spectroscopic pole ($D^{\star+}_S$ for the $K\ell\nu$ final state and $D^{\star+}$ for $\pi\ell\nu$).[31]

The form factors $f_+(q^2)$ govern the corresponding semileptonic decays. The lattice QCD calculation obtains the parameters shown in Table 3.

The FOCUS experiment[32] performed a non-parametric measurement of the shape of

Table 3. Fit parameters in Eq. (8), decay rates and CKM matrix elements. The first errors are statistical; the second systematic.[24]

P	F	α	β
π	0.64(3)(6)	0.44(4)(7)	1.41(6) (13)
K	0.73(3)(7)	0.50(4)(7)	1.31(7)(13)

the form factor in $D \to K\mu\nu_\mu$. Fig. 4 shows the lattice QCD predictions for $D \to K\ell\nu$ with the FOCUS data points superimposed. In addition, they studied the shape of the form factors $f_+(q^2)$ for $D \to K\mu\nu_\mu$ and $D \to \pi\mu\nu_\mu$ with two different fitting functions: the single pole, traditionally used because of the conventional ansatz of several quark models,[23] and the BK parametrization discussed before. Table 4 shows the fit results obtained from FOCUS and CLEO III,[33] compared to the lattice QCD predictions. Both experiments obtain very good fits also with simple pole form factors, however the simple pole fit does not yield the expected spectroscopic mass. For example, FOCUS obtains $m_{pole}(D^0 \to K\mu\nu_\mu) = (1.93 \pm 0.05 \pm 0.03)$ GeV/c^2 and $m_{pole}(D^0 \to \pi\mu\nu_\mu) = (1.91^{+0.30}_{-0.15} \pm 0.07)$ GeV/c^2, while the spectroscopic poles are, respectively, 2.1121 ± 0.0007 GeV/c^2 and 2.010 ± 0.0005. This may hint that other higher order resonances are contributing to the form factors.[31] It has been argued,[35] that even the BK parametrization is too simple and that a three parameter form factor is more appropriate. However, this issue can be resolved only by much larger data samples, with better sensitivity to the curvature of the form factor near the high recoil region.

By combining the information of the measured leptonic and semileptonic width, a ratio independent of $|V_{cd}|$ can be evaluated: this is a pure check of the theory. We evaluate the ratio $R \equiv \sqrt{\Gamma(D^+ \to \mu\nu_\mu)/\Gamma(D \to \pi e^+\nu_e)}$. We assume isospin symmetry, and thus $\Gamma(D \to \pi e^+\nu_e) = \Gamma(D^0 \to \pi^- e^+\nu_e) = 2\Gamma(D^+ \to \pi^0 e^+\nu_e)$. For

Table 4. Measured shape parameter α compared to lattice QCD predictions.

$\alpha(D^0 \to K\ell\nu)$	
lattice QCD[24]	$0.5 \pm 0.04 \pm 0.07$
FOCUS[32]	$0.28 \pm 0.08 \pm 0.07$
CLEOIII[33]	$0.36 \pm 0.10^{+0.03}_{-0.07}$
Belle[34]	$0.40 \pm 0.12 \pm 0.09$
$\alpha(D^0 \to \pi\ell\nu)$	
lattice QCD[24]	$0.44 \pm 0.04 \pm 0.07$
CLEOIII[33]	$0.37^{+0.20}_{-0.31} \pm 0.15$
Belle[34]	$0.03 \pm 0.27 \pm 0.13$

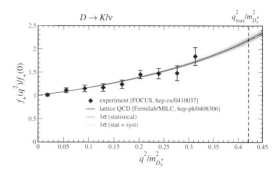

Figure 4. Shape of the form factor for $D \to K\ell\nu$:[7] MILC-Fermilab calculation compared with the non parametric data from FOCUS.

the theoretical inputs, we use the recent unquenched lattice QCD calculations in three flavors,[22,24] as they reflect the state of the art of the theory and have been evaluated in a consistent manner. The result is:

$$R_{sl}^{th} = \sqrt{\frac{\Gamma^{th}(D^+ \to \mu\nu_\mu)}{\Gamma^{th}(D \to \pi e \nu_e)}} = 0.212 \pm 0.028, \quad (9)$$

The quoted error is evaluated through a careful study of the theory statistical and systematic uncertainties, assuming Gaussian errors. The corresponding experimental quantity is calculated using the CLEO-c f_D and isospin averaged $\Gamma(D \to \pi e^+ \nu_e)$; we obtain:

$$R_{sl}^{exp} = \sqrt{\frac{\Gamma^{exp}(D^+ \to \mu\nu)}{\Gamma^{exp}(D \to \pi e \nu_e)}} = 0.249 \pm 0.022. \quad (10)$$

The theory and data are consistent at 28% confidence level, that represents a good agreement.

4 The CKM Matrix

An important goal of the next generation of precision experiments is to perform direct measurements of each individual parameter. This will enable us to perform additional unitarity checks with precision similar to the one achieved now with the first row.[12] In particular, V_{cd} and V_{cs} are now determined with high precision, but using unitarity constraints.[11]

The most recent results from LEP II, using the $W \to \ell\nu$ branching fraction, and additional inputs from other CKM parameter measurement is $V_{cs} = 0.976 \pm 0.014$.[36] The unitarity constraint implies $V_{cd} \sim V_{us} = 0.2227 \pm 0.0017$.[12]

If we use the theoretical form factors as inputs, we can extract $|V_{cs}|$ and $|V_{cd}|$ from the branching fractions reported in this paper. The results, obtained using the form factors from the unquenched lattice QCD calculation[24] and the isospin averaged semileptonic widths from CLEO-c[28] are:

$$|V_{cs}| = 0.957 \pm 0.017(exp) \pm 0.093(th) \quad (11)$$

$$|V_{cd}| = 0.213 \pm 0.008(exp) \pm 0.021(th) \quad (12)$$

A unitarity check derived uniquely from these measurements yields:

$$1 - |V_{cs}|^2 + |V_{cd}|^2 + |V_{cb}|^2 = 0.037 \pm 0.181(tot). \quad (13)$$

The mean V_{ci} and their errors have been derived from careful application of the theoretical quantities and their stated statistical and systematic errors.

These determinations are not yet competitive, but it will be interesting to see the results of future estimates, when the accuracy is comparable to the one achieved in the first row.

5 Charm as a probe for New Physics

The study of charm decays provides a unique opportunity for indirect searches for physics beyond the Standard Model. In several dynamical models, the effects of new particles observed in c, s and b transitions are correlated.[4,37] Possible new physics manifestations involve three different facets: $D^0\bar{D}^0$ oscillations, CP violation and rare decays.

5.1 $D^0\bar{D}^0$ oscillations

Two main processes contribute to $D^0\bar{D}^0$ oscillations. The short distance physics effects are depicted by higher order Feynman diagrams, such box or loop diagrams that influence the mass difference ΔM. These diagrams are sensitive to new physics, through the interference with contributions with similar topology including exotic particles in place of the d, s, b quarks present in the Standard Model loop. In addition, there is a coupling between D^0 and $\bar{D}^0$ induced by common final states such as $K\bar{K}$, $\pi\pi$ and $K\pi$. As the intermediate states are real, one conjectures that only the difference in lifetime $\Delta\Gamma$ is affected by this coupling. Thus, $\Delta\Gamma$ is expected to be dominated by Standard Model processes.

$D^0\bar{D}^0$ mixing haw been studied with a variety of different experimental methods, several of which suffer from a variety of additional complications.

The first approach, which has been pursued by a variety of experiments,[38–41] is the study of the "wrong-sign" hadronic decays such as $D^0 \to K^+\pi^-$. These decays occur via two paths: oscillation of D^0 into $\bar{D}^0$, followed by the Cabibbo favored $\bar{D}^0 \to K^+\pi^-$, or doubly Cabibbo suppressed decays $D^0 \to K^+\pi^-$. The two channels interfere and thus there is an additional parameter that affects the wrong-sign rate: the strong phase δ between $D^0 \to K^+\pi^-$ and $K^-\pi^+$ decays. Moreover it has been argued[42] that CP violation may

Table 5. $D^0 \to K^+\pi^-$ analysis. Only results of the fits allowing for CP violation are included.

Experiment	Fit Result ($\times 10^3$)
CLEO[38]	$0 < x'^2 < 0.82$
CLEO[38]	$-58 < y' < 10$
FOCUS[39]	$0 < x'^2 < 8$
FOCUS[39]	$-112 < y' < 67$ -
Belle[40]	$0 < x'^2 < 0.89$
Belle[40]	$-30 < y' < 27$
BaBar[41]	$0 < x' < 2.2$
BaBar[41]	$-56 < y' < 39$

Table 6. Summary of y_{CP} results.

Experiment	$y_{CP}(\%)$
FOCUS[44]	$3.4 \pm 1.4 \pm 0.7$
CLEO[43]	$-1.2 \pm 2.5 \pm 1.4$
Belle, untagged[45]	$-0.5 \pm 1.0 \pm 0.8$
Belle, tagged[46]	$1.2 \pm 0.7 \pm 0.4$
BaBar[47]	$0.8 \pm 0.4^{+0.5}_{-0.4}$

have non negligible effects too. Thus experiments typically perform a variety of fits for the modified variables $x' \equiv x\cos\delta + y\sin\delta$ and $y' \equiv -x\sin\delta + y\cos\delta$, under different CP violation assumptions. Table 5 summarizes the results of the most generic fit, allowing for a CP violating term.

A second class of measurements involves the study of y_{CP}: namely the normalized lifetime difference of $D^0\bar{D}^0$ CP eigenstates. In presence of CP violation, y_{CP} is a linear combination of x and y involving the CP violation phase ϕ. Table 6 summarizes experimental data on y_{CP}. The average is positive, although still consistent with 0.

The study of semileptonic D decays allows the determination of another combination of mixing parameters. Experiments study the ratio r_M defined as:

$$r_M = \frac{\int_0^\infty \mathcal{P}(D^0 \to \bar{D}^0 \to X^+\ell\bar{\nu})}{\int_0^\infty \mathcal{P}(D^0 \to X^+\ell\bar{\nu})} \approx \frac{x^2 + y^2}{2}. \tag{14}$$

Table 7 summarizes the sensitivity achieved

Table 7. Summary of mixing limits (95 % cl) from D^0 semileptonic decay studies.

Experiment	R_M	$\sqrt{x^2 + y^2}$
CLEO[48]	0.0091	0.135
BaBar[49]	0.0046	0.1
Belle[50]	0.0016	0.056

by present experiments to r_M.

Finally, a very interesting analysis method has been implemented by the CLEO experiment: they have studied the channel $D^0 \to K_S^0 \pi^+ \pi^-$. Cabibbo favored final states, such as $K^{\star-}\pi^+$, and doubly-Cabibbo suppressed channels, such as $K^{\star+}\pi^-$ interfere. They generalize the methodology that they used to identify the resonance substructure of this decay[51] to the case where the time-dependent state is a mixture of D^0 and $\bar{D}^0$.[52] In this case, the parameters x and y affect the time-dependent evolution of this system. This time-dependent Dalitz plot analysis can be used to extract the mixing and CP violation parameters. They obtain $(-4.5 < x < 9.3)\%$ and $(-6.4 < y < 3.6)\%$, It is interesting to note that this constraint has sensitivity comparable to other limits obtained from a much larger data sample.

5.2 CP violation

Within the Standard Model, CP violation effects in D decays are expected to be negligible small, as they are introduced by box diagrams or penguin diagrams containing a virtual b quark: thus they involve a strong CKM suppression ($V_{cb}V_{ub}^\star$). In contrast with the $D^0\bar{D}^0$ mixing case, where the vast theoretical effort devoted to pin down the Standard Model predictions did not yield a clear-cut result, there is a wide consensus that observing CP violation in D decays at a level much higher than $\mathcal{O}(10^{-3})$ will constitute an unambiguous signal of new physics. There is a vast array of studies that can be undertaken:[4] exploring CP violation effects

on mixing observables, searching for direct CP violation effects in D^0, D^+ and D_S^+ decays and, finally, studies of $D\bar{D}$ pairs near threshold, that exploit the quantum coherence of these states.

In general, experimental sensitivity is $\mathcal{O}(1)\%$.[4] Recent results from BaBar,[53] Belle[54], and CLEO[55] have explored CP violation in 3-body D decays. Babar obtains $\mathcal{A}(D^+ \to K^- K^+ \pi^+) = (1.4 \pm 1.0 \pm 0.8)\%$. CLEO obtains $\mathcal{A}(D^0 \to \pi^+\pi^-\pi^0) = (1 \pm 8^{+9}_{-7})\%$. Belle obtains $\mathcal{A}(D^0 \to K^+\pi^-\pi^0) = (-0.6 \pm 5.3\%$ and $\mathcal{A}(D^0 \to K^+\pi^-\pi^+\pi^-) = (-1.8 \pm 4.4\%$.

A complementary approach involves the study of observables that are sensitive to T violation,[56] such as triple product correlations in 4-body decays of D^0 and D^+. This technique has been pioneered by FOCUS,[57] through the study of triple product correlations in $D^0 \to K^+K^-\pi^+\pi^-$, $D^+ \to K_s^0 K^-\pi^+\pi^-$, $D_S^0 \to K_S^0 K^-\pi^+\pi^-$. Their present sensitivity is at the level of several percent, dominated by the statistical error. A significant improvement in the sensitivity of this technique is expected in future measurements.

6 Charm as a facet of beauty

The study of b decays has been one of our richest sources of information about the Standard Model, as well as a very powerful constraint on new physics.

As the dominant tree level diagram includes the $b \to c$ transition, the precision of our knowledge of the D decay phenomenology affects quantities associated with B decays in a variety of ways. For example, the accuracy of the determination of D hadronic branching fractions has an obvious impact on the absolute determination of B hadronic branching fractions. Moreover, the study of specific CP violation observables can be made more precise through ancillary information coming from D decays. Finally, a precise knowledge of the particle yields in D decays, allow a more precise modelling of inclusive B decays.

6.1 D absolute branching fractions

Absolute measurements of D meson branching fractions affect our knowledge of several many D and B meson decays, from which CKM parameters are extracted.

CLEO-c has employed tagged samples to obtain new values for the branching fractions $D^0 \to K^-\pi^+$, $D^+ \to K^-\pi^+\pi^+$, and other modes.[58] This powerful technique, combined with careful efficiency studies based on data, resulted in an accuracy comparable to the one of present world averages. They obtain:

$$\mathcal{B}(D^0 \to K^-\pi^+) = (3.91 \pm 0.08 \pm 0.09)\%,$$

and

$$\mathcal{B}(D^+ \to K^-\pi^+\pi^+) = (9.5 \pm 0.2 \pm 0.3)\%$$

Corrections for final state radiation are included in these branching fractions.

6.2 $D \to K_S\pi^+\pi^-$ Dalitz plot analysis and the determination of the CKM phase γ.

The decay $B^\pm \to DK^\pm$ has been the subject of intense theoretical effort to devise optimal strategies to measure the CKM angle γ. The original proposal by Gronau, London and Wyler[59] uses D decays to CP eigenstates. Subsequently Atwood, Dunietz and Soni[60] critiqued this approach and proposed a method based on D decays to flavor eigenstates. Finally, there is one method that has received a lot of attention recently,[61] the extraction of γ from a Dalitz plot analysis of $B^\pm \to D^{(\star)}K^\pm \to K^\pm K_S\pi^+\pi^-$. Charm factories can help this measurement in a variety of manners: they can provide information on $D^0\bar{D}^0$ mixing, and measure the strong phase δ between the Cabibbo favored and doubly-Cabibbo suppressed $D^0 \to K^-\pi^+$

and $\bar{D}^0 \to K^-\pi^+$, and perform unique D Dalitz plot studies.

The Dalitz plot technique illustrates the contributions that CLEO-c and, later, BE-SIII can provide to reduce the uncertainty in this determination of the angle γ. This method is attractive because it involves a D decay with a relatively large branching fraction. Moreover this three body final state comprises a very rich resonance substructure, that leads to the expectation of large strong phases. Recently both BaBar[62] and Belle[63] reported measurements on γ (BaBar) $- \phi_3$ (Belle) with this method. They obtain:

$$\phi_3 = 77^{\circ+17}_{-19}(\text{stat}) \pm 13^\circ(\text{sys}) \pm 11^\circ(\text{mod}),$$
$$\gamma = 70^\circ \pm 26^\circ(\text{stat}) \pm 10^\circ(\text{sys}) \pm 10^\circ(\text{mod}).$$

In both cases, the error labeled "mod," refers to uncertainties on the resonance substructure of the $K_S\pi^+\pi^-$ Dalitz plot. Both collaborations find that to achieve a good fit they need to include two ad-hoc $\pi\pi$ s-wave resonances that describe about 10% of the data. The study of CP tagged Dalitz plots[64] allows a model dependent determination of the D^0 and $\bar{D}^0$ phase across the Dalitz plot. Using data samples where the CP eigenstate ($\mathcal{S}_\pm$) of the D can be tagged, CLEO-c is studying the Dalitz plots $\mathcal{S}_-K_S\pi^+\pi^-$, $\mathcal{S}_-K_S\pi^+\pi^-$, as well as flavor tagged $K_S\pi^+\pi^-$ Dalitz plots. A simultaneous fit to these three Dalitz plots can validate Dalitz plot models and reduce the model dependence of these results significantly. Alternatively, a model independent result can be obtained from a binned analysis of the three CP or flavor tagged Dalitz plot. This work is under way[65] and should eventually reduce the model dependence to a couple of degrees.

7 Conclusions

Charm decays provide a rich phenomenology for a variety of important studies that improve our knowledge of several facets of the

Standard Model, and probe for signatures of new physics.

The experimental study of beauty and charm decays is prospering through vibrant experimental activity taking place in several ongoing experiments. The next few years will see an opening up of our vistas on these decays with the upcoming turn on of LHC and of a dedicated charm and beauty experiment at a hadron collider, LHCb. This experiment bears the promise of precision studies that are poised to explore thoroughly all the possible new physics manifestations alluded to in this paper.

Acknowledgments

I would like to thank the organizers for their tremendous effort that lead to a very enjoyable and productive conference. I would also like to acknowledge interesting discussions and scientific input from D. Asner, A. Kronfeld, M. Okamoto, and S. Stone. This work was supported by the United States National Science Foundation.

References

1. J. Adler *et al.* [Mark III] *Phys. Rev. Lett.* **60**, 89 (1988).
2. R. A. Briere *et al.*, CLNS-01-1742
3. W. G. Li [BES], *Int. J. Mod. Phys. A* **20**, 1560 (2005).
4. A. A. Petrov, *Nucl. Phys. Proc. Suppl.* **142**, 333 (2005) [arXiv:hep-ph/0409130].
5. L. Wolfenstein, *Phys. Rev. Lett.* **51**, 1945 (1983).
6. P. Lepage and C. Davies, *Int. J. Mod. Phys. A* **19**, 877 (2004) .
7. A. S. Kronfeld *et al.*, arXiv:hep-lat/0509169.
8. J.L. Rosner, in *Proceedings of the 1988 Banff Summer Institute*, edited by A.N. Kamal and F.C. Khanna (World Scientific, Singapore, 1989) p. 395.
9. G. Bonvicini *et al.* [CLEO], *Phys. Rev. D* **70**, 112004 (2004); [arXiv:hep-ex/0411050].
10. M. Artuso *et al.* [CLEO], arXiv:hep-ex/0508057.
11. S. Edelman *et al.*, *Phys. Lett. B* **592** 1 (2004).
12. U. Nierste, *these proceedings.*
13. J. Adler *et al.* [Mark III], *Phys. Rev. Lett.* **60**, 1375 (1998); erratum-ibid, **63**,1658 (1989); J.Z. Bai *et al.* [BES], *Phys. Lett. B* **429**, 188 (1998); M. Ablikim *et al.* [BES], *Phys. Lett. B* **610**, 183 (2005).
14. T. W. Chiu *et al.*, [hep-ph/0506266] (2005).
15. L. Lellouch and C.-J. Lin (UKQCD), *Phys. Rev. D* **64**, 094501 (2001).
16. D. Becirevic *et al.*, *Phys. Rev. D* **60**, 074501(1999).
17. S. Narison, in *QCD as a Theory of Hadrons: From Partons to Confinement, Monograph series in Physics*, S. Narison, T. Ericson editor; Cambridge Univ. Press (2003) [hep-ph/0202200].
18. A. Penin and M. Steinhauser, *Phys. Rev. D* **65**, 054006 (2002).
19. D. Ebert *et al.*, *Mod. Phys. Lett. A* **17**, 803 (2002).
20. Z. G. Wang *et al.*, *Nucl. Phys. A* **744**, 156 (2004); L. Salcedo *et al.*, *Braz. J. Phys.* **34**, 297 (2004).
21. J. Amundson *et al.*, *Phys. Rev. D* **47** 3059 (1993).
22. C. Aubin *et al.*, arXiv:hep-lat/0506030.
23. S. Stone in *Heavy Flavours*, ed. A.J. Buras and M. Lindner (World Scientific, Singapore, 1992).
24. C. Aubin *et al.*, *Phys. Rev. Lett.* **94**, 011601 (2005); [arXiv:hep-ph/0408306].
25. N. Isgur and M.B. Wise, *Phys. Rev. D* **42**, 2388 (1990).
26. B. Grinstein and D. Pirjol, *Phys. Rev. D* **70**, 114005 (2004) [arXiv:hep-ph/0404250].
27. M. Ablikim *et al.* [BES], *Phys. Lett. B*

608, 24 (2005) [arXiv:hep-ex/0410030]; M. Ablikim *et al.* [BES], *Phys. Lett. B* **597**, 39 (2004) [arXiv:hep-ex/0406028].

28. T. E. Coan *et al.* [CLEO], arXiv:hep-ex/0506052; G. S. Huang *et al.* [CLEO], arXiv:hep-ex/0506053.

29. Q. He *et al.*, *CLEO-CONF 05-3*; LP2005-429 (2005).

30. D. Becirevic and A. B. Kaidalov, *Phys. Lett. B* **478**,417 (2000); [arXiv:hep-ph/9904490].

31. S. Fajfer and J. Kamenik, arXiv:hep-ph/0509166; and references therein.

32. J.M. Link *et al.* [FOCUS], *Phys. Lett. B* **607**, 233 (2005) [arXiv:hep-ex/0410037].

33. G. S. Huang *et al.* [CLEO], *Phys. Rev. Lett.* **94**, 011802 (2005) [arXiv:hep-ex/0407035].

34. K. Abe *et al.* [Belle], arXiv:hep-ex/0510003 (2005).

35. R. J. Hill, arXiv:hep-ph/0505129.

36. http://lepewwg.web.cern.ch/ LEPEWWG/lepww/4f/Winter05/

37. Y. Nir, arXiv:hep-ph/9911321.

38. R. Godang *et al.* [CLEO], *Phys. Rev. Lett.* **84**, 5038 (2000) [arXiv:hep-ex/0001060].

39. J. M. Link *et al.* [FOCUS], *Phys. Lett. B* **618**, 23 (2005) [arXiv:hep-ex/0412034].

40. K. Abe *et al.* [BELLE], *Phys. Rev. Lett.* **94**, 071801 (2005) [arXiv:hep-ex/0408125].

41. B. Aubert *et al.* [BABAR], *Phys. Rev. Lett.* **91**, 171801 (2003) [arXiv:hep-ex/0304007].

42. L. Wolfenstein, *Phys. Lett. B* **144** 425 (1984).

43. S.E. Csorna *et al.* [CLEO], *Phys. Rev. D* **65**, 092001 (2002).

44. J. Link *et al.* [FOCUS], *Phys. Lett. B* **485**, 62 (2000).

45. K. Abe *et al.* [Belle], *Phys. Rev. Lett.* **88**, 162001 (2002).

46. K. Abe *et al.*

47. B. Aubert *et al.* [BaBar], *Phys. Rev. Lett.* **91**, 171801 (2003).

48. C. Cawlfield *et al.* [CLEO], *Phys. Rev. D* **71**, 077101 (2005) [arXiv:hep-ex/0502012].

49. B. Aubert *et al.* [BABAR], *Phys. Rev. D* **70**, 091102 (2004) [arXiv:hep-ex/0408066].

50. K. Abe *et al.* [Belle], arXiv:hep-ex/0507020.

51. H. Muramatsu *et al.* [CLEO], *Phys. Rev. Lett.* **89**, 251802 (2002) [Erratum-ibid. **90**, 059901 (2003)] [arXiv:hep-ex/0207067].

52. D. M. Asner [CLEO], *Phys. Rev. D* **72**, 012001 (2005) [arXiv:hep-ex/0503045].

53. B. Aubert *et al.* [BABAR], *Phys. Rev. D* **71**, 091101 (2005) [arXiv:hep-ex/0501075].

54. X. C. Tian *et al.* [Belle], arXiv:hep-ex/0507071 (2005).

55. D. Cronin-Hennessy *et al.* [CLEO], *Phys. Rev. D* **72**, 031102 (2005) [arXiv:hep-ex/0503052].

56. I.I. Bigi, in *KAON2001: International Conference on CP Violation*, 2001; [arXiv:hep-ph/0107102].

57. J. M. Link *et al.* [FOCUS], *Phys. Lett. B* **622**, 239 (2005) [arXiv:hep-ex/0506012].

58. Q. He *et al.* [CLEO], *Phys. Rev. Lett.* **95**, 121801 (2005) [arXiv:hep-ex/0504003].

59. M. Gronau and D. Wyler, *Phys. Lett. B* **265**, 172 (1991); M. Gronau and D. London, *Phys. Lett. B* **253**, 483 (1991).

60. D. Atwood, I. Dunietz and A. Soni, *Phys. Rev. Lett.* **78**. 3257 (1997); D. Atwood, I. Dunietz and A. Soni, *Phys. Rev. D* **63**, 036005 (2001).

61. A. Giri, Y. Grossman, A. Soffer and J. Zupan, *Phys. Rev. D* **68**, 054018 (2003).

62. B. Aubert *et al.* [BABAR], arXiv:hep-ex/0507101.

63. K. Abe *et al.* [Belle], arXiv:hep-ex/0504013 (2005).

64. D. Asner, CLEO CONF 05-12 (2005).

65. J. Rosner [CLEO], arXiv:hep-ex/ 0508024 (2005).

DISCUSSION

Vera Luth (SLAC):

What are the plans for CLEO-c to study D_s mesons?

Marina Artuso: The study of D_s mesons is one of the key components of the CLEO-c program and will start soon.

Nikolai Uraltsev (INFN Milano):

Have you tried, or is it in your plan, to measure the difference in the inclusive lepton spectra for charge and neutral D? It can come only from the Cabibbo-suppressed decays, therefore, alternatively, can you measure the similar inclusive spectra from only Cabibbo-suppressed decays?

Marina Artuso: We have not yet pursued this measurement, but it is within our capabilities and it is in our plans.

HEAVY FLAVOR OSCILLATIONS AND LIFE TIMES

RICK JESIK

Physics Department, Imperial College, Prince Consort Road, London SW7 2AZ, UK
E-mail: r.jesik@imperial.ac.uk

NO CONTRIBUTION RECEIVED

HEAVY FLAVOR, QUARKONIUM PRODUCTION AND DECAY

XIAOYAN SHEN

Institute of High Energy Physics, Chinese Academy of Science
Beijing 100049, P. R. China
E-mail: shenxy@ihep.ac.cn

Recent experimental results on quarkonium physics are reviewed. In particular, the new observed particles since last one or two years, such as $X(1835)$, $X(3872)$, $X(3940)$, $Y(3940)$ and $Y(4260)$ are discussed, the latest data on double charmonium production, heavy hadron spectroscopy and quarkonia decays are presented.

1 Introduction

The simplest QCD potential, the so called "Cornell" potential, can be written as:

$$V(r) = -\frac{4}{3}\frac{\dot{\alpha_s}}{r} + k\dot{r}.$$

The first term describes the one-gluon exchange, dominating at short distances (< 0.1 fm) with large momentum transfers. This is the asymptotically "free" regime where the α_s is small and the perturbative calculations can be performed. The second term, important at large distances (> 1 fm) with low momentum transfers, leads to the "confinement" and is in a regime where the α_s is large, making the calculation non-perturbative.

QCD has been tested extensively at the high momentum transfer by lots of high precision experiments. At low energy, it is difficult to be tested due to the non-perturbative nature. Quarkonia, the bound states of quark and its antiquark, are the QCD equivalents of positronium (e^+e^-) in QED. Quarkonia form the simplest strongly interacting systems with only two constituents (unlike baryons) and identical flavor (unlike mesons with "open" flavor). Light quarkonia are highly relativistic. They also contain mixtures of quarks of different flavors and so can be easily fall apart into other mesons. Charmonium ($c\bar{c}$) is the first heavy quarkonium discovered and is less relativistic. Bottomonium ($b\bar{b}$) is heavier, therefore is less non-relativistic and has a large number of long-lived states. The toponium system would have been completely non-relativistic. However, the weak decays will be dominant over the strong decays in such systems. Therefore, charmonium and bottomonium play a special role in probing the strong interactions. The properties of charmonia and bottomonia, and their productions and decays are good labs. for QCD in both perturbative and non-perturbative regimes.

Firstly, the new observations of $X(1835)$, $X(3940)$, $Y(3940)$ and $Y(4260)$ are reported. Then, the latest data on heavy quarkonium production, in particular the big discrepancy on double charmonium production between data and theory are presented. Recent experimental heavy hadron spectroscopy results, including the results from η'_c, h_c and $X(3872)$ are reviewed. Also reported are some latest experimental results in quarkonium decays, such as the production of σ and κ, as well as the non-$D\bar{D}$ decays of $\psi(3770)$.

2 New observations

There are long-standing predictions of baryonium states[1], multi-quark states, $q\bar{q}$-gluon hybrids and glueballs. These states have been searched for by many experiments for many years. However, none of them is identified after all the efforts. Recently there has been a revival of interest in the possible existence of the states of non-$q\bar{q}$ or non-qqq. Reported below are the new observations in searching for such states.

2.1 The observation of $X(1835)$ in $J/\psi \to \gamma\eta'\pi^+\pi^-$ at BESII

An anomalous enhancement near the mass threshold in the $p\bar{p}$ invariant mass spectrum from $J/\psi \to \gamma p\bar{p}$ decays was reported by the BESII experiment[2]. This enhancement was fitted with a sub-threshold S-wave Breit-Wigner resonance function with a mass $M = 1859^{+3+5}_{-10-25}$ MeV/c^2, a width $\Gamma < 30$ MeV/c^2 (at the 90% C.L.) and a product branching fraction (BF) $B(J/\psi \to \gamma X) \cdot B(X \to p\bar{p}) = (7.0 \pm 0.4(stat)^{+1.9}_{-0.8}(syst)) \times 10^{-5}$. This surprising experimental observation has stimulated a number of theoretical speculations[3,4,5,6,7,8] and motivated the subsequent experimental observation of a strong $p\bar{\Lambda}$ mass threshold enhancement in $J/\psi \to pK^-\bar{\Lambda}$ decay[9]. Among various theoretical interpretations of the $p\bar{p}$ mass threshold enhancement, the most intriguing one is that of a $p\bar{p}$ bound state (or barionium)[3,1,6], which has been the subject of many experimental searches[10].

If such a structure is interpreted as a $p\bar{p}$ bound states, it is desirable to observe this state in other decay modes. Possible decay modes for a $p\bar{p}$ bound state, suggested in Ref.[5,6], include $\pi^+\pi^-\eta'$.

The $J/\psi \to \gamma\pi^+\pi^-\eta'$ decay, with η' being tagged in its two decay modes, $\eta' \to \pi^+\pi^-\eta(\eta \to \gamma\gamma)$ and $\eta' \to \gamma\rho$, is analyzed based on 5.8×10^7 J/ψ events collected at BESII.

Figure 1(a) shows the $\pi^+\pi^-\eta$ invariant mass distribution. The η' signal is clearly seen. The $\pi^+\pi^-\eta'$ invariant mass spectrum for the selected events is shown in Fig. 1(b), where a peak at a mass around 1835 MeV/c^2 is observed (named $X(1835)$). The $\gamma\rho$ invariant mass distribution also shows a clear η' signal (Fig. 1 (c)). A peak near 1835 MeV/c^2 is evident in the $\pi^+\pi^-\eta'$ invariant mass spec-

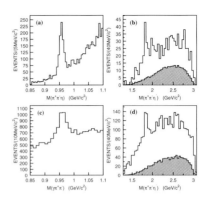

Figure 1. Invariant mass distributions for selected $J/\psi \to \gamma\pi^+\pi^-\eta'$ candidate events: (a) The $\pi^+\pi^-\eta$ invariant mass distribution. (b) The $\pi^+\pi^-\eta'$ invariant mass distribution with $\eta' \to \pi^+\pi^-\eta$. (c) The $\gamma\rho$ invariant mass distribution. (d) The $\pi^+\pi^-\eta'$ invariant mass distribution with $\eta' \to \gamma\rho$. The open histograms are data and the shaded histograms represent $J/\psi \to \gamma\pi^+\pi^-\eta'$ phase-space MC events (with arbitrary normalization).

trum (Fig. 1 (d)).

The combined $\pi^+\pi^-\eta'$ spectrum with $(\eta' \to \pi^+\pi^-\eta)$ and $(\eta' \to \gamma\rho)$ is fitted with a Breit-Wigner (BW) function convoluted with a Gaussian mass resolution function and a smooth polynomial background function. The BW mass and width of $X(1835)$ obtained from the fit (shown in Fig. 2) are $M = 1833.7 \pm 6.1$ MeV/c^2 and $\Gamma = 67.7 \pm 20.3$ MeV/c^2, respectively. The statistical significance for the signal is 7.7 σ. The mass and width of the $X(1835)$ are not consistent with any known particle[11]. The product branching fractions is determined to be

$$B(J/\psi \to \gamma X(1835)) \cdot B(X(1835) \to \pi^+\pi^-\eta')$$

$$= (2.2 \pm 0.4 \pm 0.4) \times 10^{-4}$$

The measured $X(1835)$ mass is consistent with the mass obtained from the $J/\psi \to \gamma p\bar{p}$ channel, while the width is higher by 1.9σ than the upper limit on the width reported in Ref.[2]. However, the refitted mass and width of $p\bar{p}$ threshold enhancement, after including the final state interaction effect, are consistent with those of $X(1835)$.

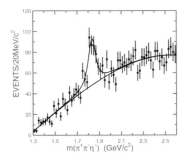

Figure 2. The $\pi^+\pi^-\eta'$ invariant mass distribution for selected events from the $J/\psi \to \gamma\pi^+\pi^-\eta'(\eta' \to \pi^+\pi^-\eta, \eta \to \gamma\gamma)$ and $J/\psi \to \pi^+\pi^-\eta'(\eta' \to \gamma\rho)$ analyses. The solid curve is the fit and the dashed curve indicates the background function.

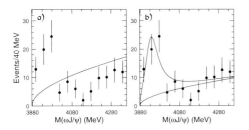

Figure 3. $B \to K\omega J/\psi$ signal yields vs $M(\omega J/\psi)$. The curve in (a) indicates the result of a fit that includes only a phase-space-like threshold function. The curve in (b) shows the result of a fit that includes an S-wave Breit-Wigner resonance term.

2.2 Near $\omega J/\psi$ threshold enhancement $Y(3940)$ in $B \to K\omega J/\psi$ at Belle

B meson decays are a prolific source of $c\bar{c}$ pairs and the large B meson samples produced at B-factories are providing opportunities to search for missing $c\bar{c}$ charmonium mesons as well as more complex states.

Based on a 253 fb^{-1} data sample that contains 275 million $B\bar{B}$ pairs collected with the Belle detector, a study of the $\omega J/\psi$ system produced in exclusive $B \to K\omega J/\psi$ decays is performed.

The B-meson signal yields from the binned one-dimensional fits to the M_{bc} and ΔE distributions for events in different $M(\omega J/\psi)$ intervals are plotted in Figs. 3(a) and (b). Here, M_{bc}, the beam constrained

mass, is equal to $\sqrt{E_{\text{beam}}^2 - p_B^2}$ and ΔE, the energy difference, is $E_{\text{beam}} - E_B$, with E_{beam} representing the cms beam energy, p_B the vector sum of the cms momenta of the B meson decay products and E_B their cms energy sum.

An enhancement, denoted as $Y(3940)$, is evident around $M(\omega J/\psi) = 3940$ MeV. The curve in Fig. 3(a) is the result of a fit with a threshold function of the form $f(M) = A_0 q^*(M)$, where $q^*(M)$ is the momentum of the daughter particles in the $\omega J/\psi$ rest frame. This functional form accurately reproduces the threshold behavior of Monte Carlo simulated $B \to K\omega J/\psi$ events that are generated uniformly distributed over phase-space. The fit quality to the observed data points is poor ($\chi^2/d.o.f. = 115/11$), indicating a significant deviation from phase-space.

Figure. 3(b) shows the results of a fit where an S-wave Breit-Wigner (BW) function is included to represent the enhancement. The fit, which has $\chi^2/d.o.f. = 15.6/8$ (CL = 4.8%), yields the mass and width of the signal to be $M = 3943 \pm 11 \pm 13$ MeV and $\Gamma = 87 \pm 22 \pm 26$ MeV. The statistical significance of $Y(3940)$ is 8.1σ and the product branching fraction is

$$\mathcal{B}(B \to KY(3940))\mathcal{B}(Y(3940) \to \omega J/\psi)$$

$$= (7.1 \pm 1.3 \pm 3.1) \times 10^{-5}$$

This $Y(3940)$ peaks above $D\bar{D}^*$. It is expected that a $c\bar{c}$ charmonium meson with this mass would dominantly decay to $D\bar{D}$ and/or $D\bar{D}^*$. While for the $c\bar{c}$-gluon hybrid charmonium states, which were first predicted in 1978[12] and are expected to be produced in B meson decays[13], their decays to $D^{(*)}\bar{D}^{(*)}$ meson pairs are forbidden or suppressed. Therefore, whether $Y(3940)$ can decay to $D\bar{D}$ and/or $D\bar{D}^*$ or not is crucial to identify its being a hybrid charmonium or a conventional charmonium state.

2.3 Observation of X(3940) in e^+e^- annihilation at $\sqrt{s} \approx 10.6$ GeV at Belle

The double charmonium production process of $e^+e^- \to J/\psi + X$ is investigated using the integrated luminosity of $350\,\mathrm{fb}^{-1}$ data sample collected by the Belle detector at the $\Upsilon(4S)$ resonance and nearby continuum. The J/ψ is reconstructed from its $\ell^+\ell^-$ decays. A partial correction for final state radiation and bremsstrahlung energy loss is performed by including the four-momentum of every photon detected within a 50 mrad cone around the electron direction in the e^+e^- invariant mass calculation.

The recoil mass spectrum of J/ψ in inclusive $e^+e^- \to J/\psi X$ for the data is shown in Fig. 4. In addition to the three previously observed peaks, the η_c, χ_{c0}, $\eta_c(2S)$, another significant peak can be seen around a mass of 3.94 GeV/c^2.

A fit to this spectrum that includes three known (η_c, χ_{c0}, $\eta_c(2S)$) and one new ($X(3940)$) charmonium states is performed. In this fit, the mass positions for the η_c, χ_{c0}, $\eta_c(2S)$ and $X(3940)$ are free parameters. The signal function for the $X(3940)$ is a convolution of the Monte Carlo line shape, with assumed zero width, with a Breit-Wigner function. The background is parametrized by a second order polynomial function and a threshold term ($\sqrt{M_{\mathrm{recoil}}(J/\psi) - 2M_D}$) to account for a possible contribution from $e^+e^- \to J/\psi D^{(*)}\overline{D}^{(*)}$. The mass of $X(3940)$ is measured to be $3.936 \pm 0.006 \pm 0.006$ GeV/c^2 and the width less than 52 MeV/c^2 at 90% C.L.. The statistical significance of the signal is 5σ.

A search for $X(3940)$ decaying into $D\overline{D}$ and $D^*\overline{D}$ final states is performed. The decay $X(3940) \to D^*\overline{D}$ is found to be the dominant decay mode with a measured branching fraction of $96^{+45}_{-32} \pm 22\%$. The upper limits for the decay $X(3940) \to D\overline{D}$ and $X(3940) \to J/\psi\omega$ are set to be less than 41% and 26% at 90%

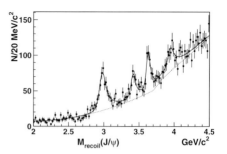

Figure 4. The distribution of masses recoiling against the reconstructed J/ψ in inclusive $e^+e^- \to J/\psi X$ events. The enhancements correspond to the η_c, χ_{c0}, $\eta_c(2S)$ and a new state, $X(3940)$. The curve represents the fit.

C.L., respectively.

2.4 Observation of Y(4260) in the $J/\psi\pi^+\pi^-$ Mass Spectrum around 4.26 GeV/c^2

The initial-state radiation events, $e^+e^- \to \gamma_{ISR}\pi^+\pi^- J/\psi$ are studied at BaBar, using an integrated luminosity of 211 fb^{-1} data collected at $\sqrt{s} = 10.58$ GeV/c^2, near the peak of the $\Upsilon(4S)$ resonance and $22 fb^{-1}$ data collected approximately 40 MeV/c^2 below this energy at the SLAC PEP-II asymmetric-energy e^+e^- storage ring. The candidate J/ψ is reconstructed via its decay to e^+e^- and $\mu^+\mu^-$.

The $\pi^+\pi^- J/\psi$ invariant mass spectrum for candidates passing all criteria is shown in Fig. 5 as points with error bars. Events that have an e^+e^- ($\mu^+\mu^-$) mass in the J/ψ sidebands [2.76, 2.95] or [3.18, 3.25] ([2.93, 3.01] or [3.18, 3.25]) GeV/c^2 but pass all the other selection criteria are represented by the shaded histogram after being scaled by the ratio of the widths of the J/ψ mass window and sideband regions. An enhancement near 4.26 GeV/c^2 is clearly observed. The Fig. 5 inset includes the ψ' region with a logarithmic scale for comparison.

An unbinned likelihood fit to the $\pi^+\pi^- J/\psi$ mass spectrum is performed using a single relativistic Breit-Wigner signal

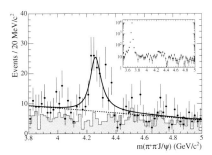

Figure 5. The $\pi^+\pi^- J/\psi$ invariant mass spectrum in the range $3.8-5.0$ GeV/c^2 and (inset) over a wider range that includes the ψ'. The points with error bars represent the selected data and the shaded histogram represents the scaled data from neighboring e^+e^- and $\mu^+\mu^-$ mass regions. The solid curve shows the result of the single-resonance fit; the dashed curve represents the background component.

function and a second-order polynomial background. The signal function is multiplied by a phase space factor and convoluted with a resolution function. The fit gives 125 ± 23 events with a mass of $4259 \pm 8(stat)^{+2}_{-6}(syst)$ MeV/c^2 and a width of $88\pm23(stat)^{+6}_{-4}(syst)$ MeV/c^2. The statistical significance of the signal is larger than $8\,\sigma$. At the present level of the statistics, the existence of additional narrow resonance(s) in this region cannot be excluded. The broad stucture at around 4.26 GeV/c^2 signifies the presence of one or more previously unobserved $J^{PC} = 1^{--}$ states containing hidden charm.

3 Heavy (flavor) quarkonium production

There are many challenging problems remaining unsolved in heavy quarkonium physics. The effective field theory NRQCD factorization approach provides a systematic method for calculating quarkonium decay and production rates. NRQCD has been very successful in describing many production data, such as the inclusive P-wave quakonium decays, quarkonium production at Tevatron and quarkonium production in deep inelastic scattering at HERA. However

it is problematic in describing the quarkonium polarization data at the Tevatron, as well as the data from double charmonium production at B factories. One of the most challenging open problems in heavy quarkonium is the large discrepancy of the double charmonium production cross sections measured in e^+e^- annihilation at B factories and the theoretical calculations from NRQCD. Here, I'll only focus on this topic.

Belle[14] observed the double $c\bar{c}$ production in e^+e^- annihilation at $\sqrt{s} \sim 10.6$ GeV. The production cross sections measured by Belle are about one order of magnitude higher than those predicted by non-relativistic QCD (NRQCD) calculations[15,16,17] for $e^+e^- \rightarrow \gamma^* \rightarrow J/\psi\, c\bar{c}$ reactions, where $c\bar{c}$ is a charmonium state with even C-parity. There have been attempts[18,19,20,21,22] to reconcile the large discrepancy between the observed cross section and predictions, and the validity of NRQCD approximations has been questioned[23,24]. It has also been suggested that at least part of the double charmonium production might be due to two virtual-photon interactions[20], i.e., $e^+e^- \rightarrow \gamma^*\gamma^* \rightarrow J/\psi\, c\bar{c}$, where odd C-parity states could be produced. However, Belle's updated results show that the contamination from $e^+e^- \rightarrow \gamma\gamma \rightarrow J/\psi J/\psi$ is small[25].

Recently, Belle and Babar presented the new measurements of the cross sections for double charmonium productions, using their $155 fb^{-1}$ and $124 fb^{-1}$ data taken at around $\Upsilon(4S)$ peak, respectively.

The recoil mass distribution for events in the J/ψ mass window is shown as points with error bars in Fig. 6. The upper plot is from Belle and the lower from BaBar. The fits to the recoil mass distribution are represented by the solid curves and the backgrounds are shown as the dashed curves.

The fit results are given in Table 1. The cross sections measured by both experiments are much larger than those predicted by many NRQCD calculations.

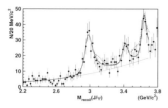

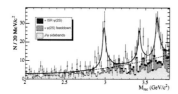

Figure 6. The points with error bars refer to the events in the J/ψ mass window. The solid curves represent the fit and the dashed curves are for background. The upper plot is for Belle and the lower for Babar.

Table 1. Comparison of experimental cross-sections with theoretical expectations (fb).

$J/\psi\,c\bar{c}$	η_c	χ_{c0}
BaBar	$17.6 \pm 2.8^{+1.5}_{-2.1}$	$10.3 \pm 2.5^{+1.4}_{-1.8}$
Belle	$25.6 \pm 2.8 \pm 3.4$	$6.4 \pm 1.7 \pm 1.0$
NRQCD[16]	2.31 ± 1.09	2.28 ± 1.03
NRQCD[15]	5.5	6.9

4 Heavy hadron spectroscopy

4.1 (Re)discovery of $\eta_c(2S)$ and $h_c(1P)$ states

The two $c\bar{c}$ states that are expected to be below open charm threshold are still not well established. They are the radially excited $n = 2$ singlet S state, the $\eta_c(2S)$ meson, and the $n = 1$ singlet P state, the $h_c(1P)$. The observation of these states and the determination of their masses would complete the below-threshold charmonium particle spectrum and provide useful information about the spin-spin part of the charmonium potential.

In 1982, the Crystal Ball collaboration[26] reported the observation of a small enhancement at $E_\gamma \sim 91$ MeV in the inclusive photon spectrum from $\psi' \to \gamma X$ decays, and interpreted it as due to $\eta_c(2S)$ with mass

Table 2. The measured $\eta_c(2S)$ mass

Exp.	reactions	Mass (MeV/c^2)
Belle(02)	$B \to KK_s K\pi$	$3654 \pm 6 \pm 8$
Belle(03)	$e^+e^- \to J/\psi X$	3622 ± 12
BaBar(03)	$\gamma\gamma \to K_s K^+\pi^-$	$3630.8 \pm 3.4 \pm 1.0$
CLEO(03)	$\gamma\gamma \to K_s K^+\pi^-$	$3642.9 \pm 3.1 \pm 1.5$
Average		3637.7 ± 4.4

3594 ± 5 MeV/c^2. This result implies a $\psi(2S)$-η'_c mass splitting that is considerably larger than heavy-quark potential model expectations. The result was not confirmed by other experiments[27].

The Belle experiment reported their observation of the $\eta_c(2S)$ in exclusive $B \to KK_s K\pi$ decays in 2002[28] and then in double charmonium production process $e^+e^- \to J/\psi X$ [29]. Later, both BaBar[30] and CLEO[31] confirmed the observation of $\eta_c(2S)$ in $\gamma\gamma \to K_s K^+\pi^-$ reaction. The measured masses of $\eta_c(2S)$ are listed in Table 2. Recent four measurements give consistent results on $\eta_c(2S)$ mass, and the averaged mass of $\eta_c(2S)$ from these experiments is significantly higher than that reported by Crystal Ball Collaboration. The deduced mass splitting of $m_{\psi(2S)} - m_{\eta_c(2S)} = 48.4 \pm 4.5$ MeV/c^2 is consistent with the heavy quark potential model calculations (42 -103 MeV/c^2) [32] and recent quenched LQCD calculations (40 -74 MeV/c^2) [33].

As mentioned above, QCD-based potential models have been quite successful in predicting masses, widths, and dominant decays of several charmonium states. The theoretical calculations also made the prediction that the hyperfine splitting $\Delta M_{hf}(\langle M(^3P_J)\rangle - M(^1P_1))$ for P-wave states should be zero. Higher-order corrections are expected to provide no more than a few MeV deviation from this result[34,35]. Lattice QCD calculations[33] predict $\Delta M_{hf}(1P) = +1.5$ to $+3.7$ MeV, but with uncertainties at the few-MeV level. Larger values of $\Delta M_{hf}(1P)$ could result if the confinement potential had a vector component or if coupled channel effects were important. In order to discriminate between these possibilities, it is necessary to identify the $h_c(^1P_1)$ state and to precisely measure its mass.

About 20 years ago, R704 experiment at CERN observed a cluster of 5 events in an exclusive scan for $p\bar{p} \to J/\psi + X$[36]. The mass of this cluster is 3525.4 ± 0.8 MeV/c^2.

The Crystal Ball Collaboration searched for h_c in the reaction $\psi(2S) \to \pi^0 h_c, h_c \to \gamma\eta_c$, but did not see it in the mass range $M(h_c) = (3515 - 3535)$ MeV [37]. The FNAL E760 Collaboration searched for h_c in the reaction $p\bar{p} \to h_c \to \pi^0 J/\psi, J/\psi \to e^+e^-$, and reported a statistically significant enhancement with $M(h_c) = 3526.2 \pm 0.2 \pm 0.2$ MeV, $\Gamma(h_c) \leq 1.1$ MeV [38]. The measurement was repeated twice by the E835 experiment with $\sim 2\times$ and $\sim 3\times$ larger luminosity, but no confirming signal for h_c was observed in $h_c \to \pi^0 J/\psi$ decay[39]. E835 experiment also searched for h_c state by a scan of $p\bar{p}$ annihilation cross section for $p\bar{p} \to h_c \to \gamma\eta_c \to \gamma\gamma\gamma$. An excess of $\eta_c\gamma$ events is observed, with a mass of $M = 3525.8 \pm 0.2 \pm 0.2$MeV. It has a probability $\mathcal{P} \sim 0.001$ to arise from background fluctuations[40].

Using 3.08×10^6 $\psi(2S)$ events accumulated with CLEO III and CLEO-c detector at the Cornell Electron Storage Ring, the isospin-violating reaction $e^+e^- \to \psi(2S) \to \pi^0 h_c$, $h_c \to \gamma\eta_c$, $\pi^0 \to \gamma\gamma$ is studied in exclusive and inclusive η_c decays, and h_c signal is observed in both cases. The statistical significance of h_c is larger than 5σ using a variety of methods to evaluate this quantity.

Table 3 shows the results for inclusive and exclusive analyses for the decay $\psi(2S) \to \pi^0 h_c \to \pi^0 \gamma\eta_c$. The combined inclusive and exclusive results give $M(h_c) = 3524.4 \pm 0.6 \pm 0.4$ MeV, which is consistent with the spin-weighted average of the χ_{cJ} states.

Table 3. Results of the inclusive and exclusive analyses for the reaction $\psi(2S) \to \pi^0 h_c \to \pi^0 \gamma\eta_c$.

	Inclusive	Exclusive
Counts	150 ± 40	17.5 ± 4.5
Significance	$\sim 3.8\sigma$	6.1σ
$M(h_c)$ (MeV)	$3524.9 \pm 0.7 \pm 0.4$	$3523.6 \pm 0.9 \pm 0.5$
$\mathcal{B}_\psi\mathcal{B}_h$ (10^{-4})	$3.5 \pm 1.0 \pm 0.7$	$5.3 \pm 1.5 \pm 1.0$

4.2 X(3872)

The $X(3872)$ was first observed by Belle[41] in $B^\pm \to K^\pm(J/\psi\pi^+\pi^-)$ and then confirmed by CDF[42], D0[43] and BaBar[44] experiments. The four experiments give consistent mass values of $X(3872)$. The averaged mass of $X(3872)$ is 3871.9 ± 0.6 MeV/c^2, which is just above $D^0\bar{D}^{*0}$ threshold (3871.3 ± 1.0 MeV/c^2).

Numerous theoretical explanations have been proposed for this high-mass, narrow-width state decaying into $J/\psi\pi^+\pi^-$. The possibilities include a $c\bar{c}$ charmonium state[45], $D^0\bar{D}^{*0}$ molecular state[46], dominant $1^{++}(2P)$ $c\bar{c}$ component with $D^0\bar{D}^{*0}/D^{*0}\bar{D}^0$ continuum component[47], $c\bar{c}g$ hybrid [48], vector glueball with a small mixture of $c\bar{c}$ [49], S-wave threshold enhancement in $D^0\bar{D}^{*0}$ scattering [50] and diquark-diquark bound state $cu\bar{c}\bar{u}$[51]. The search for more decay modes of $X(3872)$ as well as the determination of its J^{PC} will be helpful in understanding the nature of $X(3872)$.

A more detailed examination of the $X(3872)$ indicates that the $\pi^+\pi^-$ mass distributions peak near the kinematic upper limit and are consistent with the decay $\rho^0 \to \pi^+\pi^-$. If the observed decay is $X(3872) \to J/\psi\rho^0$ and if these states and their decays obey isospin symmetry, then there must be a $X(3872)^-$, which decays to $J/\psi\rho^-$, and the rate for $B \to X^-K$ should be twice of that for $B \to X^0K$. For this purpose, BaBar has performed a search for the decays of $B^0 \to X^-K^+$ and $B^- \to X^-K_s$, where $X^- \to J/\psi\pi^-\pi^0$[52], using 234M $B\bar{B}$ events. No charged signal, $X^- \to J/\psi\pi^-\pi^0$, is evident at 3.872 GeV/c^2.

BaBar also searched for $X(3872) \rightarrow \eta J/\psi$ with 90M $B\bar{B}$ events and no signal is seen[53].

CLEO searched for $X(3872)$ in $\gamma\gamma$ fusion and radiative production data and no evidence is found[54].

In order to examine the experimental constraints on the possible J^{PC} of $X(3872)$, Belle searched for $B \rightarrow K\gamma J/\psi$ and $K\pi^+\pi^-\pi^0 J/\psi$ decays in a 275M $B\bar{B}$ events sample, accumulated at a center-of-mass system energy of $\sqrt{s} = 10.58$ GeV. In the $\gamma J/\psi$ mass spectrum, a peak around 3872 MeV/c^2 can be seen, which corresponds to a statistical significance of 4.0σ. The ratio of the decay widths is obtained from the fit as:

$$\Gamma(X \rightarrow \gamma J/\psi)/\Gamma(X \rightarrow \pi^+\pi^- J/\psi) = 0.14 \pm 0.05$$

. The $\pi^+\pi^-\pi^0$ mass spectrum in $B \rightarrow \pi^+\pi^-\pi^0 J/\psi$ also shows the evidence for subthreshold decay $X(3872) \rightarrow \omega J/\psi$ and the statistical significance is 4.3σ. The ratio of the branching fractions is determined to be:

$$\frac{B(X \rightarrow \pi^+\pi^-\pi^0 J/\psi)}{B(X \rightarrow \pi^+\pi^- J/\psi)} = 1.0 \pm 0.4 \pm 0.3.$$

The evidence of these two decay modes suggests the C parity of $X(3872)$ to be $+1$.

In a search for $X(3872) \rightarrow D^0\bar{D}^{*0}$ decay in $B \rightarrow KD^0\bar{D}^{*0}$, with $\bar{D}^{*0} \rightarrow \bar{D}^0\pi^0$, there is events excess at around 3872 MeV/c^2 in $D^0\bar{D}^0\pi^0$ spectrum. This implies that the 2^{++} is not favored.

The angular distributions of a sample of $X(3872)$, produced in $B \rightarrow KX(3872)$ from 256 pb^{-1} data are analyzed. From the χ^2 test, a 1^{++} $X(3872)$ is favored.

All the results seem to favor a 1^{++} $X(3872)$. However, the statistics is low for the evidences of $X(3872) \rightarrow \gamma J/\psi$, $X(3872) \rightarrow \omega J/\psi$ and $X(3872) \rightarrow D^0\bar{D}^{*0}$. In the analyses of the angular distributions, the statistics is low to use χ^2 test. Therefore, more data are needed and the confirmation from BaBar is required to finally determine the spin-parity of $X(3872)$.

5 Selected topics from quarkonium decays

5.1 The σ production

There has been evidence for a low mass pole in the early DM2[55] and BESI[56] data on $J/\psi \rightarrow \omega\pi^+\pi^-$. A huge event concentration in the $I = 0$ S-wave $\pi\pi$ channel was seen in the region of $m_{\pi\pi}$ around 500-600 MeV in a pp central production experiment[57]. This peak is too large to be explained as background[58]. There have been many studies on the possible resonance structure in $\pi\pi$ elastic scattering[59]. It was later proved that the σ resonance is unavoidable in chiral perturbation theory in order to explain the $\pi\pi$ scattering phase shift data[60]. E791 experiment from FNAL shows the existence of a σ pole with $M = 478^{+24}_{-23} \pm 17$ MeV, $\Gamma = 324^{+42}_{-40} \pm 21$ MeV in $D^+ \rightarrow \pi^+\pi^-\pi^+$[61].

BES studied $J/\psi \rightarrow \omega\pi^+\pi^-$ decays based on 58M J/ψ events collected at BESII. Figure 7 shows the $\pi^+\pi^-$ invariant mass spectrum recoiling against ω. In addition to the well known $f_2(1270)$, a broad bump in the low mass region is clearly seen. Two independent partial wave analyses (PWA) are performed and different parametrizations of σ amplitude are used. All give the consistent results for σ pole position. The averaged pole is determined to be $(541 \pm 39 - i (252 \pm 42))$ MeV/c^2.

Based on 14M $\psi(2S)$ events collected at BESII, a partial wave analysis is performed to $\psi(2S) \rightarrow \pi^+\pi^- J/\psi$, with $J/\psi \rightarrow \mu^+\mu^-$. A severe suppression of the $\pi^+\pi^-$ invariant mass near the $\pi\pi$ threshold is distinctively different from the phase space shape, which suggests the σ production. Using different parametrizations of σ amplitude, the data can be fitted well through a strong cancellation between σ and a contact term. The obtained pole posision is $(554 \pm 14 \pm 53) - i (242 \pm 5 \pm 24))$ MeV/c^2, which is consistent with that from $J/\psi \rightarrow \omega\pi^+\pi^-$.

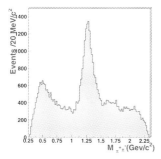

Figure 7. The $\pi^+\pi^-$ invariant mass recoiling against ω

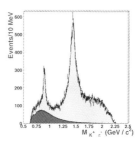

Figure 8. The $K\pi$ invariant mass recoiling against K^*. The crosses are data and histograms represent the PWA fit projection. The shaded area shows the κ contribution.

BaBar made a full amplitude analysis on $B^\pm \to \pi^\pm\pi^\pm\pi^\mp$ with $210.3 fb^{-1}$ data sample and found that the decays of $B^\pm \to \pi^\pm\pi^\pm\pi^\mp$ are dominated by decays through the intermediate resonances, such as $\rho^0(770)$ and $\rho^0(1450)$. There is no evidence for the exisitence of σ.

5.2 The κ production

The κ has been observed in the analyses on $K\pi$ scattering phase shifts by several groups. The observed mass and width are scattered in the ranges from 700 to 900 MeV/c^2 and 550 to 800 MeV/c^2, respectively, depending on the model used. However, some analyses on $K\pi$ scattering data don't need a κ pole. The E791 experiment at Fermilab reported the evidence of κ in the $D^+ \to K^+\pi^+\pi^-$ [62] with the mass and width of $797 \pm 19 \pm 43$ MeV/c^2 and $410\pm43\pm87$ MeV/c^2. However, a slightly lower statistics of CLEO $D^0 \to K^-\pi^+\pi^0$ data finds no evidence of κ[63]. The FOCUS experiment presented evidence for the existence of a coherent $K\pi$ S-wave contribution to $D^+ \to K^-\pi^+\mu^+\nu$[64].

BES performed partial wave analyses to both $J/\psi \to K^*K\pi$[65] and direct $J/\psi \to K^+K^-\pi^+\pi^-$ processes. The κ are needed in both fits. Fig. 8 is the $K\pi$ invariant mass spectrum which recoils against $K^*(892)$. The crosses are data and histograms represent the PWA fit projection. The shaded area shows the κ contribution. The pole of κ from $J/\psi \to K^*K\pi$ is determined to be $(841 \pm 30^{+81}_{-73}) - i(309 \pm 45^{+48}_{-72})$ MeV/c^2.

Both BaBar and Belle made the Dalitz plot analyses of $B \to K\pi\pi$, but no κ is included in the fit[66].

5.3 $\psi(3770)$ non-$D\bar{D}$ decays

$\psi(3770)$ was considered to decay almost entirely to pure $D\bar{D}$[67] because its mass is above the open charm-pair threshold and its width is two orders of magnitude larger than that of $\psi(2S)$. Since $\psi(3770)$ is also believed to be a mixture of the 1^3D_1 and 2^3S_1 states[68], other $\psi(2S)$-like decays for $\psi(3770)$ are expected. Many theoretical calculations[69,70,71,72,73] estimated the partial width of $\Gamma(\psi(3770) \to \pi\pi J/\psi)$. Recently, Kuang[73] obtained the partial width for $\psi(3770) \to J/\psi\pi^+\pi^-$ to be in the range of 25 to 113 keV, using Chen-Kuang potential model.

Based on 27.7 pb^{-1} data sample taken in the center-of-mass (c.m.) energy region of 3.738 GeV to 3.885 GeV, BES reported the first evidence of $\psi(3770) \to J/\psi\pi^+\pi^-$ with $J/\psi \to e^+e^-$ and $\mu^+\mu^-$[74]. Figure 9 shows the dilepton masses determined from the fitted lepton momenta of the accepted events. Two peaks are clearly seen. The lower peak mostly comes from $\psi(3770) \to J/\psi\pi^+\pi^-$, while the higher one is produced by the radiative return to the $\psi(2S)$ peak.

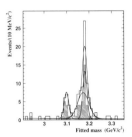

Figure 9. The distribution of the fitted dilepton masses for the events of $l^+l^-\pi^+\pi^-$ from the data; the hatched histogram is for $\mu^+\mu^-\pi^+\pi^-$, while the open one is for $e^+e^-\pi^+\pi^-$; the curves give the best fit to the data.

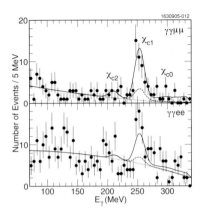

Figure 10. Energy of the lower energy photon for the selected $e^+e^- \to \gamma\gamma J/\psi$, $J/\psi \to \mu\mu$ (top) and $J/\psi \to e^+e^-$ (bottom) events at the $\psi(3770)$ resonance. The solid line shows the fit. The dotted line shows the smooth background. The dashed line shows the total background including the expected background-peaks from radiatively produced tail of the $\psi(2S)$ resonance.

A maximum likelihood fit gives 17.8 ± 4.8 $J/\psi \to l^+l^-$ signal events. After subtracting the background, $(11.8 \pm 4.8 \pm 1.3)$ $\psi(3770) \to J/\psi\pi^+\pi^-$ non-$D\bar{D}$ decay events are observed, leading to a branching fraction of $BF(\psi(3770) \to J/\psi\pi^+\pi^-) = (0.34 \pm 0.14 \pm 0.09)\%$, and a partial width $\Gamma(\psi(3770) \to J/\psi\pi^+\pi^-) = (80 \pm 33 \pm 23)$ keV, which is consistent with Kuang's estimations[73].

With a much larger data sample (an integrated luminosity of $\mathcal{L} = (280.7 \pm 2.8)$ pb^{-1}), CLEO confirmed the non-$D\bar{D}$ decays of $\psi(3770) \to \pi\pi J/\psi$ with the statistical significancies of 13σ and 3.8σ for $\pi^+\pi^-$ and $\pi^0\pi^0$ cases, respectively[75]. More precise branching fractions are obtained to be: $BF(\psi(3770) \to J/\psi\pi^+\pi^-) = (0.214 \pm 0.25 \pm 0.22)\%$ and $BF(\psi(3770) \to J/\psi\pi^0\pi^0) = (0.097 \pm 0.035 \pm 0.020)\%$. The partial widthes are $\Gamma(\psi(3770) \to J/\psi\pi^+\pi^-) = (50 \pm 6 \pm 8)$ keV and $\Gamma(\psi(3770) \to J/\psi\pi^0\pi^0) = (23 \pm 8 \pm 5)$ keV. The $\psi(3770) \to J/\psi\pi^+\pi^-$ results are consistent with those from BES but with higher precision.

Using the same data sample, CLEO first observed $\psi(3770) \to \gamma\chi_{c1} \to \gamma\gamma J/\psi$, with $J/\psi \to \mu^+\mu^-$ and e^+e^-[76]. Figure 10 shows the energy of the lower energy photon for the selected $\psi(3770) \to \gamma\gamma J/\psi$, $J/\psi \to \mu^+\mu^-$ (top) and $J/\psi \to e^+e^-$ (bottom). The excess in the χ_{c1} peak represents evidence for $\psi(3770) \to \gamma\chi_{c1}$ transitions. The fitted signal amplitudes (for the sum of the dimuon and dielectron samples) are $0.3^{+7.7}_{-0.3}$, 62 ± 11 and 24^{+12}_{-15} events for χ_{c2}, χ_{c1} and χ_{c0}, respectively. The statistical significance of the evidence for χ_{c1} signal is $6.8\ \sigma$ and the decay width of $\psi(3770)$ to $\gamma\chi_{c1}$ is $75 \pm 14 \pm 13$ (keV).

Both CLEO and BES also performed many searches for $\psi(3770)$ non-$D\bar{D}$ decays[77,78,79,80]. A statistically significant signal is found for $\phi\eta$[77] by CLEO and a suggestive suppression of $\pi^+\pi^-\pi^0$ and $\rho\pi$ is observed by CLEO and BES[77,79].

6 Summary

Many new discoveries and new results on heavy quarkonium productions and decays have been recently experimentally revitalized by BES, Belle, BaBar, CLEO, as well as other experiments. This report represents the results including the new observed particles, such as $X(1835)$, $X(3940)$, $Y(3940)$ and $Y(4260)$, further studies of $X(3872)$, the observation of h_c state, heavy flavor production, in particular the double charmonium production from BaBar and Belle, the production of

σ and κ, as well as $\psi(3770)$ non-$D\bar{D}$ decays.

More results on quarkonium are comming in the near future from BESII, CLEO-c, Belle, Babar and the experiments from Fermilab. The BESIII/BEPCII is being constructing and it will start taking physics data in the year of 2008. There is also a proposal for a new dedicated $p\bar{p}$ machine (PANDA at GSI) to explore charmonium physics. We expect more results on quarkonium production and decay to probe the strong interactions.

Acknowledgments

I would like to thank many colleagues from BaBar, Belle, BES, CLEO and the experiments at FNAL for providing me with information contained in this report. I apologize for not being able to discuss all the recent experimental results related to this talk due to the length limitation. I would also like to thank Prof. K.T. Chao and Y.P. Kuang for discussions concerning theoretical aspects.

References

1. I.S. Shapiro, *Phys. Rept.* **35**, 129 (1978); C.B. Dover, M. Goldhaber, *Phys. Rev. D* **15**, 1997 (1977).

2. BES Collaboration, J.Z. Bai *et al.*, *Phys. Rev. Lett.* **91**, 022001 (2003).

3. A. Datta, P.J. O'Donnell, *Phys. Lett.* **B567**, 273 (2003); M.L. Yan *et al.*, hep-ph/0405087; B. Loiseau, S. Wycech, hep-ph/0502127.

4. J. Ellis, Y. Frishman and M. Karliner, *Phys. Lett.* **B566**, 201 (2003); J.L. Rosner, *Phys. Rev. D* **68**, 014004 (2003).

5. C.S. Gao and S.L. Zhu, *Commun. Theor. Phys.* 42, 844 (2004), hep-ph/0308205.

6. G.J. Ding and M.L. Yan, *Phys. Rev. C* **72**, 015208 (2005).

7. B.S. Zou and H.C. Chiang, *Phys. Rev. D* **69**, 034004 (2003).

8. A. Sibirtsev *et al.*, *Phys. Rev. D* **71**, 054010 (2005).

9. BES Collaboration, M. Ablikim *et al.*, *Phys. Rev. Lett.* **93**, 112002 (2004).

10. For recent reviews of this subject, see E. Klempt *et al.*, *Phys. Rep.* **368**, 119 (2002) and J-M. Richard, *Nucl. Phys. Proc. Suppl.* **86**, 361 (2000).

11. Particle Data Group, S. Eidelman *et al.*, *Phys. Lett.* **B592**, 1 (2004).

12. D. Horn and J. Mandula, *Phys. Rev. D* **17**, 898 (1978).

13. F.E. Close, I. Dunietz, P.R. Page, S. Veseli and H. Yamamoto, *Phys. Rev. D* **57**, 5653 (1987); G. Chiladze, A.F. Falk and A.A. Petrov *Phys. Rev. D* **58**, 034013 (1988); and F.E. Close and S. Godfrey, *Phys. Lett. B* **574**, 210 (2003).

14. Belle Collaboration, K. Abe et al, *Phys. Rev. Lett.* **89**, 142001 (2002).

15. Kui-Yong Liu, Zhi-Guo He and Kuang-Ta Chao, *Phys. Lett. B* **557**, 45 (2003).

16. E. Braaten and J. Lee, *Phys. Rev. D* **67**, 054007 (2003).

17. K. Hagiwara, E. Kou and C.-F. Qiao, *Phys. Lett. B* **570**, 39 (2003).

18. Y. J. Zhang, Y. J. Gao and K. T. Chao, hep-ph/0506076.

19. J.P. Ma and Z.G. Si, *Phys. Rev. D* **70**, 074007 (2004)

20. G. T. Bodwin, J. Lee and E. Braaten, *Phys. Rev. D* **67**, 054023 (2003).

21. G. T. Bodwin, J. Lee and E. Braaten, *Phys. Rev. Lett.* **90**, 162001 (2003).

22. S. J. Brodsky, A. S. Goldhaber and J. Lee, *Phys. Rev. Lett.* **91**, 112001 (2003).

23. A. E. Bondar and V. L. Chernyak, *Phys. Lett. B* **612**, 215 (2005).

24. N. Brambilla *et al*, hep-ph/0412158.

25. Belle Collaboration, K. Abe *et al*, *Phys. Rev. D* **70**, 071102 (2004).

26. C. Edwards *et al* (Crystal Ball Collab.), *Phys. Rev. Lett.* **48**, 70 (1982).

27. T.A. Armstrong *et al* (E760 Collab.), *Phys. Rev.* **D52**, 4839 (1995); M. Ma-

138

suzawa, Ph.D. Thesis, Northwestern Univ. report UMI-94-15774 (1993), unpublished; M. Ambrogiani *et al.*, *Phys. Rev.* **D64**, 052003 (2001); P. Abreu *et al.* (DELPHI), *Phys. Lett.* **B441**, 479 (1998); and M.Acciarri *et al.* (L3), *Phys. Lett.* **B461**, 155 (1999).

28. Belle Collaboration, S.-K. Choi *et al,* *Phys. Rev. Lett.* 89 (2002) 102001; Erratum-ibid. **89** 129901 (2002).

29. Belle Collaboration, K. Abe *et al,* *Phys. Rev. Lett.* **89** 142001 (2002).

30. BaBar Collaboration, *Phys. Rev. Lett.* **92** 142002 (2004).

31. CLEO Collaboration, *Phys. Rev. Lett.* **92** 142001 (2004).

32. E. Eichten and F. Feinberg, *Phys. Rev.* **D23**, 2724 (1981); W. Buchmüller and S. -H. H. Tye, *Phys. Rev.* **D24**, 132 (1981); S. Godfrey and N. Isgur, *Phys. Rev.* **D32**, 189 (1985); Y. Q Chen and Y. P. Kuang, *Phys. Rev.* **D37**, 1210 (1988); E. J. Eichten and C. Quigg, *Phys. Rev.* **D49**, 5845 (1994); D. Ebert, R. N. Faustov and V. O. Galkin, *Phys. Rev.* **D62**, 034014 (2000).

33. K. Okamoto, *Phys. Rev.* **D65**, 094508 (2002)

34. T. Appelquist, R. M. Barnett and K. D. Lane, Ann. Rev. *Nucl. Part. Sci.* **28**, 387 (1978).

35. S. Godfrey and J. L. Rosner, *Phys. Rev. D* **66**, 014012 (2002).

36. R704 Collaboration, *Phys. Lett. B* **171**, 135 (1986).

37. Proceedings of the 17th Rencontre de Moriond Workshop on New Flavors, Les Arcs, France, p. 27 (1982); E. D. Bloom and C. W. Peck, *Ann. Rev. Nucl. Part. Sci.*, **33**, 143 (1983).

38. E760 Collaboration, *Phys. Rev. Lett.*, **69**, 2337 (1992).

39. D. N. Joffe, Ph.D. thesis, Northwestern University, hep-ex/0505007

40. Claudia Patrignani, hep-ex/0410085

41. Belle Collaboration, *Phys. Rev. Lett.,* **91**, 262001 (2003).

42. CDF-II Collaboration, *Phys. Rev. Lett.,* **93**, 072001 (2004).

43. D0 Collaboration, *Phys. Rev. Lett.,* **93**, 162002 (2004).

44. BaBar Collaboration, *Phys. Rev. D 71*, 071103 (2005).

45. E. Eichten, K. Lane, and C. Quigg, *Phys. Rev. Lett.* **89**, 162002 (2002); T. Barnes and S. Godfrey, *Phys. Rev. D* **69**, 054008 (2004).

46. N. Tornqvist, *Phys. Lett. B* **590**, 209 (2004); M. B. Voloshin, *Phys. Lett. B* **579**, 316 (2004); F. Close and P. Page, *Phys. Lett. B* **578**, 119 (2004); C.Y. Wong, *Phys. Rev. C* **69**, 055202 (2004); E. Braaten and M. Kusunoki, *Phys. Rev. D 69*, 074005 (2004); E. Swanson, *Phys. Lett. B* **588**, 189 (2004).

47. C. Meng, Y.J. Gao and K.T. Chao, hep-ph/0506222

48. F. Close and S. Godfrey,*Phys. Lett. B* **574**, 210 (2003) B.A. Li,*Phys. Lett. B* **605**, 306 (2005).

49. K.K. Seth, hep-ph/0411122

50. D.V. Bugg, *Phys. Rev. D* **71**, 016006 (2005).

51. L. Maiani, F. Piccinini, A.D. Polosa, and V. Riquer, *Phys. Rev. D* **71**, 014028 (2005).

52. BaBar Collaboration,*Phys. Rev. Lett.,* **93**, 041801 (2004).

53. BaBar Collaboration, *Phys. Rev. D* **71**, 031501 (2005).

54. CLEO Collaboration, *Phys. Rev. Lett.,* **94**, 032004 (2005).

55. J.E. Augustin et al.,*Nucl. Phys.* **B320** 1 (1989).

56. Ning Wu (BES Collaboration), Proceedings of the XXXVIth Rencontres de Moriond, Les Arcs, France, March 17-24, 2001.

57. D. Alde et al.,*Phys. Lett.* **B397** 350 (1997).

58. T. Ishida et al., Proceedings of Int. Conf. Hadron'95, Manchester, UK,

World Scientific, 1995.

59. V.E. Markushin and M.P. Locher,*Frascati Phys. Ser.* **15** 229 (1999).

60. Z. Xiao, H. Q. Zheng, *Nucl. Phys.* **A695** 273 (2001).

61. E791 Collaboration, *Phys. Rev. Lett.* **86** 770 (2001).

62. E791 Collaboration, *Phys. Rev. Lett.* **89** 121801 (2002).

63. CLEO Collaboration, *Phys. Rev.* **D63**, 0900001 (2001)

64. FOCUS Collaboration,*Phys. Lett.* **B535** 430 (2002)

65. BES Collaboration, hep-ex/0506055

66. Belle Collaboration, *Phys. Rev. D* **71**, 092003 (2005), BaBar Collaboration, BABAR-PUB-05/027

67. W. Bacino *et al.*, DELCO Collaboration, *Phys. Rev. Lett.* **40** 671 (1978).

68. J.L. Rosner,*Phys. Rev. D* **64**, 094002 (2001).

69. H.J. Lipkin,*Phys. Lett.* **B179** 278 (1986).

70. K. Lane, Harvard Report No. HUTP-86/A045, (1986).

71. Y.P. Kuang and T.M. Yan, *Phys. Rev. D* **24** 2874 (1981).

72. Y.P. Kuang and T.M. Yan, *Phys. Rev. D* **41** 155 (1990).

73. Y.P. Kuang, *Phys. Rev. D* **65** 094024 (2002).

74. BES Collaboration, hep-ex/0307028, *Phys. Lett. B* **605**, 63 (2005).

75. CLEO collaboration, hep-ex/0508023

76. CLEO collaboration, hep-ex/0509030

77. CLEO collaboration, hep-ex/0509011

78. CLEO collaboration, CLNS 05/1921 (LP2005-443)

79. BES Collaboration, *Phys. Rev.* **D70** 077101 (2004).

80. BES Collaboration, hep-ex/0507092

140

DISCUSSION

Ulrich Wiedner (Uppsala University):
Could you explain in more detail why the enhanced double charmonium production rate could not be explained by J/ψ + glueball production?

Xiaoyan Shen: Belle[1] measured the production and J/ψ helicity angular distributions, $(1 + \alpha_{prod}\cos^2\theta_{prod})$ and $(1 + \alpha_{heli}\cos^2\theta_{heli})$, for $e^+e^- \to J/\psi\eta_c$. The α_{prod} and α_{heli} are measured to be: $\alpha_{prod} = \alpha_{heli} = 0.93^{+0.57}_{-0.47}$. It is consistent with the expectations for the production of $J/\psi\eta_c$ via a single virtual photon ($\alpha_{prod} = \alpha_{heli} = 1$). The prediction for a spin-0 glueball contribution[2] ($e^+e^- \to J/\psi\,G$) to the $J/\psi\eta_c$ peak, $\alpha_{prod} = \alpha_{heli} = -0.87$, is disfavored.

Tord Ekelof (Uppsala University):
I have a question on the X(1834). You said that the X(1860) has 10% branching ratio to $\bar{p}p$ which was an evidence that it is a $\bar{p}p$ bound state. The X(1834) you discovered in this specific decay channel - I don't know how well these 10% can be put into a model but if this lower state the X(1834) should also then have some small branching ratio to $\bar{p}p$, how small would it be-would it be measurable?

Xiaoyan Shen: The large branching ratio of $X(1860)$, $p\bar{p}$ threshold enhancement from $J/\psi \to \gamma p\bar{p}$, to $p\bar{p}$ indicates $X(1860)$ has a large coupling to $p\bar{p}$. If it is a $p\bar{p}$ bound state below $p\bar{p}$ mass threshold, the theory predicts that its $\eta'\pi^+\pi^-$ decay mode would be the most favorable one[3,4].

The mass and width of $X(1835)$ from $J/\psi \to \gamma\eta'\pi^+\pi^-$ are consistent with those of $X(1860)$. Therefore we think these two states could be the same state.

Jonathan L. Rosner (Uni. of Chicago):
The $\Upsilon(4260)$ is just above $D_s^*\overline{D_s^*}$ threshold in the same way that the $\Upsilon(4030)$ is just above $D^*\overline{D^*}$ threshold. It might therefore be useful to look for $\Upsilon(4260) \to D_s^*\overline{D_s^*}$.

Xiaoyan Shen: It is a good suggestion. BaBar might be able to check it.

References

1. Belle Collaboration, K. Abe *et al*, *Phys. Rev. D* **70**, 071102 (2004).
2. S. J. Brodsky, A. S. Goldhaber and J. Lee, *Phys. Rev. Lett.* **91**, 112001 (2003).
3. C.S. Gao and S.L. Zhu, *Commun. Theor. Phys.* **42**, 844 (2004), hep-ph/0308205.
4. G.J. Ding and M.L. *Yan, Phys. Rev. C* **72**, 015208 (2005).

QUARK MIXING AND CP VIOLATION — THE CKM MATRIX

ULRICH NIERSTE

Institut für Theoretische Teilchenphysik, Universität Karlsruhe, 76128 Karlsruhe, Germany
E-mail: nierste@particle.uni-karlsruhe.de

I summarize the theoretical progress in the determination of the CKM elements since *Lepton-Photon 2003* and present the status of the elements and parameters of the Cabibbo-Kobayashi-Maskawa (CKM) matrix. One finds $|V_{us}| = 0.2227 \pm 0.0017$ from K and τ decays and $|V_{cb}| = (41.6 \pm 0.5) \cdot 10^{-3}$ from inclusive semileptonic B decays. The unitarity triangle can now be determined from tree-level quantities alone and the result agrees well with the global fit including flavour-changing neutral current (FCNC) processes, which are sensitive to new physics. From the global fit one finds the three CKM angles $\theta_{12} = 12.9° \pm 0.1°$, $\theta_{23} = 2.38° \pm 0.03°$ and $\theta_{13} = 0.223° \pm 0.007°$ in the standard PDG convention. The CP phase equals $\delta_{13} \simeq \gamma = (58.8 \,^{+5.3}_{-5.8})°$ at 1σ CL and $\gamma = (58.8 \,^{+11.2}_{-15.4})°$ at 2σ CL. A major progress are first results from fully unquenched lattice QCD computations for the hadronic quantities entering the UT fit. I further present the calculation of three-loop QCD corrections to the charm contribution in $K^+ \to \pi^+ \nu \bar{\nu}$ decays, which removes the last relevant theoretical uncertainty from the $K \to \pi \nu \bar{\nu}$ system. Finally I discuss mixing-induced CP asymmetries in $b \to s \bar{q} q$ penguin decays, whose naive average is below its Standard Model value by 3σ.

1   Flavour in the Standard Model

In the Standard Model transitions between quarks of different generations originate from the Yukawa couplings of the Higgs field to quarks. The non-zero vacuum expectation value v of the Higgs field leads to quark mass matrices M^u and M^d for the up-type and down-type quarks, respectively. The transformation to the physical mass eigenstate basis, in which the mass matrices are diagonal, involves unitary rotations in flavour space. The rotation of the left-handed down-type quarks relative to the left-handed up-type quarks is the physical Cabibbo-Kobayashi-Maskawa (CKM) matrix V. It appears in the couplings of the W boson to quarks and is the only source of transitions between quarks of different generations. V contains one physical complex phase, which is the only source of CP violation in flavour-changing transitions.

Flavour physics first aims at the precise determination of CKM elements and quark masses, which are fundamental parameters of the Standard Model. The second target is the search for new physics, pursued by confronting high precision data with the predictions of the Standard Model and its extensions. To this end it is useful to distinguish between charged-current weak decays and flavour-changing neutral current (FCNC) processes. The determination of CKM elements from the tree-level charged-current weak decays, discussed in Sect. 2, is practically unaffected by possible new physics.[a] By contrast, FCNC processes are very sensitive to virtual effects from new particles with masses at and above the electroweak scale, even beyond 100 TeV in certain models of new physics. FCNC processes are discussed in Sect. 3.

V can be parameterized in terms of three mixing angles θ_{12}, θ_{23}, θ_{13} and one complex phase δ_{13}, which violates CP . Adopting the PDG convention[1], in which V_{ud}, V_{us}, V_{cb} and V_{tb} are real and positive, these parameters can be determined through

$$V_{us} = \sin\theta_{12}\cos\theta_{13}, \quad V_{ub} = \sin\theta_{13}\, e^{-i\delta_{13}},$$
$$V_{cb} = \sin\theta_{23}\cos\theta_{13}. \tag{1}$$

---

[a]Still new physics can be revealed if the 3×3 CKM matrix V is found to violate unitarity: One may then infer the existence of new (for example iso-vector) quarks which mix with the known six quarks. Further leptonic decays of charged mesons are tree-level, but sensitive to effects from charged Higgs bosons.

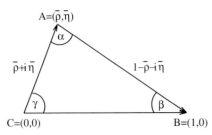

Figure 1. Unitarity triangle (UT).

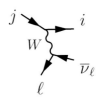

Figure 2. $|V_{uj}|$ and $|V_{cj}|$, $j = d, s, b$, are determined from semileptonic decays.

The Wolfenstein parameterization[2]

$$
V = \begin{pmatrix} 1 - \frac{\lambda^2}{2} & \lambda & A\lambda^3(\rho - i\eta) \\ -\lambda & 1 - \frac{\lambda^2}{2} & A\lambda^2 \\ A\lambda^3(1 - \rho - i\eta) & -A\lambda^2 & 1 \end{pmatrix} \quad (2)
$$

is an expansion of V in terms of $\lambda \simeq 0.22$ to order λ^3. It shows both the hierarchy of the CKM elements and their correlations, like $|V_{us}| \simeq |V_{cd}|$ and $|V_{cb}| \simeq |V_{ts}|$. The apex of the standard unitarity triangle (UT), which is shown in Fig. 1 is defined by[3]

$$
\bar{\rho} + i\bar{\eta} \equiv -\frac{V_{ub}^* V_{ud}}{V_{cb}^* V_{cd}} = \left| \frac{V_{ub}^* V_{ud}}{V_{cb}^* V_{cd}} \right| e^{i\gamma} \quad (3)
$$

$(\bar{\rho}, \bar{\eta})$ coincide with (ρ, η) up to corrections of order λ^2. With Eq. (3) and

$$
\lambda \equiv \sin\theta_{12}, \quad A\lambda^2 \equiv \sin\theta_{23}, \quad (4)
$$

the Wolfenstein parameterisation can be made exact[3], that is V can be expressed in terms of $(\lambda, A, \bar{\rho}, \bar{\eta})$ to any desired order in λ. In the following I always use the PDG phase convention and the exact definitions in Eqs. (3) and (4), with one exception: I ignore the small phase of $-V_{cd}$ (see Ref.[1]), so that I can identify $\arg V_{ub}^* = \delta_{13}$ with γ and $\arg V_{td}^*$ with the angle β of the unitarity triangle. This approximation is correct to 0.1%.

The numerical results presented in the following have been prepared with the help of the *Heavy Flavor Averaging Group (HFAG)*[4] and the *CKMfitter*[5] and *UTFit*[6] groups. *CKMfitter* uses a Frequentist treatment of theoretical uncertainties, while *UTFit* pursues a Bayesian approach, using flat probability distribution functions for theoretical uncertainties.

2 CKM elements from tree-level decays

The standard way to determine the magnitudes of the elements of the first two rows of V uses semileptonic hadron decays, depicted in Fig. 2. From Eq. (2) one realizes that an accurate determination of V_{us} or V_{ud} determines V_{cd} and V_{cs} as well. Therefore measurements of semileptonic $c \to d$ and $c \to s$ decays are usually viewed as test of the computation of the hadronic form factors entering the decay amplitudes. Charm decays are covered by Iain Stewart.[7]

2.1 V_{ud}

V_{ud} can be determined from superallowed $(0^+ \to 0^+)$ nuclear β decay and from the β decays $n \to p\,\ell\,\bar{\nu}_\ell(\gamma)$ and $\pi^- \to \pi^0\,\ell\,\bar{\nu}_\ell(\gamma)$. Since no other decay channels are open, the semileptonic decay rate can be accessed through lifetime measurements. All three methods involve the hadronic form factor of the vector current:

$$
\langle f | \bar{u}\gamma_\mu d | i \rangle,
$$

where $(i, f) = (0^+, 0^+), (n, p)$ or $(\pi^\pm, \pi^0)$. The neutron β decay further involves the form factor of the axial vector current:

$$
\langle f | \bar{u}\gamma_\mu \gamma_5 d | i \rangle
$$

The form factors parameterize the long-distance QCD effects, which bind the quarks into hadrons. The normalization of the vector current is fixed at the kinematic point of zero momentum transfer $p_i - p_f$ in the limit $m_u = m_d$ of exact isospin symmetry. The Ademollo-Gatto theorem[9] assures that corrections are of second order in the symmetry breaking parameter $(m_d - m_u)/\Lambda_{had}$, where Λ_{had} is the relevant hadronic scale. No such theorem protects the axial form factor $\langle p | \bar{u}\gamma_\mu\gamma_5 d | n \rangle$, but the corresponding parameter G_A can be extracted from asymmetries in the Dalitz plot. Experimentally the highest precision in the determination of V_{ud} is achieved in the nuclear β decay, but $n \to p \ell \bar{\nu}_\ell(\gamma)$ starts to become competitive. However, there is currently a disturbing discrepancy in the measurement of the neutron lifetime among different experiments.[8] From a theoretical point of view progress in $n \to p \ell \bar{\nu}_\ell(\gamma)$ and, ultimately, in the pristine $\pi^- \to \pi^0 \ell \bar{\nu}_\ell(\gamma)$ decay are highly desirable to avoid the nuclear effects of $0^+ \to 0^+$ transitions. On the theory side QED radiative corrections must be included to match the experimental accuracy, recently even dominant two-loop corrections to $n \to p \ell \bar{\nu}_\ell(\gamma)$ have been calculated.[10]

The world average for V_{ud} reads[11]:

$$V_{ud} = 0.9738 \pm 0.0005 \qquad (5)$$

2.2 V_{us}

V_{us} can be determined from Kaon and τ decays. The most established method uses the so-called $K\ell3$ decays $K^0 \to \pi^-\ell^+\nu_\ell$, $K^0 \to \pi^-\mu^+\nu_\ell$, $K^+ \to \pi^0\ell^+\nu_\ell$ and $K^+ \to \pi^0\mu^+\nu_\ell$. The decay rates schematically read

$$\Gamma(K \to \pi\ell^+\nu_\ell) \propto$$
$$V_{us}^2 \left| f_+^{K^0\pi^-}(0) \right|^2 \left[1 + 2\Delta_{SU(2)}^K + 2\Delta_{em}^{K\ell} \right].$$

The hadronic physics is contained in

$$\langle \pi^-(p_\pi) | \bar{s}\gamma^\mu u | K^0(p_K) \rangle =$$
$$f_+^{K^0\pi^-}(0)(p_K^\mu + p_\pi^\mu) + \mathcal{O}(p_K - p_\pi)$$
$$\Delta_{SU(2)}^{K^+} = \frac{f_+^{K^+\pi^0}(0)}{f_+^{K^0\pi^-}(0)} - 1, \quad \Delta_{SU(2)}^{K^0} = 0.$$

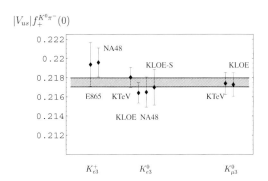

$|V_{us}| f_+^{K^0\pi^-}(0)$

Figure 3. $V_{us} f_+^{K^0\pi^-}(0)$. The horizontal band is the range quoted in Eq. (6). Courtesy of Vincenzo Cirigliano.[11]

and QED corrections are contained in $\Delta_{em}^{K\ell}$. The Ademollo-Gatto theorem[9] ensures $f_+^{K^0\pi^-}(0) = 1 + \mathcal{O}((m_s - m_d)^2/\Lambda_{had}^2)$. $f_+^{K^0\pi^-}(0) - 1$ can be calculated with the help of Chiral Perturbation Theory (χPT)[12], which exploits the fact that the pseudoscalar mesons are Goldstone bosons of a dynamically broken chiral symmetry of QCD. χPT amounts to a systematic expansion in p/Λ_{had}, M/Λ_{had}, m_ℓ/Λ_{had} and the electroweak coupling e. Here p and M denote meson momenta and masses and m_ℓ is the lepton mass. There has been a substantial progress in the calculation of both $\Delta_{em}^{K\ell}$ [13] and $f_+^{K^0\pi^-}(0)$ [14] since $LP'03$. Significant effects of $\mathcal{O}(e^2p^2)$ QED corrections on differential distributions were found; they must be included in Monte Carlo simulations. The value for $V_{us} f_+^{K^0\pi^-}(0)$ extracted from various experiments is shown in Fig. 3. The world average reads:[11]

$$f_+^{K^0\pi^-} V_{us} = 0.2175 \pm 0.0008. \qquad (6)$$

Combining the results from χPT at order p^6 and quenched lattice gauge theory (new) to[14]

$$f_+^{K^0\pi^-} = 0.972 \pm 0.012$$

one arrives at

$$V_{us} = 0.2238 \pm 0.0029 \qquad (7)$$

from $K\ell3$.

V_{us} can also be determined from the $K\mu2$ decay $K^+ \to \mu^+\nu_\mu(\gamma)$.[15] The hadronic quantity entering this decay is the Kaon decay constant F_K. Uncertainties can be better controlled in the ratio F_K/F_π and one considers

$$\frac{\Gamma(K^+ \to \mu^+\nu_\mu(\gamma))}{\Gamma(\pi^+ \to \mu^+\nu_\mu(\gamma))} =$$

$$\frac{V_{us}^2}{V_{ud}^2}\frac{F_K^2}{F_\pi^2}\frac{M_K^2 - m_\mu^2}{M_\pi^2 - m_\mu^2}\left[1 - \frac{\alpha}{\pi}(C_\pi - C_K)\right]$$

with QED corrections $C_\pi - C_K = 3.0 \pm 1.5$. Using the result[16] $F_K/F_\pi = 1.210 \pm 0.004 \pm 0.013$ computed by the MILC collaboration with 2+1 dynamical quarks, one finds (with V_{ud} from Eq. (5)):

$$V_{us} = 0.2223 \pm 0.0026 \qquad (8)$$

from the $K\mu2$ decay. This is astonishingly precise and $K^+ \to \mu^+\nu_\mu(\gamma)$ may constrain mass and couplings of a charged Higgs boson, which can mediate this decay as well.[15]

The third possibility to measure V_{us} used hadronic τ decays to the inclusive final state with strangeness $|S| = 1$. The experimental inputs are the ratios

$$R_{\tau s} = \frac{\Gamma^{\Delta S=1}(\tau \to \text{hadrons } \nu_\tau(\gamma))}{\Gamma(\tau \to e\bar\nu_e\nu_\tau(\gamma))} \propto V_{us}^2$$

$$R_{\tau d} = \frac{\Gamma^{\Delta S=0}(\tau \to \text{hadrons } \nu_\tau(\gamma))}{\Gamma(\tau \to e\bar\nu_e\nu_\tau(\gamma))} \propto V_{ud}^2$$

Here S is the strangeness. The optical theorem allows to relate $R_{\tau s,d}$ to the QCD current-current correlators $\Pi_{s,d}^T$ and $\Pi_{s,d}^L$:

$$R_{\tau s,d} = 12\pi \int_0^1 dz(1-z)^2 \times$$

$$\left[(1 + 2z)\,\text{Im}\,\Pi_{s,d}^T(z) + \text{Im}\,\Pi_{s,d}^L(z)\right]$$

with $z = s/M_\tau^2 = (p_\tau - p_{\nu_\tau})^2/M_\tau^2$. This relationship is depicted in Fig. 4. $\Pi_{s,d}^{T,L}$ can be computed through an operator product expansion (OPE). The leading term is massless perturbative QCD, subleading operators entering $\Pi_s^{T,L}$ are m_s^2 and $m_s\langle\bar q q\rangle$. The OPE amounts to an expansion in Λ_{QCD}/M_τ, m_s/M_τ and $\alpha_s(m_\tau)$. In the limit $m_s = 0$ of

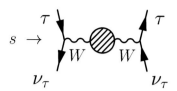

Figure 4. The optical theorem relates $\Gamma^{\Delta S=0,1}(\tau \to$ hadrons $\nu_\tau(\gamma))$ to $\Pi_{s,d}^{T,L}$. The blob denotes the hadronic states contributing to $\Pi_{s,d}^{T,L}$. The leading term in the OPE is obtained by replacing the blob by a (u,d) or (u,s) quark loop and gluons to the desired order in α_s.

exact $SU(3)_F$ symmetry the ratio $R_{\tau s}/R_{\tau d}$ would directly determine V_{us}^2/V_{ud}^2. Hence it suffices to compute the (small) $SU(3)_F$ breaking quantity[17]

$$\delta R_\tau \equiv \frac{R_{\tau d}}{V_{ud}^2} - \frac{R_{\tau s}}{V_{us}^2}.$$

With $\delta R_\tau = 0.218 \pm 0.026$[18] and experimental data from OPAL[19] one finds $R_{\tau d} = 3.469 \pm 0.014$, $R_{\tau s} = 0.1694 \pm 0.0049$ and finally:[18]

$$V_{us} = \sqrt{\frac{R_{\tau s}}{R_{\tau d}/|V_{ud}|^2 - \delta R_\tau}}$$

$$= 0.2219 \pm 0.0033_{\text{exp}} \pm 0.0009_{\text{th}}$$

$$= 0.2219 \pm 0.0034. \qquad (9)$$

The dominant source of uncertainty in δR_τ, which enters Eq. (9) as a small correction, is from m_s. In the near future it should be possible to improve on V_{us} with data from BaBar and BELLE.

In summary one finds an excellent consistency of the three numbers for V_{us} from $K\ell3$, $K\mu2$ and τ decays. This is remarkable, since the three methods use very different theoretical tools to address the strong interaction: Chiral perturbation theory, lattice gauge theory and the operator product expansion. The result nicely reflects the tremendous progress of our understanding of QCD at low energies. Averaging the results of Eqs. (7), (8) and (9) one finds:

$$V_{us} = 0.2227 \pm 0.0017 \qquad (10)$$

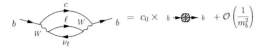

Figure 5. OPE for $\overline{B} \to X_c \ell \overline{\nu}_\ell$. The leading operator $\overline{b}b$ has dimension 3.

With V_{ud} in Eq. (5) one can perform the first-row unitarity check

$$V_{us}^2 + V_{ud}^2 + |V_{ub}|^2 - 1 \simeq V_{us}^2 + V_{ud}^2 - 1$$
$$= -0.0021 \pm 0.0012$$

The Cabibbo matrix is unitary at the 1.8σ level, just as at $LP'03$:[20]

$$V_{us}^2 + V_{ud}^2 - 1 = -0.0031 \pm 0.0017$$

2.3 V_{cb}

V_{cb} can be determined from inclusive or exclusive $b \to c\ell\nu_\ell$ decays. Exclusive decays are not discussed here. The analysis of the inclusive decay employs an OPE[21], similarly to the determination of V_{us} from τ decay discussed in Sect. 2.2. The optical theorem relates the inclusive decay rate $\overline{B} \to X_c \ell \overline{\nu}_\ell$ to the imaginary part of the B meson self energy, depicted on the LHS of Fig. 5. The OPE matches the self energy diagram to matrix element of effective operators, whose coefficients contain the short-distance information associated with the scale m_b and can be calculated perturbatively. Increasing dimensions of the operators on the RHS of Fig. 5 correspond to decreasing powers of m_b in the coefficient functions, so that the OPE amounts to a simultaneous expansion in Λ_{QCD}/m_b and $\alpha_s(m_b)$. Since $\langle B | \overline{b}b | B \rangle = 1 + \mathcal{O}(\Lambda_{QCD}^2/m_b^2)$ and there are no dimension-4 operators, non-perturbative parameters first occur at order Λ_{QCD}^2/m_b^2. They are

$$\mu_\pi^2 \propto -\langle B | \overline{b} D_\perp^2 b | B \rangle$$
$$\mu_G^2 \propto \langle B | \overline{b} i \sigma_{\mu\nu} G^{\mu\nu} b | B \rangle$$

μ_G^2, which parameterizes the matrix element of the chromomagnetic operator, can be determined from spectroscopy. Hence to order

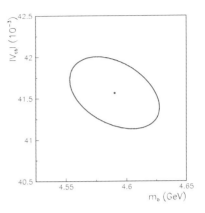

Figure 6. Fit result for V_{cb} vs. m_b, which is defined in the kinetic scheme.[22] Fit and plot are courtesy of Oliver Buchmüller and Henning Flächer. See also[23].

Λ_{QCD}^2/m_b^2 one only has to deal with the three quantities m_b, m_c and μ_π^2, which quantifies the Fermi motion of the b quark inside the B meson.

The OPE can further be applied to certain spectral moments of the $B \to X\ell\nu_\ell$ decay, the distributions of the hadron invariant mass M_X and of the lepton energy. Further the same parameters govern different inclusive decays, for instance also $B \to X_s\gamma$. Therefore there is a lot of redundancy in the determination of V_{cb}, providing powerful checks of the theoretical framework. The state of the art are fits to order Λ_{QCD}^3/m_b^3, which involve 7 parameters.[22] The result of a global fit to hadron and lepton moments in $B \to X\ell\nu_\ell$ and photon energy moments in $B \to X_s\gamma$ from BaBar, BELLE, CDF, CLEO, DELPHI[23] can be seen in Fig. 6. It gives

$$V_{cb} = 41.6 \pm 0.3_{\text{exp}} \pm 0.3_{\text{OPE moments}}$$
$$\pm 0.3_{\text{OPE } \Gamma_{\text{sl}}}$$
$$= (41.6 \pm 0.5) \cdot 10^{-3} \qquad (11)$$

from inclusive $B \to X\ell\nu_\ell$.

2.4 $|V_{ub}|$

I discuss the determination of $|V_{ub}|$ from inclusive $B \to X_u \ell \nu_\ell$ decays. Exclusive decays are discussed in.[7] In principle one could determine $|V_{ub}|$ in the same way as V_{cb}, if there were no background from $B \to X_c \ell \nu_\ell$ decays. Its suppression forces us to impose cuts on the lepton energy E_ℓ, the hadronic energy E_X, the hadron invariant mass M_X or a judiciously combination of them. M_X is too small for an OPE in the portion of phase space passing these cuts. Still some components of the hadron momentum $\vec{P}_X$ are large. The description of inclusive B decays in this region involves the non-perturbative shape function S, which is a parton distribution function of the B meson. At leading order in $1/m_b$ the same S governs the photon spectrum in $B \to X_s \gamma$ and differential decay rates in $B \to X_u \ell \overline{\nu}_\ell$. This allows us to extract S from $B \to X_s \gamma$ for the use in $B \to X_u \ell \overline{\nu}_\ell$. The goal to reduce the theoretical uncertainty below 10% requires to understand corrections in both expansion parameters α_s and Λ_{QCD}/m_b. For a correct treatment of radiative QCD corrections one must properly relate the differential decay rate $d\Gamma$ to the shape function S. This is achieved by a factorization formula, which has the schematic form:[24]

$$d\Gamma \propto H \int_0^{P_+} d\omega \, J\left(m_b(P_+ - \omega)\right) S(\omega)$$

Here H contains the hard QCD, associated with scales of order m_b. The jet function J and the shape function S contain the physics from scales of orders $M_X \sim \sqrt{m_b \Lambda_{QCD}}$ and Λ_{QCD}, respectively. P_+ and P_- are defined as $P_\pm = E_X \mp |\vec{P}_X|$. From $\Lambda_{QCD} \ll P_+ \sim M_X \sim \sqrt{m_b \Lambda_{QCD}} \ll P_- \leq m_b$ one realizes that one has to deal with a multi-scale problem, which is more complicated than $B \to X_c \ell \nu_\ell$. The second frontier of research in $B \to X_u \ell \nu_\ell$ deals with subleading shape functions s_i, which occur at order $1/m_b$. They are different in $B \to X_u \ell \nu_\ell$ and

$B \to X_s \gamma$, but their moments can be related to OPE parameters like μ_π^2, which gives some guidance to model these functions.[25] Meanwhile an event generator for $B \to X_u \ell \overline{\nu}_\ell$ decays is available,[26] with formulae which contain all available theoretical information and smoothly interpolate between the shape function and OPE regions. It is pointed out that a cut on the variable P_+, which is directly related to the photon energy in $B \to X_s \gamma$, makes the most efficient use of the $S(\omega)$ extracted from the radiative decay.[27,26] Alternatively one can eliminate $S(\omega)$ altogether by forming proper weighted ratios of the endpoint photon and lepton spectra in $B \to X_s \gamma$ and $B \to X_u \ell \nu_\ell$, respectively.[28] Using also the information from $B \to X_c \ell \nu_\ell$ on m_b and μ_π^2 the data from CLEO[29], BELLE[30] and BaBar[31] combine to the world average[4]

$$V_{ub} = (4.39 \pm 0.20_{\text{exp}} \pm 0.27_{\text{th},m_b,\mu_\pi^2}) \cdot 10^{-3}$$
$$= (4.39 \pm 0.34) \cdot 10^{-3} \qquad (12)$$

from inclusive $B \to X_u \ell \nu_\ell$. Eq. (3) implies that $|V_{ub}/V_{cb}|$ defines a circle in the $(\overline{\rho}, \overline{\eta})$ plane which is centered around $(0,0)$. With Eqs. (11) and (12) its radius is constrained to

$$R_u \equiv \sqrt{\overline{\rho}^2 + \overline{\eta}^2} = 0.45 \pm 0.04. \qquad (13)$$

2.5 $\arg V_{ub}$

$\gamma = \arg V_{ub}^*$ can be determined from exclusive $B \to \overset{(-)}{D^0} X$ decays, where X denotes one or several charmless mesons. This method exploits the interference of the tree-level $b \to c\overline{u}q$ and $b \to u\overline{c}q$ amplitudes, where $q = d$ or $q = s$. The prototype is the Gronau-London-Wyler (GLW) method[32] shown in Fig. 7. The decays $B \to D^0 X$ and $B \to \overline{D^0} X$ interfere, if both subsequent decays $D^0 \to f$ and $\overline{D^0} \to f$ are allowed. One needs four measurements to solve for the magnitudes of the $b \to c$ and $b \to u$, their relative strong phase and their relative weak phase, which is the desired UT angle γ. For example one can combine the information of the branching fractions of $B^+ \to \overset{(-)}{D^0}[\to K^\pm \pi^\mp]K^+$ and $B^\pm \to$

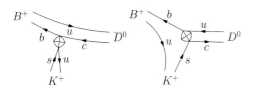

Figure 7. The Gronau-London-Wyler method combines the rates of $B^\pm \to \overset{(-)}{D^0}[\to f_i]K^\pm$ for different final states f_i.

$\overset{(-)}{D^0}[\to \pi^+\pi^-]K^\pm$. This works with untagged non-flavour-specific decays as well:[33] E.g. the final state $\overset{(-)}{D^0}\phi$ does not reveal whether the decaying meson was a B_s or $\overline{B}_s$. Still, when at least three pairs of $\overset{(-)}{B}_s \to \overset{(-)}{D^0}[\to f_i]\phi$ and $\overset{(-)}{B}_s \to \overset{(-)}{D^0}[\to \overline{f}_i]\phi$ branching fractions are measured, where $\overline{f}_i = CP f_i$ (and the f_i's are not CP eigenstates), one has enough information to solve for γ. Since no flavour tagging is involved, the Tevatron experiments may contribute to these class of γ determinations. The described determination of γ from tree-tree interference is modular, that is measurements in different decay modes can be combined, as they partly involve the same hadronic parameters. One should further first average the branching ratios from different experiments and then determine γ instead of averaging the inferred values of γ obtained from different experiments. Combining (almost) all $B^+ \to D^0 K^{+(*)}$ data gives (preliminary)[5]

$$\gamma = (70^{+12}_{-14})^\circ \qquad (14)$$

and the second solution $\gamma - 180^\circ \sim -110^\circ$. This is $\gamma = \arg V_{ub}^*$ determined from the tree-level $b \to u\bar{c}s$ amplitude.

Within the Standard Model $b \to u\bar{u}d$ decays of tagged B^0 mesons are used to determine the UT angle α. $b \to u\bar{u}d$ decays involve both a tree and a penguin amplitude. The penguin component can be eliminated, if several decay modes related by isospin are combined, as in the Gronau-London method[34]

which uses $B^+ \to \pi^+\pi^0$, $B^0 \to \pi^+\pi^-$ and $B^0 \to \pi^0\pi^0$. The $B \to \rho\pi$ and $B \to \rho\rho$ decay modes are better suited for the determination of α, because the penguin amplitude is smaller. A combined analysis of the $\pi\pi$, $\rho\pi$ and $\rho\rho$ systems gives

$$\alpha_{\exp} = (99^{+12}_{-9})^\circ \qquad (15)$$

and the second solution $\alpha_{\exp} - 180^\circ \sim -81^\circ$. The experimental result $\alpha_{\exp}$ could differ from the true $\alpha = \arg(-V_{tb}^* V_{td}/(V_{ub}^* V_{ud}))$, if new physics alters the $B_d - \overline{B}_d$ mixing amplitude. However, the influence from new physics is fully correlated in $\alpha_{\exp}$ and the CP asymmetry measured in $b \to c\bar{c}s$ decays. From the latter (see Eq. (21) below) we infer the $B_d - \overline{B}_d$ mixing phase $2\beta_{\exp} = (43.7 \pm 2.4)^\circ$. The $B_d - \overline{B}_d$ mixing phase cancels from the combination $2\gamma = 360^\circ - 2\alpha_{\exp} - 2\beta_{\exp}$, so that one obtains

$$\gamma = (59^{+9}_{-12})^\circ \qquad (16)$$

and the second solution $\gamma - 180^\circ \sim -121^\circ$. Since the isospin analysis eliminates the penguin component, this is $\gamma = \arg V_{ub}^*$ determined from the tree-level $b \to u\bar{u}d$ amplitude.

The results in Eqs. (14) and (16) are in good agreement. Their naive average is

$$\gamma = (63^{+7}_{-9})^\circ. \qquad (17)$$

The successful determination of a CP phase from a tree-level amplitude is a true novel result compared to $LP'03$. For the first time we can determine the UT from tree-level quantities alone, the result is shown in Fig. 8. This is important, because the tree-level UT can only be mildly affected by new physics and therefore likely determines the true values of $\overline{\rho}$ and $\overline{\eta}$.

3 CKM elements from FCNC processes

In the Standard Model FCNC processes are suppressed by several effects: First they only proceed through electroweak loops. Second they come with small CKM factors like

148

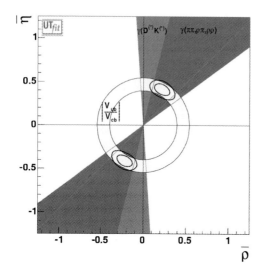

Figure 8. UT from tree quantities alone. The annulus is the constraint in Eq. (13) derived from $|V_{ub}|$. The dark shadings correspond to γ from Eqs. (14) and (16). Courtesy of Maurizio Pierini.

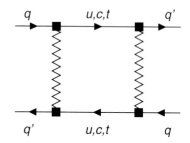

Figure 9. Meson-antimeson mixing. $(q, q') = (s, d)$, (b, d) and (b, s) for K–$\overline{\text{K}}$ mixing, B_d–$\overline{\text{B}}_\text{d}$ mixing and B_s–$\overline{\text{B}}_\text{s}$ mixing, respectively.

3.1 Meson-antimeson mixing

K–$\overline{\text{K}}$ mixing, B_d–$\overline{\text{B}}_\text{d}$ mixing and B_s–$\overline{\text{B}}_\text{s}$ mixing are all induced by box diagrams, depicted in Fig. 9. Each meson-antimeson system involves two mass eigenstates, their mass difference Δm measures the magnitude of the box diagram and therefore constrains magnitudes of CKM elements. The phase of box diagram and thereby the phases of the CKM elements involved are constrained through CP-violating quantities. Tab. 1 shows the relationship of the measurements to the CKM phenomenology. The quantities in the first two columns of Tab. 1 are well-measured and there is a lower bound on Δm_{B_s}.

3.2 ϵ_K

While ϵ_K, which quantifies indirect CP violation in $K \to \pi\pi$ decays, is measured at the percent level, its relationship to $\text{Im} V_{td}^{*2} \propto \overline{\eta}(1 - \overline{\rho})$ is clouded by hadronic uncertainties in the matrix element

$$\langle K^0 | \overline{d}s_{V-A}\overline{d}s_{V-A} | \overline{K}^0 \rangle \equiv \frac{8}{3} f_K^2 M_K^2 B_K.$$

This defines the hadronic parameter B_K, which must be computed by non-perturbative methods like lattice QCD. M_K and f_K are the well-known mass and decay constant of the Kaon. This field has experienced a major breakthrough since *LP'03*, since meanwhile fully unquenched computations with 2+1 dynamical staggered quarks are available. Us-

$|V_{ts}| \sim 0.04$ and $|V_{td}| \sim 0.01$. Loops with an internal charm quark are further suppressed by a factor of m_q^2/M_W^2 from the GIM mechanism. Radiative and leptonic decays further suffer from an additional helicity suppression, because only left-handed quarks couple to W bosons and undergo FCNC transitions. All these suppression mechanism are accidental, resulting from the particle content of the Standard Model and the unexplained smallness of most Yukawa couplings. They are absent in generic extensions of the Standard Model (like its supersymmetric generalizations) making FCNC highly sensitive to new physics, probing scales in the range of $200\,\text{GeV}$ to $100\,\text{TeV}$, depending on the model considered. This feature is a major motivation for the currently performed high-statistics experiments in flavour physics. Comparing different constraints on the UT from FCNCs processes and the tree-level constraints discussed in Sect. 2 therefore provides a very powerful test of the Standard Model.

	K−K̄ mixing	B_d−B̄_d mixing	B_s−B̄_s mixing
CP-conserving quantity:	Δm_K	Δm_{B_d}	Δm_{B_s}
CKM information:	$|V_{cs}V_{cd}|^2$	$|V_{tb}V_{td}|^2$	$|V_{tb}V_{ts}|^2$
UT constraint:	none	$R_t = \sqrt{(1-\overline{\rho})^2 + \overline{\eta}^2}$	none
CP-violating quantity:	ϵ_K	$a_{\rm CP}^{\rm mix}(B_d \to J/\psi K_S)$	$a_{\rm CP}^{\rm mix}(B_s \to J/\psi\phi)$
CKM information:	$\mathrm{Im}\,(V_{ts}V_{td}^*)^2$	$\sin(2\beta)$	$\sin(2\beta_s)$
UT constraint:	$\overline{\eta}[(1-\overline{\rho}) + \mathrm{const.}]$	$\dfrac{\overline{\eta}}{1-\overline{\rho}}$	$\overline{\eta}$

Table 1. Relationship of meson-antimeson mixing to CKM and UT parameters. $\beta = \arg V_{td}^*$ is one of the UT angles in Fig. 1 and $\beta_s = \arg(-V_{ts}) \simeq \lambda^2\overline{\eta}$.

ing MILC configurations the HPQCD collaboration reports a new result[35b]

$$B_K(\mu = 2\,\mathrm{GeV}) =$$
$$0.618 \pm 0.018_{\rm stat} \pm 0.019_{\rm chiral\ extrapolation}$$
$$\pm 0.030_{\rm discret.} \pm 0.130_{\rm pert.\ matching}$$
$$= 0.618 \pm 0.136 \tag{18}$$

in the $\overline{\rm MS}$–NDR scheme. The conventionally used renormalization scale and scheme independent parameter reads

$$\widehat{B}_K = 0.83 \pm 0.18 \tag{19}$$

The uncertainty from the perturbative lattice–continuum matching dominates over the statistical error and the errors from chiral extrapolation and discretization in Eq. (18). This matching calculation was performed in[36]. The error in Eq. (18) is a conservative estimate of the unknown two-loop contributions to this matching. If one instead takes twice the square of the one-loop result of[36] as an estimate of the uncertainty, one finds 0.036 instead of 0.130 in Eq. (18) and

$$B_K(\mu = 2\,\mathrm{GeV}) = 0.618 \pm 0.054$$
$$\widehat{B}_K = 0.83 \pm 0.07 \tag{20}$$

ϵ_K fixes $\overline{\eta}(1-\overline{\rho})$, so that it defines a hyperbola in the $(\overline{\rho},\overline{\eta})$ plane.

---

[b]In my talk I reported the preliminary value $B_K = 0.630 \pm 0.018_{\rm stat} \pm 0.015_{\rm ch.\ extr.} \pm 0.030_{\rm disc.} \pm 0.130_{\rm p.\ match.}$.

3.3 V_{td} from B_d−B̄_d mixing

The $\mathrm{B_d} - \overline{\mathrm{B}}_\mathrm{d}$ mixing mixing amplitude involves the hadronic matrix element

$$\langle B^0 |\overline{b}d_{V-A}\overline{b}d_{V-A}| \overline{B}^0 \rangle = \frac{8}{3} M_{B_d}^2 f_{B_d}^2 B_{B_d}.$$

Since the decay constant f_{B_d} is not measured, the whole combination $f_{B_d}^2 B_{B_d}$ must be obtained from lattice QCD. The hadronic matrix element, however, cancels from the "gold-plated" mixing induced CP asymmetry $a_{\rm CP}^{\rm mix}(B_d \to J/\psi K_S)$, which determines $\beta = \arg V_{td}^*$ essentially without hadronic uncertainties. Combining all data from $b \to c\overline{c}s$ modes results in[37,4]

$$\sin(2\beta) = 0.69 \pm 0.03, \quad \cos(2\beta) > 0$$
$$\Rightarrow \quad \arg(\pm V_{td}^*) = \beta = (21.8 \pm 1.2)^\circ. \tag{21}$$

The precisely measured $\Delta m_{B_d} = 0.509 \pm 0.004\,\mathrm{ps}^{-1}$ is proportional to $|V_{td}|^2 f_{B_d}^2 B_{B_d}$. The HPQCD collaboration has computed $f_{B_d} = 216 \pm 22\,\mathrm{MeV}$ with 2+1 dynamical staggered quarks.[38] This measurement is discussed in detail in[7]. Combining this with B_{B_d} from older quenched calculations results in $f_{B_d}\sqrt{\widehat{B}_{B_d}} = (246 \pm 27)\,\mathrm{MeV}$, where $\widehat{B}_{B_d} = 1.52 B_{B_d}(\mu = m_b)$ is the conventionally used scale and scheme independent variant of B_{B_d}. Then from Δm_{B_d} alone we find

$$|V_{td}| = 0.0072 \pm 0.0008,$$

where the error is reduced by a factor of 2/3 compared to the old determination from quenched lattice QCD.

3.4 $|V_{td}|/|V_{ts}|$ from $B-\bar{B}$ mixing

A measurement of the ratio $\Delta m_{B_d}/\Delta m_{B_s}$ will determine $|V_{td}|/|V_{ts}|$ via

$$\left|\frac{V_{td}}{V_{ts}}\right| = \sqrt{\frac{\Delta m_{B_d}}{\Delta m_{B_s}}}\sqrt{\frac{M_{B_s}}{M_{B_d}}}\,\xi$$

with the hadronic quantity $\xi = f_{B_s}\sqrt{\widehat{B}_{B_s}}/(f_{B_d}\sqrt{\widehat{B}_{B_d}})$ which equals $\xi = 1$ in the limit of exact SU(3)$_{\rm F}$. A new unquenched HPQCD result for f_{B_s}/f_{B_d}[38] presented in[7] can be used to refine the prediction for ξ. The lower bound $\Delta m_{B_s} \geq 14.5\,{\rm ps}^{-1}$ implies $|V_{td}/V_{ts}| \leq 0.235$ which constrains one side of the unitarity triangle:

$$R_t \equiv \sqrt{(1-\bar{\rho})^2 + \bar{\eta}^2} = \left|\frac{V_{td}}{V_{ts}\lambda}\right| \leq 1.06$$

3.5 Global fit to the unitarity triangle

The result of a global fit of $(\bar{\rho}, \bar{\eta})$ to state-of-the-art summer-2005 data is shown in Fig. 10. It uses $\widehat{B}_K = 0.85 \pm 0.02 \pm 0.07$ where the first error is Gaussian and the second is scanned over according to the standard CKMfitter method[5]. For the remaining input see[5]. The fit output is summarized in this table:

quantity	central $\pm$ CL $\equiv 1\sigma$	$\pm$ CL $\equiv 2\sigma$
$\bar{\rho}$	$0.204\,^{+0.035}_{-0.033}$	$^{+0.095}_{-0.069}$
$\bar{\eta}$	$0.336\,^{+0.021}_{-0.021}$	$^{+0.045}_{-0.060}$
α (deg)	$98.4\,^{+6.1}_{-5.6}$	$^{+16.8}_{-11.8}$
β (deg)	$22.77\,^{+0.87}_{-0.83}$	$^{+1.92}_{-2.04}$
γ (deg)	$58.8\,^{+5.3}_{-5.8}$	$^{+11.2}_{-15.4}$
$|V_{ub}|$ $[10^{-3}]$	$3.90\,^{+0.12}_{-0.12}$	$^{+0.29}_{-0.24}$
$|V_{td}|$ $[10^{-3}]$	$8.38\,^{+0.32}_{-0.44}$	$^{+0.56}_{-1.29}$

The output of the global fit agrees well with the pure tree-level determinations in Eqs. (12) and (17) and Fig. 8.

We can use Eq. (1) to determine $\theta_{13} = 0.223° \pm 0.007°$ from the fitted $|V_{ub}|$ in the table. With Eq. (1) one finds $\theta_{12} = 12.9° \pm 0.1°$ from Eq. (10) and $\theta_{23} = 2.38° \pm 0.03°$ from Eq. (11). Since $1 - \cos\theta_{13}$ is negligibly small, the Wolfenstein parameters λ and $A\lambda^2$ defined in Eq. (4) are simply given by V_{us} in Eq. (10) and V_{cb} in Eq. (11), respectively.

3.6 $K \to \pi\nu\bar{\nu}$

The rare decays $K^+ \to \pi^+\nu\bar{\nu}$ and $K_L \to \pi^0\nu\bar{\nu}$ provide an excellent opportunity to determine the unitarity triangle from $s \to d$ transitions. With planned dedicated experiments $(\bar{\rho}, \bar{\eta})$ can be determined with a similar precision as today from $b \to d$ and $b \to u$ transitions at the B factories. This is a unique and very powerful probe of the CKM picture of FCNCs. $Br(K_L \to \pi^0\nu\bar{\nu})$ is proportional to $\bar{\eta}^2$ and dominated by the top contribution. The theoretical uncertainty of the next-to-leading order (NLO) prediction[40] is below 2%. $Br(K^+ \to \pi^+\nu\bar{\nu})$ defines an ellipse in the $(\bar{\rho}, \bar{\eta})$ plane and has a sizeable charm contribution, which inflicts a larger theoretical uncertainty on the next-to-leading order (NLO) prediction[41], leading to $\mathcal{O}(5-10\%)$ uncertainties in extracted CKM parameters. Parametric uncertainties from V_{cb} and m_t largely drop out, if $\sin(2\beta)$ is calculated from $Br(K_L \to \pi^0\nu\bar{\nu})$ and $Br(K^+ \to \pi^+\nu\bar{\nu})$. Therefore the comparison of $\sin(2\beta)$ determined in Eq. (21) from the B system with $\sin(2\beta)$ inferred from $K \to \pi\nu\bar{\nu}$ constitutes a pristine test of the Standard Model.[42]

The charm contribution is expanded in two parameters: m_K^2/m_c^2 and $\alpha_s(m_c)$. The calculations of $\mathcal{O}(m_K^2/m_c^2)$ corrections was recently completed, finding a 7% increase of $Br(K^+ \to \pi^+\nu\bar{\nu})$ with a small residual uncertainty.[43] A new result are the next-to-next-to-leading order (NNLO) QCD corrections to the charm contribution.[44] This three-loop calculation reduces the theoretical error from unknown higher-order terms well below the parametric uncertainty from m_c. The branching ratio is now predicted as

$$Br(K^+ \to \pi^+\bar{\nu}\nu) = (8.0 \pm 1.1) \cdot 10^{-11}.$$

At NNLO one finds the following reduced theoretical uncertainties for parameters ex-

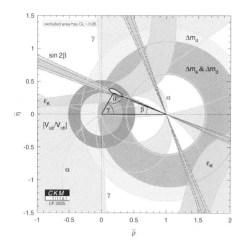
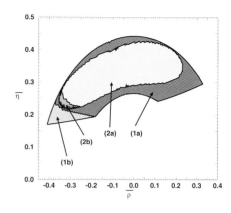

Figure 10. LHS: The UT from a global fit to summer 2005 data. RHS: The first UT fit using theoretical expressions with NLO QCD corrections, performed in 1995.[39] At that time only $|V_{ub}|$, ϵ_K and Δm_{B_d} could be used. Region 1a corresponds to a scan over 1σ ranges of the input parameters.

tracted from $Br(K_L \to \pi^0 \nu \bar{\nu})$ and $Br(K^+ \to \pi^+ \nu \bar{\nu})$:[44]

$$\frac{\delta|V_{td}|}{|V_{td}|} = 0.010, \quad \delta \sin(2\beta) = 0.006, \quad \delta\gamma = 1.2°$$

4 CP violation in $b \to s$ penguin decays

Within the Standard Model the mixing-induced CP asymmetries in $b \to s\bar{q}q$ penguin amplitudes are proportional to $\sin(2\beta)^{\text{eff}}$ which equals $\sin(2\beta)$ in Eq. (21) up to small corrections from a penguin loop with an up quark. In $b \to s\bar{u}u$ decays there is also a color-suppressed tree amplitude. In any case the corrections are parametrically suppressed by $|V_{ub}V_{us}/(V_{cb}V_{cs})| \sim 0.025$. The experimental situation is shown in Fig. 11. A naive average of the measurements of Fig. 11 gives

$$\sin(2\beta)^{\text{eff}} = 0.51 \pm 0.06,$$

which is below the value of $\sin(2\beta)$ from tree-level $b \to \bar{c}cs$ decays in Eq. (21) by 3σ. Moreover QCD factorization finds a small and positive correction to $\sin(2\beta_{\text{eff}})$–$\sin(2\beta)$ from up-quark effects.[45] While the significance of the deviation has decreased since the winter 2005 conferences, the mixing-induced CP asymme-

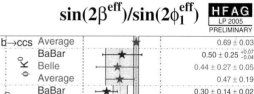
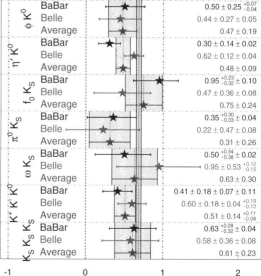

Figure 11. $\sin(2\beta)^{\text{eff}}$ from various penguin decays. The small vertical yellow band is $\sin(2\beta)$ from Eq. (21).[4]

tries in $b \rightarrow s\bar{q}q$ decays stay interesting as they permit large effects from new physics. While in B_d decays the needed interference of a B_d and $\bar{B}_d$ decay to the same final state requires a neutral K meson in the final state, $b \rightarrow s\bar{q}q$ decays of B_s mesons go to a flavourless $\bar{s}s\bar{q}q$ state, so that the desired CP effects can be studied in any final state. Hence B_s physics has the potential to become the "El Dorado" of $b \rightarrow s\bar{q}q$ penguin physics.

Acknowledgments

This presentation was made possible through the input and help from Martin Beneke, Ed Blucher, Tom Browder, Oliver Buchmüller, Christine Davies, Henning Ulrik Flächer, Paolo Franzini, Tim Gershon, Martin Gorbahn, Ulrich Haisch, Christopher Hearty, Matthias Jamin, Bob Kowalewski, Heiko Lacker, Zoltan Ligeti, Antonio Limosani, Mike Luke, Matthias Neubert, Maurizio Pierini, Jim Smith, Iain Stewart, Stéphane T'Jampens, Nikolai Uraltsev and Matthew Wingate. I am especially grateful to Vincenzo Cirigliano and Björn Lange for their thorough and patient explanations of recent progress in the fields of chiral perturbation theory and B meson shape functions. Very special thanks go to Andreas Höcker and Heiko Lacker for their day-and-night work on the presented CKMfitter plots.

References

1. S. Eidelman *et al.* [Particle Data Group], *Phys. Lett.* B **592**, 1 (2004).
2. L. Wolfenstein, *Phys. Rev. Lett.* **51**, 1945 (1983).
3. A. J. Buras, M. E. Lautenbacher, G. Ostermaier, *Phys. Rev.* D **50**, 3433 (1994).
4. *Heavy Flavor Averaging Group*, http://www.slac.stanford.edu/xorg/hfag
5. A. Höcker, H. Lacker, S. Laplace and F. Le Diberder, *Eur. Phys. J.* C **21**, 225 (2001). J. Charles *et al.* [CKMfitter Group], *Eur. Phys. J.* C **41**, 1 (2005). http://www.slac.stanford.edu/xorg/ckmfitter

6. M. Ciuchini *et al.*, *JHEP* **0107**, 013 (2001). M. Bona *et al.* [UTfit Collaboration], hep-ph/0509219. http://utfit.roma1.infn.it
7. Iain Stewart, these proceedings.
8. John Hardy, talk at *KAON 2005 Int. Workshop*, Jul 13-17, 2005, Evanston, USA.
9. M. Ademollo and R. Gatto, *Phys. Rev. Lett.* **13**, 264 (1964).
10. A. Czarnecki, W. J. Marciano and A. Sirlin, *Phys. Rev.* D **70**, 093006 (2004).
11. Vincenzo Cirigliano, talk at *KAON 2005 Int. Workshop*, Jul 13-17, 2005, Evanston, USA, and private communication.
12. J. Gasser and H. Leutwyler, *Annals Phys.* **158**, 142 (1984).
13. V. Cirigliano, M. Knecht, H. Neufeld, H. Rupertsberger and P. Talavera, *Eur. Phys. J.* C **23**, 121 (2002). V. Cirigliano, H. Neufeld and H. Pichl, *Eur. Phys. J.* C **35**, 53 (2004). T. C. Andre, arXiv:hep-ph/0406006. S. Descotes-Genon and B. Moussallam, hep-ph/0505077.
14. H. Leutwyler and M. Roos, Z. *Phys.* C **25**, 91 (1984). P. Post and K. Schilcher, *Nucl. Phys.* B **599**, 30 (2001). J. Bijnens and P. Talavera, *Nucl. Phys.* B **669**, 341 (2003). M. Jamin, J. A. Oller and A. Pich, *JHEP* **0402**, 047 (2004). V. Cirigliano, *Int. J. Mod. Phys.* A **20**, 3732 (2005). $f_+^{K^0\pi^-}$ was computed in quenched lattice QCD in: D. Becirevic *et al.*, *Eur. Phys. J.* A **24S1**, 69 (2005). V. Lubicz *et al.*, presented at *DAFNE 2004: Workshop on Physics at Meson Factories, Rome, Frascati, Italy, 7-11 Jun 2004*
15. W. J. Marciano, *Phys. Rev. Lett.* **93**, 231803 (2004).
16. C. Aubin *et al.* [MILC Collaboration], *Nucl. Phys. Proc. Suppl.* **140**, 231 (2005).
17. A. Pich and J. Prades, *JHEP* **9806**, 013 (1998). E. Gamiz, M. Jamin, A. Pich, J. Prades and F. Schwab, *JHEP* **0301**, 060 (2003).
18. E. Gamiz, M. Jamin, A. Pich, J. Prades and F. Schwab, *Phys. Rev. Lett.* **94**, 011803 (2005).
19. G. Abbiendi *et al.* [OPAL Collaboration], *Eur. Phys. J.* C **35**, 437 (2004).
20. K. R. Schubert, *Int. J. Mod. Phys.* A **19**, 1004 (2004).
21. M. A. Shifman and M. B. Voloshin, *Sov. J. Nucl. Phys.* **41**, 120 (1985) [*Yad. Fiz.* **41**, 187 (1985)]. I. I. Y. Bigi, N. G. Uraltsev

and A. I. Vainshtein, *Phys. Lett.* B **293**, 430 (1992) [Erratum-ibid. B **297**, 477 (1993)].

22. P. Gambino and N. Uraltsev, *Eur. Phys. J.* C **34**, 181 (2004). C. W. Bauer, Z. Ligeti, M. Luke, A. V. Manohar and M. Trott, *Phys. Rev.* D **70**, 094017 (2004).

23. O. Buchmüller and H. Flächer, hep-ph/0507253, and references therein.

24. C. W. Bauer and A. V. Manohar, *Phys. Rev.* D **70**, 034024 (2004). S. W. Bosch, B. O. Lange, M. Neubert and G. Paz, *Phys. Rev. Lett.* **93**, 221801 (2004).

25. K. S. M. Lee and I. W. Stewart, *Nucl. Phys.* B **721**, 325 (2005). S. W. Bosch, M. Neubert and G. Paz, *JHEP* **0411**, 073 (2004). M. Beneke, F. Campanario, T. Mannel and B. D. Pecjak, *JHEP* **0506**, 071 (2005).

26. B. O. Lange, M. Neubert and G. Paz, hep-ph/0504071.

27. T. Mannel and S. Recksiegel, *Phys. Rev.* D **63**, 094011 (2001).

28. A. K. Leibovich, I. Low and I. Z. Rothstein, *Phys. Lett.* B **513**, 83 (2001). A. K. Leibovich, I. Low and I. Z. Rothstein, *Phys. Rev.* D **61**, 053006 (2000). A. K. Leibovich, I. Low and I. Z. Rothstein, *Phys. Lett.* B **486**, 86 (2000).

29. A. Bornheim *et al.* [CLEO Collaboration], *Phys. Rev. Lett.* **88**, 231803 (2002).

30. H. Kakuno *et al.* [BELLE Collaboration], *Phys. Rev. Lett.* **92**, 101801 (2004). I. Bizjak *et al.* [Belle Collaboration], hep-ex/0505088. A. Limosani *et al.* [Belle Collaboration], *Phys. Lett.* B **621**, 28 (2005).

31. B. Aubert *et al.* [BaBar Collaboration], hep-ex/0408075. B. Aubert *et al.* [BABAR Collaboration], *Phys. Rev. Lett.* **95**, 111801 (2005). B. Aubert *et al.* [BABAR Collaboration], hep-ex/0509040. (The average quoted in Eq. (12) uses a slightly different, preliminary result.)

32. M. Gronau and D. London., *Phys. Lett.* B **253**, 483 (1991). M. Gronau and D. Wyler, *Phys. Lett.* B **265**, 172 (1991).

33. M. Gronau, Y. Grossman, N. Shuhmaher, A. Soffer and J. Zupan, *Phys. Rev.* D **69**, 113003 (2004).

34. M. Gronau and D. London, *Phys. Rev. Lett.* **65**, 3381 (1990).

35. E. Gamiz, S. Collins, C. T. H. Davies, J. Shigemitsu and M. Wingate, hep-lat/0509188.

36. T. Becher, E. Gamiz and K. Melnikov, hep-lat/0507033.

37. B. Aubert *et al.* [BABAR Collaboration], *Phys. Rev. Lett.* **94**, 161803 (2005). K. Abe *et al.* [Belle Collaboration], hep-ex/0507037. R. Barate *et al.* [ALEPH Collaboration], *Phys. Lett.* B **492**, 259 (2000). K. Ackerstaff *et al.* [OPAL collaboration], *Eur. Phys. J.* C **5**, 379 (1998). T. Affolder *et al.* [CDF Collaboration], *Phys. Rev.* D **61**, 072005 (2000).

38. A. Gray *et al.* [HPQCD Collaboration], hep-lat/0507015.

39. S. Herrlich and U. Nierste, *Phys. Rev.* D **52**, 6505 (1995).

40. G. Buchalla and A. J. Buras, *Nucl. Phys.* B **400**, 225 (1993). G. Buchalla and A. J. Buras, M. Misiak and J. Urban, *Phys. Lett.* B **451**, 161 (1999). *Nucl. Phys.* B **548**, 309 (1999).

41. G. Buchalla and A. J. Buras, *Nucl. Phys.* B **412**, 106 (1994).

42. G. Buchalla and A. J. Buras, *Phys. Lett.* B **333**, 221 (1994).

43. G. Isidori, F. Mescia and C. Smith, *Nucl. Phys.* B **718**, 319 (2005). A. F. Falk, A. Lewandowski and A. A. Petrov, *Phys. Lett.* B **505**, 107 (2001).

44. A. J. Buras, M. Gorbahn, U. Haisch and U. Nierste, hep-ph/0508165.

45. M. Beneke, *Phys. Lett.* B **620**, 143 (2005).

DISCUSSION

Luca Silvestrini (Rome and Munich):

Maybe one should rather than saying that QCD cannot explain $\sin 2\beta^{\text{eff}}$ in $b \to s$ penguins, one should say that a particular model of power suppressed corrections due to Beneke and Co. cannot do it, but this is not a model independent statement. If you just want to use data, and you say you do not know anything about power corrections, I do not think that you can infer anything from that plot.

Ulrich Nierste: The parametric suppression of the up-quark pollution by $|V_{ub}V_{us}/(V_{cb}V_{cs})| \sim 0.025$ is undisputed. Further the leading term in the $1/m_b$ expansion of $\sin 2\beta^{\text{eff}} - \sin(2\beta)$ can be reliably computed and results in the finding of Ref.[45] that $\sin 2\beta^{\text{eff}} - \sin(2\beta)$ is small and positive for the measured modes. It is true that the size of the modeled power corrections is currently widely debated. Yet I am not aware of any possible dynamical QCD effect in two-body B decays which is formally $\mathcal{O}(1/m_b)$, large in magnitude and further comes with the large strong phase needed to flip the sign of $\sin 2\beta^{\text{eff}} - \sin(2\beta)$.

QCD EFFECTS IN WEAK DECAYS

IAIN W. STEWART

Center for Theoretical Physics, Massachusetts Institute of Technology, Cambridge, MA 02139, USA
E-mail: iains@mit.edu

NO CONTRIBUTION RECEIVED

RARE DECAYS AND CP VIOLATION BEYOND THE STANDARD MODEL

LUCA SILVESTRINI

INFN, Sez. di Roma, Dip. di Fisica, Univ. di Roma "La Sapienza",
P.le A. Moro, I-00185 Rome, Italy. E-mail: Luca.Silvestrini@roma1.infn.it

We review the status of rare decays and CP violation in extensions of the Standard Model. We analyze the determination of the unitarity triangle and the model-independent constraints on new physics that can be derived from this analysis. We find stringent bounds on new contributions to $K - \bar{K}$ and $B_d - \bar{B}_d$ mixing, pointing either to models of minimal flavour violation or to models with new sources of flavour and CP violation in $b \to s$ transitions. We discuss the status of the universal unitarity triangle in minimal flavour violation, and study rare decays in this class of models. We then turn to supersymmetric models with nontrivial mixing between second and third generation squarks, discuss the present constraints on this mixing and analyze the possible effects on CP violation in $b \to s$ nonleptonic decays and on $B_s - \bar{B}_s$ mixing. We conclude presenting an outlook on Lepton-Photon 2009.

1 Introduction

The Standard Model (SM) of electroweak and strong interactions works beautifully up to the highest energies presently explored at colliders. However, there are several indications that it must be embedded as an effective theory into a more complete model that should, among other things, contain gravity, allow for gauge coupling unification and provide a dark matter candidate and an efficient mechanism for baryogenesis. This effective theory can be described by the Lagrangian

$$\mathcal{L}(M_W) = \Lambda^2 H^\dagger H + \mathcal{L}_{\text{SM}} + \frac{1}{\Lambda}\mathcal{L}^5 + \frac{1}{\Lambda^2}\mathcal{L}^6 + \dots ,$$

where the logarithmic dependence on the cutoff Λ has been neglected. Barring the possibility of a conspiracy between physics at scales below and above Λ to give an electroweak symmetry breaking scale $M_W \ll \Lambda$, we assume that the cutoff lies close to M_W. Then the power suppression of higher dimensional operators is not too severe for $\mathcal{L}^{5,6}$ to produce sizable effects in low-energy processes, provided that they do not compete with tree-level SM contributions. Therefore, we should look for new physics effects in quantities that in the SM are zero at the tree level and are finite and calculable at the quantum level. Within the SM, such quantities fall in two categories: i) electroweak precision observables (protected by the electroweak symmetry) and ii) Flavour Changing Neutral Currents (FCNC) (protected by the GIM mechanism). The first category has been discussed by S. Dawson at this conference, while the second will be analyzed here.

In the SM, all FCNC and CP violating processes are computable in terms of quark masses and of the elements of the Cabibbo-Kobayashi-Maskawa (CKM) matrix. This implies very strong correlations among observables in the flavour sector. New Physics (NP) contributions, or equivalently the operators in $\mathcal{L}^{5,6}$, violate in general these correlations, so that NP can be strongly constrained by combining all the available experimental information on flavour and CP violation.

2 The UT analysis beyond the SM

A very useful tool to combine the available experimental data in the quark sector is the Unitarity Triangle (UT) analysis. [1,2] Thanks to the measurements of the UT angles recently performed at B factories, which provide a determination of the UT comparable in accuracy with the one performed using the other available data, the UT fit is now over-constrained (see Fig. 1). It is therefore become possible to add NP contributions to all quantities entering the UT analysis and to

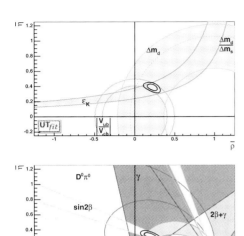

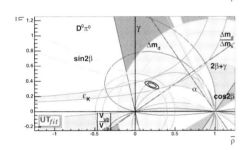

Figure 1. The UT obtained without using (top) and using only (center) the measurements of the UT angles, and the combined fit result (bottom).

perform a combined fit of NP contributions and SM parameters. In general, NP models introduce a large number of new parameters: flavour changing couplings, short distance coefficients and matrix elements of new local operators. The specific list and the actual values of these parameters can only be determined within a given model. Nevertheless, each of the meson-antimeson mixing processes is described by a single amplitude and can be parameterized, without loss of generality, in terms of two parameters, which quantify the difference between the full amplitude and the SM one.[3] Thus, for instance, in the

case of $B_q^0 - \bar{B}_q^0$ mixing we define

$$C_{B_q}\, e^{2i\phi_{B_q}} = \frac{\langle B_q^0 | H_{\text{eff}}^{\text{full}} | \bar{B}_q^0 \rangle}{\langle B_q^0 | H_{\text{eff}}^{\text{SM}} | \bar{B}_q^0 \rangle}\,,\ (q = d, s)\ \ (1)$$

where $H_{\text{eff}}^{\text{SM}}$ includes only the SM box diagrams, while $H_{\text{eff}}^{\text{full}}$ includes also the NP contributions. As far as the $K^0 - \bar{K}^0$ mixing is concerned, we find it convenient to introduce a single parameter which relates the imaginary part of the amplitude to the SM one:

$$C_{\epsilon_K} = \frac{\text{Im}[\langle K^0 | H_{\text{eff}}^{\text{full}} | \bar{K}^0 \rangle]}{\text{Im}[\langle K^0 | H_{\text{eff}}^{\text{SM}} | \bar{K}^0 \rangle]}\,. \qquad (2)$$

Therefore, all NP effects in $\Delta F = 2$ transitions are parameterized in terms of three real quantities, C_{B_d}, ϕ_{B_d} and C_{ϵ_K}. NP in the B_s sector is not considered, due to the lack of experimental information, since both Δm_s and $A_{\text{CP}}(B_s \to J/\psi\phi)$ are not yet measured.

NP effects in $\Delta B = 1$ transitions can also affect some of the measurements entering the UT analysis, in particular the measurements of α and A_{SL}.[4] However, under the hypothesis that NP contributions are mainly $\Delta I = 1/2$, their effect can be taken into account in the fit of the $B \to \pi\pi, \rho\pi, \rho\rho$ decay amplitudes. Concerning A_{SL}, penguins only enter at the Next-to-Leading order and therefore NP in $\Delta B = 1$ transitions produces subdominant effects with respect to the leading $\Delta B = 2$ contribution.

The results obtained in a global fit for C_{B_d}, C_{ϵ_K}, C_{B_d} vs. ϕ_{B_d}, and γ vs. ϕ_{B_d} are shown in Fig. 2, together with the corresponding regions in the $\bar{\rho}$–$\bar{\eta}$ plane.[4]

To illustrate the impact of the various constraints on the analysis, in Fig. 3 we show the selected regions in the ϕ_{B_d} vs. C_{B_d} and ϕ_{B_d} vs. γ planes using different combinations of constraints. The first row represents the pre-2004 situation, when only $|V_{ub}/V_{cb}|$, Δm_d, ε_K and $\sin 2\beta$ were available, selecting a continuous band for ϕ_{B_d} as a function of γ and a broad region for C_{B_d}. Adding the determination of γ (second row), only four regions in the ϕ_{B_d} vs. γ plane survive,

158

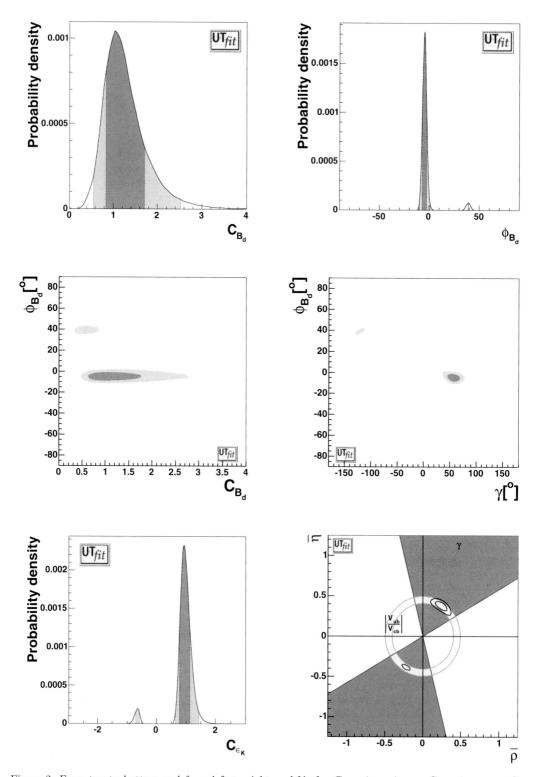

Figure 2. From top to bottom and from left to right, p.d.f.'s for C_{B_d}, ϕ_{B_d}, ϕ_{B_d} $vs.$ C_{B_d}, ϕ_{B_d} $vs.$ γ, C_{ϵ_K} and the selected region on the $\bar{\rho} - \bar{\eta}$ plane obtained from the NP analysis. In the last plot, selected regions corresponding to 68% and 95% probability are shown, together with 95% probability regions for γ (from DK final states) and $|V_{ub}/V_{cb}|$. Dark (light) areas correspond to the 68% (95%) probability region.

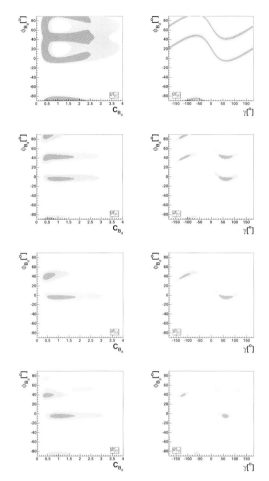

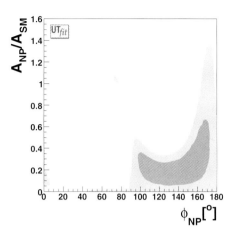

Figure 4. P.d.f. in the $(A_{\mathrm{NP}}/A_{\mathrm{SM}})$ vs. ϕ_{NP} plane for NP in the $|\Delta B| = 2$ sector (see Eq. (3)).

Figure 3. From top to bottom: distributions of ϕ_{B_d} vs. C_{B_d} (left) and ϕ_{B_d} vs. γ (right) using the following constraints: i) $|V_{ub}/V_{cb}|$, Δm_d, ε_K and $\sin 2\beta$; ii) the constraints in i) plus γ; iii) the constraints in ii) plus $\cos 2\beta$ from $B_d \to J/\psi K^*$ and β from $B \to Dh^0$; iv) the constraints in ii) plus α.

two of which overlap in the ϕ_{B_d} vs. C_{B_d} plane. Two of these solutions have values of $\cos 2(\beta + \phi_{B_d})$ and $\alpha - \phi_{B_d}$ different from the SM predictions, and are therefore disfavoured by $(\cos 2\beta)^{\mathrm{exp}}$ and by the measurement of $(2\beta)^{\mathrm{exp}}$ from $B \to Dh^0$ decays, and by α^{exp} (third and fourth row respectively). On the other hand, the remaining solution has a very large value for A_{SL} and is therefore disfavoured by $A_{\mathrm{SL}}^{\mathrm{exp}}$, leading to the final results already presented in Fig. 2. The numerical results of the analysis can be found

in ref. [4] (see ref. [2,5] for previous analyses).

Before concluding this section, let us analyze more in detail the results in Fig. 2. Writing

$$C_{B_d} e^{2i\phi_{B_d}} = \frac{A_{\mathrm{SM}} e^{2i\beta} + A_{\mathrm{NP}} e^{2i(\beta + \phi_{\mathrm{NP}})}}{A_{\mathrm{SM}} e^{2i\beta}},$$

(3)

and given the p.d.f. for C_{B_d} and ϕ_{B_d}, we can derive the p.d.f. in the $(A_{\mathrm{NP}}/A_{\mathrm{SM}})$ vs. ϕ_{NP} plane. The result is reported in Fig. 4. We see that the NP contribution can be substantial if its phase is close to the SM phase, while for arbitrary phases its magnitude has to be much smaller than the SM one. Notice that, with the latest data, the SM ($\phi_{B_d} = 0$) is disfavoured at 68% probability due to a slight disagreement between $\sin 2\beta$ and $|V_{ub}/V_{cb}|$. This requires $A_{\mathrm{NP}} \neq 0$ and $\phi_{\mathrm{NP}} \neq 0$. For the same reason, $\phi_{\mathrm{NP}} > 90°$ at 68% probability and the plot is not symmetric around $\phi_{\mathrm{NP}} = 90°$.

Assuming that the small but non-vanishing value for ϕ_{B_d} we obtained is just due to a statistical fluctuation, the result of our analysis points either towards models with no new source of flavour and CP violation beyond the ones present in the SM (Minimal Flavour Violation, MFV), or towards models in which new sources of flavour and CP violation are only present in $b \to s$ transitions. In the rest of this talk we will consider these two possibilities, starting from the former.

3   MFV models

We now specialize to the case of MFV. Making the basic assumption that the only source of flavour and CP violation is in the Yukawa couplings,[6] it can be shown that the phase of $|\Delta B| = 2$ amplitudes is unaffected by NP, and so is the ratio $\Delta m_s / \Delta m_d$. This allows the determination of the Universal Unitarity Triangle independent on NP effects, based on $|V_{ub}/V_{cb}|$, γ, $A_{CP}(B \to J/\Psi K^{(*)})$, β from $B \to D^0 h^0$, α, and $\Delta m_s / \Delta m_d$.[7] We present here the determination of the UUT, which is independent of NP contributions in the context of MFV models. The details of the analysis and the upper bounds on NP contributions that can be derived from it can be found in ref. [4]

In Fig. 5 we show the allowed region in the $\bar{\rho} - \bar{\eta}$ plane for the UUT. The corresponding values and ranges are reported in Tab. 1. The most important differences with respect to the general case are that i) the lower bound on Δm_s forbids the solution in the third quadrant, and ii) the constraint from $\sin 2\beta$ is now effective, so that we are left with a region very similar to the SM one.

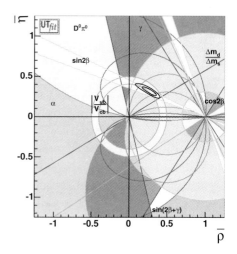

Figure 5. The selected region on $\bar{\rho}$-$\bar{\eta}$ plane obtained from the determination of the UUT.

Table 1. Results of the UUT analysis.

	UUT (68%)	UUT (95%)
$\bar{\rho}$	0.259 ± 0.068	$[0.107, 0.376]$
$\bar{\eta}$	0.320 ± 0.042	$[0.241, 0.399]$
$\sin 2\beta$	0.728 ± 0.031	$[0.668, 0.778]$
$\alpha[^\circ]$	105 ± 11	$[81, 124]$
$\gamma[^\circ]$	51 ± 10	$[33, 75]$
$[2\beta + \gamma][^\circ]$	98 ± 12	$[77, 123]$
$\Delta m_s \,[\text{ps}^{-1}]$	20.6 ± 5.6	$[10.6, 32.6]$

Starting from the determination of the UUT, one can study rare decays in MFV models.[8] In general, a model-independent analysis of rare decays is complicated by the large number of higher dimensional operators that can contribute beyond the SM.[9] The situation drastically simplifies in MFV models, where (excluding large $\tan \beta$ scenarios) no new operators arise beyond those generated by W exchange. Since the mass scale of NP must be higher than M_W, we can further restrict our attention to operators up to dimension five, since higher dimensional operators will suffer a stronger suppression by the scale of NP. In this way, we are left with NP contributions to two operators only: the FCNC Z and magnetic vertices.[a] NP contributions can be reabsorbed in a redefinition of the SM coefficients of these operators: $C = C_{SM} + \Delta C$ for the Z vertex and $C_7^{\text{eff}} = C_{7SM}^{\text{eff}} + \Delta C_7^{\text{eff}}$ for the magnetic operator.[b]

The analysis goes as follows: using

---

[a]The chromomagnetic vertex should also be considered, but this is not necessary for the analysis presented here.[8]

[b]We find it convenient to redefine the C function at the electroweak scale, and the C_7^{eff} function at the hadronic scale.

the CKM parameters as determined by the UUT analysis, one can use BR($B \to X_s\gamma$), BR($B \to X_s l^+ l^-$) and BR($K^+ \to \pi^+ \nu\bar\nu$) to constrain ΔC and ΔC_7^{eff}. Then, predictions can be obtained for all other K and B rare decays. Fig. 6 shows the constraints on the NP contributions. Three possibilities emerge: i) the SM-like solution with NP corrections close to zero; ii) the "opposite C" solution with the sign of C flipped by NP and C_7^{eff} close to the SM value; iii) the "opposite C_7" solution with the sign of C_7^{eff} flipped, which however requires a sizable deviation from the SM also in C.

The corresponding predictions for other rare decays are reported in Fig. 7, and the 95% probability upper bounds are summarized in Tab. 2, together with the SM predictions obtained starting from the UUT analysis. It is clear that, given present constraints, rare decays can be only marginally enhanced with respect to the SM, while strong suppressions are still possible. Future improvements in the measurements of BR($B \to X_s\gamma$), BR($B \to X_s l^+ l^-$) and BR($K^+ \to \pi^+ \nu\bar\nu$) will help us to reduce the allowed region for NP contributions. Another very interesting observable is the Forward-Backward asymmetry in $B \to X_s l^+ l^-$.[10] Indeed, the two solutions for ΔC_7^{eff} and the corresponding possible values of ΔC give rise to different profiles of the normalized $\bar A_{\text{FB}}$ (see eq. (3.10) of ref.[8], where more details can be found). This can be seen explicitly in Fig. 8.

4 New Physics in b → s transitions

We concluded sec. 2 pointing out two possible NP scenarios favoured by the UT analysis: the first one, MFV, was discussed in the previous section, now we turn to the second one, i.e. models with new sources of flavour and CP violation in $b \to s$ transitions. Indeed, most NP models fall in this class. Since the SM flavour $SU(3)$ symmetry is strongly broken by the top (and bottom) Yukawa cou-

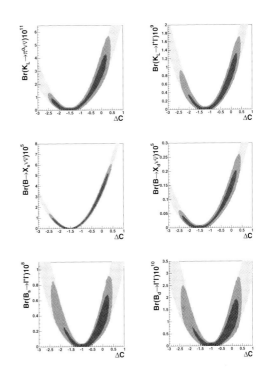

Figure 7. P.d.f.'s for the branching ratios of the rare decays $Br(K_L \to \pi^0 \nu\bar\nu)$, $Br(K_L \to \mu\bar\mu)_{\text{SD}}$, $Br(B \to X_{d,s}\nu\bar\nu)$, and $Br(B_{d,s} \to \mu^+\mu^-)$ as a function of ΔC. Dark (light) areas correspond to the 68% (95%) probability region. Very light areas correspond to the range obtained without using the experimental information.

plings, flavour models are not very effective in constraining NP contributions to $b \to s$ transitions.[12] The same happens in models of gauge-Higgs unification or composite Higgs models, due to the large coupling between the third generation and the EW symmetry breaking sector.[13] Last but not least, the large atmospheric neutrino mixing angle suggests the possibility of large NP contributions to $b \to s$ processes in SUSY-GUTs.[14]

This well-motivated scenario is becoming more and more interesting since B factories are probing NP effects in $b \to s$ penguin transitions, and the Tevatron and LHCb will probe NP effects in $B_s - \bar B_s$ mixing in the near future. For the latter process, there is a solid SM prediction which states that $\Delta m_s > 28\,(30)$ ps^{-1} implies NP at 2σ (3σ).

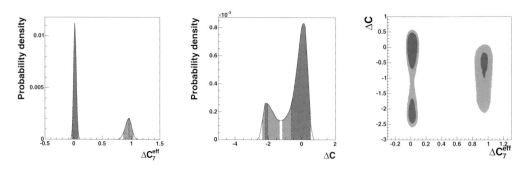

Figure 6. P.d.f.'s for ΔC_7^{eff} (left), ΔC (middle) and ΔC vs. ΔC_7^{eff} (right).

Table 2. Upper bounds for rare decays in MFV models at 95% probability, the corresponding values in the SM (using inputs from the UUT analysis) and the available experimental information.

Branching Ratios	MFV (95%)	SM (68%)	SM (95%)	exp[11]
$Br(K^+ \to \pi^+ \nu\bar{\nu}) \times 10^{11}$	< 11.9	8.3 ± 1.2	$(6.1, 10.9)$	$(14.7^{+13.0}_{-8.9})$
$Br(K_L \to \pi^0 \nu\bar{\nu}) \times 10^{11}$	< 4.59	3.08 ± 0.56	$(2.03, 4.26)$	$< 5.9 \cdot 10^4$
$Br(K_L \to \mu\bar{\mu})_{\text{SD}} \times 10^9$	< 1.36	0.87 ± 0.13	$(0.63, 1.15)$	-
$Br(B \to X_s \nu\bar{\nu}) \times 10^5$	< 5.17	3.66 ± 0.21	$(3.25, 4.09)$	< 64
$Br(B \to X_d \nu\bar{\nu}) \times 10^6$	< 2.17	1.50 ± 0.19	$(1.12, 1.91)$	$< 2.2 \cdot 10^2$
$Br(B_s \to \mu^+\mu^-) \times 10^9$	< 7.42	3.67 ± 1.01	$(1.91, 5.91)$	$< 1.5 \cdot 10^2$
$Br(B_d \to \mu^+\mu^-) \times 10^{10}$	< 2.20	1.04 ± 0.34	$(0.47, 1.81)$	$< 3.9 \cdot 10^2$

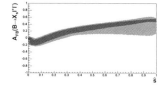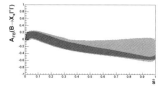

Figure 8. P.d.f. for the normalized forward-backward asymmetry in $B \to X_s l^+ l^-$ for $\Delta C_7^{\text{eff}} \sim 0$ with $\Delta C > -1$ (left), for $\Delta C_7^{\text{eff}} \sim 0$ with $\Delta C < -1$ (middle) and for $\Delta C_7^{\text{eff}} \sim 1$ (right). Dark (light) areas correspond to the 68% (95%) probability region.

For $b \to s$ penguin transitions, $B \to X_s \gamma$ and $B \to X_s l^+ l^-$ decays strongly constrain the FCNC Z and magnetic effective vertices, as already discussed in the previous section in the simplified case of MFV. On the other hand, NP contributions to the chromomagnetic $b \to s$ vertex and to dimension six operators are only mildly constrained by radiative and semileptonic decays, so that they can contribute substantially to $b \to s$ hadronic decays, although in any given model all these NP contributions are in general correlated

and thus more constrained.

As shown in the talk by K. Abe at this conference, B-factories are now probing NP in $b \to s$ transitions by measuring the coefficient $\mathcal{S}$ of the $\sin \Delta m_d t$ term in time-dependent CP asymmetries for $b \to s$ nonleptonic decays. Neglecting the doubly Cabibbo suppressed $b \to u$ contributions, one should have $\mathcal{S} = \sin 2\beta$ for all $b \to s$ channels within the SM, so that deviations from this equality would signal NP in the decay amplitude.[15] However, $b \to u$ terms may also cause deviations $\Delta \mathcal{S}$ from the equality above, so that the estimate of $\Delta \mathcal{S}$ becomes of crucial importance in looking for NP. While a detailed analysis of $\Delta \mathcal{S}$ goes beyond the scope of this talk,[16] the reader should be warned that $\Delta \mathcal{S}$ might be quite large for channels that are not pure penguins, and in particular for final states containing η' mesons. [c] In this respect, it is of fundamental importance to improve the measurement of pure penguin channels, such as ϕK_S, as well as to enlarge the sample of available $b \to s$ and $b \to d$ channels, in order to be able to use flavour symmetries to constrain ΔS.

The problem of computing $\Delta \mathcal{S}$ in any given NP model is even tougher: as is well known, in the presence of two contributions to the amplitude with different weak phases, CP asymmetries depend on hadronic matrix elements, which at present cannot be computed in a model-independent way. One has then to resort to models of hadronic dynamics to estimate $\Delta \mathcal{S}$, with the large theoretical uncertainties associated to this procedure.

With the above *caveat* in mind, let us now focus on SUSY and discuss the phenomenological effects of the new sources of flavour and CP violation in $b \to s$ processes that arise in the squark sector.[18] In general, in the MSSM squark masses are neither flavour-universal, nor are they aligned to quark masses, so that they are not flavour diagonal in the super-CKM basis, in which quark masses are diagonal and all neutral current (SUSY) vertices are flavour diagonal. The ratios of off-diagonal squark mass terms to the average squark mass define four new sources of flavour violation in the $b \to s$ sector: the mass insertions $(\delta^d_{23})_{AB}$, with $A, B = L, R$ referring to the helicity of the corresponding quarks. These δ's are in general complex, so that they also violate CP.

One can think of them as additional CKM-type mixings arising from the SUSY sector. Assuming that the dominant SUSY contribution comes from the strong interaction sector, *i.e.* from gluino exchange, all FCNC processes can be computed in terms of the SM parameters plus the four δ's plus the relevant SUSY masses: the gluino mass $m_{\tilde{g}}$, the average squark mass $m_{\tilde{q}}$ and, in general, $\tan\beta$ and the μ parameter.[d] Barring accidental cancellations, one can consider one single δ parameter, fix the SUSY masses and study the phenomenology. The constraints on δ's come at present from BR's and CP asymmetries in $B \to X_s\gamma$, $B \to X_s l^+ l^-$ and from the lower bound on Δm_s. Since gluino exchange does not generate a sizable ΔC in the notation of the previous section, the combined constraints from radiative and semileptonic decays are particularly stringent.

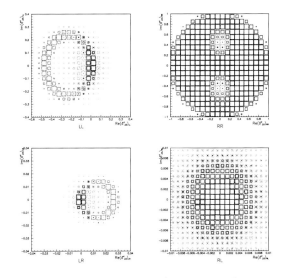

Figure 9. P.d.f.'s in the $\text{Re}(\delta^d_{23})_{AB} - \text{Im}(\delta^d_{23})_{AB}$ plane for $A, B = L, R$, as determined by $B \to X_s\gamma$ (violet), $B \to X_s l^+ l^-$ (light blue) and all constraints (dark blue).

Fixing as an example $m_{\tilde{g}} = m_{\tilde{q}} = -\mu = 350$ GeV and $\tan\beta = 10$, one obtains the

---

[c]Theoretical uncertainties might be larger than what expected even in the golden mode $B \to J/\psi K_S$, although they can be reduced with the aid of other decay modes.[17]

[d]The last two parameters are irrelevant as long as $\tan\beta$ is of $\mathcal{O}(1)$.

constraints on δ's reported in Fig. 9. Several comments are in order at this point: i) only $(\delta_{23}^d)_{\text{LL,LR}}$ generate amplitudes that interfere with the SM one in rare decays. Therefore, the constraints from rare decays for $(\delta_{23}^d)_{\text{RL,RR}}$ are symmetric around zero, while the interference with the SM produces the circular shape of the $B \to X_s \gamma$ constraint on $(\delta_{23}^d)_{\text{LL,LR}}$. ii) We recall that LR and RL mass insertions generate much larger contributions to the (chromo)magnetic operators, since the necessary chirality flip can be performed on the gluino line ($\propto m_{\tilde{g}}$) rather than on the quark line ($\propto m_{\tilde{b}}$). Therefore, the $B \to X_s \gamma$ constraint is much more effective on these insertions. iii) The $\mu \tan \beta$ flavour-conserving LR squark mass term generates, together with a flavour changing LL mass insertion, an effective $(\delta_{23}^d)_{\text{LR}}^{\text{eff}}$ that contributes to $B \to X_s \gamma$. Having chosen a negative μ, we have $(\delta_{23}^d)_{\text{LR}}^{\text{eff}} \propto -(\delta_{23}^d)_{\text{LL}}$ and therefore the circle determined by $B \to X_s \gamma$ in the LL and LR cases lies on opposite sides of the origin (see Fig. 9). iv) For LL and LR cases, $B \to X_s \gamma$ and $B \to X_s l^+ l^-$ produce bounds with different shapes on the Re δ – Im δ plane (violet and light blue regions in Fig. 9), so that applying them simultaneously only a much smaller region around the origin survives (dark blue regions in Fig. 9). This shows the key role played by rare decays in constraining new sources of flavour and CP violation in the squark sector. v) For the RR case, the constraints from rare decays are very weak, so that almost all δ's with $|(\delta_{23}^d)_{\text{RR}}| < 1$ are allowed, except for two small forbidden regions where Δm_s goes below the experimental lower bound.

Having determined the p.d.f's for the four δ's, we now turn to the evaluation of $\mathcal{S}$ as defined at the beginning of this section. We use the approach defined in ref. [19] to evaluate the relevant hadronic matrix elements, warning the reader about the large uncontrolled theoretical uncertainties that affect this evaluation. Let us focus for concreteness on the

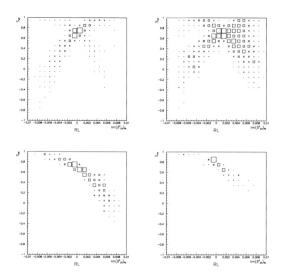

Figure 10. From top to bottom and from left to right, p.d.f.'s for $\mathcal{S}$ for B decays to ϕK_S, ωK_S, $\eta' K_S$ and πK_S as a function of Im $(\delta_{23}^d)_{\text{RL}}$.

effects of $(\delta_{23}^d)_{\text{RL}}$. Imposing that the BR's are correctly reproduced, we obtain the estimates of $\mathcal{S}$ for the ϕK_s, $\eta' K_s$, ωK_s and $\pi^0 K_s$ final states reported in Fig. 10. One can see that $(\delta_{23}^d)_{\text{RL}}$ insertions can produce sizable deviations from the SM expectations for $\mathcal{S}$ in the $\eta' K_s$ and ωK_s channels. Similar results hold for the other δ's.

Another place where δ_{23}^d mass insertions can produce large deviations from the SM is Δm_s. In this case, hadronic uncertainties are under control, thanks to the Lattice QCD computation of the relevant matrix elements,[20] and the whole computation is at the same level of accuracy as the SM one.[21] Considering for example the contribution of $(\delta_{23}^d)_{\text{RR}}$ mass insertions, starting from the constraints in Fig. 9, one obtains the p.d.f. for Δm_s reported in Fig. 11, where for comparison we also report the compatibility plot within the SM.[1] Much larger values are possible in the SUSY case, generally accompanied by large values of the CP asymmetry in $B_s \to J/\psi \phi$: both would be a clear signal of NP to be revealed at hadron colliders.

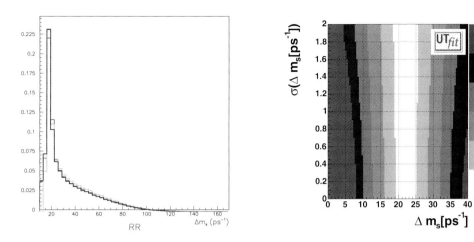

Figure 11. Left: p.d.f. for Δm_S obtained considering $(\delta_{23}^d)_{\rm RR}$ and the SUSY parameters given in the text, as determined by $B \to X_s\gamma$ (violet), $B \to X_s l^+ l^-$ (light blue) and all constraints (dark blue). Right: compatibility plot for Δm_s in the SM.

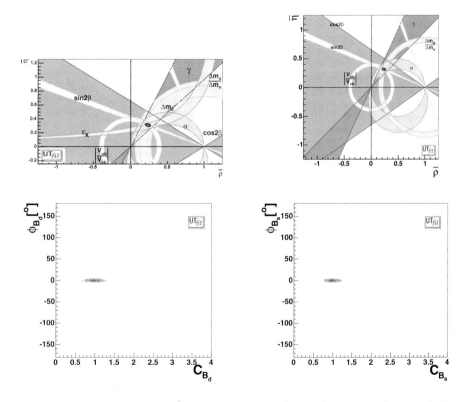

Figure 12. Outlook for Lepton-Photon 2009: the SM UT (top left), the UUT (top right), the ϕ_{B_d} vs. C_{B_d} plane (bottom left) and the ϕ_{B_s} vs. C_{B_s} plane (bottom right). See the text for details.

5 Conclusions and Outlook

Let us summarize the results presented in this talk in four messages:

1 The recent results from B factories make the UT fit overconstrained. This allows us to simultaneously fit SM CKM parameters and NP contributions to $\Delta F = 2$ transitions, in the most general scenario with NP also affecting $\Delta F = 1$ decays. With present data, the SM-like solution in the first quadrant for the UT is strongly favoured (see Fig. 2). The nonstandard solution in the third quadrant has only 4 % probability.

2 From the generalized UT analysis, we can conclude that NP contributions to $\Delta B = 2$ transitions can be of $\mathcal{O}(1)$ if they carry the same weak phase of the SM, otherwise they have to be much smaller or vanishing (see Fig. 4). New sources of flavour and CP violation must therefore be either absent (MFV) or confined to $b \to s$ transitions. The latter possibility is naturally realized in many NP scenarios.

3 In MFV models, the UUT can be determined, independently of NP contributions, with an accuracy comparable to the SM analysis. Together with the available data on $B \to X_s\gamma$, $B \to X_s l^+l^-$ and $K^+ \to \pi^+\nu\bar{\nu}$, this allows to derive stringent upper bounds on other rare K and B decays. Sizable enhancements with respect to the SM are excluded, while strong suppressions are still possible at present.

4 Although the constraints from $B \to X_s\gamma$ and $B \to X_s l^+l^-$ are becoming more and more stringent, NP in $b \to s$ transitions is still allowed to a large extent and might produce sizable deviations from the SM in the time-dependent CP asymmetries in $b \to s$ nonleptonic decays and in $B_s - \bar{B}_s$ mixing. This situation can be realized in SUSY models, where detailed computations of the deviations from the SM can be performed.

We are bound to witness further improvements in the experimental and theoretical inputs to the above analysis in the near future. In the next few years, the UUT analysis might well become the standard analysis, NP contributions to $\Delta F = 2$ transitions will be either revealed or strongly constrained, and rare decays will provide stringent bounds on NP in $\Delta F = 1$ processes or, hopefully, show some deviation from the SM expectation. In Fig. 12 I show a pessimistic view of what we might see at Lepton-Photon 2009, in the dull scenario in which everything remains consistent with the SM.[4] Also in this case, however, flavour physics will remain a crucial source of information on the structure of NP. This information is complementary to the direct signals of NP that we expect to see at the LHC.

I conclude reminding the reader that, for reasons of space, I had to omit several very interesting topics, including in particular lepton flavour violation and electric dipole moments, which might also reveal the presence of NP in the near future.

Acknowledgments

I am very much in debt to my flavour collaborators: C. Bobeth, M. Bona, A.J. Buras, M. Ciuchini, T. Everth, E. Franco, V. Lubicz, G. Martinelli, A. Masiero, F. Parodi, M. Pierini, P. Roudeau, C. Schiavi, A. Stocchi, V. Vagnoni, S. Vempati, O. Vives and A. Weiler. I acknowledge the support of the EU network "The quest for unification" under the contract MRTN-CT-2004-503369.

References

1. M. Bona *et al.* [UTfit Collaboration], *JHEP* **0507**, 028 (2005).
2. J. Charles *et al.* [CKMfitter Group], *Eur. Phys. J. C* **41**, 1 (2005).
3. J. M. Soares and L. Wolfenstein, *Phys. Rev. D* **47**, 1021 (1993); N. G. Deshpande *et al.*, *Phys. Rev. Lett.* **77**, 4499 (1996); J. P. Silva and L. Wolfenstein, *Phys. Rev. D* **55**, 5331 (1997); A. G. Cohen *et al.*, *Phys. Rev. Lett.* **78**, 2300 (1997); Y. Grossman *et al.*, *Phys.*

Lett. B **407**, 307 (1997).

4. M. Bona *et al.* [UTfit Collaboration], arXiv:hep-ph/0509219.

5. S. Laplace *et al.*, *Phys. Rev.* D **65**, 094040 (2002); M. Ciuchini *et al.*, eConf **C0304052**, WG306 (2003); Z. Ligeti, *Int. J. Mod. Phys.* A **20**, 5105 (2005); F. J. Botella *et al.*, *Nucl. Phys.* B **725**, 155 (2005); K. Agashe *et al.*, arXiv:hep-ph/0509117.

6. G. D'Ambrosio *et al.*, *Nucl. Phys.* B **645**, 155 (2002).

7. A. J. Buras *et al.*, *Phys. Lett.* B **500**, 161 (2001).

8. C. Bobeth *et al.*, *Nucl. Phys.* B **726**, 252 (2005).

9. A. Ali *et al.*, *Phys. Rev.* D **66**, 034002 (2002); G. Hiller and F. Kruger, *Phys. Rev.* D **69**, 074020 (2004); G. Buchalla *et al.*, *JHEP* **0509**, 074 (2005);

10. G. Burdman, *Phys. Rev.* D **57**, 4254 (1998).

11. V. V. Anisimovsky *et al.* [E949 Collaboration], *Phys. Rev. Lett.* **93** 031801 (2004); A. Alavi-Harati *et al.* [The E799-II/KTeV Collaboration], *Phys. Rev.* D **61** 072006 (2000); R. Barate *et al.* [ALEPH Collaboration], *Eur. Phys. J.* C **19** 213 (2001); B. Aubert *et al.* [BABAR Collaboration], *Phys. Rev. Lett.* **93** 091802 (2004); A. Abulencia *et al.* [CDF Collaboration], arXiv:hep-ex/0508036.

12. A. Masiero *et al.*, *Phys. Rev.* D **64**, 075005 (2001).

13. K. Agashe *et al.*, *Nucl. Phys.* B **719**, 165 (2005); G. Martinelli *et al.*, *JHEP* **0510**, 037 (2005).

14. T. Moroi, *Phys. Lett.* B **493**, 366 (2000); D. Chang *et al.*, *Phys. Rev.* D **67**, 075013 (2003); R. Harnik *et al.*, *Phys. Rev.* D **69**, 094024 (2004).

15. Y. Grossman and M. P. Worah, *Phys. Lett.* B **395**, 241 (1997); M. Ciuchini *et al.*, *Phys. Rev. Lett.* **79**, 978 (1997); D. London and A. Soni, *Phys. Lett.* B **407**, 61 (1997); Y. Grossman *et al.*, *Phys. Rev.* D **58**, 057504 (1998); M. Ciuchini *et al.*, arXiv:hep-ph/0407073.

16. See the session of WG4 on this topic at the CKM 2005 workshop.

17. M. Ciuchini *et al.*, arXiv:hep-ph/0507290.

18. F. Gabbiani *et al.*, *Nucl. Phys.* B **477**, 321 (1996); R. Barbieri and A. Strumia, *Nucl. Phys.* B **508**, 3 (1997); A. L. Kagan and M. Neubert, *Phys. Rev.* D **58**, 094012 (1998); S. A. Abel *et al.*, *Phys. Rev.* D **58**, 073006 (1998); A. Kagan, arXiv:hep-ph/9806266; R. Fleischer and T. Mannel, *Phys. Lett.* B **511**, 240 (2001); T. Besmer *et al.*, *Nucl. Phys.* B **609**, 359 (2001); E. Lunghi and D. Wyler, *Phys. Lett.* B **521**, 320 (2001); M. B. Causse, arXiv:hep-ph/0207070; G. Hiller, *Phys. Rev.* D **66**, 071502 (2002); S. Khalil and E. Kou, *Phys. Rev.* D **67**, 055009 (2003); *Phys. Rev. Lett.* **91**, 241602 (2003); eConf **C0304052**, WG305 (2003); *Phys. Rev.* D **71**, 114016 (2005); G. L. Kane *et al.*, *Phys. Rev.* D **70**, 035015 (2004); S. Baek, *Phys. Rev.* D **67**, 096004 (2003); K. Agashe and C. D. Carone, *Phys. Rev.* D **68**, 035017 (2003); J. F. Cheng *et al.*, *Phys. Lett.* B **585**, 287 (2004); D. Chakraverty *et al.*, *Phys. Rev.* D **68**, 095004 (2003); J. F. Cheng *et al.*, *Nucl. Phys.* B **701**, 54 (2004); E. Gabrielli *et al.*, *Nucl. Phys.* B **710**, 139 (2005); S. Khalil, *Mod. Phys. Lett.* A **19**, 2745 (2004); *Phys. Rev.* D **72**, 035007 (2005).

19. M. Ciuchini *et al.*, *Phys. Rev.* D **67**, 075016 (2003) [Erratum-ibid. D **68**, 079901 (2003)].

20. D. Becirevic *et al.*, *JHEP* **0204**, 025 (2002).

21. M. Ciuchini *et al.*, *Nucl. Phys.* B **523**, 501 (1998); A. J. Buras *et al.*, *Nucl. Phys.* B **586**, 397 (2000).

QCD AND HADRON STRUCTURE

Previous page:

The Uppsala Castle (above), built during the reign of Gustavus Wasa in 1549. The current building dates from the XVIII century, after the original was destroyed in the great fire of 1702. The symposium dinner was held in the Hall of State

The Gunilla bell tower (below), erected in front of the castle in the mid 1700s in memory of Queen Gunilla. It still tolls at 6:00 and 21:00 every day.

HAS THE QUARK-GLUON PLASMA BEEN SEEN?

JOHANNA STACHEL

Physikalisches Institut, Ruprecht-Karls-Universität Heidelberg, Philosophenweg 12,
D 69120 Heidelberg, Germany
E-mail: stachel@physi.uni-heidelberg.de

Data from the first three years of running at RHIC are reviewed and put into context with data obtained previously at the AGS and SPS and with the physics question of creation of a quark-gluon plasma in high energy heavy ion collisions. Also some very recent and still preliminary data from run4 are included

1 Introduction

Very shortly after the discovery of asymptotic freedom[1] it became apparent that, as a consequence, at high temperature and/or at high density quarks and gluons would also become deconfined[2], leading to a phase transition from confined hadronic matter to an unconfined phase. This was studied in subsequent years and since the early 1980ies this phase is called the Quark-Gluon Plasma (QGP).

The conditions for this phase transition were studied in lattice QCD and state of the art calculations[3] obtain as critical temperature for the phase transition for two light and one heavier quark flavors a value for the critical temperature of $T_c = 173 \pm 15$ MeV and for the critical energy density of $\epsilon_c = 0.7 \pm 0.2$ GeV/fm^3. It is believed since many years that in collisions of heavy atomic nuclei at high energies such conditions should be reached. This motivated an experimental program starting simultaneously in 1986 at the Brookhaven AGS and at the CERN SPS, initially with light projectile nuclei such as Si and S and from 1992 and 1994, respectively, with Au and Pb projectiles. The experimental results from this program prompted a press release from CERN[4] in February 2000 stating that the combined results from the experiments proved that a new state of matter other than ordinary hadronic matter had been created in these collisions, in which quarks were 'liberated to roam freely'. The experimental results were clearly not reconcilable with the known hadronic physics and it could be estimated that the critical temperature had been exceeded in the early phase of the collision by about 20-30 % and the critical energy density by somewhat more than a factor 2. On the other hand, from those data nothing could be said yet that would characterize the properties of the new state of matter. Hence, at that time the term QGP was not used for the new state of matter.

In the summer of 2000, RHIC as a dedicated collider for heavy ions started operation with two large experiments, PHENIX and STAR, and two smaller experiments, BRAHMS and PHOBOS. In the first 3 years of operation data for Au + Au collisions with an integrated luminosity of $85/\mu$b, for p + p collisions with 2/pb, and for d + Au collisions with 25/nb were collected and a summary of the results was recently published in a special issue of Nuclear Physics A by all four heavy ion experiments[5,6,7,8]. In the 2004 Au + Au run the 1/nb level was exceeded and data start to appear from this run. Here I will rely mostly on published data and review some of the key observations from the first 3 years including only a few of the still preliminary first run4 observations.

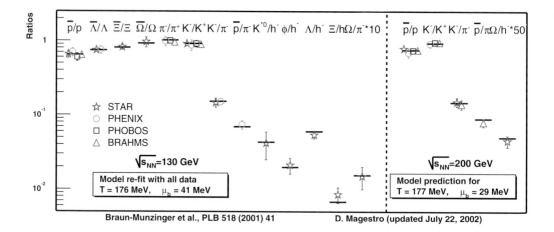

Figure 1. Hadron yield ratios measured at RHIC in comparison to calculations within a statistical model based on a grand canonical ensemble (updated version of[10], taken from[9]).

2 Experimental Results

2.1 Hadron Production and Statistical Models

Hadron yields have been measured for a large range of species at the AGS, SPS and at RHIC. It was realized already for many years that the data for central collisions of heavy nuclei can be rather accurately reproduced by calculations for a chemically equilibrated system in terms of a grand canonical ensemble (a review and complete set of references can be found in[9]). For the lower RHIC energy of $\sqrt{s} = 130$ GeV the data are final and published and for 200 GeV data are emerging currently. Figure 1 shows experimental yield ratios from all four RHIC experiments in comparison to a statistical model fit.

In the calculations, there are two free fit parameters, the temperature and the baryon chemical potential. For top RHIC energy the temperature is fitted as 177 ± 5 MeV, practically unchanged from $\sqrt{s} = 130$ and 17.3 GeV; the baryo-chemical potential is dropping continuously with increasing beam energy reflecting an increasing transparency of the nuclei at higher energies and an increas-

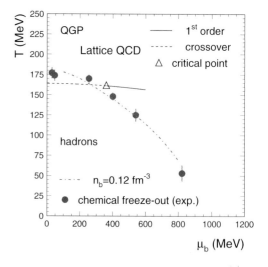

Figure 2. Phase diagram of nuclear matter in the temperature - baryon chemical potential plane. Experimental points for hadro-chemical freeze-out are shown together with a recent lattice QCD calculation[12] and a curve of constant total baryon density. Figure from[11].

ing dominance of baryon-antibaryon production. This is shown in Figure 2 where results of statistical model fits at various beam energies are summarized and shown together with recent results from lattice QCD[12].

It appears that from top SPS energy upwards the temperature at which hadrochemical equilibrium is achieved is not changing anymore and practically coincides with the lattice QCD prediction for the critical temperature, while at lower beam energies it is falling. At $\sqrt{s} = 8.8$ GeV it is only 148 ± 5 MeV. The strangeness suppression that is well established for pp and e^+e^- collisions appears to be completely lifted. This leads to an enhancement in the yields of particularly multistrange hadrons in heavy ion collisions as compared to pp results. For the Omega baryon at SPS energy this enhancement is[13] a factor 17. How hadrons like the Ω can be equilibrated on the time scales of the nuclear collision has been a puzzle for several years and there is consensus that with two-body collisions and the known hadronic cross sections this is not possible[14,15,16]. A possible explanation has been presented recently[14]. In the direct vicinity of the phase transition the densities of particles are rising very rapidly due to the increase of degrees of freedom by more than a factor of 3 between a hadron gas and a QGP. At these high densities multi-hadron collisions become dominant and can drive even the Ω yield into equilibrium in a fraction of a fm/c. Conversely, already 5 MeV below the critical temperature the densities are so low that the system falls out of equilibrium and the yields cannot follow anymore a decreasing temperature. Therefore the authors of[14] conclude that the rapid equilibration is a direct consequence of the phase transition from QGP to hadronic matter and that, at least at high beam energies, the chemical equilibration temperature is a direct experimental measure of the critical temperature.

2.2 Elliptic Flow

Momentum distributions in three dimensions are analyzed with transverse coordinates relative to the reaction plane of the collision spanned by the impact parameter vector and the beam direction and a decomposition in terms of Fourrier coefficients is performed. Already at the Bevalac sizeable anisotropies were observed for heavy colliding nuclei. In particular, the quadrupole coefficient v_2 was found to be negative, explained by shadowing of the emitted particles by the target and projectile spectator remnants[17]. At AGS energies a sign change was observed[18] by E877, i.e. the momentum spectra were harder in the reaction plane than perpendicular to it. The interpretation used a prediction from hydrodynamics[19] that, for semiperipheral collisions, in the early phase of the collision the pressure gradient was larger in this direction due to the excentricity of the nuclear overlap region[a]. From the hydrodynamic evolution it would follow, that this anisotropy in pressure gradient would evolve with time into an anisotropy in momentum space, driven by the initial condition and the equation of state of the expanding system. This was confirmed by a microscopic analysis within a transport model[21]. From this the name 'elliptic flow' originated for the quadrupole coefficient v_2.

At the higher SPS energy growing positive coefficients v_2 were found[22,23,24] and the sign change was traced to occur[25] at beam momenta per nucleon of about 4 GeV/c. At RHIC energies very large values of v_2 were observed[26,27,28,29], typically about 50 % above SPS top energy results.

This was studied differentially for different hadronic species and as function of p_t as shown for data from STAR in Figure 3. It is observed that for more massive hadrons the rise of v_2 starts at larger values of p_t. For

---

[a]The use of hydrodynamics to describe the dynamics of a hadronic collision goes back to the 1950ies[20].

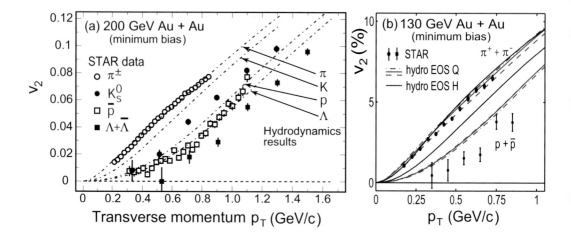

Figure 3. Left: Elliptic flow coefficient v_2 as function of p_t for different particle species[27]. Together with the experimental data results from a hydrodynamics calculation including a phase transition are shown[30]. Right: Experimental data for pions and protons at a lower RHIC energy[26]. Also shown are hydrodynamics calculations[30] with and without phase transition from QGP to hadronic phase. Figure from[7].

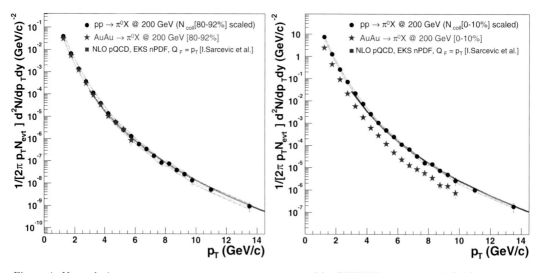

Figure 4. Neutral pion transverse momentum spectra measured by PHENIX in peripheral (left) and central (right) Au + Au collisions (stars) together with pp data from the same experiment scaled with the number of binary collisions (circles)[37,38]. Yellow band: Normalization uncertainties of the pp data. Black line: NLO pQCD calculation. Figure taken from[39].

the first time there was quantitative agreement with hydrodynamic calculations[30,31] in terms of p_t and hadronic species dependence, as also shown in Fig. 3. These hydrodynamic calculations also reproduce the overall features of the p_t spectra of differenct hadrons, although in details there are deviations stemming from the different treatment of the hadronic phase and freeze-out (see Fig. 20 in[8] and references there). It is common to all the hydrodynamics calculations that, in order to reproduce the data, a rapid initial equilibration on a time scale faster than 1 fm/c is required[30,31,32].

At p_t above 2-3 GeV/c, where hydrodynamics should no longer hold as a theoretical description, another type of scaling was discovered[33,7]: dividing both v_2 and p_t by the number of constituent quarks in a hadron all results match rather well even including multistrange baryons. It was realized that an old idea of quark coalescence[34] could be the underlying physics[35] and indeed calculations based on the assumption of coalescence of valence quarks during hadronization of a QGP reproduce this feature rather well[36].

2.3 High Momentum Suppression

One of the highlights of the RHIC experimental program is the observation of a strong suppression in the production of hadrons at high transverse momentum when compared to pp collisions. Figure 4 shows the p_t spectrum of neutral pions in Au + Au collisions as compared to a measurement in pp in the same experiment and at the same energy[37,38]. The pp spectrum compares well with a calculation in NLO pQCD. In order to compare, the pp spectrum has been scaled with the number of binary nucleon-nucleon collisions in a Au + Au collisions at a given centrality. The number N_{coll} of binary collisions is given by the collision geometry - measured in the data with some resolution -, the well known nuclear density distribu-

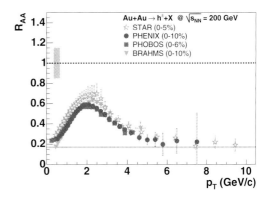

Figure 5. Ratio R_{AA} of the p_t spectrum for central Au + Au collisions normalized to the pp spectrum scaled with the number of binary collisions for charged particles from all four RHIC experiments. Figure from[40].

tion, and the inelastic pp cross section. The collision centrality in Au + Au collisions is characterized by the fraction of the geometric cross section for which events have been selected, which is related to the impact parameter. The yellow bands in Fig. 4 reflect the systematic uncertainty in this scaling. One can observe that for peripheral collisions pp and Au + Au collisions agree very well, while in central collisions the Au + Au spectrum is significantly suppressed.

This is better visualized by building the ratio R_{AA} between the Au + Au p_t spectrum and the pp spectrum scaled with N_{coll} as shown in Figure 5. All four RHIC experiments observe a suppression by about a factor of five for p_t larger than 4 GeV/c. Since not all experiments measure neutral pions, the ratio is shown here for charged hadrons, but at large p_t the data for all hadron species merge. At low p_t the ratio R_{AA} is expected to be below one because there, due to the dominance of soft processes, the appropriate scaling is with the number of participants, i.e. nucleons in the nuclear overlap region. It is expected that this ratio should rise as hard scattering becomes dominant and, in fact, due to the well known Cronin enhancement, in the region of 2-6 GeV/c values above one

176

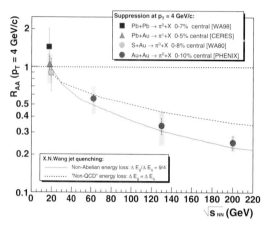

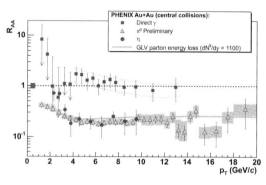

Figure 6. Suppression factor R_{AA} for pions as function of beam energy at a fixed p_t value of 4 GeV/c. The yellow line shows a calculation[45] including parton energy loss in a medium with high parton density. Figure from[44].

Figure 7. Preliminary PHENIX results for the suppression factor R_{AA} out to high p_t for π^0 and η mesons together with a calculation based on a high gluon rapidity density[43]. Also shown are the results for direct photons. Figure from[53].

are expected. Contrary to this expectation the data show a suppression.

The suppression is not unexpected. It was predicted that in a medium with high parton density the radiative energy loss of a quark or gluon should be strongly enhanced[41,42], leading to a very effective thermalization of jets in a hot color charged medium. Calculations employing a large initial gluon rapidity density of about 1100 can account[43] for the data at top RHIC energy. The beam energy dependence of the R_{AA} ratio was presented recently by d'Enterria[44] and it appears that the suppression evolves in a very smooth way from top SPS energy onwards. The $R_{AA}(p_t = 4\text{GeV/c})$ values are shown in Figure 6 for the top SPS energy and three RHIC energies. Already the values of about 1.0 measured at the SPS represent a slight suppression as compared to the normal Cronin enhancement[44]. Going from $\sqrt{s_{nn}} = 17.3$ to 62.4 to 200 GeV the gluon rapidity density needed to reproduce the data grows[43] from 400 to 650 to 1100. An alternative formulation of this in medium suppression is by increasing and large opacities of the medium

traversed[45].

The proof that this is really a final state effect probing the properties of the medium traversed by the parton is given by the observation that in d + Au collisions in the same experiments no suppression is seen, but rather the expected Cronin enhancement[46,47,48,49].

Direct photons were measured by PHENIX in pp and Au + Au collisions at top RHIC energy[50,51]. The pp spectra are rather close to a NLO pQCD calculation[52]. The Au + Au photon spectra are within errors consistent with the scaled pp result and hence the expectation from NLO pQCD. For all centralities they do not show any significant suppression. This is shown in Figure 7 where also the most recent neutral pion results from run4 [53] extending out to $p_t = 20$ GeV/c are displayed. It is remarkable that the suppression of the pion p_t spectrum remains practically constant over a large range in p_t from 4 to 20 GeV/c, close to the predicted behavior for a medium with initial gluon rapidity density of 1100 (see Fig. 7).

The high initial gluon densities correspond to an initial temperature of about twice the critical temperature and to initial energy densities $\epsilon_0 = 14$ - 20 GeV/fm^3 well

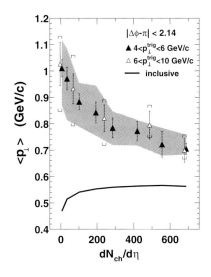

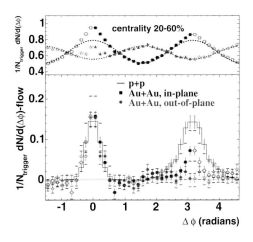

Figure 8. Mean transverse momentum in a cone of an opening angle of one radian opposite to a leading particle as a function of the collision centrality; distributions are for two ranges of leading particle transverse momentum as well as for the inclusive particle distribution. Data and figure from STAR [54].

Figure 9. Azimuthal angle correlation between a high p_t trigger particle (4-6 GeV/c) with all particles in the window p_t = 2-4 GeV/c for intermediate centrality Au + Au collisions. The distribution is shown for both particles in a ± 45 degree window around the reaction plane orientation (in-plane) or a same window perpendicular to it (out-of-plane). Figure from[56].

in line with the initial conditions needed for the hydrodynamics calculations to describe spectra and elliptic flow (see previous section) and bracketed by the estimates based on the Bjorken formula and transverse energy production.

The observed high p_t suppression pattern is different for different hadronic species[5,8,7]. In particular, a pattern appears where at intermediate values of p_t of 2-6 GeV the suppression of baryons is significantly weaker than that of mesons. The proton/pion or also the Λ/K_s^0 ratios peak at values 1.5-1.6 for p_t = 3-4 GeV/c, close to the ratio 3/2 expected in quark coalescence models.

Parton thermalization is displayed in a very clean way be recent results of the STAR collaboration[54]. Evaluating the mean transverse momentum in a cone opposite to a high p_t trigger particle as a funtion of centrality, a gradual decrease for more central Au + Au collisions is observed and in the most central

collisions a value very close to the inclusive mean p_t is reached (see Figure 8).

In azimuthal correlations of two high p_t particles it was seen that the away-side peak disappears in central Au + Au collisions for a choice of trigger p_t of 4-6 GeV and p_t of the correlated particle of 2-4 GeV/c [55]. In pp, d + Au and peripheral Au + Au collisions a clear peak opposite to the trigger particle is observed in the same type of correlation, also measured by STAR[55,48]. Recently, it was shown that the effect is very strong in case the away-side jet is emitted out of the reaction plane and much weaker for emission in the reaction plane[56] as displayed in Figure 9. This supports the strong correlation of the suppression with the length of matter traversed by the parton.

When lowering the p_t cut on the correlated hadron, a very broad structure appears on the side opposite to the trigger particle. This was shown by STAR[54] for a cut on the correlated hadron of $p_t = 0.15 - 4$ GeV/c. This calls to mind a similar observation at SPS energy by CERES[57] where for a condition $p_t \geq 1.2$ GeV/c for both particles also very strong broadening of the away-side structure with increasing collision centrality in Pb + Au collisions was observed. Recent data[58] from PHENIX display a tantalizing feature as shown in Figure 10: For a trigger particle p_t of 4-6 GeV/c and a correlated particle p_t of 1.0 - 2.5 GeV/c the away-side peak seen in peripheral Au + Au collisions develops actually into a hole at $\Delta\phi = \pi$ for more central collisions while a very broad peak appears with a maximum at $\Delta\phi = \pi - 1$ as can be seen in Fig. 10. A suggestion has been made that this could be the Mach cone due to the sonic boom of the quenched jet. A parton traversing a quark-gluon plasma with velocity larger than the velocity of sound in the QGP ($\sqrt{1/3}$ for an ideal gas) would radiate only up to a cone angle of about 1 rad [59,60]. If this could be established it would have far reaching consequences since it would be an observable linked directly to the speed of sounds of the quark-gluon plasma and thereby its equation of state. It remains an experimental challenge to establish an actual cone topology in two dimensions.

2.4 Charm Quarks and Quarkonia

Open charm has been measured indirectly from the inclusive electron p_t spectra after subtracting known contributions from photon conversions and light hadron decays by PHENIX[61]. The spectrum remaining after subtraction is dominated [b] by open charm and beauty contributions. Recently results for an elliptic flow analysis were shown[62] of the electrons dominantly from open charm decays. There is a significant nonzero value in the p_t range 0.4 - 1.6 GeV/c. This is con-

---

[b] A possible contribution to the electron spectrum from the Drell-Yan process cannot be ruled out at present, though.

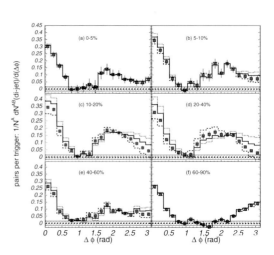

Figure 10. Azimuthal correlations of a leading particle of $p_t = 4$-6 GeV/c and any particle with $p_t = 1$-2.5 GeV/c for different centralities of a Au + Au collision. Data and figure from PHENIX[58].

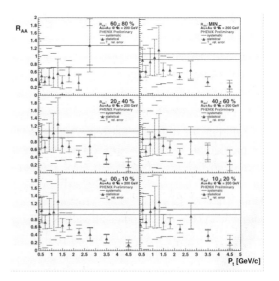

Figure 11. R_{AA} suppression factor for electrons dominantly from open charm and open beauty decay for Au + Au collisions at top RHIC energy for different collision centrality (see text) measured by PHENIX. Figure from[64].

firmed by preliminary STAR data[63] that extend the overall transverse momentum coverage by adding the range $p_t = 1.5 - 3.0$ GeV/c. Together, the data paint a consistent picture that indeed the electrons from open charm decay exhibit elliptic flow, i.e. follow the collective motion of the light quarks. This would imply that the charm quark thermalizes to a significant degree. Note that this is a necesarry prerequisite for any formation of charmed hadrons by statistical hadronization (see below).

In that case also jet quenching should be observed for charmed hadrons. Indeed, in still preliminary data it was shown recently that electron spectra, after the subtraction of contributions from conversion and light hadron decays, show high p_t suppression for central Au + Au collisions[64]. The R_{AA} factor drops practically as low as for pions at p_t of 4 GeV/c, i.e. to values of about 0.2. In a recent publication[65] the suppression for electrons from D meson decay was studied for different transport coefficients. The preliminary RHIC data would be consistent with a calculation using a transport coefficient of 14 GeV^2/fm (see Fig. 2 of[65]), at the upper end of the range needed to reproduce R_{AA} for pions. This is very surprizing, in particular also in view of the fact, that at p_t of about 4 GeV/c also the contribution of b-quarks to the electron spectrum should become sizeable.

At top SPS energy, for central Pb + Pb collisions a significant suppression of J/ψ production was observed in the NA50 experiment[66,67]. This suppression is compared to the socalled 'normal' nuclear absorption seen also in pA collisions. From analysis of all pA data, a cross section for normal nuclear absorption of 4.1±0.4 mb was extracted[67]. To this normal nuclear absorption all results from heavy ion collisions can be compared. It turns out that S + U data as well as data from peripheral Pb + Pb collisions agree with this normal nuclear absorp-

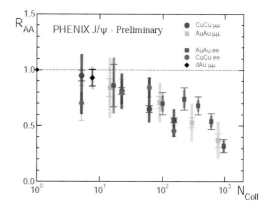

Figure 12. J/ψ yield in Au + Au and Cu + Cu collisions at top RHIC energy normalized to the measured result for pp collisions, scaled with the number of inelastic collisions. Results are shown for decays into electron and muon pairs at mid- and forward rapidities, respectively. Figure from[53].

tion curve. For transverse energies above 40 GeV or a length of nuclear matter seen by the J/ψ of L ≥ 7 fm the points from Pb + Pb collisions fall increasingly below this normal nuclear absorption curve. Theoretically, the suppression can be explained by disappearance of the J/ψ (or possibly only the charmonia states that feed it) in a hot colored medium or by interaction with comovers (chiefly pions), albeit with a very large density of more than 1/fm^3, i.e. a value not deemed achievable for a hadron gas.

The first results for J/ψ production in central Au + Au collisions at RHIC energy came very recently from PHENIX[53] (run4); they are displayed in Figure 12. As compared to pp as well as d + Au collisions there is a significant suppression. The suppression is, however, rather similar to the one observed at SPS. This is in contrast to some predictions[68] that in central Pb + Pb collisions at RHIC only J/ψ mesons from the corona should survice, i.e. order of 5 % of the normal unsuppressed yield. The actually observed yield by far exceeds this expectation. This could be

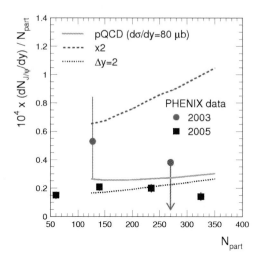

Figure 13. J/ψ yield per participating nucleon for Au + Au collisions at top RHIC energy compared to the yield expected from statistical recombination. Data from[53], statistical hadronization prediction from[71]. For statistical hadronization a standard interval of $\Delta y = 1$ is used. Calculations are shown for a $c\bar{c}$ cross section per unit rapidity of 80 μb, of 160 μb. Also shown: a calculation for 80 μb and $\Delta y = 2.0$.

seen as indication, that at RHIC the energy density in the QGP is not yet high enough to dissolve the J/ψ but rather only enough to dissolve higher $c\bar{c}$ states. Recent results from high temperature lattice QCD indicate that the J/ψ bound state may only disappear[69] in the vicinity of T = 2 T_c.

A maybe more interesting alternative has been proposed[70]: Even if the initially formed $c\bar{c}$ pairs are completely dissociated in the hot QGP, at hadronization charmed hadrons may form in a statistical fashion by the same mechanism described above for hadrons involving up, down, and strange valence quarks. This includes also the formation of charmonia and it was pointed out in[70] that the formation probability of J/ψ mesons would grow quadratically with the $c\bar{c}$ rapidity density. Such increased reformation of J/ψ by statistical hadronization could possibly account for a suppression apparently not much stronger than at SPS.

Figure 13 shows the prediction of[71] to-

gether with the recent and still preliminary RHIC data for $J/\psi \to e^+e^-$. Currently the main uncertainty in this consideration is the overall charm production yield, which enters quadratically. A calculation is shown using a cross section per unit rapidity of 80 μb (400 μb integrated) as expected from NLO pQCD[68]. The overall charm production cross section at RHIC energy has so far been measured indirectly by PHENIX[61] for $\sqrt{s} = 130$ and 200 GeV from the inclusive electron spectra in the way described above for Au + Au collisions and, at full RHIC energy, also for d + Au and pp collisions[72]. It is found that the integrated charm cross section, when scaled with the number of binary collisions, agrees for all three collision systems. The value is about 50% above the NLO pQCD calculation[68] but agrees within errors. On the other hand, in STAR, D mesons have been reconstructed via their hadronic decay to $K\pi$ in d + Au collisions and a charm cross section per nucleon nucleon collision has been extracted[73]. It is twice as large as the PHENIX value by nearly two standard deviations. The experimental situation concerning open charm production needs to be improved before the J/ψ puzzle can be better addressed. Only measurements at LHC will unambiguously clarify the role of statistical hadronisation of charm, since with this mechanism a significant J/ψ enhancement in Pb + Pb collisions at LHC was predicted[71] instead of suppression. Such an enhancement would be an unambiguous signal of deconfinement.

3 Summary and Outlook

Hadron yields are found to be in chemical equilibrium. For top SPS energy and up this can be achieved by multi-particle collisions in the direct vicinity of T_c and hence the observed chemical equilibration temperature is an experimental measure of the critical temperature for the phase transition.

At RHIC energies, spectra and azimuthal

correlations are quantitatively described by hydrodynamics. This requires rapid local thermalization and high initial energy densities more than tenfold above the calculated critical energy density for the phase transition between hadronic matter and QGP.

High p_t hadrons are suppressed in central Au + Au collisions and this is a medium effect. Jet quenching in a hot color charged medium was predicted, modelling of the data with high parton density is successful. There are some indications of valence quark coalescence in hadronic observables.

The observations that lead to the CERN press release are confirmed by the RHIC experiments. Beyond this additional features are observed that start to probe the properties of the new state of matter. Much progress in this direction is expected from the high luminosity RHIC data just starting to appear and, from 2007, from the LHC heavy ion program.

Acknowledgements

I thank Peter Braun-Munzinger for numerous enlightening discussion.

References

1. D.J. Gross, F. Wilczek, *Phys. Rev. Lett.* **30** (1973) 1343; H.D. Politzer, *Phys. Rev. Lett.* **30** (1973) 1346.
2. J.C. Collins, M.J. Perry, *Phys. Rev. Lett.* **34** (1975) 1353; N. Cabibbo, G. Parisi, *Phys. Lett.* B **59** (1975) 67.
3. F. Karsch, *Nucl. Phys.* A **698** (2002) 199c, hep-ph/0103314.
4. CERN press release Feb. 10, 2000; U. Heinz and M. Jacob, nucl-th/0002042.
5. I. Arsene et al., BRAHMS Collaboration, *Nucl. Phys.* A **757** (2005) 1, nucl-ex/0410020.
6. B.B. Back et al., PHOBOS Collaboration, *Nucl. Phys.* A **757** (2005) 28, nucl-ex/0410022.
7. J. Adams et al., STAR Collaboration, *Nucl. Phys.* A **757** (2005) 102, nucl-ex/0501009.
8. K. Adcox et al., PHENIX Collaboration, *Nucl. Phys.* A **757** (2005) 184, nucl-ex/0410003.
9. P. Braun-Munzinger, K. Redlich, and J. Stachel, in Quark-Gluon Plasma 3, eds. R.C. Hwa and X.N. Wang, (World Scientific, Singapore, 2004), 491.
10. P. Braun-Munzinger, D. Magestro, K. Redlich, and J. Stachel, *Phys. Lett.* B **518** (2001) 41.
11. updated version of Figure in P. Braun-Munzinger, J. Stachel, *J. Phys.* G **28** (2002) 1971, taken from A. Andronic, P. Braun-Munzinger, J. Stachel, to be published.
12. Z. Fodor, S.D. Katz, *JHEP* **404** (2004) 50, hep-lat/0402006.
13. E. Andersen et al., WA97 Collaboration, *Phys. Lett.* B **449** (1999) 401.
14. P. Braun-Munzinger, J. Stachel, and C. Wetterich, *Phys. Lett.* B **596** (2004) 61.
15. C. Greiner, *AIP Conf.Proc.* **644** (2003) 337.
16. P. Huovinen, J.I. Kapusta, *Phys. Rev.* C **69** (2004) 014902.
17. H. Gutbrod et al., *Phys. Rev.* C **42** (1990) 640.
18. J. Barrette et al., E877 Collaboration, *Phys. Rev.* C **55** (1997) 1420.
19. H. Stöcker, W. Greiner, *Phys. Rept.* **137** (1986) 277.
20. L.D. Landau, *Izv. Akad. Nauk. Ser. Fiz.* **17** (1953) 51.
21. H. Sorge, *Phys. Lett.* B **402** (1997) 251.
22. H. Appelshäuser for the NA45 Collaboration, Proc. QM2001; G. Agakichiev et al., NA45 Collaboration, *Phys. Rev. Lett.* **92** (2004) 032301.
23. C. Alt et al., NA49 Collaboration, *Phys. Rev.* C **68** (2003) 034903.
24. M.M. Aggarwal et al., WA98 Collaboration, *Eur. Phys. J.* C **41** (2005) 287.
25. P. Chung et al., E895 Collaboration, *Phys. Rev.* C **66** (2002) 021901.
26. C. Adler et al., STAR Collaboration, *Phys. Rev. Lett.* **87** (2001) 182301.
27. J. Adams et al., STAR Collaboration, *Phys. Rev. Lett.* **92** (2004) 052302.
28. S.S. Adler et al., PHENIX Collaboration, *Phys. Rev. Lett.* **91** (2003) 182301.
29. B.B. Back et al., PHOBOS Collaboration, *Phys. Rev. Lett.* **89** (2002) 222301.
30. P. Huovinen, P.F. Kolb, U.W. Heinz, P.V. Ruuskanen, and S.A. Voloshin, *Phys. Lett.* B **503** (2001) 58; P. Huovinen, private communication.
31. D. Teaney, J. Lauret, and E.V. Shuryak,

182

Phys. Rev. Lett. **86** (2001) 4783 and nucl-th/0110037.

32. P.F. Kolb, U. Heinz, in Quark-Gluon Plasma 3, eds. R.C. Hwa and X.N. Wang, (World Scientific, Singapore, 2004) 634.

33. X. Dong, S. Esumi, P. Sorensen, and N. Xu, *Phys. Lett.* B **597** (2004) 328.

34. K.P. Das, R.C. Hwa, *Phys. Lett.* B **68** (1977) 459; Erratum-ibid.**73** (1978) 503.

35. S.A. Voloshin, *Nucl. Phys.* A **715** (2003) 379c.

36. R.J. Fries, B. Müller, C. Nonaka, S.A. Bass, *Phys. Rev. Lett.* **90** (2003) 202303.

37. S.S. Adler et al., PHENIX Collaboration, *Phys. Rev. Lett.* **91** (2003) 072301.

38. S.S. Adler et al., PHENIX Collaboration, *Phys. Rev. Lett.* **91** (2003) 241803.

39. D. d'Enterria for the PHENIX Collaboration, inv. talk at Nato Adv. Study Inst., Kemer, Turkey, 2003, nucl-ex/0401001.

40. D. d'Enterria, Proc. Renc. de Moriond 2004, nucl-ex/0406012.

41. X.N. Wang and M. Gyulassy, *Phys. Rev. Lett.* **68** (1992) 1480.

42. H. Baier, Y.L. Dokshitzer, A.H. Mueller, S. Peigne, and D. Schiff, *Nucl. Phys.* B **483** (1997) 291 and B**84** (1997) 265.

43. I. Vitev, Proc. ICPAQGP, Kolkata 2005, *J. Phys.* G in print, hep-ph/0503221.

44. D. d'Enterria, Proc. "Hard Probes 2004", *Eur. Phys. J.* C, in print, nucl-ex/0504001.

45. Q. Wang and X.N. Wang, *Phys. Rev.* C **71** (2005) 014903.

46. B.B. Back et al., PHOBOS Collaboration, *Phys. Rev. Lett.* **91** (2003) 072302.

47. S.S. Adler et al., PHENIX Collaboration, *Phys. Rev. Lett.* **91** (2003) 072303.

48. J. Adams et al., STAR Collaboration, *Phys. Rev. Lett.* **91** (2003) 072304.

49. I. Arsene et al., BRAHMS Collaboration, *Phys. Rev. Lett.* **91** (2003) 072305.

50. S.S. Adler et al., PHENIX collaboration, *Phys. Rev.* D **71** (2005) 071102R.

51. S.S. Adler et al., PHENIX collaboration, *Phys. Rev. Lett.* **94** (2005) 232301.

52. L.E. Gordon and W. Vogelsang, *Phys. Rev.* D **48** (1993) 3136.

53. Y. Akiba for the PHENIX collaboration, Proc. Quark Matter 2005 Conference, *Nucl. Phys.* A , in print, nucl-ex/0510008.

54. J. Adams et al., STAR Collaboration, nucl-ex/0501016.

55. C. Adler et al., STAR Collaboration, *Phys. Rev. Lett.* **90** (2003) 082302.

56. J. Adams et al., STAR Collaboration, *Phys. Rev. Lett.* **93** (2004) 252301.

57. D. Adamova et al., CERES Collaboration, *Phys. Rev. Lett.* **92** (2004) 032301.

58. S.S. Adler et al., PHENIX Collaboration, nucl-ex/0507004.

59. H. Stöcker, *Nucl. Phys.* A **750** (2005) 121.

60. J. Casalderrey-Solana, E. Shuryak, and D. Teaney, hep-ph/0411315.

61. K. Adcox et al., PHENIX Collaboration, *Phys. Rev. Lett.* **88** (2002) 192303; S.S. Adler et al., PHENIX Collaboration, *Phys. Rev. Lett.* **94** (2004) 082301.

62. S.S. Adler et al., PHENIX Collaboration, nucl-ex/0503003.

63. M.A.C. Lamont for the STAR Collaboration, Proc. ICPAQGP Kolkata 2005, *J. Phys.* G in print.

64. B. Jacak for the PHENIX collaboration, Proc. ICPAQGP Kolkata 2005, *J. Phys.* G in print, nucl-ex/0508036.

65. N. Armesto, A. Dainese, C.A. Salgado, U. Wiedemann, *Phys. Rev.* D **71** (2005) 054027.

66. M.C. Abreu et al., NA50 Collaboration, *Phys. Lett.* B **521** (2001) 195.

67. B. Alessandro et al., NA50 Collaboration, *Eur. Phys. J.* C **39** (2005) 335.

68. R. Vogt, *Int. J. Mod. Phys.* E **12** (2003) 21; and hep-ph/0203151.

69. F. Karsch., Proc. Int. Conf. Hard and Electromagnetic Probes, Ericeira, Portugal 2004, hep-lat/0502014.

70. P. Braun-Munzinger, J. Stachel, *Phys. Lett.* B **490** (2000) 196.

71. A. Andronic, P. Braun-Munzinger, K. Redlich, J. Stachel, *Phys. Lett.* B **571** (2003) 36.

72. K. Adcox et al., PHENIX Collaboration, *J. Phys.* G **30** (2004) S1189.

73. J. Adams et al., STAR Collaboration, *Phys. Rev. Lett.* **94** (2005) 062301.

DISCUSSION

Igor Klebanov (Princeton University):
I cannot resist making a comment related to my talk tomorrow. One can ask why the strongly coupled plasma has low viscosity/entropy ratio. In fact string theory gives a clue about this via the AdS/CFT correspondence. It also suggests why the energy density at $T \approx 2T_C$ is only about 80 % of what it would be if the plasma were weakly coupled.

Johanna Stachel: Indeed it was pointed out by Shuryak that this ratio of viscosity to entropy may be very small indicative of a close to ideal fluid. This stems from studying the effect of viscous corrections to the elliptic flow variable; the data appear to require no such correction. However, we do not have at present a good experimental measure of viscosity yet and this question requires further study.

Adriano Di Giacomo (Pisa University and INFN): Could you comment on the experimental evidence of the critical point at $\mu \neq 0$ in the phase diagram you showed in one of your first slides?

Johanna Stachel: It has been proposed that fluctuations in various observables such as baryon number (in some rapidity interval) or possibly strangeness could grow very much in the vicinity of the critical point. Many fluctuations variables have already been studied in experiments and beyond a few tantalizing but still very preliminary hints no evidence for unusually large fluctuations has been established.

Gigi Rolandi (CERN):
ALICE was designed before RHIC data. Is there any aspect of the detector you would optimize in a different way now that you have seen RHIC data?

Johanna Stachel: While there still will be a lot of interest in soft probes, most of the new information at LHC is expected in the sector of hard probes, such as heavy quarks, jets and direct photons. The main goal will be to probe the properties of the Quark-Gluon Plasma. Already in 1996 we started to optimize ALICE in direction of hard probes. You were part of the committee that for instance recommended the addition of the TRD in order to optimize the performance for quarkonia, open charm and beauty, triggering on jets and generally improving the high momentum performance of the experiment in the central barrel. For instance the momentum resolution at 100 GeV is in present simulations close to 3 %. So, in summary we should be very well prepared for the challenges of heavy ion physics at LHC.

Günter Grindhammer (MPI Munich):
Can the fast equilibration of the quark-gluon soup be understood within QCD?

Johanna Stachel: All approaches to understand this within perturbative QCD were not successful so far. However, nonperturbative phenomena such as instabilities have been studied recently and could provide a way.

Bennie Ward (Baylor University):
In your talk you mentioned the press release about the possible discovery of a liquid state, but in the equation of state plot, you showed data agreeing with a quark gluon plasma. Which do you wish to claim, a gas or a liquid?

Johanna Stachel: As I commented already in my answer to Klebanov, a good experimental determination of the viscosity is still outstanding. Also, in lattice QCD at present no indications of a liquid behavior have been established yet. Also, the two approaches are not necessarily in contradiction. For example, in nuclear physics some phenomena are well described by the liquid drop model, for others the Fermi gas is a useful formalism.

HAVE PENTAQUARK STATES BEEN SEEN?

VOLKER D. BURKERT

Jefferson Laboratory

12000 Jefferson Avenue, Newport News, VA23606, USA

E-mail: burkert@jlab.org

The status of the search for pentaquark baryon states is reviewed in light of new results from the first two dedicated experiments from CLAS at Jefferson Lab and of new analyses from several labs on the $\Theta^+(1540)$. Evidence for and against the heavier pentaquark states, the $\Xi(1862)$ and the $\Theta_c^0(3100)$ observed at CERN and at HERA, respectively, are also discussed. I conclude that the evidence against the latter two heavier pentaquark baryons is rapidly increasing making their existence highly questionable. I also conclude that the evidence for the Θ^+ state has significantly eroded with the recent CLAS results, but still leaves room for a state with an intrinsic width of $\Gamma < 0.5$ MeV. New evidence in support of a low mass pentaquark state from various experiments will be discussed as well.

1 Introduction

The announcement in 2003 of the discovery of the $\Theta^+(1540)$, a state with flavor exotic quantum numbers and a minimum valence quark content of $(uudd\bar{s})$[1], generated a tremendous amount of excitement in both the medium-energy nuclear physics and the high energy physics communities. Within less than one year the initial findings were confirmed by similar observations in nine other experiments [2,3,4,5,6,7,8,9,10], both in high energy and in lower energy measurements. These results seemed to beautifully confirm the theoretical prediction, within the chiral soliton model by D. Diakonov, M. Petrov, and M. Polyakov [11], of the existence of state with strangeness $S = +1$, a narrow width, and a mass of about 1.53 GeV. This state was predicted as the isosinglet member of an anti-decuplet of ten states, three of which (Θ^+, Ξ^{--}, Ξ^+) with exotic flavor quantum numbers that experimentally can be easily distinguished from ordinary 3-quark baryons. Two observations of heavier pentaquark candidates at CERN [12] and at HERA [13] added to the expectation that a new avenue of research in hadron structure and strong QCD had been opened up. Yet, to this day two years after the initial announcement was made, I am here to address the question if pentaquark states have really been observed. What happened?

2 The positive sightings of the Θ^+

A summary of the published experimental evidence for the Θ^+ and the heavier pentaquark candidates is given in table 1. In most cases the published width is limited by the experimental resolution. The observed masses differ by up to more than 20 MeV. The quoted significance S in some cases is based on a naive, optimistic evaluation $S = signal/\sqrt{background}$, while for unknown background a more conservative estimate is $S = signal/\sqrt{signal + background}$, which would result in lowering the significance by one or two units. Despite this, these observations presented formidable evidence for a state at a mass of 1525-1555 MeV. A closer look at some of the positive observations begins to reveal possible discrepancies.

2.1 A problem with the width and production ratios Θ^+/Λ^*?

The analysis of $K^+ A$ scattering data showed that the observed Θ^+ state must have an intrinsically very narrow width. Two analyses of different data sets found finite width of $\Gamma = 0.9 \pm 0.3$ MeV [16,17], while others came up with upper limits of 1 MeV [18] to

186

Table 1. Initial positive observations of the Θ^+, Ξ_5, and Θ_c^0 pentaquark candidates.

Experiment	Reaction	Energy (GeV)	Mass (MeV/c^2)	significance
LEPS	$\gamma^{12}C \to K^- X$	$E_\gamma \approx 2$	1540 ± 10	4.6σ
DIANA	$K^+ Xe \to pK_s^0 X$	$E_{K^+} < 0.5$	1539 ± 2	4σ
CLAS(d)	$\gamma d \to pK^- K^+ n$	$E_\gamma < 3.8$	1542 ± 5	5.2
SAPHIR	$\gamma p \to K_s^0 K^+ n$	$E_\gamma < 2.65$	$1540 \pm 4 \pm 2$	4.4σ
CLAS(p)	$\gamma p \to \pi^+ K^- K^+ n$	$E_\gamma = 4.8 - 5.5$	1555 ± 10	7.8σ
νBC	$\nu A \to pK_s^0 X$	range	1533 ± 5	6.7σ
ZEUS	$ep \to epK_s^0 X$	$\sqrt{s} = 320$	1522 ± 1.5	4.6σ
HERMES	$ed \to pK_s^0 X$	$E_e = 27.6$	$1528 \pm 2.6 \pm 2.1$	5.2σ
COSY	$pp \to \Sigma^+ pK_s^0$	$P_p = 3$	1530 ± 5	3.7σ
SVD	$pA \to pK_s^0 X$	$E_p = 70$	$1526 \pm 3 \pm 3$	5.6σ
NA49	$pp \to \Xi^- \pi^- X$	$E_p = 158$	1862 ± 2	4σ
H1	$ep \to D^{*-} pD^{*+} \bar{p} X$	$\sqrt{s} = 320$	$3099 \pm 3 \pm 5$	5.4σ

several MeV [19,20]. When compared with the production ratio of the $\Lambda^*(1520)$ hyperon with intrinsic width of $\Gamma_{\Lambda^*} = 15.9$ MeV the rate of the total cross section for the formation of the two states is expected to be $R_{\Theta^+,\Lambda^*} \equiv \sigma_{tot}(\Theta^*)/\sigma_{tot}(\Lambda^*) = 0.014$ for a 1 MeV width of the Θ^+ [21]. Although this relationship holds strictly for resonance formation at low energies only, dynamical models for the photoproduction of the Θ^+ show that the production cross section at modest energies of a few GeV strongly depends on the width of the state[24,25,26,27]. One therefore might expect R_{Θ^+,Λ^*} to remain small in the few GeV energy range. The published data however, suggested otherwise: Much larger ratios were observed than expected from the estimate based on the Θ^+ width. These results, together with the upper limits obtained from experiment with null results, including the most recent results from CLAS are summarized in table 2.

3 Non-observations of the Θ^+.

Something else that happened in 2004 and 2005 was a wave of high energy experiments presenting high statistics data that did not

confirm the existence of the Θ^+ state. There are two types of experimental results, one type of experiments studies the decays of intermediate states produced in e^+e^- collisions, and gives limits in terms of branching ratios. The other experiments searched for the Θ^+ in fragmentation processes. Several experiments give upper limits on the R_{Θ^+,Λ^*} ratio. Detailed discussions of experiments that claimed sightings of a pentaquark candidate state, as well as those that generated null results are presented in detail in recent reviews[14,15].

4 Are these results consistent ?

It is difficult to compare the low energy experiments with the high energy experiments. Low energy experiments study exclusive processes where completely defined final states are measured, and hadrons act as effective degrees of freedom. At high energies we think in terms of quark degrees of freedom, and fragmentation processes are more relevant. How can these different processes be compared quantitatively? The only invariant quantities for a resonance are quantum numbers, mass, and intrinsic width. In the absence of a resonance signal we can only place an up-

Table 2. The ratio R_{Θ^+,Λ^*} measured in various experiments. The first 5 experiments claimed a Θ^+ signal, while the others give upper limits. The last two results are from the most recent CLAS measurements that give very small upper limits.

Experiment	Energy (GeV)	Θ^+/Λ^* (%)
LEPS	$E_\gamma \approx 2$	~ 40
CLAS(d)	$E_\gamma = 1.4 - 3.8$	~ 20
SAPHIR	$E_\gamma = 1.4 - 3$	10
ZEUS	$\sqrt{s} = 320$	5
HERMES	$E_\gamma \approx 7$	~ 200
CDF	$\sqrt{s} = 1960$	< 3
HERA-B	$\sqrt{s} = 42$	< 2
SPHINX	$\sqrt{s} = 12$	< 2
Belle	$\sqrt{s} \approx 2$	< 2.5
BaBar	$\sqrt{s} = 12$	< 3
CLAS-2(p)	$E_\gamma = 1.4 - 3.8$	< 0.2
CLAS-2(d)	$E_\gamma = 1.4 - 3.6$	
	$K^+\Lambda$ analysis	≈ 1.5

per limit on its width as a function of the invariant mass in which the resonance signal is expected to occur. It is therefore the total resonance width, or an upper limit on it, that allows us to compare processes for different reactions. It may not be unreasonable to assume that a narrow width of the Θ^+ would result in much reduced production cross section compared to broader states such as the Λ^* both at low and at high energies. A similar conclusion may be drawn if one considers quark fragmentation as the main source of hadron production at high energies, e.g. in $e^+e^- \to q\bar{q}$, where hadrons are generated through the creation of $q\bar{q}$ pairs from the vacuum via glue string breaking. In a scenario of independent creation of a number of $q\bar{q}$ pairs starting with a single $q\bar{q}$ pair created in e^+e^- annihilation, or a single quark knocked out of a target nucleon in deep inelastic scattering, four additional $q\bar{q}$ pairs from the vacuum are needed to form a $(uudd\bar{s})$ 5-quark object. This should be much less likely to occur than the creation of a 3-quark baryon such as the

Λ^*. The latter requires creation of only two additional $q\bar{q}$ pairs.

If we take the estimate $R_{\Theta^+,\Lambda^*} \sim 0.015$ for a 1 MeV width of the Θ^+ from low energy resonance formation as a guide, we have a way of relating high energy and low energy processes. Comparing the limits for that ratio from table 2 one can make several observations: 1) The first set of experiments claiming sighting of the Θ^* show very large ratios. 2) The second set of experiments quoting upper limits are not below the ratio extracted from low energy $K^+\Lambda$ analysis. 3) The recent CLAS results are an order of magnitude below that value. This is to be contrasted with the very large ratios measured in the first set of experiments in table 2. The focus of new experimental investigations should be to verify that these initial results are indeed correct.

5 New results - mostly against the existence of pentaquark states.

During the past six months much new evidence against and some in favor of the existence of pentaquark baryons have emerged. There are new high statistics results from the CLAS detector at Jefferson Lab. New analyses of previously published data have become available from ZEUS and the SVD-2 collaborations, and LEPS studied a new channel with claimed Θ^+ sensitivity. The Belle and BaBar collaborations have generated high statistics data that test the lower energy photoproduction results, and high energy experiments at Fermilab, HERA and CERN confront the claims for the Ξ_5 and Θ_c^0 pentaquark candidates. New evidence for a doubly charged Θ^{++} comes from the STAR detector at RHIC. These new results will be discussed in the following sections.

5.1 New results from CLAS.

The CLAS collaboration has recently completed the first two dedicated high statistics

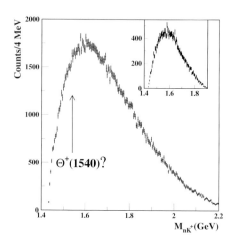

Figure 1. The invariant mass M_{nK^+} from the CLAS high statistics experiment on $\gamma p \to K_s^0 K^+ n$.

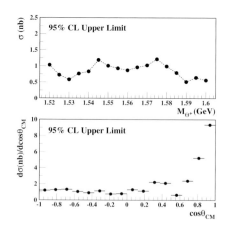

Figure 2. The CLAS upper limit on the total cross section for Θ^+ production from hydrogen (top). The bottom panel shows the limit on the differential cross section at a $K^+ n$ mass of 1540 MeV.

experiments aimed at verifying previously reported observations of the Θ^+.

The first experiment measured the reaction $\gamma p \to K_s^0 K^+ (n)$, where the neutron is reconstructed using 4-momentum conservation. The $M_{K^+ n}$ invariant mass distribution shown in Fig. 1 is structureless. An upper limit for the Θ^+ cross section is derived by fitting the data with a polynomial background distribution and a sliding Gaussian that represents the experimental resolution. The upper limit at 95% c.l. is shown in Fig. 2 versus the mass of the Θ^+. In the mass range of (1.525 to 1.555) GeV, a limit of (0.85 - 1.3) nb (95% c.l.) is derived. What does this result tell us? There are several conclusions that can be drawn from the CLAS result on the proton.

1) It directly contradicts, by two orders of magnitude in cross section, the SAPHIR experiment[4] that claimed a significant signal in the same channel and in the same energy range, and published a cross section of 300nb for Θ^+ production.

2) Together with the extracted Λ^* cross section it puts an upper limit on the Θ^+/Λ^* ratio in table 2 that is an order of magnitude lower than the value from the $K^+ A$ analysis,

and strongly contradicts the 'positive' Θ^+.

3) It puts a very stringent limit on a possible production mechanism. For example, it implies a very small coupling $\Theta^+ N K^*$ which in many hadronic models was identified as a major source for Θ^+ production.

4) If there is no large isospin asymmetry in the elementary process, the γD and γA experiments at lower statistics should not be able to see a signal. Possible mechanisms to obtain a large isospin asymmetry have been discussed in the literature following the first announcement of the new CLAS data[30,31].

The second new CLAS experiment measured the reaction $\gamma D \to p K^- K^+ (n)$, where the neutron again is reconstructed from the overdetermined kinematics. This experiment represents a dedicated measurements to verify a previous CLAS result that claimed more than 4.6σ significance for the Θ^+ in the same channel and same energy range. The aim is to measure the possibly preferred production on neutrons through $\gamma n \to K^- K^+ n$. To avoid the complication of precise neutron detection the recoil proton is measured instead, requiring momenta of greater than 0.3 GeV/c for the proton to be detected in CLAS. This reduces the acceptance for the exclusive re-

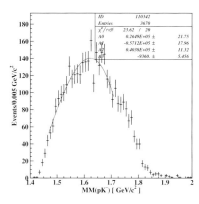

Figure 3. The missing mass M_{pK^-} from the CLAS high statistics deuterium experiment. Events are selected with the same kinematics as the previously published results.

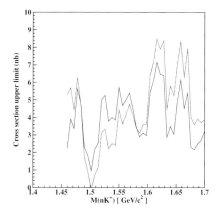

Figure 4. The upper limit (95% c.l.) of the total cross section for Θ^+ photoproduction from neutrons.

action by a large factor. The preliminary results, representing about 50% of the full statistics, are shown in Fig. 4. Again, no significant signal is seen in a data sample with about seven times the statistics of the previous result. From this result an upper limit of 5nb (95% c.l.) is derived for the elementary cross section on the neutron. The limit is somewhat model-dependent as rescattering effects in the deuteron must be taken into account. The result clearly contradicts the previous lower statistics data. In order to un-

Table 3. Limit on Θ^+ width from recent CLAS results. Upper limits for the total cross section on protons of 1.25nb, and on neutrons of 4nb are used to determine the limit on the width. The first line in each row is for γp, the second line is for γn. The cross section is computed for a $J^P = 1/2^+$ assignment of the Θ^+ and a width of 1 MeV.

Publication	$\sigma(\gamma N)$ (nbarn)	Γ_{Θ^+} (MeV)
S. Nam et al. [26]	2.7	< 0.5
	2.7	< 1.7
Y. Oh et al., [27]	~ 1.6	< 0.8
	~ 8.7	< 0.5
C.M Ko et al., [25]	15	< 0.08
	15	< 0.25
W. Roberts [24]	5.2	< 0.24
	11.2	< 0.4

derstand the discrepancy the older data have been reanalyzed with a background distribution extracted from the new high statistics data set. The results show an underestimation of the background normalization in the original analysis. A new fit with the improved background yields a signal with a significance of 3σ compared to the $(5.2\pm0.6)\sigma$ published.

What is the impact of the combined CLAS data on proton and neutron? For this we compare the cross section limits with various dynamical model calculations[24,25,26,27]. In hadro-dynamical models, the cross section is computed based on an effective Lagrangian approach. The comparison is shown in table 3 for the $J^P = 1/2^+$ assignment and $\Gamma_{\Theta^+} = 1$ MeV. The upper limit for the combined proton and neutron targets would be less than 0.5 MeV for at least one of the targets in each model.

5.2 BaBar study of quasi-real photoproduction

The BaBar collaboration also studied the quasi-real photoproduction of $e + Be \rightarrow pK_s^0 + X$ [33]. In this case electrons with energies of ~ 9 GeV resulting from small an-

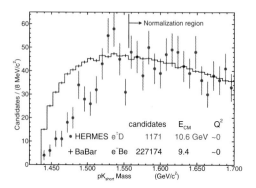

Figure 5. Comparison of BaBar results on quasi-real photoproduction of pK_s^0 from beryllium, with the HERMES results. The very high statistics BaBar data do not show any structure near 1530 MeV, while the HERMES data do. The falloff of the HERMES data near the lower mass end may indicate acceptance limitations.

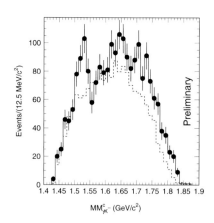

Figure 6. The LEPS data on deuterium. The missing mass distribution $MM(\gamma K^-)$ showing a peak at 1530 MeV.

gle scattering off the positron beam interact with the beryllium beam pipe. The scattered electron is not detected, and the invariant mass of final state inclusive pK_s^0 is studied for possible contributions from $\Theta^+ \to pK_s^0$. There is no evidence for a signal. The data can be directly compared to the HERMES results which were taken in quasi-real photoproduction kinematics from deuterium at higher electron beam energies. The comparison is shown in Fig. 5. While at high masses the two distributions coincide, a potential loss of acceptance at HERMES is seen for low mass pK_s^0 pairs. The HERMES peak may, at least in part, be the result of the acceptance rising up below the nominal Θ^+ mass. The absence of any signal in the high statistics BaBar data calls the signal observed by HERMES into question. BaBar also compare their null results with the ZEUS signal observed at $Q^2 > 20$ GeV^2. However, since ZEUS sees no signal at low Q^2 and BaBar only probes the quasi-real photoproduction kinematics, this comparison is indeed misleading.

5.3 LEPS at SPring-8

The LEPS experiment originally claimed the discovery of the Θ^+ in photoproduction from

a carbon target in the inclusive reaction $\gamma C \to K^- X$ plotting the Fermi-momentum corrected missing mass M_X. The experiment has been repeated with a liquid deuterium target and higher statistics. The still preliminary results are shown in Fig. 6. A peak at 1530 MeV is observed. The data also show a large ratio of Θ^+/Λ^* (see table 2). Since these data are obtained at energies similar to the new CLAS data on deuterium, they need to be confronted with the recent exclusive CLAS data taken on deuterium in the reaction $\gamma d \to K^- pK^+ n$, and the resulting cross section limit for the elementary cross section on neutrons. This will require extraction of a normalized cross section from the LEPS data. There are also new results from LEPS on the channel $\gamma d \to \Lambda^* X$. The mass distribution $M_{K^+ n}$ is shown in Fig. 7. After accounting for background from $\Lambda^*(1520)$ and sideband subtraction, a narrow peak near 1530 MeV/c^2 with 5σ significance is claimed. The signal emerges only when events are selected with $M_{K^-p} \sim M_{\Lambda^*}$, indicating that the process $\gamma D \to \Lambda^* \Theta^+$ may be observed. The mass distribution also shows an excess of events near 1600 MeV/c^2.

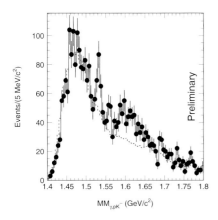

Figure 7. The LEPS data on deuterium.The missing mass distribution MM_{γ,K^-p} with events selected with the invariant mass M_{K^-p} near the Λ^*. A peak is seen at a mass of 1530 MeV, and is interpreted as the Θ^+.

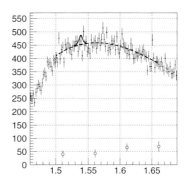

Figure 8. The Belle pK_s^0 mass spectrum.

5.4 Results from Belle

New results by the Belle collaboration have been presented recently[22]. Belle uses hadrons created in high energy e^+e^- collisions and reconstructs the hadron interaction with the vertex detector materials. The momentum spectrum is sufficiently low so that resonance formation processes such as $K^+n \rightarrow \Theta^+ \rightarrow pK_s^0$ can be studied. The high statistics pK_s^0 invariant mass shows no signal, as is seen in Fig. 8. The upper limit on the formation cross section can be used to extract an upper limit for the Θ^+ width, which is shown in Fig. 9. At a specific mass of 1539 MeV, an upper limit of $\Gamma_{\Theta^+} < 0.64$ MeV (90% c.l.) is derived. The mass corresponds to the Θ^+ mass claimed by the DIANA experiment[2]. However, if one allows the entire mass range for the Θ^+ from 1525 to 1555 MeV claimed by experiments, the upper limit would be $\Gamma_{\Theta^+} < 1$ MeV (90% c.l.). The latter value confirms the limit derived in previous analyses.

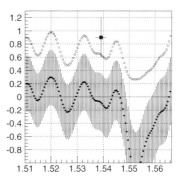

Figure 9. The Belle upper limit on the Θ^+ width. The dotted line shows the 90% c.l. limit for the width. The data point is result of the analysis of the DIANA result in $K^+Xe \rightarrow K_s^0 p + X$.

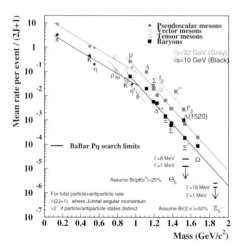

Figure 10. The BaBar baryon mass spectrum.

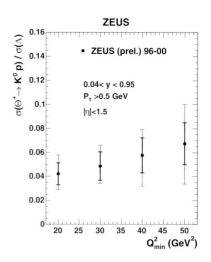

Figure 11. Q^2-dependence of the Θ^+/Λ^* ratio measured by ZEUS.

5.5 BaBar results in quark fragmentation

The BaBar collaboration at SLAC searches for the Θ^+ as well as the Ξ^{--} pentaquark states directly in e^+e^- collisions[23], mostly in the quark fragmentation region. With high statistics no signal is found for either $\Theta^+(1540)$ or $\Xi^{--}(1862)$, and upper limits are placed on their respective yields. The results are shown in Fig. 10. The limit on the production rates are 8 or 4 times lower than the rates of ordinary baryons at the respective masses. It is, however, not obvious what this result implies. The slope for the production of pseudoscalar mesons is $d(event\ rate)/d(mass) = 10^{-2}/\text{GeV}$. For 3-quark baryons it is $10^{-4}/\text{GeV}$, i.e. the rate drops by a factor of 10,000 per one unit of GeV in mass. In the quark fragmentation region, if we extrapolate from mesons where only one $q\bar{q}$ pair must be created to form a meson starting with one of the initial quarks in the e^+e^- annihilation, and baryons where two $q\bar{q}$ pairs are needed, to pentaquarks where four $q\bar{q}$ pairs are needed, the slope for pentaquark production in fragmentation would be $10^{-8}/\text{GeV}$. Since there

is no rate measured for a pentaquark state there is no normalization point available. If we arbitrarily normalize the pentaquark line at the point where baryon and meson lines intersect, the line falls one order of magnitude below the upper limit for the Θ^+ and several orders of magnitude below the upper limit for the Ξ^{--} assuming a mass of 1862 MeV for the latter. The sensitivity of quark fragmentation to 5-quark baryon states is thus questionable. Moreover, the limit for $R_{\Theta^+,\Lambda^*} < 0.02$ at 95% c.l. is not in contradiction with the ratio estimated at low energy assuming a width of $\Gamma_{\Theta^+} = 1$ MeV [17].

5.6 New Θ^+ analysis from ZEUS

The ZEUS collaboration has extended the analysis of their Θ^+ signal and studied possible production mechanisms[32]. The signal emerges at $Q^2 > 20$ GeV^2 and remains visible at $Q^2 > 50$ GeV^2. The Θ^+ and $\bar{\Theta}^-$ signals are nearly equally strong, however, the signal is present only at forward rapidity $\eta^{lab} > 0$ and not visible at backward rapidity $\eta^{lab} < 0$. There is currently no possible production mechanism that would generate such a pattern. The ZEUS collaboration extracted

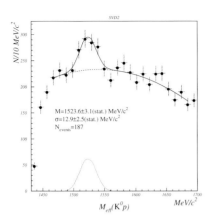

Figure 12. Results of a new analysis by the SVD-2 group of their published data.

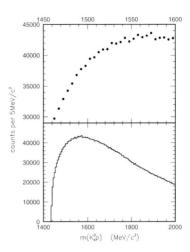

Figure 13. Invariant mass $M_{K^0 p}$ measured in the CERN WA89 hyperon experiment.

the Q^2-dependence of the ratio R_{Θ^+,Λ^*} which is shown in Fig. 11. It shows a weak dependence on Q^2 .

5.7 New results from high energy hadronic interaction experiments

The SVD-2 collaboration has reanalyzed their published data[42] with much improved event reconstruction efficiency. The experiment measured the reaction $pA \to pK_s^0 X$ using a 70 GeV incident proton beam. The main component in the detector system is the silicon vertex detector (SVD). Events are divided into two samples: events with the K_s^0 decaying inside and events with the decay outside the SVD. The two distributions both show a significant peak at the mass of 1523 MeV/c^2. One of the distributions is shown in Fig. 12. A combined significance for two independent data sets of $\sim 7.5\ \sigma$ is obtained. The strangeness assignment in the pK_s^0 channel is not unique, and could also indicate excitation of a Σ^* resonance. In this case one would expect a decay $\Sigma \to \Lambda\pi$, which is not observed. Therefore, an exotic $S = +1$ assignment of that peak is likely should it be a resonant state. The SVD-2 results have been challenged by the WA89 collaboration that measured the process $\Sigma^- A \to pK_s^0 X$ in

comparable kinematics[43]. Their mass distribution, shown in Fig. 13, does not exhibit any signal in the mass range of the Θ^+ candidate. The WA89 collaboration claims their results to be incompatible with the SVD results.

6 An isovector Θ^{++} candidate?

Inspired by the prediction of Diakonov et al., of an anti-decuplet of 5-quark states, with the Θ^+ being an isoscalar, the focus of the search for the lowest mass pentaquark was on an isoscalar baryon with $S = +1$. However, searches have also been conducted for a possible isovector baryon state with charge $Q = +2$. The final state to study is pK^+. No signal was seen in any of these searches. However, recently the STAR collaboration at RHIC presented data indicating a small but significant Θ^{++} candidate[44]. The data are shown in Fig. 14 before and after background subtraction. A peak with a significance of 5σ is seen at a mass of about 1530 MeV/c^2 in the d-Au collision sample. The Λ^* signal is also clearly visible.

If the Θ^{++} signal is real, then there must be also a signal in the singly charged channel, i.e. a Θ^+ . A small peak with relatively low significance appears in the $K_s^0 p$ invariant

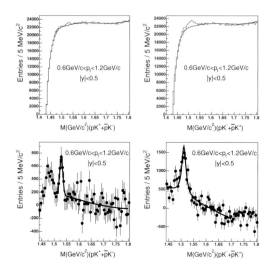

Figure 14. Invariant mass $pK^+ + \bar{p}K^-$ (left) and $K^-p + K^+\bar{p}$ (right) measured by STAR. The spectra are shown before (top) and after (bottom) subtraction of background from mixed events. The l.h.s. shows the Θ^{++} candidate signal, the r.h.s shows the Λ^* signal.

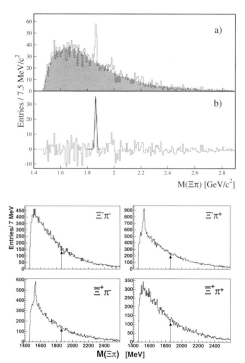

Figure 15. The NA49 results (top) on the claimed $\Xi_5(1862)$. The sum of the $\Xi^-\pi^-$ and $\Xi^-\pi^+$ distributions are shown before and after background subtraction. The COMPASS results are displayed in the bottom panel. The $\Xi(1530)$ state is clearly seen in the neutral charge combinations, while the Ξ_5 expected at 1862 MeV is absent.

mass spectrum, however shifted by about 10 MeV/c^2 to higher mass values.

So far I have focused in my talk on the Θ^+, as without the evidence for the Θ^+ there would not have been any search for other pentaquark states within the anti-decuplet. However, much effort has been put recently into the search for the two heavier pentaquark candidates claimed in two high energy experiments.

7   Status of Ξ_5 and Θ_c^0

A candidate for a 5-quark Ξ_5 has been observed in the $\Xi^-\pi^-$ final state by the CERN NA49 experiment, and a candidate Θ_c^0 for the charmed equivalent of the Θ^+ has been claimed by the H1 experiment at HERA in the channel D^*p. In contrast to the Θ^+, which has been claimed in at least ten experiments, the heavier candidates have not been seen in any other experiment. The Ξ_5 state of NA49 has been searched for by

several experiments[34,35,37,38,39,40,41]. The ratios $\Xi_5/\Xi(1530)$ determined by several experiments are shown in table 7. However, the highest energy experiments probe production through quark fragmentation, and may not be directly comparable to the NA49 results. The FOCUS photoproduction experiment and the COMPASS muon scattering experiment are close to the kinematics of NA49. The COMPASS results are shown in Fig. 15 and compared to the original NA49 results. No signal is observed. FOCUS also did not observe a signal and obtained an upper limit nearly two orders lower than the signal seen by NA49. A summary of the search for the Ξ_5 is shown in Fig. 16.

The $\Theta_c^0(3100)$ pentaquark candidate so

Table 4. Results of searches for the Ξ_5 pentaquark state.

Exp.	Initial state	Energy (GeV)	$\frac{\Xi_5^{--}}{\Xi(1530)}$
NA49	pp	$E_p = 158$	0.24
COMPASS	$\mu^+ A$	$E_\mu = 160$	< 0.046
ALEPH	e^+e^-	$\sqrt{s} = M_Z$	< 0.075
BaBar	e^+e^-	$m_{Y(4s)}$	< 0.0055
CDF	$p\bar{p}$	$\sqrt{s} = 1960$	< 0.03
E690	pp	$E_p = 800$	< 0.003
FOCUS	γp	$E_\gamma < 300$	< 0.003
HERA-B	pA	$E_p = 920$	< 0.04
HERMES	eD	$E_e = 27.6$	< 0.15
WA89	$\Sigma^- A$	$E_\Sigma = 340$	< 0.013
ZEUS	ep	$\sqrt{s} = 310$	not seen

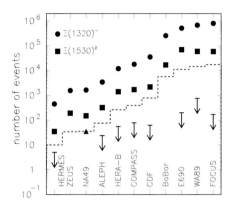

Figure 16. Summary of experimental results in the search for the exotic cascade Ξ_5. A comparison of the number of observed ground state $\Xi^-(1320)$ and excited state $\Xi^0(1530)$ is shown. The arrows indicate upper limits for the number of exotic Ξ_5 candidates. Only NA49 has observed a signal.

far has only been seen by the H1 experiment at HERA. Several other experiments came up empty-handed[45], and ZEUS and FOCUS claim incompatibility of their results with the H1 findings. The H1 results and the FOCUS results are shown in Fig. 17.

8 Summary and conclusions

Over the past year the evidence for the existence of pentaquark baryons has clearly lost much of its original significance.

The evidence for the two heavy pentaquark candidate states, the $\Xi_5(1862)$ and the $\Theta_c^0(3100)$, observed at unexpected masses, and each seen in one experiment only, has been drastically diminished. Several experiments with high sensitivity to the relevant processes have found no indication of these states. In the face of overwhelming evidence against these states, experiments claiming positive sightings should either explain why other experiments are not sensitive, or should re-evaluate their own results.

The situation with the Θ^+ state observed at masses near 1540 MeV is less clear, although evidence for the state has also diminished significantly. So far more than ten experiments claimed to have observed a nar-

row state with exotic flavor quantum number $S = +1$. Two of the initial results (SAPHIR and CLAS(d)) have been superseded by higher statistics measurements from CLAS[28,29] conducted at same energies and with same or overlapping acceptances. No signal was found in either case. In addition, the HERMES results are being challenged by new high statistics data from BaBar[33]. It is remarkable that experiments claiming a Θ^+ signal, measured Θ^+/Λ^* ratios much above values naively expected from K^+A scattering analysis (see table 2). HERMES even measures a Θ^+ cross section that is significantly higher than the cross section for Λ^* production.

The Belle experiment[22] studying K^+A scattering, is beginning to challenge the DIANA results, the second experiment claiming observation of the Θ^+. Belle extracted an upper limit of 0.64 MeV (90% c.l.) for the width of any Θ^+ signal at a mass of 1539 MeV. In the larger mass range of 1525-1555 an upper limit of 1 MeV has been extracted. This is to

MeV are obtained. Much smaller limits of < 0.1 MeV are obtained for $J^P = 3/2^-$, while for $J^P = 1/2^-$, limits of 1 to 2.5 MeV are extracted. Although these limits are model-dependent, taken together they still present formidable constraints on the Θ^+ width. Hadronically decaying resonances with total decay widths of less than a few MeV would seem unusual, but widths of a few hundred keV or less would make the existence of the state highly unlikely. In order to have quantitative tests of the LEPS results, which is the only remaining low energy photoproduction experiment with a positive signal, the old and new results from LEPS should be turned into normalized cross sections and compared to the CLAS data on deuterium.

When the dust will have settled on the issue of narrow pentaquark baryons, we will have learned a lot about the physics of hadrons, no matter what the final outcome will be.

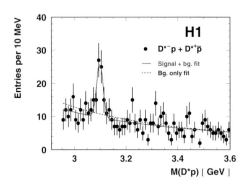

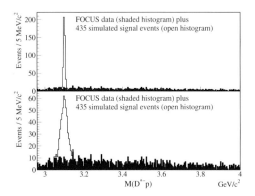

Figure 17. Top: The results of H1 on a charmed $\Theta_c^0(3100)$ candidate. Bottom: Results of searches on the Θ_c^0 candidate by FOCUS. The colored area shows the measured distributions, while the solid line indicates the expected signal extrapolated from the H1 measurement.

be contrasted with the width extracted from the DIANA experiment as well as from K^+D scattering of $\Gamma_{\Theta^+} = 0.9 \pm 0.3$ MeV. However, the Belle limit is not (yet) in strict contradiction to the DIANA results at this point. It will be interesting to see if the Belle limit can be further reduced with higher statistics.

The new CLAS results on protons and neutrons also challenge the value of the Θ^+ width. Using hadronic models, upper limits of 0.1 to 0.6 MeV are obtained for proton targets, and the $J^P = 1/2^+$ assignment. For neutron targets limits from 0.26 and 1.7

References

1. T. Nakano *et al.* (LEPS), *Phys. Rev. Lett.* 91:012002 (2003)
2. V.V. Barmin *et al.* (DIANA), *Phys. Atom. Nuclei* 66, 1715 (2003)
3. S. Stepanyan *et al.* (CLAS), *Phys. Rev. Lett.* 91:25001 (2003)
4. J. Barth *et al.* (SAPHIR), *Phys. Lett.* B 572, 127 (2003)
5. A.E. Asratyan, A.G. Dolgolenko, M.A. Kubantsev, *Phys. Atom. Nuclei* 67, 682 (2004)
6. V. Kubarovsky *et al.* (CLAS), *Phys. Rev. Lett.* 92:032001 (2004)
7. A. Airapetian, *et al.* (HERMES), *Phys. Lett.* B 585, 213 (2004)
8. S. Chekanov *et al.*, (ZEUS), *Phys. Lett.* B 591, 7 (2004)
9. M. Abdel-Barv *et al.* (COSY-TOF), *Phys. Lett.* B 595, 127 (2004)
10. A. Aleev *et al.* (SVD), hep-ex/0401024
11. D. Diakonov, P. Petrov, M. Polyakov, *Z. f. Phys.* A 359, 305 (1997)
12. C. Alt *et al.* (NA49), *Phys. Rev. Lett.* 92:042003 (2004)
13. A. Aktas *et al.* (H1), *Phys. Lett.* B 588, 17

(2004)

14. K. Hicks, hep-ph/0504027

15. M. Danilov, hep-ex/0509012

16. R.N. Cahn and G.H. Trilling, *Phys. Rev. D* 69:011501 (2004)

17. W. Gibbs, *Phys. Rev. C* 70:045208 (2004)

18. A. Sibirtsev *et al.*, *Eur. Phys. J. A* 23:491 (2005)

19. R. Arndt, I. Strakovsky, R. Workman, *Phys. Rev. C* 68:042201, 2003, Erratum-ibid.C69:019901 (2004)

20. R. Gothe and S. Nussinov, hep-ph/0308230

21. W. Gibbs, private communications

22. K. Abe *et al.* (BELLE), hep-ex/0507014

23. Tetiana Berger-Hryn'ova (BaBar), hep-ex/0510043

24. W. Roberts; *Phys. Rev. C* 70, 065201 (2004)

25. C.M. Ko and W. Liu, nucl-th/0410068

26. S. Nam *et al.*; hep-ph/0505134

27. Y. Oh, K. Nakayama, and T.-S.H. Lee, hep-ph/0412363

28. M. Battaglieri *et al.* (CLAS), hep-ex/05-10061, submitted to *Phys. Rev. Lett.*

29. S. Stepanyan *et al.* (CLAS), paper contributed to this conference

30. S. Nam, A. Hosaka, and H. Kim, hep-ph/0508220

31. Marek Karliner and Harry J. Lipkin, hep-ph/0506084

32. ZEUS collaboration, submitted to XXII International Symposium on Lepton Photon Interactions at High Energies, July 2005, Uppsala, Sweden.

33. K. Goetzen (BaBaR), hep-ex/0510041; J. M. Izen, Talk presented at the EINN workshop on "New hadrons: Facts and Fancy", Milos, Greece, September 19-20, 2005

34. B. Aubert *et al.* (BaBar), hep-ex/0502004

35. D.O. Litvintsev (CDF), hep-ex/0410024

36. E.S. Ageev *et al.*(COMPASS), *Eur. Phys. J. C* 41, 469 (2005);

37. K. Stenson (FOCUS), hep-ex/0412021

38. K.T. Knöpfle *et al.* (HERA-B), *J. Phys. G* 30:S871 (2004); J. Spengler (HERA-B), hep-ex/0504038

39. A. Airapetian *et al.* (HERMES), *Acta Phys. Polon.* B 36:2213 (2005)

40. M.I. Adamavich *et al.* (WA89), *Phys. Rev. C* 70:022201 (2004)

41. S. Chekanov *et al.* (ZEUS), paper submitted to LP2005, DESY-05-18 (2005)

42. A. Aleev *et al.* (SVD-2), hep-ex/0509033

43. M.I. Adamovich *et al.* (WA89), hep-ex/05100013

44. H. Z. Huang, nucl-ex/0509037

45. J.M. Link *et al.* (FOCUS); hep-ex/0506013; S. Chekanov *et al.* (ZEUS), paper submitted to LP2005, DESY-04-164(2004); A. Airapetian *et al.* (HERMES), *Phys. Rev. D* 71:032004 (2005)

DISCUSSION

Michael Danilov (ITEP):

(1) What is the fate of the $7.8\,\sigma$ CLAS signal in γp collisions?

(2) Is there any explanation of comparable Θ^+ and Λ^* rates in experiments which claim Θ^+ observation and limits of a few % on this ratio in experiments which do not see Θ^+?

Volker Burkert :

(1) The CLAS signal in $\gamma p \rightarrow \pi^+ K^- K^+ n$ was observed with the π^+ at forward angles. This result still stands, although the significance maybe lower than what is quoted in the paper. This process was measured at rather high photon energies, and the same kinematics can not be reached with the new high statistics data taken at lower energies. An experiment at 6 GeV is planned for 2006/7 to check this result with high statistics.

(2) With regard to the Θ^+ and $\Lambda(1530)$ ratio, we know from the narrow width of 1 MeV or less, that the ratio for the formation process of the two states is 1.4% or less. I would not expect to observe much larger ratios in low energy processes. In the two cases I discussed, the SAPHIR results on hydrogen, and the 2003 CLAS(d) results on deuterium, the large ratio of Θ^+ and $\Lambda(1530)$ observed in both experiments turned out to be incorrect. The high statistics 2005 CLAS(p) results on protons give an upper limit (95% c.l.) of 0.16%, while SAPHIR had a ratio of about 10%. The 2005 CLAS(d) data on deuterium also give only an upper limit for that ratio. Most high energy experiments give upper limits of a few % for that ratio. ZEUS measured a ratio of about 5%. HERMES quotes a Θ^+ cross section that is several times larger than the one for Λ^* production. The HERMES result is now also being challenged by the latest results from BaBar that access a similar phase space in quasi-real photoproduction on light nuclei but has much higher statistics and sees no signal.

Vincenzo Cavasinni (Pisa):

Besides possible statistical fluctuations, are there possible experimental systematic effects which could produce an enhancement in the Θ^+ mass region.

Volker Burkert :

There are a number of effects that can generate enhancements and even narrow structures. Kinematical reflections due to heavy meson production that decay to $K^+ K^-$ final states may appear as shoulders or enhancements in mass distributions when projected on the meson-baryon axis of the Dalitz plot if the phase space is truncated by selecting events in some region of phase space. There can also be issues of particle misidentification that can result in false peaks. False peaks can also be generated by so-called ghost tracks. Usually the analyzers are aware of such pitfalls and check their analysis to protect against such effects.

Barbara Badelek (Uppsala/Warsaw):

(1) Concerning the cascade pentaquark: also COMPASS does not see it (published on www in March '05) which is important due to the kinematics being almost identical to that of NA49 and the statistics is ten times larger.

(2) Concerning your last remark in the conclusions: the situation resembles that of the formed "A2 split" which has indeed split the scientific community at that time. Effects of $20\,\sigma$ were observed. Now A2 is for sure not split and we do not bother about those experiments that saw the effect.

Volker Burkert :

Thank you for these comments.

DEVELOPMENTS IN PERTURBATIVE QCD

GAVIN P. SALAM

LPTHE, Universities of Paris VI and VII and CNRS UMR 7589, Paris, France.

A brief review of key recent developments and ongoing projects in perturbative QCD theory, with emphasis on conceptual advances that have the potential for impact on LHC studies. Topics covered include: twistors and new recursive calculational techniques; automation of one-loop predictions; developments concerning NNLO calculations; the status of Monte Carlo event generators and progress in matching to fixed order; analytical resummation including the push to NNLL, automation and gap between jets processes; and progress in the understanding of saturation at small x.

1 Introduction

A significant part of today's research in QCD aims to provide tools to help better constrain the standard model and find what may lie beyond it. For example one wishes to determine, as accurately as possible, the fundamentals of the QCD and electroweak theories, such as α_s, quark masses, and the elements of the CKM matrix. One also needs precise information about 'pseudo-fundamentals,' quantities such as parton distribution functions (PDFs) that could be predicted if we knew how to solve non-perturbative QCD, but which currently must be deduced from experimental data. Finally, one puts this information together to predict the QCD aspects of both backgrounds and signals at high energy colliders, particularly at the Tevatron and LHC, to help maximize the chance of discovering and understanding any new physics.

Other facets of QCD research seek to extend the boundaries of our knowledge of QCD itself. The underlying Yang-Mills field theory is rich in its own right, and unexpected new perturbative structures have emerged in the past two years from considerations of string theory. In the high-energy limit of QCD it is believed that a new state appears, the widely studied colour glass condensate, which still remains to be well understood. And perhaps the most challenging problem of QCD is that of how to relate the partonic and hadronic degrees of freedom.

Given the practical importance of QCD for the upcoming LHC programme, this talk will concentrate on results (mostly since the 2003 Lepton-Photon symposium) that bring us closer to the well-defined goals mentioned in the first paragraph. Some of the more explorative aspects will also be encountered as we go along, and one should remember that there is constant cross-talk between the two. For example: improved understanding of field theory helps us make better predictions for multi-jet events, which are important backgrounds to new physics; and by comparing data to accurate perturbative predictions one can attempt to isolate and better understand the parton-hadron interface.

The first part of this writeup will be devoted to results at fixed order. At tree level we will examine new calculational methods that are much more efficient than Feynman graphs; we will then consider NLO and NNLO calculations and look at the issues that arise in going from the Feynman graphs to useful predictions.

One of the main uses of fixed-order order predictions is for understanding rare events, those with extra jets. In the second part of the writeup we shall instead turn to resummations, which help us understand the properties of typical events.

Throughout, the emphasis will be on the conceptual advances rather than the detailed phenomenology. Due to lack of space, some active current topics will not be covered, in particular exclusive QCD. Others are discussed elsewhere in these proceedings.[1]

2 Fixed-order calculations

2.1 Tree-level amplitudes and twistors

Many searches for new physics involve signatures with a large number of final-state jets. Even for as basic a process as $t\bar{t}$ production, the most common decay channel (branching ratio of 46%), $t\bar{t} \to b\bar{b}W^+W^- \to b\bar{b}q\bar{q}q\bar{q}$ involves 6 final-state jets, to which there are large QCD multi-jet backgrounds.[2] And at the LHC, with $10\,\text{fb}^{-1}$ (1 year) of data, one expects of the order of 2000 events with 8 or more jets[3] ($p_t(\text{jet}) > 60\,\text{GeV}$, $\theta_{ij} > 30\,\text{deg}$, $|y_i| < 3$).

For configurations with such large numbers of jets, even tree-level calculations become a challenge — for instance $gg \to 8g$ involves 10525900 Feynman diagrams (see ref.[4]). In the 1980's, techniques were developed to reduce the complexity of such calculations.[5] Among them colour decomposition,[6] where one separates the colour and Lorentz structure of the amplitude,

$$\mathcal{A}^{\text{tree}}(1, 2, \ldots, n) = g^{n-2} \sum_{perms}$$

$$\underbrace{\text{Tr}(T_1 T_2 \ldots T_n)}_{\text{colour struct.}} \underbrace{A^{\text{tree}}(1, 2, \ldots, n)}_{\text{colour ordered amp.}}; \quad (1)$$

the use[7] of spinor products $\langle ij \rangle \equiv \langle i^-|j^+ \rangle = \overline{u_-(k_i)}u_+(k_j)$ and $[ij] = \langle i^+|j^- \rangle$ as the key building blocks for writing amplitudes; and the discovery[8] and subsequent proof[9] of simple expressions for the subset of amplitudes involving the maximal number of same-helicity spinors (maximum helicity violating or MHV amplitudes), i.e. $n - 2$ positive helicity spinors for an n-gluon amplitude:

$$A^{\text{tree}}(--++\ldots) = \frac{i\langle 12 \rangle^4}{\langle 12 \rangle \langle 23 \rangle \ldots \langle n1 \rangle}. \quad (2)$$

Ref.[9] also provided a computationally efficient recursion relation for calculating amplitudes with arbitrary numbers of legs, and with these and further techniques,[10] numerous programs (e.g. MadEvent,[11] ALPGEN,[12] HELAC/PHEGAS,[13] CompHEP,[14] GRACE,[15] Amegic[16]) are able to provide results for processes with up to 10 legs.

The past two years have seen substantial unexpected progress in the understanding of multi-leg tree-level processes. It was initiated by the observation (first made by Nair[17]) that helicity amplitudes have a particularly simple form in 'twistor' space, a space where a Fourier transform has been carried out with respect to just positive helicity spinors. In twistor space a duality appears[18] between the weakly-coupled regimes of a topological string theory and $\mathcal{N} = 4$ SUSY Yang-Mills. This has led to the postulation, by Cachazo, Svrcek and Witten (CSW),[19] of rules for deriving non-MHV $\mathcal{N} = 4$ SUSY amplitudes from MHV ones, illustrated in fig. 1. For purely gluonic amplitudes the results are identical to plain QCD,[20] because tree level SUSY amplitudes whose external legs are gluons have only gluonic propagators.

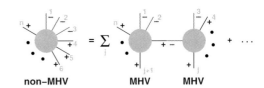

Figure 1. Graphical illustration of the CSW rules: by joining together two MHV amplitudes with an off-shell scalar propagator one obtains an amplitude with an extra negative helicity (NMHV).

Recently a perhaps even more powerful set of recursion relations was proposed by Britto, Cachazo and Feng (BCF),[21] which allows one to build a general n-leg diagram by joining together pairs of on-shell sub-diagrams. This is made possible by continuing a pair of reference momenta into the complex plane. The proof[22] of these relations (and subsequently also of the CSW rules[23]) is remarkably simple, based just on the planar nature of colour-ordered tree diagrams, their analyticity structure, and the asymptotic behaviour of known MHV amplitudes.

The discovery of the CSW and BCF recursion relations has spurred intense activity, about 150 articles citing the original papers[18,19] having appeared in the 18 months following their publication. Questions addressed include the derivation of simple expressions for specific amplitudes,[24] the search for computationally efficient recursive formulations,[25] extensions to processes with fermions and gluinos,[26] Higgs bosons,[27] electroweak bosons,[28] gravity,[29] and the study of multi-gluon collinear limits.[30] This list is necessarily incomplete and further references can be found in a recent review[31] as well as below, when we discuss applications to loop amplitudes.

2.2   One-loop amplitudes

For quantitatively reliable predictions of a given process it is necessary for it to have been calculated to next-to-leading order (NLO). A wide variety of NLO calculations exists, usually in the form of publicly available programs,[32] that allow one to make predictions for arbitrary observables within a given process. The broadest of these programs are the MCFM,[33] NLOJET[34] and PHOX[35] families.

For a $2 \to n$ process the NLO calculation involves the $2 \to n + 1$ tree-level diagram, the $2 \to n$ 1-loop diagram, and some method for combining the tree-level matrix element with the loop contributions, so as to cancel the infrared and collinear divergences present in both with opposite signs. We have seen above that tree-level calculations are well understood, and dipole subtraction[36] provides a general prescription for combining them with the corresponding 1-loop contributions. The bottleneck in such calculations remains the determination of the 1-loop contribution. Currently, $2 \to 3$ processes are feasible, though still difficult, while as yet no full $2 \to 4$ 1-loop QCD calculation has been completed.

In view of the difficulty of these loop calculations, a welcome development has been the compilation, by theorists and experimenters at the Les Houches 2005 workshop, Physics at TeV Colliders,[37] of a realistic prioritized wish-list of processes. Among the most interesting remaining $2 \to 3$ processes one has $pp \to WW + \mathrm{jet}$, $pp \to VVV$ ($V = W$ or Z) and $pp \to H + 2\,\mathrm{jets}$. The latter can be considered a 'background' to Higgs production via vector boson fusion, insofar as the isolation of the vector-boson fusion channel for Higgs production would allow relatively accurate measurements of the Higgs couplings.[38] A number of $2 \to 4$ processes are listed as backgrounds to $t\bar{t}H$ production ($pp \to t\bar{t}q\bar{q}, t\bar{t}b\bar{b}$), $WW \to H \to WW$ or to general new physics ($pp \to V + 3\,\mathrm{jets}$) or specifically SUSY ($pp \to VVV + \mathrm{jet}$).

Two broad classes of techniques have been used in the past for 1-loop calculations: those based directly on the evaluation of the Feynman diagrams (sometimes for the 1-loop-tree interference[39]) and those based on unitarity techniques to sew together tree diagrams (see the review[40]). Both approaches are still being actively pursued.

Today's direct evaluations of 1-loop contributions have, as a starting point, the automated generation of the full set of Feynman diagrams, using tools such as QGRAF[41] and FeynArts.[42] The results can be expressed in terms of sums of products of group-theoretic (e.g. colour) factors and tensor one-loop integrals, such as

$$I_{n;\mu_1..\mu_i} = \int d^{4+2\epsilon}\ell \frac{\ell_{\mu_1}\ldots\ell_{\mu_i}}{(\ell + k_1)^2 \ldots (\ell + k_n)^2} \tag{3}$$

Reduction procedures exist (e.g.[43,44]) that can be applied recursively so as to express the $I_{n;\mu_1..\mu_i}$ in terms of known scalar integrals. Such techniques form the basis of recent proposals for automating the evaluation of the integrals, where the recursion relations are solved by a combination of analytical and numerical methods,[45] or purely numerically,[46,47,48] in some cases with special care as regards divergences that appear in the coefficients of individual terms of the recursion relation but vanish in the sum.[47,49] Results from these approaches include a new compact form for $gg \rightarrow g\gamma\gamma$ at 1 loop[50] and the 1-loop contribution to the 'priority' $pp \rightarrow H + 2\,$jets process.[51] Related automated approaches have also been developed in the context of electroweak calculations,[52] where recently a first $2 \rightarrow 4$ 1-loop result ($e^+e^- \rightarrow 4$ fermions) was obtained.[53]

Another approach[54] proposes subtraction terms for arbitrary 1-loop graphs, such that the remaining part of the loop integral can carried out numerically in 4 dimensions. The subtraction terms themselves can be integrated analytically for the sum over graphs and reproduce all infrared, collinear and ultraviolet divergences.

The above methods are all subject to the problem of the rapidly increasing number of graphs for multi-leg processes. On the other hand, procedures based on 'sewing' together tree graphs to obtain loop graphs[40] (cut constructibility approach) can potentially benefit from the simplifications that emerge from twistor developments for tree graphs. This works best for $\mathcal{N} = 4$ SUSY QCD, where cancellations between scalars, fermions and vector particles makes the 'sewing' procedure simplest and for example all gluonic (and some scalar and gluino) NMHV 1-loop helicity amplitudes are now known;[55] also, conjectures for $\mathcal{N} = 4$ SUSY n-leg MHV planar graphs, at any number of loops, based on 4-gluon two and three-loop calculations,[56] have now been explicitly verified for 5 and 6 gluon two-loop amplitudes.[57] For $\mathcal{N} = 1$ SUSY QCD, known results for all MHV amplitudes[58] have been reproduced[59] and new results exist for some all-n NMHV graphs[60] as well as full results for 6 gluons.[61] For plain QCD, progress has been slower,

though all finite 1-loop graphs ($+ + + + \ldots$, $- + + + \ldots$) were recently presented[62] and understanding has also been achieved for divergent graphs,[63] including the full result for all 1-loop ($- - + + + \ldots$) MHV graphs.[64] The prospects for the twistor-inspired approach are promising and one can hope that it will soon become practically competitive with the direct evaluation methods.

2.3 NNLO jet calculations

NNLO predictions are of interest for many reasons: in those processes where the perturbative series has good convergence they can help bring perturbative QCD predictions to the percent accuracy level. In cases where there are signs of poor convergence at NLO they will hopefully improve the robustness of predictions and, in all cases, give indications of the reliability of the series expansion. Finally they may provide insight in the discussion of the relative importance of hadronisation and higher-order perturbative corrections.[65,66,67]

So far, NNLO predictions are available mostly only for processes with 3 external legs, such as the total cross section for $Z \rightarrow$ hadrons or the inclusive $pp \rightarrow W, Z$[68,69,70] or Higgs[69,71,72] cross sections. A full list is given in table 1 of Stirling's ICHEP writeup.[73]

Most current effort is being directed to the $e^+e^- \rightarrow 3\,$jets process, where NLO corrections are often large, and where one is free of complications from incoming coloured particles. The ingredients that are needed are the squared 5-parton tree level (M_5) and 3-parton 1-loop (M_{3a}) amplitudes, and the interference between 4-parton tree and 1-loop (M_4), and between 3-parton tree and 2-loop amplitudes (M_{3b}), all of which are now known (for references see introduction of[74]). A full NNLO prediction adds the integrals over phase-space of these contributions, multiplied by some jet-observable function J that depends on the momenta p_i

$$J_{NNLO} = \int d^D \Phi_5 M_5 J(p_{1..5}) +$$

$$\int d^D \Phi_4 M_4 J(p_{1..4}) + \int d^D \Phi_3 (M_{3a+b}) J(p_{1..3}).$$

Each of the terms is infrared and collinear (IRC) divergent, because of the phase space integration ($M_{4,5}$) and/or the loop integral ($M_{3,4}$). The current bottleneck for such calculations is in canceling these divergences for an arbitrary (IRC safe) J.

A standard approach at NLO is to introduce subtraction terms (e.g. in the dipole[36] formalism), schematically,

$$J_{NLO}^{4-jet} = \int d^D \Phi_5 (M_5 J(p_{1..5}) - S_5 J(\tilde{p}_{1..4}))$$

$$+ \int d^D \Phi_4 (M_4 J(p_{1..4}) + S_4 J(p_{1..4}))$$

such that each integral is separately finite and that, alone, the S_5 and S_4 terms integrate to equal and opposite divergent contributions (they both multiply the same 4-parton jet function $J(p_{1..4})$ and the $\tilde{p}_{1..4}$ are a specifically designed function of the $p_{1..5}$). At NNLO a similar method can be envisaged, and considerable work has gone towards developing a general formalism.[75,76,77,78,79,74]

An alternative approach, sector decomposition,[80,81,82,83] introduces special distributions f_{-i} (involving plus-functions, like those in splitting functions), which isolate the ϵ^{-i} divergent piece of a given integral ($\epsilon = (D - 4)/2$),

$$\int d^D \Phi_5 M_5 J(p_{1..5}) = \frac{1}{\epsilon^4} \int d^4 \Phi_5 f_{-4} M_5 J(p_{1..5})$$

$$+ \frac{1}{\epsilon^3} \int d^4 \Phi_5 f_{-3} M_5 J(p_{1..5}) + \dots,$$

where the integration is performed in transformed variables that simplify the separation of divergences. In such an approach one thus obtains separate results for each power of ϵ in both real and virtual terms, making it easy to combine them.

A useful testing ground for a number of these approaches has been $e^+ e^- \rightarrow$

2 jets.[83,84,78] Fundamentally new results are the differential distributions for W, Z, H production[85] in the sector-decomposition approach and the $(\alpha_s C_F / 2\pi)^3$ contribution to $\langle 1 - \text{Thrust} \rangle$ in $e^+ e^- \rightarrow 3$ jets, -20.4 ± 4, in the 'antenna' subtraction approach.[74]

2.4 NNLO splitting functions

The landmark calculation of 2004 was probably that of the NNLO splitting functions by Moch, Vermaseren and Vogt[86] (MVV). These are important for accurate DGLAP evolution of the parton distributions, as extracted from fixed target, HERA and Tevatron data, up to LHC scales.

The results for the splitting functions take about 15 pages to express, though the authors have also provided compact approximations for practical use. These are gradually being adopted in NNLO fits.[87,88] Mostly the NNLO splitting functions are quite similar to the estimates obtained a few years ago[89] based on a subset of the moments and known asymptotic limits. In particular it remains true that the NNLO corrections are in general small, both compared to NLO and in absolute terms. The worst region is that of small x for the singlet distributions, shown in fig. 2, where the NLO corrections were large and there is a significant NNLO modification as well. Studies of small-x resummation suggest that further high-order effects should be modest,[90,91] though so far only the gluon sector has been studied in detail.

Another potentially dangerous region is that of large x: the splitting functions converge well, but the coefficient functions have $[\alpha_s^n \ln^{2n-1}(1 - x)]/(1 - x)$ enhancements. There are suggestions that the all-order inclusion of these enhancements via a threshold resummation may help improve the accuracy of PDF determinations.[92] Fresh results from the MVV group,[93] for the third order electromagnetic coefficient functions, quark and gluon form factors, and new threshold

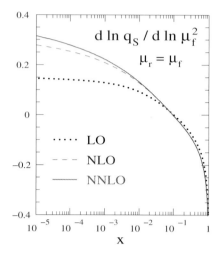

Figure 2. Impact[86] of NNLO DGLAP corrections on the derivative of a toy singlet quark distribution $q_S = \sum_i (q_i + \bar{q}_i)$.

resummation coefficients should provide the necessary elements for yet higher accuracy at large x. Together with the large-x part of the NNLO splitting functions these have been used as inputs to the N^3LO soft-collinear enhanced terms for the Drell-Yan and Higgs cross sections.[94,95]

Finally, as splitting-function calculations approach accuracies at the 1% level, one should consider also the relevance of QED corrections.[87,96]

2.5 Other accuracy-related issues

In view of the efforts being devoted to improving the accuracy of theoretical predictions, it is disheartening to discover that there are still situations where accuracy is needlessly squandered through incorrect data-theory comparisons. This is the case for the inclusive jet cross sections in the cone algorithm at the Tevatron, where a parameter $R_{sep} = 1.3$ is introduced in the cone algorithm used for the NLO calculation (see e.g.[97]), but not in the cone algorithm applied to the data. R_{sep} is the multiple of the cone radius beyond which a pair of partons is not recombined. It dates to early theoretical

work[98] on NLO corrections to the cone algorithm in hadron collisions, when there was no public information on the exact jet algorithm used by the experiments: R_{sep} was introduced to parametrize that ignorance.

The use of R_{sep} in just the theory introduces a spurious NLO correction (at the $5 - 10\%$ level[98]), meaning that the data-theory comparisons are only good to LO. The size of the discrepancy is comparable to the NLO theory uncertainty. As this is smaller than experimental errors, for now the practical impact is limited. However, as accuracies improve it is essential that theory-experiment comparisons be done consistently, be it with properly used cone algorithms[99] or with the (more straightforward and powerful) k_t algorithm,[100] which is finally starting to be investigated.[101]

An accuracy issue that is easily overlooked when discussing QCD developments is the non-negligible impact of electroweak effects at large scales. The subject has mostly been investigated for leptonic initial and final states at a linear collider (ILC). The dominant contributions go as $\alpha_{EW} \ln^2 P_t/M_W$. Since the LHC can reach transverse momenta (P_t) an order of magnitude larger than the ILC the electroweak effects are very considerably enhanced at the LHC, being up to $30 - 40\%$.[102,103] Among the issues still to be understood in such calculations (related to the cancellation of real and virtual corrections) is the question of whether experiments will include events with W and Z's as part of their normal QCD event sample, or whether instead such events will be treated separately.

3 All-order calculations

All-order calculations in QCD are based on the resummation of logarithmically dominant contributions at each order. Such calculations are necessary if one is to investigate the properties of *typical* events, for which each extra power of α_s is accompanied by large soft and collinear logarithms.

The two main ways of obtaining all-order resummed predictions are with exclusive Monte Carlo event generators, and with analytical resummations. The former provide moderately accurate (leading log (LL) and some parts of NLL) predictions very flexibly for a wide variety of processes, with hadronisation models included. The latter are able to provide the highest accuracy (at parton level), but usually need to be carried out by hand (painfully) for each new observable and/or process. All-order calculations are also used in small-x and saturation physics, which will be discussed briefly at the end of this section.

3.1 Monte Carlo event generators (MC)

Various issues are present in current work on event generators — the switch from Fortran to C++; improvements in the showering algorithms and the modelling of the underlying event; and the inclusion of information from fixed-order calculations.

The motivation for moving to C++ is the need for a more modern and structured programming language than Fortran 77. C++ is then a natural choice, in view both of its flexibility and its widespread use in the experimental community.

Originally it was intended that Herwig[104], Pythia[105] and Ariadne/LDC[106,107] should all make use of a general C++ event generator framework known as ThePEG.[108] Herwig++,[109] based on ThePEG, was recently released for e^+e^- and work is in progress for a hadron-hadron version. Pythia 7 was supposed to have been the Pythia successor based on the ThePEG, however instead a standalone C++ generator Pythia 8 is now being developed,[110] perhaps to be interfaced to ThePEG later on. Another independent C++ event generator has also recently become available, SHERPA,[111] whose showering and hadroni-

sation algorithms are largely based on those of Pythia, and which is already functional for hadron collisions.

In both the Pythia and Herwig 'camps' there have been developments on new showering algorithms. Herwig++ incorporates an improved angular-ordered shower[112] in which the 'unpopulated' phase space regions have been shrunk. Pythia 6.3 has a new parton shower[113] based on transverse momentum ordering (i.e. somewhat like Ariadne) which provides an improved description of e^+e^- event-shape data and facilitates the modelling of multiple interactions in hadron collisions. Separately, investigations of alternatives to standard leading-log backward evolution algorithms for initial-state showers are also being pursued.[114,115]

3.2 Matching MC & fixed order

Event generators reproduce the emission patterns for soft and collinear gluons and also incorporate good models of the transition to hadrons. They are less able to deal with multiple hard emissions, which, as discussed above, are important in many new particle searches. There is therefore a need to combine event generators with fixed-order calculations.

The main approach for this is the CKKW[116] proposal. Events are generated based on the n-parton tree matrix elements (for various n), keeping an event only if its n partons are sufficiently well separated to be considered as individual jets (according to some threshold jet-distance measure, based e.g. on relative k_t). Each event is then assigned a 'best-guess' branching history (by jet clustering). This defines a scale for each hard branching, which enters into the running coupling as well as Sudakov form factors that account for virtual corrections. The normal parton showering is then added on to each event at scales below the threshold jet-distance measure. Over the past two years

206

this method has become widely adopted and is available in all major generators.[117]

As well as seeking to describe more jets it is also important to increase the accuracy of event generators for limited number of jets, by including NLO corrections. Here too there is one method that has so far dominated practical uses, known as MC@NLO.[118] Very roughly, it takes the standard MC and modifies it according to

$$\mathrm{MC} \to \mathrm{MC}(1 + \mathrm{NLO} - \mathrm{NLO_{MC}}), \quad (4)$$

where $\mathrm{NLO_{MC}}$ represents the effective NLO corrections present by default in the MC. Since the MC usually has the correct soft and collinear divergences, the combination $\mathrm{NLO} - \mathrm{NLO_{MC}}$ should be finite. Eq. (4) is therefore a well-behaved way of introducing exactly the correction needed to guarantee NLO correctness. A prediction from MC@NLO for b-production,[119] figure 3, is compared to data and a purely analytical approach,[120] and one notes the good agreement between all three.

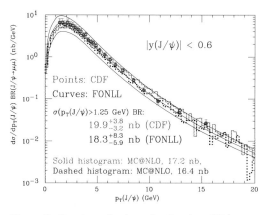

Figure 3. Spectrum for b-production ($\to J/\Psi$) compared to MC@NLO and an analytical prediction.[120]

A bottleneck in widespread implementation of the MC@NLO approach is the need to know $\mathrm{NLO_{MC}}$, which is different for each generator, and even process. Furthermore MC@NLO so far guarantees NLO correctness only for a fixed number of jets — e.g. it can provide NLO corrections to W pro-

duction, but then $W + 1$ jet is only provided to LO. An approach to alleviate both these problems proposes[121] to combine, for example, W, $W + 1$ jet, $W + 2$ jets, etc. with a procedure akin to CKKW. It seeks to alleviate the problem of needing to calculate $\mathrm{NLO_{MC}}$ as follows: when considering $W + m$ jets the $(m + 1)^{\mathrm{th}}$ emission (that needed for NLO accuracy) is generated not by the main MC, but by a separate mini well-controlled generator, designed specifically for that purpose and whose NLO expansion is easily calculated (as is then the analogue of eq.(4)). The only implementation of this so far (actually of an earlier, related formalism[122]) has been for e^+e^-.[123]

3.3 Analytical resummations

It is in the context of analytical resummations that one can envisage the highest resummation accuracies, as well as the simplest matching to fixed order calculations. Rather than directly calculating the distribution $d\sigma(V)/dV$ of an observable V, one often considers some integral transform $F(\nu)$ of the distribution so as to reduce $F(\nu)$ to the form

$$\ln F(\nu) = \sum_n (\underbrace{\alpha_s^n L^{n+1}}_{\mathrm{LL}} + \underbrace{\alpha_s^n L^n}_{\mathrm{NLL}} + \ldots), \quad (5)$$

with $L = \ln \nu$.

Much of the information for certain $\mathrm{N^3LL}$ threshold resummations was recently provided by the MVV group.[93] The highest accuracy for full phenomenological distributions is for the recently calculated Higgs transverse momentum spectrum[124] at the LHC and the related[125] energy-energy-correlation (EEC) in e^+e^-,[126] both of which have also been matched to NLO fixed order.

Boson transverse momentum spectra and the EEC are among the simplest observables to resum. For more general observables and processes, such as event shapes or multi-jet events, the highest accuracy obtained so far has been NLL, and the calculations are both tedious and error-prone. This has prompted work on understanding general features of re-summation. One line of research[127] examines the problem of so-called 'factorisable' observables, for an arbitrary process. This extends the understanding of large-angle soft colour evolution logarithms, whose resummation was originally pioneered for 4-jet processes by the Stony Brook group,[128,129] in which unexplained hidden symmetries were recently discovered,[130,131] notably between kinematic and colour variables.

Separately, the question of how to treat general observables has led to a procedure for automating resummations for a large class of event-shape-like observables.[132] It avoids the need to find an integral transform that factorizes the observable and introduces a new concept, recursive infrared and collinear safety, which is a sufficient condition for the exponentiated form eq. (5) to hold. Its main application so far has been to hadron-collider dijet event shapes,[133] (see also[134]) which provide opportunities for experimental investigation of soft-colour evolution and of hadronisation and the underlying event at the Tevatron and LHC.

The above resummations apply to 'global' observables, those sensitive to radiation everywhere in an event. For non-global observables, such as gap probabilities[135] or properties of individual jets,[136,137] a new class of enhanced term appears, non-global logarithms $\alpha_s^n L^n$ (NGL), fig. 4. Their resummation has so far only been possible in the large-N_C limit,[136,138] though the observation of structure related to BFKL evolution,[139] has inspired proposals for going to finite N_C.[140]

Non-global and soft colour resummations (in non-inclusive form[129]) come together when calculating the probability of a gap between a pair of jets at the Tevatron/LHC, relevant as a background to the W fusion process for Higgs production.[141] Recently it has been pointed out that contributions that are sub-leading in $1/N_C^2$ play

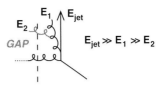

Figure 4. Diagram giving NGL: to calculate the probability of there being no emission into the gap, one should resum a large-angle energy-ordered cascade of emissions rather than just direct emission from the original hard partons.

an important role for large gaps,[142] and also that there are considerable subtleties when using a k_t jet algorithm to define the gap.[143]

3.4 Small x and saturation

The rise of the gluon at small x, as predicted by BFKL,[144] leads eventually to such high gluon densities that a 'saturation' phenomenon should at some point set in. Usually one discusses this in terms of a saturation scale, $Q_s^2(x)$, below (above) which the gluon distribution is (un)saturated. From HERA data, it is believed that at $x = x_0 \simeq 10^{-4} - 10^{-5}$, Q_s is of order $1\,\mathrm{GeV}$ and that it grows as $Q_s^2(x) = (x/x_0)^{-\lambda}\,\mathrm{GeV}^2$ with $\lambda \simeq 0.3$.[145] Q_s may be of relevance to the LHC because the typical transverse scale E_m of minimum bias minijets should satisfy the relation $Q_s^2(s/E_m^2) \sim E_m^2$, whose solution gives $E_m \sim (sx_0)^{\frac{\lambda}{2+\lambda}}\,\mathrm{GeV}^{\frac{2-\lambda}{2+\lambda}} \simeq 2.7 - 3.6\,\mathrm{GeV}$, or if one doesn't trust the normalization, a factor 1.7 relative to the Tevatron. The exact phenomenology is however delicate.[146,147]

The theoretical study of the saturation scale has seen intense activity these past 18 months, spurred by two observations. Firstly it was pointed out[148] that the Balitsky-Kovchegov (BK) equation, often used to describe the onset of saturation and the evolution $d \ln Q_s^2 / d \ln x$, is in the same universality class as the Fisher Kolmogorov Petrovsky Piscounov (FKPP) reaction-diffusion equation, much studied in statistical physics[149] and whose travelling wave solutions relate to the evolution of the saturation scale. Secondly, large corrections were discovered[150] when going beyond the BK mean-field approximation: to LO, $\lambda_{BK} \propto \alpha_s$, while the non mean-field corrections go as $\alpha_s / \ln^2 \alpha_s^2$. Such corrections turn out to be a familiar phenomenon in stochastic versions of the FKPP equation, with α_s^2 in QCD playing the role of the minimum particle density in reaction-diffusion systems with a finite number of particles.[151] The stochastic corrections also lead to a large event-by-event dispersion in the saturation scale.

These stochastic studies are mostly based on educated guesses as to the form of the small-x evolution beyond the mean-field approximation. There has also been extensive work on finding the full equation that replaces BK beyond the mean-field approximation, a number of new formulations having been proposed.[152,153,154,155] It will be interesting to examine how their solutions compare to the statistical physics related approaches.

4 Concluding remarks

Of the topics covered here, the one that has been the liveliest in the past year is that of 'twistors'.[a] Its full impact cannot yet be gauged, but the dynamic interaction between the QCD and string-theory communities on this subject will hopefully bring further important advances.

More generally one can ask if QCD is on track for the LHC. Progress over the past years, both calculationally and phenomenologically (e.g. PDF fitting) has been steady. Remaining difficulties, for example in high (NNLO) and moderate (NLO) accuracy calculations are substantial, however the considerable number of novel ideas currently being discussed encourages one to believe that significant further advances will have been made by the time LHC turns on.

---

[a]Second place goes to saturation.

Acknowledgments

Many people have helped with the preparation of this talk, through explanations of material I was unfamiliar with, suggestions of topics to include, and useful comments. Among them: Z. Bern, T. Binoth, J. Butterworth, M. Cacciari, M. Ciafaloni, P. Ciafaloni, D. Comelli, D. Dunbar, Yu. L. Dokshitzer, R. K. Ellis, J.-P. Giullet, D. Kosower, L. Lonnblad, A. H. Mueller, G. Marchesini, S. Moretti, S. Munier, M. H. Seymour, A. Vogt, B. R. Webber, G. Zanderighi. I wish also to thank the organisers of the conference for the invitation, financial support, and warm hospitality while in Uppsala.

References

1. I. Stewart, these proceedings; G. Ingelman, these proceedings.
2. A. Juste, these proceedings.
3. P. D. Draggiotis, R. H. P. Kleiss and C. G. Papadopoulos, *Eur. Phys. J.* C **24**, 447 (2002).
4. E. W. N. Glover, 'Precision Phenomenology and Collider Physics', talk given at Computer Algebra and Particle Physics workshop, DESY Zeuthen, April 2005.
5. Reviewed in L. J. Dixon, hep-ph/9601359.
6. F. A. Berends and W. Giele, *Nucl. Phys.* B **294**, 700 (1987); M. L. Mangano, S. J. Parke and Z. Xu, *Nucl. Phys.* B **298** 653 (1988); M. L. Mangano, *Nucl. Phys.* B **309**, 461 (1988).
7. J. F. Gunion and Z. Kunszt, *Phys. Lett.* B **161** 333 (1985). R. Kleiss and W. J. Stirling, *Nucl. Phys.* B **262**, 235 (1985).
8. S. J. Parke and T. R. Taylor, *Phys. Rev. Lett.* **56**, 2459 (1986).
9. F. A. Berends and W. T. Giele, *Nucl. Phys.* B **306**, 759 (1988).
10. F. Caravaglios and M. Moretti, *Phys. Lett.* B **358**, 332 (1995).
11. F. Maltoni and T. Stelzer, *JHEP* **0302**, 027 (2003).
12. M. L. Mangano *et al.*, *JHEP* **0307**, 001 (2003).
13. A. Kanaki and C. G. Papadopoulos, hep-ph/0012004.

14. A. Pukhov *et al.*, hep-ph/9908288.

15. F. Yuasa *et al.*, *Prog. Theor. Phys. Suppl.* **138**, 18 (2000).

16. F. Krauss, R. Kuhn and G. Soff, *JHEP* **0202**, 044 (2002)[hep-ph/0109036].

17. V. P. Nair, *Phys. Lett.* B **214**, 215 (1988).

18. E. Witten, *Commun. Math. Phys.* **252**, 189 (2004); R. Roiban, M. Spradlin and A. Volovich, *JHEP* **0404**, 012 (2004).

19. F. Cachazo, P. Svrcek and E. Witten, *JHEP* **0409**, 006 (2004).

20. S. J. Parke and T. R. Taylor, *Phys. Lett.* B **157**, 81 (1985) **174B**, *465 (1986)]*; Z. Kunszt, *Nucl. Phys.* B **271**, 333 (1986).

21. R. Britto, F. Cachazo and B. Feng, *Nucl. Phys.* B **715**, 499 (2005).

22. R. Britto *et al.*, *Phys. Rev. Lett.* **94**, 181602 (2005).

23. K. Risager, hep-th/0508206.

24. D. A. Kosower, *Phys. Rev.* D **71**, 045007 (2005); R. Roiban, M. Spradlin and A. Volovich, *Phys. Rev. Lett.* **94**, 102002 (2005); M. x. Luo and C. k. Wen, *Phys. Rev.* D **71**, 091501 (2005); R. Britto *et al.*, *Phys. Rev.* D **71**, 105017 (2005);

25. I. Bena, Z. Bern and D. A. Kosower, *Phys. Rev.* D **71**, 045008 (2005).

26. J. B. Wu and C. J. Zhu, *JHEP* **0407**, 032 (2004), *JHEP* **0409**, 063 (2004); G. Georgiou, E. W. N. Glover and V. V. Khoze, *JHEP* **0407**, 048 (2004).

27. L. J. Dixon, E. W. N. Glover and V. V. Khoze, *JHEP* **0412**, 015 (2004); S. D. Badger, E. W. N. Glover and V. V. Khoze, *JHEP* **0503**, 023 (2005).

28. Z. Bern *et al.*, *Phys. Rev.* D **72**, 025006 (2005).

29. J. Bedford *et al.*, *Nucl. Phys.* B **721**, 98 (2005).

30. T. G. Birthwright *et al.*, *JHEP* **0507**, 068 (2005); *JHEP* **0505**, 013 (2005).

31. F. Cachazo and P. Svrcek, hep-th/0504194.

32. http://www.cedar.ac.uk/hepcode

33. J. M. Campbell and R. K. Ellis, *Phys. Rev.* D **62**, 114012 (2000).

34. Z. Nagy, *Phys. Rev. Lett.* **88**, 122003 (2002).

35. T. Binoth *et al.*, *Eur. Phys. J.* C **16**, 311 (2000).

36. S. Catani and M. H. Seymour, *Nucl. Phys.* B **485**, 291 (1997) **510**, *503 (1997)]*. S. Catani *et al.*, *Nucl. Phys.* B **627**, 189 (2002).

37. http://lappweb.in2p3.fr/conferences/LesHouches/Houches2005/

38. D. Zeppenfeld *et al.*, *Phys. Rev.* D **62**, 013009 (2000).

39. J. M. Campbell, E. W. N. Glover and D. J. Miller, *Phys. Lett.* B **409**, 503 (1997).

40. Z. Bern, L. J. Dixon and D. A. Kosower, *Ann. Rev. Nucl. Part. Sci.* **46**, 109 (1996).

41. P. Nogueira, *J. Comput. Phys.* **105**, 279 (1993).

42. T. Hahn, *Comput. Phys. Commun.* **140**, 418 (2001).

43. G. 't Hooft and M. J. G. Veltman, *Nucl. Phys.* B **153**, 365 (1979); G. Passarino and M. J. G. Veltman, *Nucl. Phys.* B **160**, 151 (1979).

44. A. I. Davydychev, *Phys. Lett.* B **263**, 107 (1991); O. V. Tarasov, *Phys. Rev.* D **54**, 6479 (1996).

45. T. Binoth *et al.*, hep-ph/0504267.

46. W. T. Giele and E. W. N. Glover, *JHEP* **0404**, 029 (2004).

47. F. del Aguila and R. Pittau, *JHEP* **0407**, 017 (2004).

48. A. van Hameren, J. Vollinga and S. Weinzierl, *Eur. Phys. J.* C **41**, 361 (2005).

49. W. Giele, E. W. N. Glover and G. Zanderighi, *Nucl. Phys. Proc. Suppl.* **135**, 275 (2004).

50. T. Binoth *et al.*, *JHEP* **0503**, 065 (2005).

51. R. K. Ellis, W. T. Giele and G. Zanderighi, hep-ph/0506196, hep-ph/0508308.

52. A. Ferroglia *et al.*, *Nucl. Phys.* B **650**, 162 (2003); A. Aleksejevs, S. Barkanova and P. Blunden, Contributed Paper 30 (see also nucl-th/0212105); G. Belanger *et al.*, hep-ph/0308080; A. Denner, S. Dittmaier, hep-ph/0509141.

53. A. Denner *et al.*, *Phys. Lett.* B **612**, 223 (2005).

54. Z. Nagy and D. E. Soper, *JHEP* **0309**, 055 (2003).

55. Z. Bern *et al.*, *Phys. Rev.* D **71**, 045006 (2005); R. Britto, F. Cachazo and B. Feng, *Phys. Rev.* D **71**, 025012 (2005); Z. Bern, L. J. Dixon and D. A. Kosower, *Phys. Rev.* D **72**, 045014 (2005); K. Risager, S. J. Bidder and W. B. Perkins, hep-th/0507170 (and references therein).

56. Z. Bern, J. S. Rozowsky and B. Yan, *Phys. Lett.* B **401**, 273 (1997); C. Anastasiou *et al.*, *Phys. Rev. Lett.* **91**, 251602 (2003); Z. Bern, L. J. Dixon and V. A. Smirnov, *Phys. Rev.* D **72**, 085001 (2005).

57. E. I. Buchbinder and F. Cachazo, hep-th/0506126.

58. Z. Bern *et al.*, *Nucl. Phys.* B **435**, 59 (1995).

59. C. Quigley and M. Rozali, *JHEP* **0501**, 053 (2005). J. Bedford *et al.*, *Nucl. Phys.* B **706**, 100 (2005).

60. S. J. Bidder *et al.*, *Phys. Lett.* B **612**, 75 (2005).

61. R. Britto *et al.*, hep-ph/0503132.

62. Z. Bern, L. J. Dixon and D. A. Kosower, hep-ph/0505055.

63. A. Brandhuber *et al.*, hep-th/0506068; Z. Bern, L. J. Dixon and D. A. Kosower, hep-ph/0507005.

64. D. Forde and D. A. Kosower, hep-ph/0509358.

65. Y. L. Dokshitzer and B. R. Webber, *Phys. Lett.* B **352**, 451 (1995).

66. E. Gardi and G. Grunberg, *JHEP* **9911**, 016 (1999).

67. J. Abdallah *et al.* [DELPHI Collaboration], *Eur. Phys. J.* C **29**, 285 (2003).

68. R. Hamberg, W. L. van Neerven and T. Matsuura, *Nucl. Phys.* B **359**, 343 (1991) **644**, *403 (2002)]*.

69. R. V. Harlander and W. B. Kilgore, *Phys. Rev. Lett.* **88**, 201801 (2002).

70. V. Ravindran, J. Smith and W. L. van Neerven, *Nucl. Phys.* B **682**, 421 (2004).

71. C. Anastasiou and K. Melnikov, *Nucl. Phys.* B **646**, 220 (2002).

72. V. Ravindran, J. Smith and W. L. van Neerven, *Nucl. Phys.* B **665**, 325 (2003); *Nucl. Phys.* B **704**, 332 (2005).

73. W. J. Stirling, hep-ph/0411372.

74. A. Gehrmann-De Ridder, T. Gehrmann and E. W. N. Glover, hep-ph/0505111.

75. D. A. Kosower, *Phys. Rev.* D **67**, 116003 (2003).

76. S. Weinzierl, *JHEP* **0303**, 062 (2003).

77. W. B. Kilgore, *Phys. Rev.* D **70**, 031501 (2004).

78. S. Frixione and M. Grazzini, *JHEP* **0506**, 010 (2005).

79. G. Somogyi, Z. Trocsanyi and V. Del Duca, *JHEP* **0506**, 024 (2005).

80. M. Roth and A. Denner, *Nucl. Phys.* B **479**, 495 (1996); T. Binoth and G. Heinrich, *Nucl. Phys.* B **585**, 741 (2000).

81. G. Heinrich, *Nucl. Phys. Proc. Suppl.* **116**, 368 (2003). *ibid.* **135**, 290 (2004).

82. C. Anastasiou, K. Melnikov and F. Petriello, *Phys. Rev.* D **69**, 076010 (2004).

83. T. Binoth and G. Heinrich, *Nucl. Phys.* B **693**, 134 (2004).

84. C. Anastasiou, K. Melnikov and F. Petriello, *Phys. Rev. Lett.* **93**, 032002 (2004).

85. C. Anastasiou *et al.*, *Phys. Rev.* D **69**, 094008 (2004); C. Anastasiou, K. Melnikov and F. Petriello, *Phys. Rev. Lett.* **93**, 262002 (2004).

86. S. Moch, J. A. M. Vermaseren and A. Vogt, *Nucl. Phys.* B **688**, 101 (2004). *ibid.* **691**, 129 (2004).

87. A. D. Martin *et al.*, *Eur. Phys. J.* C **39**, 155 (2005).

88. S. Alekhin, hep-ph/0508248.

89. W. L. van Neerven and A. Vogt, *Phys. Lett.* B **490**, 111 (2000).

90. M. Ciafaloni *et al.*, *Phys. Rev.* D **68**, 114003 (2003); *Phys. Lett.* B **587**, 87 (2004).

91. G. Altarelli, R. D. Ball and S. Forte, *Nucl. Phys. Proc. Suppl.* **135**, 163 (2004). hep-ph/0310016.

92. G. Corcella and L. Magnea, hep-ph/0506278.

93. S. Moch, J. A. M. Vermaseren and A. Vogt, *Phys. Lett.* B **606**, 123 (2005); hep-ph/0504242; *Nucl. Phys.* B **726**, 317 (2005); *JHEP* **0508**, 049 (2005); *Phys. Lett.* B **625**, 245 (2005).

94. S. Moch and A. Vogt, hep-ph/0508265.

95. E. Laenen and L. Magnea, hep-ph/0508284.

96. B. F. L. Ward and S. A. Yost, hep-ph/0509003.

97. M. Wobisch [D0 Collaboration], *AIP Conf. Proc.* **753**, 92 (2005).

98. S. D. Ellis, Z. Kunszt and D. E. Soper, *Phys. Rev. Lett.* **69**, 3615 (1992).

99. G. C. Blazey *et al.*, hep-ex/0005012.

100. S. Catani *et al.*, *Nucl. Phys.* B **406**, 187 (1993); S. D. Ellis and D. E. Soper, *Phys. Rev.* D **48**, 3160 (1993).

101. M. Martinez [CDF Collaboration], *eConf* **C0406271**, MONT05 (2004); V. D. Elvira [D0 Collaboration], *Nucl. Phys. Proc. Suppl.* **121**, 21 (2003).

102. E. Maina, S. Moretti and D. A. Ross, *Phys. Lett.* B **593**, 143 (2004) **614**, *216 (2005)]* ; S. Moretti, M. R. Nolten and D. A. Ross, hep-ph/0503152.

103. J. H. Kuhn *et al.*, hep-ph/0508253; hep-ph/0507178.

104. G. Corcella *et al.*, *JHEP* **0101**, 010 (2001).

105. T. Sjostrand *et al.*, hep-ph/0308153.

106. L. Lonnblad, *Comput. Phys. Commun.* **71**, 15 (1992).

107. H. Kharraziha and L. Lonnblad, *JHEP* **9803**, 006 (1998).

108. M. Bertini, L. Lonnblad and T. Sjostrand,

Comput. Phys. Commun. **134**, 365 (2001); http://www.thep.lu.se/ThePEG/

109. S. Gieseke *et al.*, *JHEP* **0402**, 005 (2004).

110. http://www.thep.lu.se/~torbjorn/ pythiaaux/future.html

111. T. Gleisberg *et al.*, *JHEP* **0402**, 056 (2004).

112. S. Gieseke, P. Stephens and B. Webber, *JHEP* **0312**, 045 (2003).

113. T. Sjostrand and P. Z. Skands, *Eur. Phys. J.* C **39**, 129 (2005).

114. H. Tanaka, *Prog. Theor. Phys.* **110**, 963 (2003).

115. S. Jadach and M. Skrzypek, *Acta Phys. Polon.* B **36**, 2979 (2005).

116. S. Catani *et al.*, *JHEP* **0111**, 063 (2001).

117. S. Mrenna and P. Richardson, *JHEP* **0405**, 040 (2004). A. Schalicke and F. Krauss, *JHEP* **0507**, 018 (2005). N. Lavesson and L. Lonnblad, *JHEP* **0507**, 054 (2005).

118. S. Frixione and B. R. Webber, *JHEP* **0206**, 029 (2002).

119. S. Frixione, P. Nason and B. R. Webber, *JHEP* **0308**, 007 (2003).

120. M. Cacciari *et al.*, *JHEP* **0407**, 033 (2004).

121. Z. Nagy and D. E. Soper, hep-ph/0503053.

122. M. Kramer and D. E. Soper, *Phys. Rev.* D **69**, 054019 (2004).

123. M. Kramer, S. Mrenna and D. E. Soper, hep-ph/0509127.

124. G. Bozzi *et al.*, hep-ph/0508068. *Phys. Lett.* B **564**, 65 (2003).

125. Y. L. Dokshitzer, D. Diakonov and S. I. Troian, *Phys. Rept.* **58**, 269 (1980).

126. D. de Florian and M. Grazzini, *Nucl. Phys.* B **704**, 387 (2005).

127. R. Bonciani *et al.*, *Phys. Lett.* B **575**, 268 (2003).

128. J. Botts and G. Sterman, *Nucl. Phys.* B **325**, 62 (1989); N. Kidonakis and G. Sterman, *Phys. Lett.* B **387**, 867 (1996); N. Kidonakis and G. Sterman, *Nucl. Phys.* B **505**, 321 (1997); N. Kidonakis, G. Oderda and G. Sterman, *Nucl. Phys.* B **531**, 365 (1998);

129. G. Oderda, *Phys. Rev.* D **61** 014004 (2000). C. F. Berger, T. Kucs and G. Sterman, *Phys. Rev.* D **65**, 094031 (2002).

130. Y. L. Dokshitzer and G. Marchesini, hep-ph/0509078; hep-ph/0508130.

131. M. H. Seymour, hep-ph/0508305.

132. A. Banfi, G. P. Salam and G. Zanderighi, *Phys. Lett.* B **584**, 298 (2004); *JHEP* **0503**, 073 (2005).

133. A. Banfi, G. P. Salam and G. Zanderighi, *JHEP* **0408**, 062 (2004).

134. G. Sterman, hep-ph/0501270.

135. M. Dasgupta and G. P. Salam, *JHEP* **0203**, 017 (2002).

136. M. Dasgupta and G. P. Salam, *Phys. Lett.* B **512**, 323 (2001);

137. A. Banfi and M. Dasgupta, *JHEP* **0401**, 027 (2004).

138. A. Banfi, G. Marchesini and G. Smye, *JHEP* **0208**, 006 (2002).

139. G. Marchesini and A. H. Mueller, *Phys. Lett.* B **575**, 37 (2003).

140. H. Weigert, *Nucl. Phys.* B **685**, 321 (2004).

141. R. B. Appleby and M. H. Seymour, *JHEP* **0309**, 056 (2003).

142. J. R. Forshaw, A. Kyrieleis and M. H. Seymour, *JHEP* **0506**, 034 (2005).

143. R. B. Appleby and M. H. Seymour, *JHEP* **0212**, 063 (2002); A. Banfi and M. Dasgupta, hep-ph/0508159.

144. L.N. Lipatov, *Sov. J. Nucl. Phys.* **23** 338 (1976); E.A. Kuraev, L.N. Lipatov and V.S. Fadin, *Sov. Phys. JETP* **45** 199 (1977); I.I. Balitsky and L.N. Lipatov, *Sov. J. Nucl. Phys.* **28** 822 (1978); V. S. Fadin and L. N. Lipatov, *Phys. Lett.* B **429**, 127 (1998); M. Ciafaloni and G. Camici, *Phys. Lett.* B **430**, 349 (1998).

145. J. Bartels, K. Golec-Biernat and H. Kowalski, *Phys. Rev.* D **66**, 014001 (2002).

146. V. A. Khoze *et al.*, *Phys. Rev.* D **70**, 074013 (2004).

147. R. S. Thorne, *Phys. Rev.* D **71**, 054024 (2005).

148. S. Munier and R. Peschanski, *Phys. Rev.* D **69**, 034008 (2004); *Phys. Rev. Lett.* **91**, 232001 (2003).

149. U. Ebert and W. van Saarloos, *Physica* D **146**, 1 (2000).

150. A. H. Mueller and A. I. Shoshi, *Nucl. Phys.* B **692**, 175 (2004).

151. E. Iancu, A. H. Mueller and S. Munier, *Phys. Lett.* B **606**, 342 (2005).

152. E. Iancu and D. N. Triantafyllopoulos, *Nucl. Phys.* A **756**, 419 (2005); *Phys. Lett.* B **610**, 253 (2005).

153. E. Levin and M. Lublinsky, hep-ph/0501173.

154. A. Kovner and M. Lublinsky, *JHEP* **0503**, 001 (2005); *Phys. Rev. Lett.* **94**, 181603 (2005).

155. Y. Hatta *et al.*, hep-ph/0504182. hep-ph/0505235.

DISCUSSION

Bennie Ward (Baylor University):
To what extent are the deduced MHV rules using twistors now proven?

Gavin Salam: an outline of a field-theoretic proof for the MHV (CSW) rules was given in the same article[22] as the proof of the BCF rules. Recently (after the Lepton Photon Symposium) a more detailed version of the proof has appeared.[23]

QUANTUM CHROMODYNAMICS AT COLLIDERS

JON M. BUTTERWORTH

Department of Physics & Astronomy, University College London, WC1E 6BT, London, UK
E-mail: J.Butterworth@ucl.ac.uk

QCD is the accepted (that is, the effective) theory of the strong interaction; studies at colliders are no longer designed to establish this. Such studies can now be divided into two categories. The first involves the identification of observables which can be both measured and predicted at the level of a few percent. Such studies parallel those of the electroweak sector over the past fifteen years, and deviations from expectations would be a sign of new physics. These observables provide a firm "place to stand" from which to extend our understanding. This links to the second category of study, where one deliberately moves to regions in which the usual theoretical tools fail; here new approximations in QCD are developed to increase our portfolio of understood processes, and hence our sensitivity to new physics. Recent progress in both these aspects of QCD at colliders is discussed.

1 The Data and the Experiments

QCD studies at colliders involve measurements of the hadronic final state in e^+e^-, lepton-hadron and hadron-hadron collisions. The lepton colliders also allow the study of effective photon-photon, lepton-photon and photon-hadron collisions, due to the almost-on-shell photon beam which accompanies lepton beams. In collisions involving these photons, the photon may participate directly in the hard process, or it may act as a source of partons much like a hadron. Together, this array of different colliding beams provides us with many data and rich opportunities to learn from cross-comparison between experiments.

Data presented at this meeting include precise measurements of a great number of properties of the final state, and these measurement are used to demonstrate and improve our understanding of the physics. With the confidence that this is understood, it then becomes possible to infer, from an increasing number of measurements, information about the initial state; that is, quarks and gluons in their natural habitat inside hadrons. This in turn enables us to predict effects at future colliders, particularly the Large Hadron Collider under construction at CERN[1].

In sections 2-5, the final state measurements are discussed. In the subsequent section, some experimental advances in the current knowledge of parton densities within the Dokshitzer-Gribov-Lipatov-Altarelli-Parisi (DGLAP) paradigm are presented. Following that, some measurements in regions of phase space where DGLAP evolution is not applicable are discussed. This includes low x and diffractive effects, at which point I conclude this contribution and hand over to the next speaker[2].

2 Fragmentation and Hadron Production

An obvious observable to start with in looking at QCD final states is the charged particle multiplicity. This has been measured as a function of the energy scale of the interaction by many experiments. A summary[3] is shown in Fig. 1. The energy scale dependence is seen to be universal to within a few percent for reasonable definitions of the energy scale in e^+e^- and DIS, and the proton data from ISR also lies close to the same curve. This is well modelled by the current Monte Carlo (MC) models. The shape is also described by next-to-leading-order (NLO) QCD (not shown), where local parton-hadron duality is assumed to give an arbitrary constant normalisation factor.

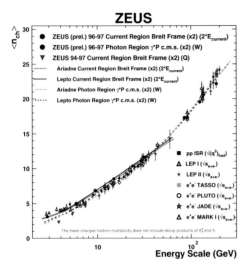

Figure 1. The charged particle multiplicity as a function of energy scale for a selection of experiments.

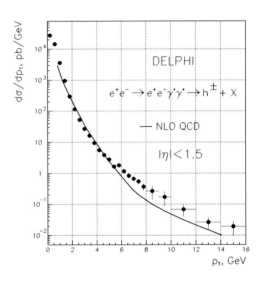

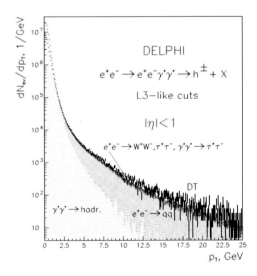

Figure 2. The charged particle cross section in $\gamma\gamma$ collisions as a function of particle transverse momentuym (p_T) as measured by DELPHI. The upper plot is the DELPHI measurement of the cross section compared to NLO QCD. The lower plot is the DELPHI data analysed using cuts close to those used by L3 (see text).

To make more precise statements about QCD fragmentation, measurements can be designed specifically to suit precise calculations. Accurate calculations for quark and gluon fragmentation exist for hemispheres of a fragmenting diquark of di-gluon system. In the case of quarks, this is a natural configuration for comparison with e^+e^- data. Obtaining a comparable configuration for gluons, however, is more difficult. In a contribution from OPAL[4] the jet boost algorithm is employed to do this. Precise agreement is observed for $0.06 < x < 0.8$. Because of this level of agreement, fundamental parameters of the theory can be extracted with confidence. An impressive recent example is the measurement of the ration of the gluon and quark colour factors, $C_A/C_F = 2.261 \pm 0.014 \pm 0.036 \pm 0.066$, by DELPHI[5], where the first error is statistical, the second the experimental systematic error and the the third the theoretical uncertainty. This agrees well with the QCD expectation of 2.25.

One assumption employed in such measurements is that the soft, hadronization stage can be controlled and seperated from the hard QCD process. This assumption has been tested in many measurements, and several new results from HERA[6] have tested it in the case of charm quarks. Here it has been shown that the fraction of charm quarks fragmenting to the various charmed hadrons is the same (to within the measurement accuracy of a few %) in DIS and photoproduction at HERA as it is in e^+e^- annihilation. Comparisons between the fragmentation function at HERA, LEP and CLEO also show qualitative agreement. A fit of the fragmentation function using NLO calculations would allow a more quantitative statement to be made here, and would be of great interest; as would more accurate measurements from HERA II.

The claim is that for some QCD observables the theoretical understanding is so good that deviations in the data really do mean new physics. This claim was challenged by two results from the L3 collaboration, where in $\gamma\gamma$ events, both the charged particle and jet cross sections lie above the NLO QCD prediction, with a discrepancy which increases as the scale increases[7]. This discrepancy seems impossible to reconcile with QCD; yet the scale is so low ($p_T \approx 5$GeV for the charged hadrons) that some beyond-the-standard-model explanation seems unlikely. The charged particle measurement has been repeated by DELPHI[8], however, and no such discrepancy is seen (Fig.2 - note that no theoretical uncertainty is shown). To their great credit, DELPHI have gone further, solving the puzzle by mimicking the L3 analysis and showing that for the L3 selection cuts there is a large background from annihilation, which has the correct charactierstics to explain the discrepancy. This is also shown in Fig.2; it is then a victory for some kind of precision QCD. It is tempting to speculate that the e^+e^- background may also contribute to the excess seen by L3 in the jet cross section.

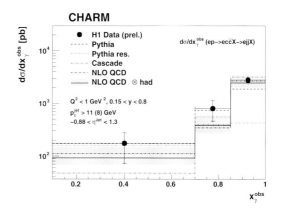

Figure 3. The x_γ^{OBS} distribution in charm photoproduction.

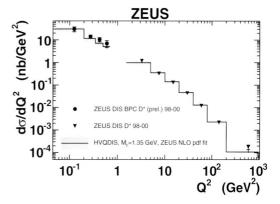

Figure 4. The inclusive charm cross section as a function of Q^2.

3   Charm and Beauty Production

Recent data on fragmentation properties of charm have been briefly discussed above. The production cross sections for both charm and bottom quarks also represent an important investigative tool for QCD, and since bottom in particular is often used as a tag in searches for new physics, the QCD production mechanism is of particular importance. An understanding of the production dynamics as well as inclusive rates is needed. Results continue to be produced from $p\bar{p}$, ep DIS and photoproduction.

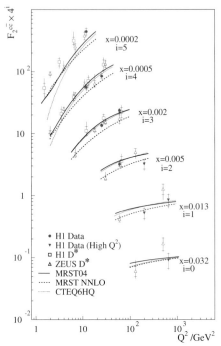

Figure 5. The charm tagged structure function of the proton.

3.1 Charm cross sections

Photoproduction of charm has been measured using tagged D^*+ jets and via lifetime tagging[9,10,11]. Changing the fraction of the photon's momentum seen in the jets, x_γ^{OBS}, from values near one to lower values allows one to move from so-called direct processes, dominated by point-like photons, to resolved processes, in which the photon acts as a source of partons similar to a hadron. Both regions are well described by NLO QCD calculations (Fig.3). In addition, the inclusive cross section is well understood in both the photoproduction and DIS regimes, from photon virtualities of near zero up to 1000GeV^2 (Fig. 4). Expressed as the charm structure function $F_2^{c\bar{c}}$, the data is already quite precise and is still being accumulated. Again, NLO QCD describes it well (Fig. 5)[12].

On a related topic, inelastic J/ψ production, the debate about colout octet terms is not yet resolved. NLO QCD corrections to the colour singlet term are very large[11].

3.2 Beauty cross sections

Inclusive measurements of bottom-tagged cone dijets from the CDF II have been measured[17] and compared to PYTHIA[13], HERWIG[14] and MC@NLO[15] (Fig. 6a). The normalisation of the LO MCs has a large uncertainty associated with it due to higher order terms. However, it is significant that PYTHIA describes the shape of the data very well for $E_T^{\text{jet}} > 40\text{GeV}$. MC@NLO is in good agreement with the cross section at high transverse momenta but falls below the data at $E_T^{\text{jet}} < 70\text{GeV}$. Apart from the NLO terms, one difference between the two programs is that PYTHIA includes a multiparton interaction model to describe the underlying event. Adding such a model to MC@NLO in the shape of JIMMY[16], leads to good agreement between MC@NLO and the data for $E_T^{\text{jet}} > 40\text{GeV}$ (Fig. 6b).

There are also measurements from D0 of muon-tagged jets[18], where within 50% errors NLO calculations describe the data. At HERA, DIS and direct photoproduction measurements are reasonably well described, though there is a tendency for the data to be above the calculations. This seems particularly pronounced at low x_γ^{OBS} (see Fig.7), where it is possible that non-perturbative effects such as the underlying event may play some role. Precision data from HERA II will hopefully clarify the situation.

Finally, the first measurements of the beauty stucture function $F_2^{b\bar{b}}$ have now been made[12], shown in Fig. 8. These lag the similar charm measurements in statistical precision, but there are many more data to come, and it will be an important challenge for the theory to describe such inclusive measurements well.

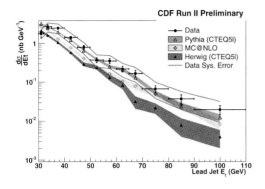

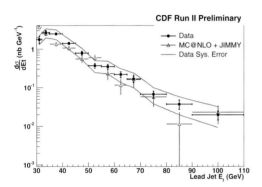

Figure 6. Bottom-quark jet cross sections from CDF II.

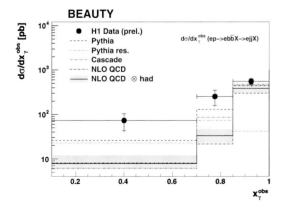

Figure 7. The x_γ^{OBS} distribution in bottom photoproduction.

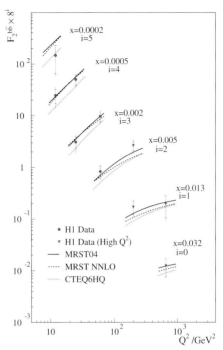

Figure 8. The bottom-quark tagged structure of the proton.

3.3 Charm and Bottom production dynamics

The charm statistics at HERA are sufficient that the production dynamics may be measured. Several measurements already exist[11], and there are new measurements now of the azimuthal correlation of dijets in charm events[10], as well as jet shapes for charm jets[9]. Both are sensitive to QCD radiation in these processes. The azimuthal decorrelation is well described by leading-logarithmic parton shower models for both resolved and direct photoproduction; NLO calculations for massive charm quarks (e.g. in which the charm is not an active quark in the photon or proton) describe the direct case well, but fail to describe the low-x_γ^{OBS} decorrelation (see Fig.9. The jet shapes are well described by PYTHIA's parton showers for high x_γ^{OBS}, but the jets are narrower in the data than in the MC at low x_γ^{OBS}.

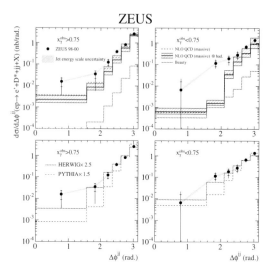

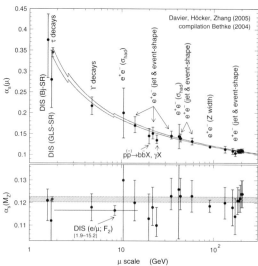

Figure 9. The $\Delta\phi$ distribution for charm jets.

Figure 10. A selection of α_s measurements.

In the case of beauty, the Tevatron data allow studies of such properties in bottom quark events. The dijet correlation is reasonably well described by MC@NLO, but the addition of multiparton interactions does again improve the agreement. PYTHIA also does a reasonable job.

Finally, a beautiful new measurement of the ratio of bottom- to light-quark jet rates from DELPHI[19] leads to an accurate measurement of the running b-mass $m_b(Q) = 4.25 \pm 0.11$GeV at threshold.

In summary of this section, it does seem that in general charm and bottom production are well described by NLO QCD, but that there is a need to combine state-of-art non-perturbative models with the best perturbative calculations in order to get this level of agreement. This is true particularly for measurements in hadronic collisions spanning a large range in transverse energy.

4 Jet Structure and Event Shapes

Measurements of jet cross sections and event shapes continue to improve in precision, as do calculations of such properties. This means that the strong coupling, α_s, may be extracted from a large number of final states in many processes. At this conference, new results from e^+e^- (JADE, OPAL, ALEPH) and ep (H1, ZEUS) were presented[20,21]. A particularly interesting measurement is the ALEPH extraction from τ decays, shown in Fig. 10, which greatly improves the accuracy at low scales[21]. In general, none of the others is a great leap forward in itself, but all steadily improve accuracy of the world average, and build confidence in our understanding of QCD.

Behind this achievement lies an increasing number of well-understood QCD processes. Perhaps particularly noteworthy this year are the new inclusive jet measurements from Tevatron Run II and HERA, where the use of well-controlled jet algorithms and the impressive level of knowledge of the energy scale and resolution in the experiments means that the data really lay down a strong challenge for the theoretical predictions. Some of the CDF II results are shown Fig. 11; here the $K_\perp$ algorithm has been used with different distance parameters; this is an important technique, in that any new physics effect seen in such cross sections should be present for all reasonable choices, whereas the sensitivites to some non-perturbative effects will vary between different algorithms and parameters. Another interesting process with new data is prompt photon production, where both HERA and Tevatron have new data[22,23]. The D0 data in particular now show impressive agreement with QCD over a wide range of transverse energy.

5   Production of jets with bosons

When the LHC starts delivering data, an unprecedented number of W and Z particles will be produced, usually in association with jets. They feature in many "standard candle" cross sections which will be used to extract parton densities and calibrate the detectors, as well as in many exotic signatures for new physics. It is imperative to understand as far as possible equivalent processes at existing colliders, particularly the Tevatron. The dijet correlation[24] at D0 is shown in Fig 12. It is well described by NLO QCD in the important wide-angle area where the fixed-order tree-level diagrams are most significant, and is described by parton shower MC in the low angle regions, as expected. Importantly, the SHERPA program matches these two types of calculations and describes the whole shape well[25].

A related cross section is the diphoton decorrelation, measured by CDF[23], shown in Fig. 13. The angle between the two photons is well described by NLO QCD as contained in the DIPHOX[26] program. The RESBOS[27] calculation does not include NLO fragmentation contributions and falls below the data at high angles.

Run II measurements of Z cross sections are now coming out, and both the inculsive Z rapidity[28] and the N-jet rate in Z events[29] are in good agreement with NLO QCD (Fig.14).

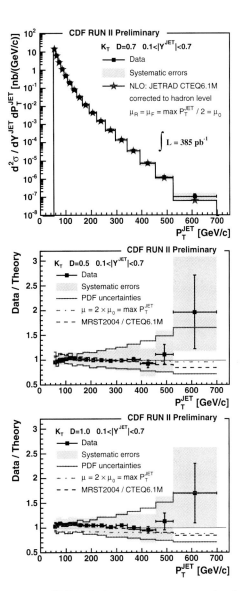

Figure 11. CDF inclusive jet measurements using the $K_\perp$ algorithm. The top plot shows the measured differential cross section $d\sigma/dp_T^{\text{jet}}$ compared to NLO QCD for $R = 0.7$. The lower two plots show the ratio of data/theory for similar cross sections measured with $R = 0.5$ and $R = 1.0$.

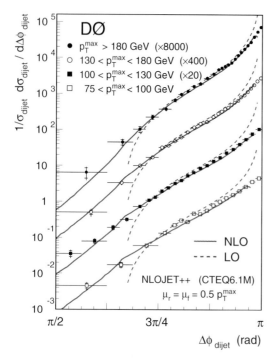

Figure 12. Dijet decorrelation from D0.

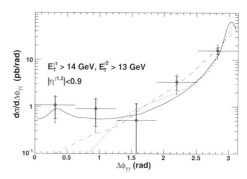

Figure 13. Diphoton decorrelation from CDF. The points are the data, the solid line is the DIPHOX calculation and the dashed is RESBOS (see text).

6 Parton Densities

There has been major theoretical progress in this area, as discussed in the previous contribution[30]. There have also been some notable experimental advances, which are discussed below.

6.1 High x

The kinematic plane at the LHC is shown in Fig. 15, along with the regions where LHC and other data will be able to constrain the gluon density in the proton. There is an urgent need more information about the gluon at high x (say 0.05 and above) and at Q^2 between 100 and 10000 GeV^2, so that reliable predictions may be made for the highest energy cross sections at LHC. In addition there is a strong correlation between α_s and the gluon for intermediate x values (0.001 to 0.05) in fits to F_2.

Including DIS jet cross sections in the fit constrains the coupling, but these cross sections are dominantly quark initiated and depend only weakly on the gluon density. Jet photoproduction, on the other hand, is dominantly gluon initiated over a wide kinematic range, as can reach very high x. ZEUS have included both in a fit[35], with their latest inclusive cross section data, and see a significant improvement in the accuracy of both α_s and the gluon at high x. Perhaps most excitingly, the jet data used was a fraction (around a tenth) of the total expected by the end of HERA II. There are major improvements expected[36].

HERA II is also now producing high luminosities of electron-proton collisions (rather than positron-proton), and early measurements were shown at this conference. The large increase of statistics, matching or bettering that achieved with positrons, and coupled with lepton polarization, brings several benefits. One is the ability to measure the electroweak structure of quark coupling (see a previous contribution[31]). The measurement of charged and neutral currents will also allow constraints on flavour composition of proton to be made from HERA data alone, avoiding nuclear correction uncertainties from fixed target data. These data also

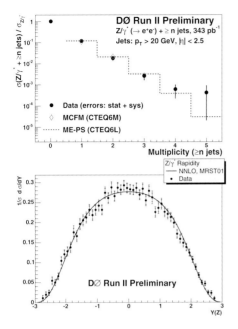

Figure 14. The N-jet rate in Z events, compared to MC calculation (upper plot) and the Z rapidity distribution (lower plot).

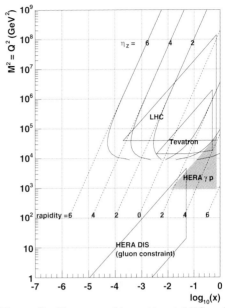

Figure 15. The parton kinematics at LHC and elsewhere. The curved lines show the region where Z+jet production might be used to constrain the gluon at the LHC. The HERA and Tevatron regions shown are those where the gluon may be constrained from F_2 fits and jet production.

reach up to high x.

At lower Q^2 it is still in principle possible to reach high x, since the scattered electron may be measured. However, the radiative corrections are such in this region that while reconstruction of Q^2 from electron is good, it is very poor for x. A new measurement from ZEUS[32] uses the hadronic jet to reconstruct x. As x increases, the jet moves forward and will at some point be lost down the forward beampipe. However, in this case it is possible to set a minimum x based on the fact that the hadronic jet escaped, and integrate above this. The measurement gives a good sensitivity to the high x structure function, as shown in Fig. 16.

Finally in this subsection, the W asymmetry measurements from tevatron run II are now appearing[33]. They are sensitive to flavour composition in proton at high x and will be important input to new fits.

6.2 Low Q^2

Measuring inclusive lepton-proton cross-sections in the low Q^2 region probes the transition from a region where perturbative calculations are valid to a region where non-perturbative techniques must be used to make any prediction. It also provides the lowest reach in x, and thus sensitivity to high density QCD. Two new measurements from H1 have been presented in this area[34]. In the first, QED Compton events, with a high virtuality exchanged electron, are used. In this case the electron virtuality means that the final state electron can be detected even when the virtuality of the exchanged photon is very low. In the second such measurement, initial state photon radiation is tagged, which implies a low virtuality incoming electron with an energy lower than the beam energy. This incoming electron energy is measured from the longitudinal energy imbalance in the cen-

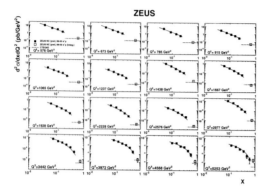

Figure 16. New measurements of F_2 at high x.

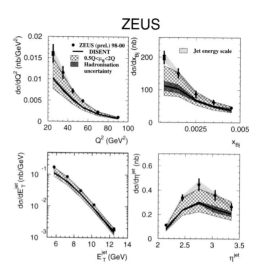

Figure 17. Forward jet cross sections at HERA.

tral detector. This allows the measurement to be made at lower Q^2 while keeping x moderately high. Both of these measurements provide new data in the transition region between DIS and photoproduction.

7 Peripheral Collisions, Low x and Diffraction

The low Q^2 region discussed above is an example of a measurement where we deliberately extend into a region where the usual theoretical tools are expected to fail. Moving into such regions allow the investigation of new approximations in QCD such as clever resummations, new evolution equations, new perturbative expansions, high parton densities and correlated parton distributions. Using the data to verify or falsify such tools extends our portfolio of understood QCD phenomena. There is a large overlap in this area with both the previous[30] and following[2] speakers, and I will concentrate on the topics least aligned with theirs.

7.1 New resummations and evolutions

The parton density fits discussed above all use the DGLAP evolution equations, which are strongly ordered in the scale, $Q_1 \gg Q_2 \gg Q_3$. For inclusive properties, this is the dominant configuration. However, it is of course possible to select kinematic configurations in which a large evolution in x (or equivalently in rapidity) is required, but where this evolution takes place at a Q^2 which is both in the perturbative regime and approximately constant. New measurements have been made in forward jet production (Fig.17) in DIS and other related processes at HERA[38].

In such a region the DGLAP evolution is not applicable. Thus if NLO fixed order QCD with DGLAP parton densities is used to try and predict such cross sections, the predictions have large uncertainties. It is also seen that they usually lie below the data. Leading-logarithmic Monte Carlos can do better than this, and in particular, the CCFM-based MC CASCADE[37] probably has the ability to describe such cross sections. However, it has a strong dependence on the unintegrated gluon density, which is extracted from fits to data. The new data should be used to constrain this further.

Such effects may also be studied in vector meson and photon production. The vector mesons I leave to the next speaker[2], but will mention here the new data from DELPHI $\gamma^*\gamma^*$ collisions, where a signifcant x evolution can occur along the exchanged quark line.

Again, calculations (BFKL-based) which resum $\log(x)$ terms seem to have the best chance of describing the data.

A consistent, and reasonably precise, description of high rapidity/low x data seems to be within reach. This would give a real boost to the credibility of this approach, and would be a great help for predicting forward jet rates at LHC.

8 Conclusion

In an increasing number of important processes at high energy colliders, perturbative QCD calculations, and the data, are rather precise, and in rather good agreement with each other. New data from Tevatron and HERA, and (re)analysis of old data from PETRA and LEP, continue to improve the situation, as do theoretical advances. There is still room for improvement of course, but for some important processes QCD is now very precisely understood, and there have been recent significant advances in measurement and theory. As an aside, the point is now being reached where for some observables, electroweak effects are comparable to QCD uncertainties[39]. For other processes, while QCD is becoming better understood, there is still experimental and theoretical work to do. A list of such processes, in approximate decreasing order of how well they are understood, could be:

- Parton density functions at high Q^2 and intermediate x, ideal jet fragmentation.

- Multijet processes, Boson+jets; Heavy flavour production.

- Parton density functions at low and high x.

- High rapidities and rapidity gaps.

- Diffraction, absorption and total cross sections.

- Off-diagonal and unintegrated parton density functions.

- Underlying events (a topic hardly touched on here, but where there is lots of work on tuning to Tevatron, HERA, SPS and other data[40,41]).

In all these areas existing data, as well as data still to come from Tevatron run II, HERA II and RHIC, provide a challenge. Data from LHC will make great use of such developments, and will also challenge the theory further.

Acknowledgments

My thanks to D. Alton, R. Field, G. Ingelman, E. Perez, G. Salam, J. Schiek, P. Wells and M. Wing for discussions and material, as well as to those others whose work I have presented here. I also thank the organisers for a conference which was stimulating, well run and fun.

References

1. L. Rolandi, F. Gianotti, these proceedings.
2. G. Ingelman, these proceedings.
3. ZEUS Coll., ZEUS-prel-05-016; LP2005 paper 283.
4. OPAL Coll., G. Abbiendi *et al.*, *Phys. Rev.* D **69** 032002 (2004).
5. DELPHI Coll., M. Seibert, K. Hamacher and J. Drees DELPHI-2005-12 CONF 732; LP2005 paper 135.
6. H1 Coll., A. Aktas *et al. Eur. Phys. J.* C **38** 447 (2005); H1prelim-05-074, LP2005 paper 407.
   ZEUS Coll., S. Chekanov *et al.* hep-ex/0508019; ZEUS-prel-05-007, LP2005 paper 266.
7. L3 Coll., P. Achard *et al. Phys. Lett.* B **554** 105 (2003).*Phys. Lett.* B **602** 157 (2004).
8. DELPHI Coll., M. Chapkin *et al.*, DELPHI-2005-018 CONF 738. LP2005 paper 90.
9. H1 Coll., H1prelim-05-074, H1prelim-05-077; LP2005 papers 405, 409.

224

10. ZEUS Coll., S. Chekanov *et al.*, hep-ex/0505008; hep-ex/0507089; ZEUS-prel-04-024, LP2005 paper 265.

11. For a recent review, see J. M. Butterworth and M. Wing, "High energy photoproduction", *Rep. Prog. Phys.* 68 2773-2828 (2005).

12. H1 Coll., A. Aktas *et al.*, hep-ex/0507081; *Eur. Phys. J.* C **40** 349 (2005).

13. T. Sjöstrand *et al.* *Comput. Phys. Commun.* **135** 238 (2001).

14. G. Corcella *et al.*, *JHEP* **0101** 010 (2001).

15. S. Frixione and B. R. Webber, *JHEP* **0206** (2002) 029; S. Frixione, P. Nason and B. R. Webber, *JHEP* **0308** 007 (2003).

16. J. M. Butterworth, J. R. Forshaw and M. H. Seymour, *Z. Phys.* C **72** 637 (1996).

17. CDF Collaboration Note 6985, available from `http://www-cdf.fnal.gov`

18. D0 Collaboration D0 note 4754-CONF.

19. DELPHI Coll., P. Bambade *et al*, LP2005 paper 57.

20. JADE Coll., C. Pahl, *et al* hep-ex/0408123; J. Schieck, *et al* hep-ex/0408122. H1 Coll., H1prelim-05-033, LP2005 paper 390. ZEUS Coll., S. Chekanov *et al.* arXiv:hep-ex/0502007; LP2005 paper 281, ZEUS-prel-05-024 OPAL Coll, G. Abbiendi *et al.*, *Eur. Phys. J.* C **40** 287 (2005); hep-ex/0507047.

21. ALEPH Coll., S. Schael *et al.*, hep-ex/0506072; M. Davier, A. Höcker and Z. Zhang, hep-ph/0507078.

22. H1 Coll., A. Aktas *et al.*, *Eur. Phys. J.* C **38** 437 (2005). D0 Coll., D0-4859-CONF.

23. CDF Collaboration, D. Acosta *et al.*, *Phys. Rev. Lett.* **95** 022003 (2005).

24. D0 Coll., V. M. Abazov *et al.* *Phys. Rev. Lett.* **94** 221801 (2005).

25. S. Schumann, LP2005 poster 50. T. Gleisberg et al, *JHEP* **0402** 056 (2004).

26. T. Binoth *et al.*, *Eur. Phys. J.* C **16** 311 (2000).

27. C. Balazs, E. L. Berger, S. Mrenna and C. P. Yuan, *Phys. Rev.* D **57** 6934 (1998).

28. D0 Coll., "Measurement of $Z \to e^+ e^-$ rapidity distribution", `http://www-d0.fnal.gov/Run2Physics/WWW/results/ew.htm`

29. D0 Coll., D0-4794-CONF.

30. G. Salam, these proceedings.

31. C. Diaconu, these proceedings.

32. ZEUS Coll., ZEUS-prel-05-005, LP2005 paper 261.

33. CDF Coll., D. Acosta *et al. Phys. Rev.* D **71** 051104 (2005).

34. H1 Coll., A. Aktas *et al.*, *Phys. Lett.* B **598** 159 (2004); H1prelim-04-042, LP2005 paper 389.

35. ZEUS Coll., S. Chekanov *et al.*, *Eur. Phys. J.* C42, 1 (2005).

36. C. Gwenlan, A. Cooper-Sarkar and C. Targett-Adams, hep-ph/0509220, prepared for the HERA-LHC workshop.

37. H. Jung and G. P. Salam, *Eur. Phys. J.* C **19** 351 (2001).

38. ZEUS Coll., LP2005 paper 25, 278. H1 Coll., LP2005 paper 378, 379, 381, 384.

39. E. Maina *et al.*, LP2005 paper 359; A. D. Martin, R. G. Roberts, W. J. Stirling and R. S. Thorne, *Eur. Phys. J.* C **39** 155 (2005).

40. HERA-LHC workshop, `http://www.desy.de/~heralhc/`

41. TEV4LHC workshop, `http://conferences.fnal.gov/tev4lhc/`

DISCUSSION

Tord Ekelöf (Uppsala University):

Given the history of the discrepancy for the inclusive jet cross-section at high p_T found in the early days of the Tevatron, which eventually came down to be compatible with QCD, and taking into account the extrapolation of QCD needed for the LHC — how confident can we be that it would be a signal of new physics to see an excess at the yet much higher p_T that we will reach with LHC, if we do not couple that to high p_T leptons, missing energy or other weak probes?

Jon Butterworth: Certainly there are cases (for instance the L3 high p_T data mentioned in the talk) where QCD cannot be "tweaked" to explain away discrepancies. How much tweaking is allowed, and therefore ultimately our confidence in any such new physics signal, depends largely on how carefully we measure and calculate QCD at colliders now, and eventually at LHC itself. I would add that weak probes at hadron colliders don't escape such effects entirely - they still depend upon the parton densities, for instance, which turned out to be a large uncertainty affecting the Tevatron excess.

Lorenzo Magnea (University of Torino):

In b production cross section, MC@NLO shows large improvements in fitting the data with respect to HERWIG (MC@LO) — PYTHIA succeeds in fitting the same data by tuning internal parameters. This is not necessarily good news: one of the two Monte Carlos is getting the right answer for the wrong reason! MC@NLO signals that hard radiation is important, PYTHIA assigns the same effect to the underlying event or leading-log evolution.

Jon Butterworth: You are right in the sense that both effects have to be taken seriously, and seem to be important. And it is certainly true that by tuning "internal parameters" we can muddy the picture. My approach would be to implement what we know first, then use well grounded models to see how close we are to understanding what is happening. There are certainly cases (as in my previous answer) where no "internal parameters" can help! Also - Monte Carlos should be treated as best estimators of physics, not as black boxes with internal parameters!

HARD DIFFRACTION — 20 YEARS LATER

GUNNAR INGELMAN

High Energy Physics, Uppsala University, Box 535, SE-75121 Uppsala, Sweden
E-mail: gunnar.ingelman@tsl.uu.se

The idea of diffractive processes with a hard scale involved, to resolve the underlying parton dynamics, was published 1985 and experimentally verified 1988. Today hard diffraction is an active research field with high-quality data and new theoretical models. The trend from Regge-based pomeron models to QCD-based parton level models has given insights on QCD dynamics involving perturbative gluon exchange mechanisms, including the predicted BFKL-dynamics, as well as novel ideas on non-perturbative colour fields and their interactions. Extrapolations to the LHC include the interesting possibility of diffractive Higgs production.

1 Introduction

'Gaps – in my understanding after 20 years' could have been an appropriate title on this talk, which will focus on today's understanding of hard diffraction based on models that include both hard perturbative and soft non-perturbative QCD. I will start, however, with some of the 'historical' milestones that have established this new research field. Recent data have revealed problems with models based on the pomeron in 'good old' Regge phenomenology. The key issue is whether the pomeron is a part of the proton wave function or diffraction is an effect of the scattering process. The latter seems more appropriate in today's QCD-based models.

Recently, a new kind of 'hard gap' events, having a large momentum transfer across the gap, have been observed in terms of a rapidity gap between two high-$p_\perp$ jets in $p\bar{p}$ and diffractive vector meson production at large momentum transfer in γp. This has given the first real evidence for the BFKL-dynamics predicted since long by QCD.

Thus, hard diffraction provides a 'QCD laboratory' where several aspects of QCD dynamics can be investigated. Remember that QCD, in particular in its non-perturbative domain, still has major unsolved problems.

Based on our current theoretical understanding, interesting predictions are made for hard diffraction in future experiments, *e.g.* at the LHC. Here, much interest is presently focused on the possibility to produce the Higgs boson in diffractive events, which may even provide a potential for Higgs discovery!

2 'Historical' milestones

Going back to 'ancient' pre-QCD history, Regge theory provided a phenomenology for total and diffractive cross-sections with a dominant contribution from the exchange of a pomeron with vacuum quantum numbers. Hadronic scattering events were then classified as elastic, single and double diffraction, double pomeron exchange and totally inelastic depending on the observable distribution of final state particles in rapidity.[a] For example, single diffraction is then characterised by a leading proton (or other beam hadron) separated by a large rapidity gap (*i.e.* without particles) to the X-system of final state particles (Fig. 1a). One should note that in this Regge approach, there is no hard scale involved. In particular, there is no large momentum transfer across the gap which may therefore be called a 'soft gap'.

The basic new idea introduced 20 years ago by myself and Peter Schlein[1], was to consider a hard scale in the X-system in order to resolve an underlying parton level

---

[a]Rapidity $y = \frac{1}{2}\ln\frac{E+p_z}{E-p_z} \approx -\ln\tan\frac{\theta}{2} = \eta$ pseudo-rapidity for a particle with $(E, \vec{p}_\perp, p_z)$ and polar angle θ to beam axis z.

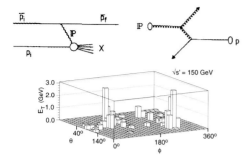

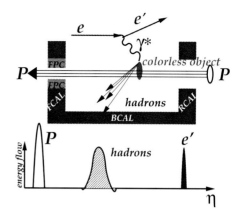

Figure 1. (a) Single diffractive $p\bar{p}$ scattering via pomeron exchange giving a leading beam particle separated by a rapidity gap to the X-system. (b) Hard parton level scattering in the X-system producing high-$p_\perp$ jets. (c) Jets in the central calorimeter of an event triggered by a leading proton in the UA8 Roman pot detectors giving the discovery of hard diffraction.

Figure 2. Schematics of diffractive deep inelastic scattering in the ZEUS detector and the final state rapidity distribution.

interaction and thereby be able to investigate the process in a modern QCD-based framework. We formulated this in a model with an effective pomeron flux in the proton, $f_{I\!\!P/p}(x_{I\!\!P}, t)$, and parton distributions in the pomeron $f_{q,g/I\!\!P}(z, Q^2)$, such that cross-sections for hard diffractive processes could be calculated from the convolution

$$d\sigma \sim f_{I\!\!P/p} \, f_{q,g/I\!\!P} \, f_{q,g/p} \, d\hat{\sigma}_{\text{pert. QCD}} \quad (1)$$

of these functions with a perturbative QCD cross-section for a hard parton level process (Fig. 1ab). This enabled predictions of diffractive jet production at the CERN $p\bar{p}$ collider and also of diffractive deep inelastic scattering. Although this seems quite natural in today's QCD language, it was rather controversial at the time.

It was therefore an important breakthrough when the UA8 experiment at the CERN $p\bar{p}$ collider actually discovered[2] hard diffraction by triggering on a leading proton in their Roman pot detectors and finding jets in the UA2 central calorimeter (Fig. 1c) in basic agreement with our model implemented in a 'Lund Monte Carlo' event generator. These observed jets had normal jet properties and by investigating their longitudinal momentum distribution one could infer

that the partons in the pomeron have rather a hard distribution, $f_{q,g/I\!\!P}(z) \sim z(1-z)$ and also indications[3] of a superhard component $\sim \delta(1-z)$.

In spite of this discovery, hard diffraction was not fully recognised in the whole particle physics community. It was therefore a surprise to many when diffractive deep inelastic scattering (DDIS) was discovered by ZEUS[4] and H1[5] at HERA in 1993. These events were quite spectacular with the whole forward detector empty (Fig. 2), *i.e.* a large rapidity gap as opposed to the abundant forward hadronic activity in normal DIS events. A surprisingly large fraction $\sim 10\%$ of all DIS events were diffractive. Moreover, they showed the same Q^2 dependence and were not suppressed with increasing Q^2, demonstrating that DDIS is *not* a higher twist process but leading twist.

The diffractive DIS cross-section can be written [6]

$$\frac{d\sigma}{dx \, dQ^2 \, dx_{I\!\!P} \, dt} = \frac{2\pi\alpha^2}{xQ^4} \left(1 + (1-y)^2\right) F_2^{D(4)} \quad (2)$$

where fractional energy loss $x_{I\!\!P}$ and four-momentum transfer t from the proton define the diffractive conditions. For most of the data, the leading proton is not observed and hence t is effectively integrated out giv-

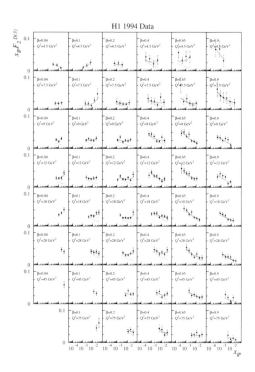

Figure 3. H1 data[7] on the diffractive structure function $F_2^{D(3)}(x_{I\!P}; \beta, Q^2)$ with fits based on the Regge models with pomeron and reggeon exhange.

ing the structure function $F_2^{D(3)}(x_{I\!P}, \beta, Q^2)$ which has now been obtained from rapidity gap events with high precision (Fig. 3). The variables

$$\beta = \frac{-q^2}{2q \cdot (p_p - p_Y)} = \frac{Q^2}{Q^2 + M_X^2 - t} \quad (3)$$

$$x_{I\!P} = \frac{q \cdot (p_p - p_Y)}{q \cdot p_p} = \frac{Q^2 + M_X^2 - t}{Q^2 + W^2 - M_p^2} = \frac{x}{\beta}$$

are model-independent invariants.

In $p\bar{p}$, UA8 has provided more information on diffractive jet production through analyses of such cross-sections[8]. Several different diffractive hard scattering processes have been observed in $p\bar{p}$ at the Tevatron. Events with jets, W, Z, $b\bar{b}$ or J/ψ have a large rapidity gap, and are thus diffractive, in about 1% of the cases, as shown in Table 1. This is an order of magnitude smaller relative rate of hard diffraction compared to the 10% in DIS at HERA.

$$R_{\text{hard}} = \frac{1}{\sigma_{\text{hard}}^{\text{tot}}} \int_{x_{F\,\text{min}}}^{1} dx_F \, \frac{d\sigma_{\text{hard}}}{dx_F}$$

$R_{\text{hard}}[\%]$	Exp.	observed	SCI
dijets	CDF	0.75 ± 0.10	0.7
W	CDF	1.15 ± 0.55	1.2
W	DØ	$1.08^{+0.21}_{-0.19}$	1.2
$b\bar{b}$	CDF	0.62 ± 0.25	0.7
Z	DØ	$1.44^{+0.62}_{-0.54}$	$1.0^\star$
J/ψ	CDF	1.45 ± 0.25	$1.4^\star$

Table 1. Tevatron data on the ratio in % of diffractive hard processes to all such hard events, where diffraction is defined by a rapidity gap corresponding to a leading proton with large x_F. For comparison results of the SCI model discussed in section 4 (* denote predictions in advance of data).

3 Pomeron approach

Using Regge factorisation, the diffractive DIS structure function can be written

$$F_2^{D(4)}(x, Q^2, x_{I\!P}, t) = f_{I\!P/p}(x_{I\!P}, t) F_2^{I\!P}(\beta, Q^2) \quad (4)$$

in terms of a pomeron flux and a pomeron structure function, where $x_{I\!P} \simeq p_{I\!P}/p_p$ is interpreted as the momentum fraction of the pomeron in the proton and $\beta \simeq p_{q,g}/p_{I\!P}$ is the momentum fraction of the parton in the pomeron.

Good fits with data can be obtained (Fig. 3), provided that also a Reggeon exchange contribution is included. Factoring out the fitted $x_{I\!P}$ dependence, one obtains the diffractive structure function $F_2^{D(2)}(\beta, Q^2)$ (or $F_2^{I\!P}$) shown in Fig. 4. The Q^2 dependence is rather weak and thus shows approximate scaling indicating scattering on pointlike charges. There is, however, a weak $\log Q^2$ dependence which fits well with conventional perturbative QCD evolution.

The β dependence is quite flat, which can be interpreted as hard parton distributions in the pomeron. This is borne out in a full next-to-leading order (NLO) QCD fit giving the parton distributions shown in Fig. 5, which demonstrate the dominance of the gluon distribution.

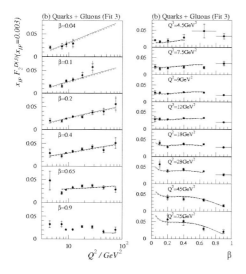

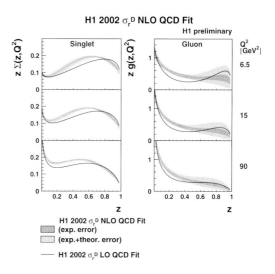

Figure 4. H1 data[7] on the diffractive structure function $F_2^{D(2)}(\beta, Q^2)$ with QCD fits.

Figure 5. Diffractive quark (singlet $z\Sigma(z, Q^2)$) and gluon ($zg(z, Q^2)$) momentum distributions ('in the pomeron') obtained[7] from a NLO DGLAP fit to H1 data on diffractive DIS.

This framework can now be used to calculate various processes. In contributions[9] to this conference, new HERA data on diffractive $D^\star$ and dijets are compared to calculations using these diffractive parton densities folded with the corresponding perturbative QCD matrix elements in NLO. For DIS the model calculation agree well with the data, both in absolute normalization and in the Q^2 and $p_\perp$ dependences. In photoproduction, the shapes of distributions agree with data, but the model normalization is too large by about a factor 2 (Table 2). This fits with the theoretical knowledge that QCD factorization has been proven for diffraction in DIS[10], but not in photoproduction (*i.e.* low Q^2) or hadronic interactions. Remember that the photon state $|\gamma\rangle = |\gamma\rangle_0 + |q\bar{q}(g)\rangle + |\rho\rangle \ldots$ has not only the direct component, but also hadronic components.

Using this model with diffractive parton densities from HERA to calculate diffractive hard processes at the Tevatron, one obtains cross-sections which are an order of magnitude larger than observed, as illustrated in Fig. 6. This problem can be cured by modifications of the model, in particular, by intro-

$\dfrac{\sigma(\text{data})}{\sigma(\text{theory})}$ for	H1	ZEUS
$D^\star$ in diffr. DIS	~ 1	~ 1
dijets in diffr. DIS	~ 1	~ 1
$D^\star$ in diffr. γp	—	~ 0.4
dijets in diffr. γp	~ 0.5	~ 0.5

Table 2. Ratio of HERA data on diffractive production of $D^\star$ and dijets to model calculation based on diffractive parton distributions and NLO perturbative QCD matrix elements.

ducing some kind of damping[11] at high energies, such as a pomeron flux 'renormalization'. It is, however, not clear whether this is the right way to get a proper understanding.

Thus, there are problems with the pomeron approach. The pomeron flux and the pomeron parton densities do not seem to be universal quantities. They cannot be separately well defined since only their product is experimentally measurable. Moreover, it may be improper to think of the pomeron as 'emitted' from the proton, because the soft momentum transfer t at the proton-pomeron vertex imply a long space-time scale such that they move together for an extended time

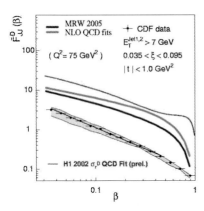

Figure 6. D0 data on diffractive dijets at the Tevatron compared with model calculations based on diffractive parton densities from DIS at HERA.

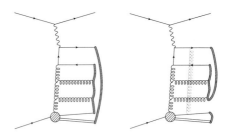

Figure 7. Gluon-induced DIS at small x with colour flux tube, or string, configuration in (a) the conventional Lund string model connection of partons and (b) after a soft colour-octet exchange (dashed gluon line) between the remnant and the hard scattering system resulting in a phase space region without a string leading to a rapidity gap after hadronisation.

which means that there should be some crosstalk between such strongly interacting objects. In order to investigate these problems, alternative approaches have been investigated where the pomeron is not in the initial state, *i.e.* not part of the proton wave function but an effect of the QCD dynamics of the scattering process.

4 QCD-based approaches

A starting point can here be the standard hard perturbative interactions, since they should not be affected by the soft interactions. On the other hand we know that there should be plenty of soft interactions, below the cut-off Q_0^2 for perturbation theory, because α_s is then large giving a large interaction probability (*e.g.* unity for hadronisation). Soft colour exchange may then very well have a strong influence on the colour topology of the event and thereby on the final state via hadronisation. Ideally one would like to have a single model describing both diffractive gap events and non-diffractive events.

A simple, but phenomenologically successful attempt in this spirit is the soft colour interaction (SCI) model[12]. Consider DIS at small x, which is typically gluon-initiated

leading to perturbative parton level processes as illustrated in Fig. 7. The colour order of the perturbative diagram has conventionally been used to define the topology of the resulting non-perturbative colour string fields between the proton remnant and the hard scattering system (Fig. 7a).

Hadronisation, *e.g.* described by the Lund model[13], will then produce hadrons over the full rapidity region. One should remember, however, that we have proper theory to rely on only for the hard perturbative part of the event, which is separated from the soft dynamics in both the initial and the final parts by the QCD factorization theorem. This hard part is above a perturbative QCD cut-off $Q_0^2 \sim 1 \text{GeV}^2$ with an inverse giving a transverse size which is small compared to the proton diameter. Thus, the hard interactions can be viewed as being embedded in the colour field of the proton and hence one can consider interactions of the outgoing partons with this 'background' field. Fig. 7b illustrates a soft gluon exchange that rearranges colour so that the hard scattering system becomes a colour singlet and the proton remnant another singlet. These systems hadronise independently of each other and are separated by a rapidity gap. The gap can be large because the primary gluon has a small momentum fraction x_0 given by the gluon density $g(x_0, Q_0^2) \sim x_0^{-\alpha}(1-x_0)^5$, leaving a large momentum fraction $(1-x_0)$ to the remnant which can form a leading proton.

The soft gluon exchange is non-perturbative and hence its probability cannot be calculated theoretically. The model, therefore, introduces a single parameter P for the probability of exchanging such a soft gluon between any pair of partons, where one of them should be in the remnant representing the colour background field. Applying this on the partonic state, including remnants, in the Lund Monte Carlo generators LEPTO for ep and PYTHIA for hadronic interactions, e.g. $p\bar{p}$, leads to variations of the string topologies and thereby different final states after hadronisation.

The definition of diffraction through the *gap-size is a highly infrared sensitive observable*, as demonstrated in Fig. 8 for DIS at HERA. At the parton level, even after perturbative QCD parton showers, it is quite common to have large gaps. Hadronising the conventional string topology from the pQCD phase, leads to an exponential suppression with the gap-size, i.e. a huge non-perturbative hadronisation effect. Introducing the soft colour interactions causes a drastic effect on the hadron level result, with a gap-size distribution that is not exponentially suppressed but has the plateau characteristic for diffraction. The result of the SCI model is remarkably stable with respect to variation of the soft gluon exchange probability parameter, illustrating that the essential effect arises when allowing the possibility to rearrange the colour string topology. The gap events are in this approach nothing special, but a fluctuation in the colour topology of the event.

Selecting the gap events in the Monte Carlo one can extract the diffractive structure function and the model (choosing $P \approx 0.5$) describes quite well[14] the main features of $F_2^{D(3)}(x_{I\!P}, \beta, Q^2)$ observed at HERA (Fig. 3). This is not bad for a one-parameter model!

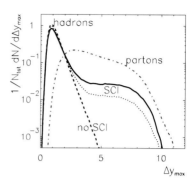

Figure 8. Distribution of the size Δy_{max} of the largest rapidity gap in DIS events at HERA simulated using LEPTO (standard small-x dominated DIS event sample with $Q^2 \geq 4$ GeV^2 and $x \geq 10^{-4}$). The dashed-dotted curve represents the parton level obtained from hard, perturbative processes (matrix elements plus parton showers). The dashed curve is for the hadronic final state after standard Lund model hadronisation, whereas adding the Soft Color Interaction model results in the full curve. The dotted curve is when the SCI probability parameter P has been lowered from its standard value 0.5 to 0.1.

By moving the SCI program code from LEPTO to PYTHIA, exactly the same model can be applied to $p\bar{p}$ at the Tevatron. Using the value of the single parameter P obtained at HERA, one obtains the correct overall rates of diffractive hard processes as observed at the Tevatron, see Table 1. Differential distributions are also reproduced[15] as exemplified in Fig. 9, which also demonstrates that the pomeron model is far above the data and PYTHIA without the SCI mechanism is far below. The SCI model also reproduces the observed two-gap events (conventionally called double pomeron exchange) with a central hard scattering system[15].

The phenomenological success of the SCI model indicates that it captures the most essential QCD dynamics responsible for gap formation. It is therefore interesting that recent theoretical developments provide a basis for this model. As is well known, the QCD factorization theorem separates the hard and

232

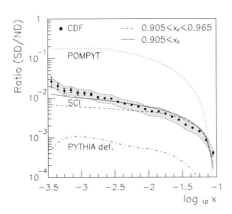

Figure 9. Ratio of diffractive to non-diffractive dijet events versus momentum fraction x of the interacting parton in $\bar{p}$. CDF data (with error band) compared to the POMPYT pomeron model, default PYTHIA and the SCI model.[15]

soft dynamics and is the basis for the definition of the parton density functions (pdf)

$$f_{q/N} \sim \int dx^- e^{-ix_B p^+ x^-/2}\langle N(p)\,|\,\bar{\psi}(x^-)$$
$$\gamma^+ W[x^-;0]\,\psi(0)\,|\,N(p)\rangle_{x^+=0}$$

where the nucleon state sandwiches an operator including the Wilson line $W[x^-;0] =$ $\mathrm{P}\exp\left[ig\int_0^{x^-} dw^- A_a^+(0,w^-,0_\perp)t_a\right]$ which is a path-ordered exponential of gluon fields. The physical interpretation becomes transparent if one expands the exponential giving[16]

$$W[x^-;0] \sim 1 + g\int \frac{dk_1^+}{2\pi}\frac{\tilde{A}^+(k_1^+)}{k_1^+ - i\varepsilon} + \quad (5)$$
$$g^2\int \frac{dk_1^+ dk_2^+}{(2\pi)^2}\frac{\tilde{A}^+(k_1^+)\tilde{A}^+(k_2^+)}{(k_1^+ + k_2^+ - i\varepsilon)(k_2^+ - i\varepsilon)} + \ldots$$

with terms of different orders in the strong coupling g. As illustrated in Fig. 10, the first term is the scattered 'bare' quark and the following terms corresponds to rescattering on the target colour field via 1,2...gluons. This rescattering[16] has leading twist contributions for longitudinally polarised gluons, which are instantaneous in light-front time

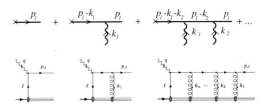

Figure 10. Diagram representation of Eq. 6 with the scattered quark from a hard vertex (marked by the cross) and having 0,1,2... rescatterings on the gauge field of the target (upper part) and its application in DIS (lower part).

$x^+ = t + z$ and occurs within Ioffe coherence length $\sim 1/m_p x_{Bj}$ of the hard DIS interaction.

This implies a rescattering of the scattered quark with the spectator system in DIS (Fig. 10). Although one can choose a gauge such that the scattered quark has no rescatterings, one can not 'gauge away' all rescatterings with the spectator system[16]. The sum of the couplings to the $q\bar{q}$-system in Fig. 11a gives the same result in any gauge and is equivalent to the colour dipole model in the target rest system (discussed below). Thus, there will always be such rescatterings and their effects are absorbed in the parton density functions obtained by fitting inclusive DIS data.

This has recently been used[17] as a basis for the SCI model. As illustrated in Fig. 11a, a gluon from the proton splits into a $q\bar{q}$ pair that the photon couples to. Both the gluon and its splitting are mostly soft since this has higher probability ($g(x)$ and α_s). The produced $q\bar{q}$ pair is therefore typically a large colour dipole that even a soft rescattering gluon can resolve and therefore interact with. The discussed instantaneous gluon exchange can then modify the colour topology before the string-fields are formed such that colour singlet systems separated in rapidity arise producing a gap in the final state after hadronisation as described by the SCI model.

Similarly, the initial gluon may also split softly into a gluon pair followed by perturbative $g \to q\bar{q}$ giving a small $q\bar{q}$ pair (Fig. 11b). Soft rescattering gluons can then not resolve the $q\bar{q}$, but can interact with the large-size $q\bar{q}$–g colour octet dipole and turn that into a colour singlet system separated from the target remnant system that is also in a colour singlet state. Higher order perturbative emissions do not destroy the gap, since it occurs in the rapidity region of the hard system and not in the gap region.

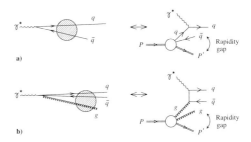

Figure 12. Diffractive DIS in the semi-classical approach[18] where the photon fluctuates into a $q\bar{q}$ or $q\bar{q}g$ system that interacts non-perturbatively with the proton colour field in the proton rest frame (left) and the corresponding Breit frame interpretations (right).

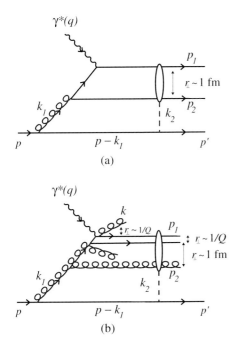

(a)

(b)

Figure 11. Low-order rescattering correction to DIS in the parton model frame where the virtual photon momentum is along the negative z-axis with $q = (q^+, q^-, \vec{q}_\perp) \simeq (-m_p x_B, 2\nu, \vec{0})$ and the target is at rest. The struck parton absorbs nearly all the photon momentum giving $p_1 \simeq (0, 2\nu, \vec{p}_{1\perp})$ (aligned jet configuration). In (a) the virtual photon strikes a quark and the diffractive system is formed by the $q\bar{q}$ pair (p_1, p_2) which rescatters coherently from the target via 'instantaneous' longitudinal (A^+) gluon exchange with momentum k_2. In (b) the $Q\bar{Q}$ quark pair which is produced in the $\gamma^* g \to Q\bar{Q}$ subprocess has a small transverse size $r_\perp \sim 1/Q$ and rescatters like a gluon. The diffractive system is then formed by the $(Q\bar{Q})\,g$ system. The possibility of hard gluon emission close to the photon vertex is indicated. Such radiation (labeled k) emerges at a short transverse distance from the struck parton and is not resolved in the rescattering.

Moreover, the rescattering produces on-shell intermediate states having imaginary amplitudes[16], which is a characteristic feature of diffraction. This theoretical framework implies the same Q^2, x, W dependencies in both diffractive and non-diffractive DIS, in accordance with the observation at HERA.

In this approach the diffractive structure function is a convolution $F_2^D \sim g_p(x_{I\!P}) f(\beta)$ between the gluon density and the gluon splitting functions $f(\beta) \sim \beta^2 + (1 - \beta)^2$ for $g \to q\bar{q}$ and $f(\beta) \sim (1 - \beta(1-\beta))/(\beta(1-\beta))$ for $g \to gg$ (assuming that these perturbative expressions provide reasonable approximations also for soft splittings). Connecting to the pomeron language, this means that the gluon density replaces the pomeron flux and the gluon splitting functions the pomeron parton densities. One can note that the weak β dependence observed in the diffractive structure function (Fig. 4), here gets a natural explanation since $F_2^{I\!P}(\beta) = F_2^{D(4)}/f_{I\!P/p} \sim \frac{x}{\beta} F_2^{D(4)}(x)$, $i.e.$ the applied factor x/β essentially cancels the increase of the proton structure function F_2 at small x.

Another approach, which has some similarities, is the so-called semi-classical approach[18]. The analysis is here made in the proton rest frame where the incoming photon fluctuates into a $q\bar{q}$ pair or a $q\bar{q}$–g system that traverses the proton (Fig. 12). The soft interactions of these colour dipoles with

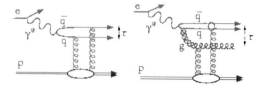

Figure 13. Diffractive DIS in the model[19] based on two-gluon exchange between the non-perturbative proton and the perturbative fluctuation of the photon to $q\bar{q}$ and $q\bar{q}g$ colour dipoles.

the non-perturbative colour field of the proton is estimated using Wilson lines describing the interaction of the energetic partons with the soft colour field of the proton. The colour singlet exchange contribution to this process has been derived and shown to give leading twist diffraction when the dipole is large. This corresponds to a dipole having one soft parton (as in the aligned jet model), which is dominantly the gluon in Fig. 12b. One is thus testing the large distances in the proton colour field. This soft field cannot be calculated from first principles and therefore modelled involving parameters fitted to data. This theoretical approach is quite successful in describing the data (in Fig. 3) on $F_2^{D(3)}(x_{I\!P}; \beta, Q^2)$.

Yet another approach starts from perturbative QCD and attempts to describe diffractive DIS as a two-gluon exchange[19]. Again the photon fluctuates into a $q\bar{q}$ or a $q\bar{q}$-g colour dipole. The upper part of the corresponding diagrams shown in Fig. 13 can be calculated in perturbative QCD giving essentially the β and Q^2 dependencies of the diffractive structure function

$$x_{I\!P} F_2^{D(3)} = F_{q\bar{q}}^T + F_{q\bar{q}g}^T + F_{q\bar{q}}^L \qquad (6)$$

with contributions of $q\bar{q}$ and $q\bar{q}g$ colour dipoles from photons with transverse (T) and longitudinal (L) polarization given by

$$F_{q\bar{q}}^T = A\left(\frac{x_0}{x_{I\!P}}\right)^{n_2} \beta(1-\beta)$$

$$F_{q\bar{q}g}^T = B\left(\frac{x_0}{x_{I\!P}}\right)^{n_2} \alpha_s \ln\left(\frac{Q^2}{Q_0^2} + 1\right)(1-\beta)^\gamma$$

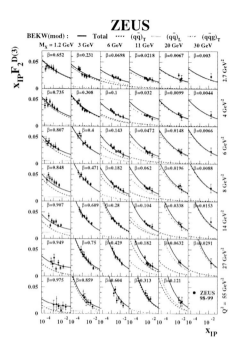

Figure 14. ZEUS data[20] on the diffractive structure function $F_2^{D(3)}(x_{I\!P}; \beta, Q^2)$ with fits of the model[19] for perturbative QCD two-gluon exchange.

$$F_{q\bar{q}}^L = C\left(\frac{x_0}{x_{I\!P}}\right)^{n_4} \frac{Q_0^2}{Q^2}\left[\ln\left(\frac{Q^2}{4Q_0^2\beta} + 1.75\right)\right]^2$$
$$\beta^3(1-2\beta)^2$$

where $n_\tau = n_{\tau 0} + n_{\tau 1}\ln\left[\ln\frac{Q^2}{Q_0^2} + 1\right]$ for twist $\tau = 2, 4$.

The lower part of the diagram, with the connection to the proton, cannot be calculated perturbatively. This soft dynamics is introduced through a parameterisation where one fits the $x_{I\!P}$ dependence, which introduces parameters for the absolute normalization A, B, C as well as $n_{\tau 0}, n_{\tau 1}, \gamma$. The result[20] is a quite good fit to the data as shown in Fig. 14. The different contributions from the two dipoles and photon polarizations are also shown, which provide interesting information on the QCD dynamics described by this approach.

This perturbative two-gluon exchange mechanism is theoretically related to the processes in the following section.

5   Gaps between jets and BFKL

In the processes discussed so far, the hard scale has not involved the gap itself since the leading proton has only been subject to a soft momentum transfer across the gap. A new milestone was therefore the observation[21] at the Tevatron of events with a gap between two high-$p_\perp$ jets. This means that there is a large momentum transfer across the gap and perturbative QCD should therefore be applicable to understand the process. This is indeed possible by considering elastic parton-parton scattering via hard colour singlet exchange in terms of two gluons as illustrated in Fig. 15. In the high energy limit $s/|t| \gg 1$, where the parton cms energy is much larger than the momentum transfer, the amplitude is dominated by terms $\sim [\alpha_s \ln(s/|t|)]^n$ where the smallness of α_s is compensated by the large logarithm. These terms must therefore be resummed leading to the famous BFKL equation describing the exchange of a whole gluon ladder (including virtual corrections and so-called reggeization of gluons).

This somewhat complicated equation has been solved numerically[22], including also some non-leading corrections which turned out to be very important at the non-asymptotic energy of the Tevatron. This gave the matrix elements for an effective $2 \to 2$ parton scattering process, which was implemented in the Lund Monte Carlo PYTHIA such that parton showers and hadronisation could be added to generate complete events. This reproduces the data, both in shape and absolute normalization, which is not at all trivial, as demonstrated in Fig. 16. The non-leading corrections are needed since the asymptotic Mueller-Tang result has the wrong E_T dependence. A free gap survival probability parameter, which in other models is introduced to get the correct overall normalization, is not needed in this approach. Amazingly, the correct gap rate results from the complete model including parton show-

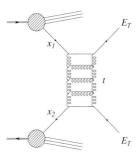

Figure 15. Hard colour singlet exchange through a BFKL gluon ladder giving a rapidity gap between two high-$p_\perp$ jets.

ers, parton multiple scattering and hadronisation through PYTHIA together with the above discussed soft colour interaction model. The latter must be included, since the rescatterings are always present as explained above and without them an *ad hoc* 15% gap survival probability factor would have to be introduced.

This process of gaps between jets provides strong evidence for the BFKL dynamics as predicted since long by QCD, but which has so far been very hard to establish experimentally.

Related to this is the new results from ZEUS[23] on the production of J/ψ at large momentum transfer t in photoproduction at HERA. The data, shown in Fig. 17, agree well with perturbative QCD calculations[24] (based on the hard scales t and $m_{c\bar{c}}$) for two-gluon BFKL colour singlet exchange. As illustrated in Fig. 18, not only the simple two-gluon exchange is included, but also the full gluon ladder in either leading logarithm approximation or with non-leading corrections. Although the conventional DGLAP approximation can provide a good description in part of the t-region, in order to describe the full t-region and the energy dependence of this process one needs the BFKL formalism. Thus, this provides another evidence for BFKL dynamics.

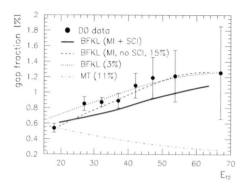

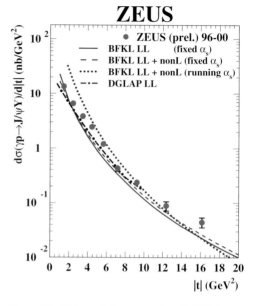

Figure 16. Fraction of jet events having a rapidity gap in $|\eta| < 1$ between the jets versus the second-highest jet-E_T. D0 data compared to the colour singlet exchange mechanism[22] based on the BFKL equation with non-leading corrections and with the underlying event treated in three ways: simple 3% gap survival probability, PYTHIA's multiple interactions (MI) and hadronisation requiring a 15% gap survival probability, MI plus soft colour interactions (SCI) and hadronisation with no need for an overall renormalisation factor. Also shown is the Mueller-Tang (MT) asymptotic result with a 11% gap survival probability.

Figure 17. Differential cross-section $d\sigma/d|t|$ for the process $\gamma + p \rightarrow J/\psi + Y$. ZEUS data compared[23] to BFKL model calculations using leading log (LL) with fixed α_s, and including non-leading (non-L) corrections with fixed or running α_s as well as simple leading log DGLAP.

6 Future

The discussions above illustrate that hard diffraction is a 'laboratory' for QCD studies. This obviously includes small-x parton dynamics and two- or multi-gluon exchange processes, such as gluon ladders and the BFKL equation. Moreover, high gluon densities lead to the concept of continuous colour fields and the interactions of a high energy parton traversing a colour field as described by the soft colour interaction model and the semiclassical approach discussed above. This has natural connections to the quark-gluon plasma and the understanding of the jet quenching phenomenon as an energy loss for a parton moving through the plasma colour field.

Fundamental aspects of hadronisation also enters concerning gap formation and the production of leading particles from a small mass colour singlet beam remnant system.

The latter is not only a problem in diffraction, but of more general interest, e.g. to understand how one should map a small-mass $c\bar{c}$ pair onto the discrete charmonium states[25].

We can still not exclude the possibility that the pomeron is some special kind of non-perturbative colour singlet gluonic system in the proton wave function. If so, this could be connected with the recently developed three-dimensional proton structure[26] $f_{q,g/p}(x, b, Q^2)$, where the quarks and gluons in the proton do not only have the normal x, Q^2 variables, but also an impact parameter b giving the transverse distance of the struck parton from the proton center as obtained by Fourier transformation of measured transverse momentum. Such analyses have led to speculations that the proton has a core of valence quarks surrounded by a cloud of sea partons. It is conceivable that such a cloud contains special gluon configurations that correspond to the pomeron.

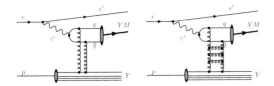

Figure 18. Diffractive vector meson production at large momentum transfer as described by perturbative QCD hard colour singlet exchange via two gluons and a gluon ladder in the BFKL framework[24].

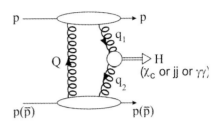

Figure 19. Diagram for the exclusive process $pp \rightarrow pHp$, where the scales $M_H \gg Q \gg \Lambda_{QCD}$ motivate the use of perturbative QCD.

As long as QCD has important unsolved problems, diffraction will continue to be an interesting topic. This brings us to the LHC, where we will have all of the above diffractive processes but also some new ones.

A topical and controversial issue currently is diffractive Higgs production at LHC. The first idea was here to exploit the fact that diffractive events are cleaner due to less hadronic activity and that it should therefore be easier to reconstruct a Higgs through its decay products in such an environment. This has even been considered as a Higgs discovery channel. At the Tevatron the cross-sections turns out to be very small due to the low energy available when the leading proton has only lost a small energy fraction[27]. At the LHC, however, the cross-sections are quite reasonable, but Monte Carlo studies[27] show that the events are not as clean as expected because the energy is so large that one can have a leading proton and a large gap (typically outside the detector) and still have plenty of energy to produce a lot of hadronic activity together with the Higgs.

The exclusive process $pp \rightarrow pHp$ is, however, quite interesting. By measuring the protons one can here calculate the Higgs mass with missing mass techniques, without even looking at the central system. Therefore, plans are in progress for adding such leading proton spectrometers some hundred meters downstream in the LHC beam line (*e.g.* the '420 m' project). On the theoretical side there has been several model calculations for this exclusive process. The cross-section estimates vary by some orders of magnitude[28], so some models must be substantially wrong.

The most reliable state-of-the-art calculation is by the Durham group[29] based on the diagram in Fig. 19. The basic process is calculated in perturbative QCD using Sudakov factors to include the requirement of no gluon radiation that would destroy the gap. There are, of course, also soft processes that might destroy the gap and these are taken into account by a non-trivial estimate of the gap survival probability. This gives a cross-section $\sigma \sim 3~fb$ for $M_H = 120$ GeV at LHC giving ~ 90 events for $\int \mathcal{L} \sim 30~fb^{-1}$, which is certainly of experimental interest. At the Tevatron the cross-section is $\sigma \sim 0.2~fb$, which is too small to be of interest. Here, however, one can make important tests of this model calculation by instead of the Higgs consider similar exclusive production of smaller mass systems such as χ_c, jet-jet and $\gamma\gamma$ which have larger cross-sections[29].

7 Conclusions

After 20 years of hard diffraction it is obvious that there has been great progress. Most importantly, this phenomenon has been discovered both in $p\bar{p}$ and ep resulting in a lot of high-quality data, much more than could be presented here. The developments of theory and models have provided working phenomenological descriptions, but we do not have solid theory yet. Hard diffraction has become an important part of QCD research where, in particular, the interplay of hard and soft dynamics can be investigated.

In the new QCD-based models, emphasized above, the pomeron is not a part of the proton wave function, but diffraction is an effect of the scattering process. Models based on interactions with a colour background field are here particularly intriguing, since they provide an interesting approach which avoids conceptual problems of pomeron-based models, such as the pomeron flux, but also provide a basis for constructing a common theoretical framework for all final states, diffractive gap events as well as non-diffractive events.

But, there are still gaps in our understanding. This is not altogether bad, because it means that we have an interesting future first a few years at HERA and the Tevatron and then at the LHC.

References

1. G. Ingelman and P.E. Schlein, *Phys. Lett. B* **152**, 256 (1985).
2. R. Bonino *et al.*, *Phys. Lett. B* **211**, 239 (1988).
3. A. Brandt *et al.*, *Phys. Lett. B* **297**, 417 (1992).
4. M. Derrick *et al.*, *Phys. Lett. B* **315**, 481 (1993). M. Derrick *et al.*, *Z. Phys. C* **68**, 569 (1995).
5. T. Ahmed *et al.*, *Nucl. Phys. B* **429**, 477 (1994). T. Ahmed *et al.*, *Phys. Lett. B* **348**, 681 (1995).
6. G. Ingelman and K. Prytz, *Z. Phys. C* **58**, 285 (1993).
7. C. Adloff *et al.*, *Z. Phys. C* **76**, 613 (1997).
8. A. Brandt *et al.*, *Phys. Lett. B* **421**, 395 (1998). A. Brandt *et al.*, *Eur. Phys. J. C* **25**, 361 (2002).
9. ZEUS collaboration, LP2005 papers 268, 293, 295.H1 collaboration, LP2005 papers 396, 397
10. J. C. Collins, *Phys. Rev. D* **57**, 3051 (1998) [*Erratum-ibid. D* **61**, 019902 (2000)].
11. K. Goulianos, *Phys. Lett. B* **358**, 379 (1995). K. Goulianos and J. Montanha, *Phys. Rev. D* **59**, 114017 (1999). S. Erhan and P. E. Schlein, *Phys. Lett. B* **427**, 389 (1998).
12. A. Edin, G. Ingelman and J. Rathsman, *Phys. Lett. B* **366**, 371 (1996). A. Edin, G. Ingelman and J. Rathsman, *Z. Phys. C* **75**, 57 (1997).
13. B. Andersson, G. Gustafson, G. Ingelman and T. Sjöstrand, *Phys. Rept.* **97**, 31 (1983).
14. A. Edin, G. Ingelman and J. Rathsman, in proc. 'Monte Carlo generators for HERA physics', DESY-PROC-1999-02, arXiv:hep-ph/9912539.
15. R. Enberg, G. Ingelman and N. Timneanu, *Phys. Rev. D* **64**, 114015 (2001).
16. S. J. Brodsky, P. Hoyer, N. Marchal, S. Peigne and F. Sannino, *Phys. Rev. D* **65**, 114025 (2002).
17. S. J. Brodsky, R. Enberg, P. Hoyer and G. Ingelman, *Phys. Rev. D* **71**, 074020 (2005).
18. W. Buchmuller, T. Gehrmann and A. Hebecker, *Nucl. Phys. B* **537**, 477 (1999). A. Hebecker, *Phys. Rept.* **331**, 1 (2000).
19. J. Bartels, J. R. Ellis, H. Kowalski and M. Wusthoff, *Eur. Phys. J. C* **7**, 443 (1999).
20. S. Chekanov *et al.*, *Nucl. Phys. B* **713**, 3 (2005).
21. F. Abe *et al.*, *Phys. Rev. Lett.* **80**, 1156 (1998). B. Abbott *et al.*, *Phys. Lett. B* **440**, 189 (1998).
22. R. Enberg, G. Ingelman and L. Motyka, *Phys. Lett. B* **524**, 273 (2002).
23. ZEUS collaboration, LP2005 paper 291
24. R. Enberg, L. Motyka and G. Poludniowski, *Eur. Phys. J. C* **26**, 219 (2002).
25. C. B. Mariotto, M. B. Gay Ducati and G. Ingelman, *Eur. Phys. J. C* **23**, 527 (2002).
26. L. Frankfurt, M. Strikman and C. Weiss, *Annalen Phys.* **13**, 665 (2004). M. Diehl, arXiv:hep-ph/0509170.
27. R. Enberg, G. Ingelman, A. Kissavos and N. Timneanu, *Phys. Rev. Lett.* **89**, 081801 (2002). R. Enberg, G. Ingelman and N. Timneanu, *Phys. Rev. D* **67**, 011301 (2003).
28. V.A. Khoze, A.D. Martin and M.G. Ryskin, *Eur. Phys. J. C* **26**, 229 (2002).
29. A.D. Martin, V.A. Khoze and M.G. Ryskin, arXiv:hep-ph/0507305 and references therein

DISCUSSION

Jon Butterworth (UCL):

Is there a connection between the soft colour interactions which give rapidity gaps, and those which can form the J/ψ in colour octet calculations?

Gunnar Ingelman: Yes, I think so. These are different theoretical techniques for describing the same, or similar, basic soft interactions. The colour octet model provides a nice theoretical framework with proper systematics, but one still need to fit uncalculable non-perturbative matrix elements to data. The SCI model started as a very simple phenomenological model and is now connected to QCD rescattering theory giving it more theoretical support. Also I think one should not see these different approaches, including the QCD-based models for diffraction that I have discussed, as excluding each other. It is not so that one is right and the others wrong. None is fully right, but they can be better or worse for describing and understanding different aspects of these poorly understood soft phenomena and all of them are likely to contribute to an improved understanding.

Klaus Mönig (LAL/DESY):

If the Pomeron is in fact soft gluon exchange, shouldn't there be rapidity gaps also between quark and gluon jets at LEP?

Gunnar Ingelman: When using the SCI model for e^+e^- we observed a very low rate of gaps, as is also the case in the data. This is due to the lack of an initial hadron giving a spectator system. The large gaps in ep and $p\bar{p}$ depend on having a hadron remnant system taking a large fraction of the beam momentum when a soft gluon interacts. This remnant is then far away in rapidity from the rest

and when it emerges as a colour singlet there will be large gap in rapidity after hadronisation.

Stephen L Olsen (Hawaii):

At the B-factories we see an anomalously strong cross-section for exclusive $e^+e^- \rightarrow J/\psi\,\eta_c$ (at $\sqrt{s} = 10.6$ GeV). This is as large a "gap" as possible at this energy. Can your methods be used to adress processes like these?

Gunnar Ingelman: In principle yes, since the SCI model is implemented in the Lund Monte Carlos one could try it on anything that those Monte Carlos can simulate. However, there is a technical complication here when hadronising into an exclusive two-body final state, which the Lund hadronisation model is not constructed for and not suitable for. In spite of this, we have made recent progress in applying the model to exclusive B-meson decays, such as $B \rightarrow J/\psi K$, were the SCI mechanism increases the rate for such decay modes which are colour suppressed in the conventional theory. It may therefore be possible to apply it also for your process, but it cannot be obtained by just running the Monte Carlo straightforwardly—a dedicated study of the problem would be required.

POLARIZATION IN QCD

JEAN-MARC LE GOFF

DAPNIA/SPhN CEA-Saclay, F91191 Gif-sur-Yvette, France

E-mail: jmlegoff@cea.fr

We first deal with the muon anomalous magnetic moment which is found to differ from the standard model prediction by 2.7 σ. A new proposal will aim at reducing the error by a factor 2. The main part of the paper then deals with the spin structure of the nucleon which can be studied in terms of quark and gluon helicity distributions, quark transversity distributions and generalized parton distributions. The main recent results are first indications that the total gluon spin in the nucleon might be small and a first measurement of the Collins fragmentation function which is needed to extract transversity distributions from semi-inclusive DIS data.

1 The muon anomalous magnetic moment

The muon anomalous magnetic moment $a_\mu = (g_\mu - 2)/2$ can be both measured and computed very accurately. Deviations from the standard model prediction are expected according to many models: SUSY, leptoquark, muon substructure or anomalous W coupling.

The E821 experiment[1] in Brookhaven makes use of a muon beam obtained by pion weak decay. This beam is naturally polarized due to parity violation in pion decay. The muons are injected in a ring where their spins rotate slightly faster than their momentum. This precession is proportional to a_μ. The rate of positrons from muon decay is recorded in 24 electromagnetic calorimeters. Due to parity violation in muon decay this rate is a decreasing exponential modulated at the precession frequency, as illustrated in Fig. 1. The rate is fitted by $N(t) = N_0 \, e^{-t/\tau}[1 + A\cos(\omega_a t + \phi)]$ which provides the precession frequency ω_a. The magnetic field is measured by a NMR probe in units of the free proton precession frequency, ω_p. Knowing the ratio of muon-to-proton magnetic moment from muonium experiments, the ratio ω_a/ω_p provides $a_\mu = (11,659,208 \pm 5 \pm 3) \times 10^{-10}$. This is an accuracy of 0.5 ppm, which is 15 times better than the previous experiment.

The calculation of a_μ in the standard model involves several contributions dis-

Table 1. World averaged experimental measurement and various theoretical contributions to a_μ in units of 10^{-10}, see text

QED	$11,659,471.94 \pm 0.14$
Had LO	693.4 ± 6.4
Had LBL	12.0 ± 3.5
Had HO	-10.0 ± 0.6
weak	15.4 ± 0.22
total	$11,659,182.7 \pm 7.3$
exp	$11,659,208 \pm 6$
exp-the	25.3 ± 9.4

played in table 1. The QED contribution is by far the largest. It is computed up to 4 loops and is then very accurately known. The weak contribution is small; it is computed to 2 loops and is also very accurately known. The hadronic contribution is split into three terms: leading order (LO), light by light term (LBL) and higher order (HO). The evaluation of the leading order term (hadronic vacuum polarization) is based on e^+e^- data. An evaluation from tau decay data is also possible but it requires several delicate corrections and it does not give a result compatible with those from e^+e^-.

As can be seen in the table, the world averaged experimental result differ from the standard model prediction by 25.3 ± 9.4, i.e. 2.7 σ. A new experiment, E969, is proposed in order to reach $\delta a_\mu \approx 2 \cdot 10^{-10}$. The main ideas and the ring will be kept, the num-

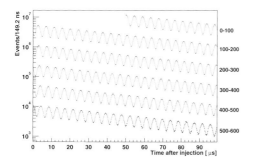

Figure 1. The positron rate in E821 as a function of time exhibits an exponential shape (note logarithmic vertical scale) modulated at the precession frequency.

ber of muons will be increased by a factor 5 and the systematics decreased. Together with expected improvements in the theory a gain by a factor 2 is expected in the error on the difference between experiment and theory. This should clarify whether the 2.7 σ discrepancy observed is a real effect or not.

2 Quark polarized distributions

The spin $1/2$ of the nucleon can be decomposed in the contributions from its constituents as[2]

$$\frac{1}{2} = \frac{1}{2}\Delta\Sigma + \Delta G + L_q + L_g, \qquad (1)$$

where $\Delta\Sigma$ is the total number of quarks with spin parallel to the spin of the nucleon minus the number of quarks with spin antiparallel; once weighted by the spin of the quark, i.e. $1/2$, this is the contribution from the quark spins to the nucleon spin. Similarly, ΔG is the contribution from the spin of the gluons (which have spin 1). Finally, L_q and L_g are the contributions from the orbital angular momentum of quarks and gluons, respectively.

The contribution $\Delta\Sigma$ can be measured in inclusive deep inelastic scattering (DIS) experiments where a high energy lepton scatters on a nucleon target through the exchange of

a virtual photon. There exist only two independent Lorentz invariants which can be chosen as $Q^2 = -q_\mu^2$, the virtuality of the photon, and $x_{bj} = Q^2/2M\nu$, where ν is the photon energy and M the nucleon mass. Q^2 gives the probe resolution, while x_{bj} can be interpreted as the fraction of the nucleon momentum carried by the quark which absorbed the virtual photon. In unpolarized DIS one measures cross sections which involve two structure functions depending only on the two invariants x and Q^2. To first order (i.e. in the parton model) the Q^2 dependence vanishes, indicating scattering on point-like particles (the quarks) and we have e.g. $F_1(x) = \frac{1}{2}\sum e_q^2 q(x)$. Here the quark distribution function $q(x)$ gives the probability density for finding inside the nucleon a quark of flavor q and momentum fraction x.

In polarized DIS one measures cross-section spin differences, $\Delta\sigma = \sigma^{\uparrow\downarrow} - \sigma^{\uparrow\uparrow}$, which involve the polarized structure functions g_1 and g_2. In the parton model g_1 reads $g_1(x) = \frac{1}{2}\sum e_q^2 \Delta q(x)$, where $\Delta q(x) = q^+(x) - q^-(x)$ counts the number of quarks with spin parallel to the spin of the nucleon minus the number of quarks with spin antiparallel.

In 1988 the EMC experiment measured $\Gamma_1 = \int_0^1 g_1(x)dx = \frac{1}{2}\sum e_q^2\Delta q$, where $\Delta q = \int_0^1 q(x)dx$. The sum $\Delta u + \Delta d + \Delta s$ is then the total number of quarks with spin parallel minus those with spin antiparallel, whatever their momentum and flavor, i.e. the quark spin contribution to the nucleon spin, $\Delta\Sigma$. Combining the measurement of Γ_1 with hyperon beta-decay data and using flavor SU(3) symmetry, they got $\Delta\Sigma = 12 \pm 9 \pm 14\%$. Due to a theoretical expectation of $\approx 60\%$ this came as a big surprise, which was advertised as the "spin crisis", and the corresponding paper[3] is still one of the 6 most cited experimental papers on the SLAC SPIRES database. The result was confirmed by SMC[4], SLAC[5] and HERMES[6], giving $\Delta\Sigma = 20$ to 30 % depending on the analysis. The

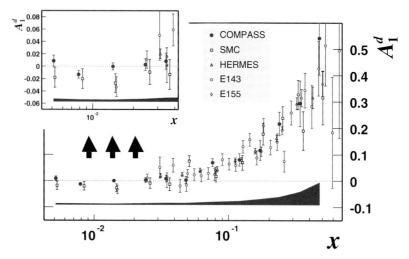

Figure 2. The virtual-photon deuteron asymmetry $A_1^d \approx g_1/F_1$ measured by COMPASS. The error bars represent the statistical error and the band the systematic error.

main uncertainty in this result arises from the extrapolation in the unmeasured region at $x < 0.003$.

The new g_1^d data[7] from COMPASS presented in fig 2 cover the range $0.004 < x < 0.5$. Below a few 10^{-2} they have much smaller errors than earlier data as can be seen in the insert. Including these data in a QCD analysis performed by COMPASS changes $\Delta\Sigma$ from $0.202^{+0.042}_{-0.077}$ to $0.237^{+0.024}_{-0.029}$, where we note a reduction of the error by about a factor 2.

The final g_1 data of HERMES[8] on proton, deuteron and neutron, presented in Fig. 3, exhibit a very good statistical accuracy. The smearing between x bins, due to spectrometer resolution and radiative effects, was taken into account in the analysis, resulting in a correlation between the measurements in the different bins.

Jefferson laboratory has obtained the first accurate virtual-photon neutron asymmetries $A_1^n \approx g_1^n/F_1^n$ at high x [9], showing for the first time that A_1^n becomes positive for $x > 0.5$. Combined with world averaged A_1^p data, this provides a measurement of $\Delta u/u$ and $\Delta d/d$ as illustrated in Fig. 4. If orbital angular momentum is neglected, per-

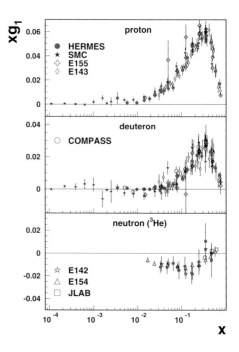

Figure 3. The final set of HERMES g_1 data compared to published data.

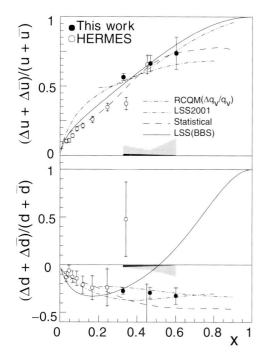

Figure 4. $\Delta u/u$ and $\Delta d/d$ as a function of x from JLab (full circles) and HERMES (open circles) compared to various models.

turbative QCD predicts that both $\Delta u/u$ and $\Delta d/d$ should go to 1 when x goes to 1. In the figure we note that this seems indeed to be the case for $\Delta u/u$ but that $\Delta d/d$ remains negative up to the highest measured x, 0.6. This might indicate that orbital angular momentum cannot be neglected, at least in the particular domain of high x.

As a conclusion of this section devoted to quark contributions, we should say that the measurement of Γ_1 does not exactly provide $\Delta\Sigma$, but rather the singlet axial matrix element a_0. Naively, the two quantities are identified. However, due to the axial anomaly in QCD, they are related by

$$a_0 = \Delta\Sigma - \frac{3\alpha_s}{2\pi}\Delta G. \qquad (2)$$

If $\Delta G = 0$ then $\Delta\Sigma = a_0 \approx 0.2$, but if ΔG is large, on the order of 2.5, we get $\Delta\Sigma$ on the order of 0.6, which solves the spin crisis.

3 Gluon polarized distribution

3.1 With a lepton beam

Although they do not carry an electric charge the gluons can be probed with a lepton beam using the photon gluon fusion (PGF) process, $\gamma^* g \rightarrow q\bar{q}$, where the gluon interacts with the virtual photon through the exchange of a quark and a $q\bar{q}$ pair is produced in the final state. The cross section for this process is of course much smaller than for the leading order absorption of the virtual photon by a quark. Therefore a tagging is needed.

The cleanest tagging is provided by requiring the $q\bar{q}$ pair to be a $c\bar{c}$ pair, since the presence of c quarks inside the nucleon (intrinsic charm) is negligible. The c quark is identified in the form of a D^0 meson which decays to $K\pi$ with a 4% branching ratio. Due to multiple scattering inside the thick polarized target the D^0 decay vertex cannot be distinguished from the primary vertex and the D^0 is seen as a peak in the reconstructed $K\pi$ mass above a combinatorial background. This background can be reduced by selecting D^0 coming from the decay $D^* \rightarrow D^0\pi_s$. Because the difference of masses $M_{D^*} - M_{D^0} - M_\pi$ is only 6 MeV the π is soft and there is little phase space for the background. Due to the background and the small cross section, this channel is statistically limited: using 2002 and 2003 data COMPASS gets $\Delta G/G = -1.08 \pm 0.76$. There is twice as much data in the 2004 run and the data taking is resuming in 2006 with an improved apparatus.

In the absorption of the virtual photon by a quark, the produced hadrons go in the direction of the virtual photon with a small transverse momentum p_t relative to it. An alternative tagging of PGF is then obtained by requiring a pair of hadrons with large p_t. However, there are several sources of physical background and the measured asymmetry can be written as $A_{\parallel} = R_{PGF}\, a_{LL}^{PGF}(\Delta G/G) + A_{bckg}$. Here R_{PGF}

is the fraction of PGF event in the sample; a_{LL}^{PGF} is the analyzing power, i.e. the spin asymmetry of the PGF process which can be computed in perturbative QCD; and A_{bckg} is the asymmetry of the background. The background includes the QCD Compton process, $\gamma^*q \rightarrow qg$, and at low Q^2 the resolved photon processes where the photon fluctuates to a hadronic state and one of the partons from the photon interacts with one of the partons from the nucleon.

Analyzing 2002 and 2003 data at $Q^2 < 1$ GeV^2, using Pythia to estimate R_{PGF} and A_{bckg}, COMPASS obtains[10] :

$$\frac{\Delta G}{G}(x = 0.10) = 0.024 \pm 0.089 \pm 0.057. \quad (3)$$

Such a low systematic error is obtained due to the fact that a large part of the error is proportional to the asymmetry, which is small. This result is presented in Fig. 5, together with earlier measurements. The HERMES measurement[11] is dominated by low Q^2 data where the resolved photon contribution is important. The background asymmetry was however neglected in this analysis, which may cast some doubt on the result. The SMC result[12] and the COMPASS result at $Q^2 > 1$ come from a region where the background is easier do deal with. However, they suffer from low statistics.

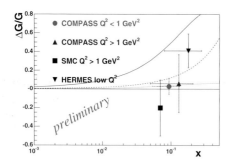

Figure 5. Various measurements of $\Delta G/G$ compared to three parametrisations from GRSV corresponding to integrals $\Delta G = \int_0^1 \Delta G(x)dx = 0.2$ (dots), 0.6 (dashes) and 2.5 (continuous).

The experimental results are compared to three parametrisations from GRSV[13] corresponding to integrals $\Delta G = \int_0^1 \Delta G(x)dx = 0.2$, 0.6 (standard) and 2.5 (max). We can see that the new COMPASS result is not compatible with the "max" parametrization with $\Delta G = 2.5$. This favors lower values of ΔG, but it is still possible that $\Delta G(x)$ has a more complicated shape, e.g. crossing zero around $x = 0.1$.

3.2 With a pp collider

The gluon polarization can also be measured in a polarized proton-proton collider, such as RHIC. The golden channel is $pp \rightarrow \gamma+$ jet $+X$. This occurs through the partonic process $qg \rightarrow q\gamma$, so it provides a convolution of ΔG in one proton with Δq in the other proton. There is a physical background, $q\bar{q} \rightarrow \gamma g$, but its contribution is small in RHIC kinematics. RHIC has not yet accumulated enough luminosity to use this channel.

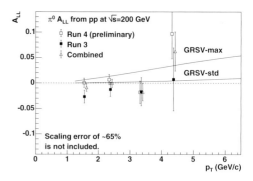

Figure 6. Longitudinal spin asymmetry A_{LL} for $\vec{p}\vec{p} \rightarrow \pi^0 X$ measured by Phenix.

Another possibility is to look at π^0 production. At the partonic level this involves $gg \rightarrow gg$, $gq \rightarrow gq$ and $qq \rightarrow qq$. So the measured asymmetry includes contributions proportional to $\Delta G \otimes \Delta G$, $\Delta q \otimes \Delta G$ and $\Delta q \otimes \Delta q$. Fig. 6 presents the results obtained[14] by the Phenix collaboration at RHIC out of runs 3 and 4, compared to two of the previous GRSV parametrizations[13]. The error bars are still large but the result tends to disfavor the max parametrization. Run 5 was finished in June 2005 and it is expected to provide a gain by a factor ≈ 100 in terms of the factor of merit, $\mathcal{L}P^4$.

4 Transversity

At leading twist there exist three parton distribution functions for the nucleon[15]. The unpolarized pdf, $q(x)$, the longitudinally polarized pdf or helicity pdf, $\Delta q(x) = q^+(x) - q^-(x)$, and the transversity pdf, $\Delta_T q(x) = q^{\uparrow\uparrow}(x) - q^{\uparrow\downarrow}(x)$. Here $q^{\uparrow\uparrow}(x)$ is the probability density function for finding a quark with spin parallel to the spin of the nucleon but when the direction of the spin is observed perpendicularly to the momentum. Because rotations do not commute with Lorentz boosts transversity and helicity distributions differ. For instance, due to gluons having spin 1, there is no gluon transversity distribution, while there is a gluon helicity distribution.

Transversity is a chiral odd function. Therefore, since all hard processes conserve helicity, it decouples from DIS as illustrated in Fig. 7. In order to measure transversity, one needs to introduce a second soft object in addition to transversity, to allow for helicity flip. The simplest solution is to introduce a second proton and to measure Drell-Yan process, $pp \to l^+l^-X$. This happens through the hard process $q\bar{q} \to \gamma^* \to l^+l^-$ and provides the convolution of $\Delta_T q$ in one proton by $\Delta_T \bar{q}$ in the other proton.

It is also possible to measure transversity in semi-inclusive DIS. In this case the second soft object is a polarized fragmentation function and one measures the product of transversity times this polarized fragmentation function. Two kinds of polarized fragmentation have been used so far. The Collins fragmentation function tells how much the transverse spin of the fragmenting quark is reflected in the azimuthal distribution of the produced hadrons. It results in an azimuthal asymmetry in terms of the Collins angle,

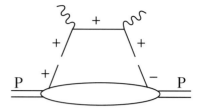

Figure 7. A pdf represents the probability to emit a parton from the nucleon. The square of the corresponding amplitude is then the lower part of the diagram. In the case of transversity the helicity of the emitted and reabsorbed partons are opposite whereas all hard processes involved e.g. in DIS conserve helicity. So the hard (upper) part of the diagram cannot be connected to the soft (lower) part, i.e. transversity cannot be measured in DIS.

$\phi_{col} = \phi_h + \phi_s - \pi$, where ϕ_h is the azimuthal angle of the produced hadron and ϕ_s that of the target spin. Note that an azimuthal asymmetry in terms of the Sivers angle, $\phi_{siv} = \phi_h - \phi_s$, is also possible. It is not related to transversity but to quark momentum distribution in the plane transverse to the nucleon momentum (intrinsic k_T).

The so-called interference fragmentation function is due to some interference effects in the production of two hadrons. It results in an azimuthal distribution in terms of the angle $\phi_{RS} = \phi_R + \phi_S - \pi$. In order to define ϕ_R we must introduce the momentum of the two hadrons $\vec{p}_1$ and $\vec{p}_2$, $\vec{p}_h = \vec{p}_1 + \vec{p}_2$, $\vec{R} = \vec{p}_1 - \vec{p}_2$ and $\vec{R}_\perp$ which is the component of $\vec{R}$ perpendicular to $\vec{p}_h$. Then ϕ_R is the azimuthal angle of $\vec{R}_\perp$.

Fig. 8 presents the Collins asymmetry measured on a proton target by the HERMES collaboration[16]. A clear evidence for non-zero asymmetry is seen both for the production of π^+ and π^-. Fig. 9 presents the Sivers asymmetry measured also by HERMES[16]. The asymmetry for π^- is compatible with zero, but for π^+ it is clearly positive. Such a non-zero Sivers asymmetry should in some way be related with orbital angular momentum.

Fig. 10 shows the Collins and Sivers asymmetries measured by COMPASS on a

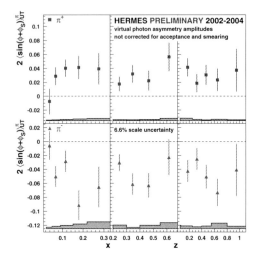

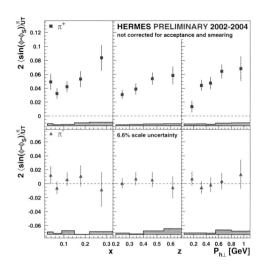

Figure 8. The Collins asymmetry measured by HER-MES on a proton target for π^+ (upper plot) and π^- (lower plot) as a function of x, z and p_t.

Figure 9. The Sivers asymmetry measured by HER-MES on a proton target for π^+ (upper plot) and π^- (lower plot) as a function of x, z and p_t.

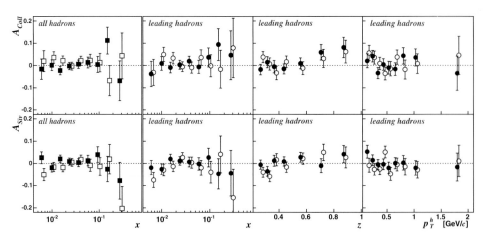

Figure 10. The Collins (upper plot) and Sivers (lower plot) asymmetries measured by COMPASS on a deuteron target for π^+ (black squares) and π^- (open squares) as a function of x, z and p_t.

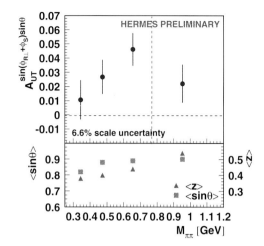

Figure 11. Interference asymmetry measured by HERMES on the proton as a function of the two pion mass. The vertical dashed line indicates the position of the ρ mass.

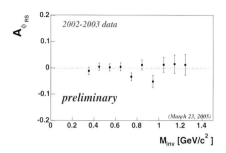

Figure 12. Interference asymmetry measured by COMPASS on the deuteron as a function of the two pion mass.

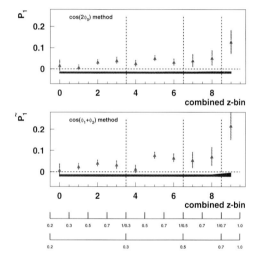

Figure 13. The azimuthal asymmetry measured by Belle as a function of z_1 and z_2.

deuteron target[17]. The data cover a wider range of x than HERMES data but, in spite of statistical errors comparable to those of HERMES, no Collins nor Siver asymmetry is observed, neither for positive nor negative hadrons. This is ascribed to a possible cancellation between proton and neutron asymmetries. Three times more statistics is available on tape and a run with a proton target is planned in 2006.

The azimuthal asymmetries in terms of ϕ_{RS} (interference asymmetry), measured by HERMES on a proton target, is shown in Fig. 11 as a function of the two pion mass[18]. It clearly shows a positive asymmetry. No change of sign is observed at the ρ mass in contrast with the prediction of a model by Jaffe[19].

The corresponding asymmetries measured by COMPASS on a deuteron target[20] are shown in Fig. 12. Once again no asymmetry is observed, probably due to a cancellation between proton and neutron asymmetries.

The Collins asymmetry, as measured by HERMES and COMPASS, is the product of transversity times a polarized fragmentation function, which is called the Collins function. This Collins function has to be measured independently. This can be done in an e^+e^- collider. If we consider the plane defined by the beam and jet axes, the cross section can be written as $\sigma = A + B\cos(\phi_1 + \phi_2)\Delta_T D_q^h(z_1)\Delta_T D_q^h(z_2)$, where ϕ_1 (ϕ_2) is the azimuthal angle of the produced hadron around the axis of jet 1 (jet 2) with respect to the beam and jet plane. Hadrons h_1 and h_2 have different momentum, so $\Delta_T D_q^h(z)$ is probed at two different z.

The results obtained by Belle[21] are presented in Fig. 13. The first 4 points correspond to $0.2 < z_2 < 0.3$ and increasing z_1. We see a non-zero asymmetry increasing with z_1. The next 3 points correspond to $0.3 < z_2 < 0.5$ and increasing z_1, we see again an asymmetry increasing with z_1. The next two points correspond to $0.5 < z_2 < 0.7$. In the last point, where both z_1 and z_2 are larger than 0.7, we see a pretty large asymmetry. So we clearly see a non-zero asymmetry and it is increasing with z_1 and z_2 as expected. Ten times more statistics is available, which should allow for a deconvolution and real extraction of the Collins function as a function of z.

5 Generalized parton distributions

Generalized parton distributions[22] (GPD) appear in deeply virtual Compton scattering, which is the exclusive process $\gamma^* p \rightarrow \gamma p$. In the deep region, i.e. large Q^2 and $-t \ll Q^2$ where t is the transfer to the nucleon, this process can be factorized at all orders in QCD. At leading order, as illustrated in Fig. 14, a quark of longitudinal momentum fraction $x - \xi$ is emitted by the nucleon, the quark absorbs the virtual photon, emits a real photon and is reabsorbed with a momentum fraction $x + \xi$. The factorization involves new soft objects, the generalized parton distributions which are real functions of x, ξ and t. There exist four of them, labelled H, $\tilde{H}$, E and $\tilde{E}$.

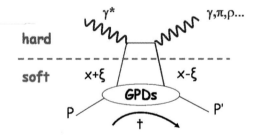

Figure 14. The DVCS diagram at leading order.

Due to the optical theorem the DIS cross-section is proportional to the imaginary part of forward Compton scattering, $\gamma^* p \rightarrow \gamma^* p$. The diagram for the DVCS amplitude is then very similar to the diagram for the DIS cross-section. The difference is that DVCS probes correlations $\Psi(x - \xi)\Psi^*(x + \xi)$ while DIS probes $|\Psi(x)|^2$ and that DVCS implies a non-zero momentum transfer to the nucleon, t. Two of the GPDs correspond to the same nucleon spin in the initial and final states, in the limit $t = 0$ and $\xi = 0$ they give back pdf, namely $H(x,0,0) = q(x)$ and $\tilde{H}(x,0,0) = \Delta q(x)$. The two other GPDs have $s \neq s'$ and disconnect from the cross section in the limit $t = 0$.

The GPDs are also related to the form factors $F(t)$, which are the Fourier transforms of the spatial distributions of quarks inside the nucleon. We have the sum rules $\int H(x,\xi,t)dx = F_1(t)$ and $\int E(x,\xi,t)dx = F_2(t)$, where F_1 and F_2 are the Dirac and Pauli form factors. In the limit $\xi = 0$ the GPDs can be shown to provide a 3D view of the nucleon in terms of the momentum fraction and the impact parameter, $f(x,d_\perp)$. It is therefore intuitive that GPDs have something to do with the orbital angular momentum. This relation is formalized in the so-called Ji sum rule.

We can see that the DVCS diagram involves a loop over x. Therefore the amplitude is an integral of GPD over x:

$$T_{DVCS} \propto \int \frac{dx}{x - \xi + i\epsilon} H(x,\xi,t) =$$

$$\mathcal{P} \int \frac{dx}{x - \xi} H(x,\xi,t) - i\pi H(x = \xi, \xi, t). \quad (4)$$

In addition DVCS interferes with the Bethe-Heitler process which is just an elastic scattering on the nucleon accompanied by the radiation of a real photon, either by the initial or the final lepton.

This interference can actually be useful. Since the Bethe-Heitler amplitude is known, the measurement of the interference provides a measurement of the DVCS amplitude. The interference can be measured in single spin asymmetry (polarized beam)

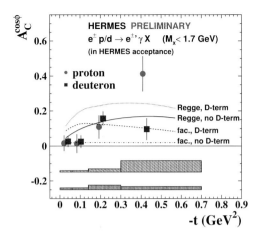

Figure 15. Beam charge asymmetry for the DVCS process as a function of $-t$ for the proton and the deuteron. HERMES data are compared to the predictions of different models of proton GPDs

which selects the imaginary part of the DVCS amplitude, i.e. $H(x = \xi, \xi, t)$; while the beam charge asymmetry (e^+ versus e^-) selects the real part of the DVCS amplitude, i.e. $\int \frac{dx}{x-\xi} H(x, \xi, t)$.

Measurements of single spin asymmetries and beam charge asymmetries have been performed by Jefferson Lab[23] and HERMES[24]. Fig. 15 presents the beam charge asymmetries obtained recently by HERMES[25]. We see that the error bars are still too large in these exploratory data to discriminate the models. However, much more precise data are currently being analyzed at Jefferson Lab and new precise data should be taken by HERMES in the coming years. After 2010 DVCS experiments at COMPASS and at a 12 GeV upgraded Jefferson lab should cover a wider kinematic range.

6 Conclusions

The E821 experiment at Brookhaven gives a muon anomalous momentum which is 2.7 σ from the standard model prediction. A new experiment and progress in theory are expected to reduce the error by a factor 2.

The spin structure of the nucleon is a very active field. More topics than could be discussed here are being studied, like the tensor structure function of the deuteron at HERMES.

There are some first indications that the gluon spin contribution to the nucleon, ΔG, could be smaller than the value of about 2.5 which was somewhat expected because it would solve the "spin crisis". If this is indeed the case, we are left with the crisis, i.e. with a small contribution of the quark spin, $\Delta \Sigma = 0.2 - 0.3$.

Non-zero asymmetries related to transversity have been measured by HERMES. The extraction of transversity then requires to know a polarized fragmentation function, the Collins function. Asymmetries related to the Collins function were measured in $e^+ e^-$ collisions by Belle and the Collins function itself should be extracted soon.

Generalized parton distributions (GPD) represent an opening field. Exploratory measurements have been performed and new much more precise data are expected soon.

Several new projects are developing over the world. An upgraded version of COMPASS should allow for the measurements of GPD in a wide kinematical range. An upgraded Jefferson Lab at 12 GeV will also probe GPDs. The PAX project at the $p\bar{p}$ collider at GSI will be the ideal tool to study transversity in the cleanest channel, i.e. in Drell-Yan processes. The most complete program is certainly that of the eRHIC project of a polarized electron-proton collider. This will give the possibility to reach low x, to have enough level arm in Q^2 for an accurate NLO analysis providing Δq and ΔG, to measure directly $\Delta G(x)$ and to further study GPDs.

References

1. Muon (g-2) Collaboration, G. W. Bennett et al, *Phys. Rev. Lett.* **92**, 161802 (2004); David W. Hertzog and William M. Morse, *Annu. Rev. Nucl. Part. Sci.* **54**, 141-174 (2004).
2. Reviews: M. Anselmino, A. Efremov and E. Leader, *Phys. Rep.* **261**, 1 (1995), hep-ph/9501369; B. Lampe and E. Reya, *Phys. Rep.* **332**, 1 (2000), hep-ph/9810270
3. J. Ashman et al., EMC coll., *Phys. Lett.* **B206**, 364 (1988).
4. SMC, D. Adams et al., *Phys. Rev.* **D58**, 112001 (1998).
5. E142, Anthony et al. *Phys. Rev.* **D54**, 6620 (1996); E143, Abe et al., *Phys. Rev.* **D58**, 112003 (1998); E154, Abe et al., *Phys. Lett.* **B405**, 180 (1997); E155, Anthony et al., *Phys. Lett.* **B493**, 19 (2000).
6. HERMES, A. Airapetian et al., *Phys. Lett.* **B442**, 484 (1998).
7. COMPASS, E.S. Ageev et al., *Phys. Lett.* **B612**, 154 (2005).
8. HERMES: D. Reggiani, AIP Conference Proceedings 792, DIS 2005, p 913, Springer Verlag (2005).
9. X. Zheng et al., *Phys. Rev. Lett.* **92**, 012004 (2004).
10. COMPASS: C. Bernet, AIP Conference Proceedings 792, DIS 2005, p 949, Springer Verlag (2005); hep-ex/0507049.
11. HERMES, A. Airapetian et al., *Phys. Rev. Lett.* **84**, 2584 (2000).
12. SMC, B. Adeva et al., *Phys. Rev.* **D70**, 012002 (2004).
13. M. GLuck, E. Reya, M. Stratmann, W. Vogelsang, *Phys. Rev.* **D63**, 094005 (2001).
14. A. Deshpande, AIP Conference Proceedings 792, DIS 2005, p 1001, Springer Verlag (2005).
15. Review: V. Barone, A. Drago and P.G. Ratcliffe, *Phys. Rep.* **359**, 1 (2002); hep-ph/0104283.
16. HERMES: A. Airapetian *et al.*, *Phys. Rev. Lett.* **94**, 012002 (2005); hep-ex/0408013.
17. COMPASS: V. Y. Alexakhin *et al.*, *Phys. Rev. Lett.* **94**, 202002 (2005); hep-ex/0503002.
18. HERMES: P.B. Van der Nat, AIP Conference Proceedings 792, DIS 2005, p 953, Springer Verlag (2005).
19. R. L. Jaffe, X. m. Jin and J. Tang, *Phys. Rev. Lett.* **80**, 1166 (1998); hep-ph/9709322.
20. COMPASS: R. Joosten, AIP Conference Proceedings 792, DIS 2005, p 957, Springer Verlag (2005).
21. K. Abe et al., the Belle Collaboration, contributing paper to this conference, hep-ex/0507063.
22. Reviews: M. Diehl, *Phys. Rep.* **388**, 41 (2003); A.V. Belitsky, D. Müller and A. Kirchner, *Nucl. Phys.* **B629**, 323 (2002).
23. CLAS, S. Stepanyan et al., *Phys. Rev. Lett.* **87**, 182002 (2001).
24. A. Airapetian et al., *Phys. Rev. Lett.* **87**, 182001 (2001), hep-ex/0106068
25. M. Kopytin, AIP Conference Proceedings 792, DIS 2005, p 949, Springer Verlag (2005).

DISCUSSION

Volker Burkert (Jefferson Lab):

If ΔG remains small does that mean that the orbital angular momentum contributions to the nucleon spin must be large? If so, does this put more emphasis on programs to measure the generalized parton distributions (GPD) to allow access to L_z ?

Jean-Marc Le Goff: Actually I can see two levels of ΔG being small. The axial anomaly gives

$$\Delta\Sigma = a_0 + \frac{3\alpha_s}{2\pi}\Delta G, \qquad (5)$$

with a_0 0.2 to 0.3. If ΔG is on the order of 2 or 3 then we get $\Delta\Sigma$ on the order of 0.6 as expected, which "solves the spin crisis". So the first level of being small is relative to the large values of 2 or 3. And the data tend to disfavor such a scenario, even if they cannot exclude it completely.

On the other hand we have the momentum sum rule

$$\frac{1}{2} = \frac{1}{2}\Delta\Sigma + \Delta G + L_q + L_g. \qquad (6)$$

Here, with a_0 0.2 to 0.3, we need $\Delta G \approx 0.4$ to fill the sum rule. So the other level of being small is relative to this value of 0.4. And according to the current data a value of 0.4 is as likely as a null value. So I do not think that the current ΔG data can be used to claim that L_z should be important.

However, there are several indications that L_z should not be neglected. I mentioned in the talk the observation of a non-zero Sivers asymmetry and also the finding that $\Delta d/d$ is negative at high x, which both point to orbital angular momentum. In addition GPDs are a very interesting topic besides the issue of L_z. They generalized the notion of pdf and form factor, providing a 3d view of the nucleon. So I think we have very strong motivations to study GPD and I am looking forwards seeing the new accurate Jefferson lab DVCS data.

QCD AND STRING THEORY

IGOR R. KLEBANOV

Joseph Henry Laboratories, Princeton University, Princeton, NJ 08544, USA
E-mail: klebanov@princeton.edu

This talk begins with some history and basic facts about string theory and its connections with strong interactions. Comparisons of stacks of Dirichlet branes with curved backgrounds produced by them are used to motivate the AdS/CFT correspondence between superconformal gauge theory and string theory on a product of Anti-de Sitter space and a compact manifold. The ensuing duality between semi-classical spinning strings and long gauge theory operators is briefly reviewed. Strongly coupled thermal SYM theory is explored via a black hole in 5-dimensional AdS space, which leads to explicit results for its entropy and shear viscosity. A conjectured universal lower bound on the viscosity to entropy density ratio, and its possible relation to recent results from RHIC, are discussed. Finally, some available results on string duals of confining gauge theories are briefly reviewed.

1 Introduction

String theory[a] is well known to be the leading prospect for quantizing gravity and unifying it with other interactions[1,2]. One may also take a broader view of string theory as a description of string-like excitations that arise in many different physical systems, such as the superconducting flux tubes, cosmic strings, and of course the chromo-electric flux tubes in non-Abelian gauge theories, which are the subject of my talk. You could object that these string-like excitations are "emergent" rather than fundamental phenomena. We will see, however, that there is no sharp distinction between "emergent" and fundamental strings. We will exhibit examples, stemming from the AdS/CFT correspondence[3,4,5], where the "emergent" and fundamental strings are dual descriptions of the same theory. Besides being of great theoretical interest, such gauge/string dualities are becoming a useful tool for studying strongly coupled gauge theories. A developing connection that is highlighted in this talk is with the new results at RHIC:[6] there are indications that a rather strongly coupled Quark-Gluon Plasma (sQGP) has been ob-

served.

2 Some early history

String Theory was born out of attempts to understand the Strong Interactions. Empirical evidence for a string-like structure of hadrons comes from arranging mesons and baryons into approximately linear Regge trajectories. Studies of πN scattering prompted Dolen, Horn and Schmid[7] to make a duality conjecture stating that the sum over s-channel exchanges equals the sum over t-channel ones. This posed the problem of finding the analytic form of such dual amplitudes. Veneziano[8] found the first, and very simple, expression for a manifestly dual 4-point amplitude:

$$A(s,t) \sim \frac{\Gamma(-\alpha(s))\Gamma(-\alpha(t))}{\Gamma(-\alpha(s)-\alpha(t))} \quad (1)$$

with an exactly linear Regge trajectory $\alpha(s) = \alpha(0) + \alpha's$. Soon after, Nambu[9], Nielsen[10] and Susskind[11] independently proposed its open string interpretation. This led to an explosion of interest in the early 70's in string theory as a description of strongly interacting particles. The basic idea is to think of a meson as an open string with a quark at one end-point and an anti-quark at another. Then various meson states arise as different excitations of such an open string. The split-

---

[a]Due to a strict length limit, I did not include figures in this manuscript. The figures are included in the PowerPoint version of this talk, available at the 2005 Lepton-Photon Symposium web site.

ting of a string describes the decay of a meson into two mesons, for example.

The string world sheet dynamics is governed by the Nambu-Goto area action

$$S_{\mathrm{NG}} = -T \int d\sigma d\tau \sqrt{-\det \partial_a X^\mu \partial_b X_\mu} \ , \quad (2)$$

where the indexes a, b take two values ranging over the σ and τ directions on the world sheet. The string tension is related to the Regge slope through $T = \frac{1}{2\pi\alpha'}$. The quantum consistency of the Veneziano model requires that the Regge intercept is $\alpha(0) = 1$, so that the spin 1 state is massless but the spin 0 is a tachyon. But the ρ meson is certainly not massless, and there are no tachyons in the real world. This is how the string theory of strong interactions started to run into problems.

Calculation of the string zero-point energy gives

$$\alpha(0) = \frac{d - 2}{24} \ . \quad (3)$$

Hence the model has to be defined in 26 space-time dimensions. Attempts to quantize such a string model directly in 3+1 dimensions led to tachyons and problems with unitarity. Consistent supersymmetric string theories were discovered in 10 dimensions, but their relation to the strong interactions was initially completely unclear. Most importantly, the Asymptotic Freedom of strong interactions was discovered[12], singling out Quantum Chromodynamics (QCD) as the exact field theory of strong interactions. At this point most physicists gave up on strings as a description of strong interactions. Instead, since the graviton appears naturally in the closed string spectrum, string theory emerged as the leading hope for unifying quantum gravity with other forces[13].

3   QCD gives strings a chance

Now that we know that a non-Abelian gauge theory is an exact description of strong interactions, is there any room left for string theory in this field? Luckily, the answer is positive. At short distances, much smaller than 1 fermi, the quark anti-quark potential is Coulombic, due to Asymptotic Freedom. At large distances the potential should be linear due to formation of a confining flux tube[14]. When these tubes are much longer than their thickness, one can hope to describe them, at least approximately, by semi-classical Nambu strings[15]. This appears to explain the existence of approximately linear Regge trajectories: a linear relation between angular momentum and mass-squared

$$J = \alpha' m^2 + \alpha(0) \ , \quad (4)$$

is provided by a semi-classical spinning relativistic string with massless quark and anti-quark at its endpoints. A semi-classical string approach to the QCD flux tubes is widely used, for example, in jet hadronization algorithms based on the Lund String Model[16].

Semi-classical quantization around a long straight Nambu string predicts the quark anti-quark potential[17]

$$V(r) = Tr + \mu + \frac{\gamma}{r} + O(1/r^2) \ . \quad (5)$$

The coefficient γ of the universal Lüscher term depends only on the space-time dimension d and is proportional to the Regge intercept: $\gamma = -\pi(d - 2)/24$. Recent lattice calculations of the force vs. distance for probe quarks and anti-quarks[18] produce good agreement with this value in $d = 3$ and $d = 4$ for $r > 0.7 fm$. Thus, long QCD strings appear to be well described by the Nambu-Goto area action. But quantization of short, highly quantum QCD strings, that could lead to a calculation of light meson and glueball spectra, is a much harder problem.

The connection of gauge theory with string theory is strengthened by 't Hooft's generalization of QCD from 3 colors ($SU(3)$ gauge group) to N colors ($SU(N)$ gauge group)[19]. The idea is to make N large, while keeping the 't Hooft coupling $\lambda = g_{\mathrm{YM}}^2 N$

fixed. In this limit each Feynman graph carries a topological factor N^χ, where χ is the Euler characteristic of the graph. Thus, the sum over graphs of a given topology can perhaps be thought of as a sum over world sheets of a hypothetical "QCD string." Since the spheres (string tree diagrams) are weighted by N^2, the tori (string one-loop diagrams) – by N^0, etc., we find that the closed string coupling constant is of order N^{-1}. Thus, the advantage of taking N to be large is that we find a weakly coupled string theory. In the large N limit the gauge theory simplifies in that only the planar diagrams contribute. But directly summing even this subclass of diagrams seems to be an impossible task. From the dual QCD string point of view, it is not clear how to describe this string theory in elementary terms.

Because of the difficulty of these problems, between the late 70's and the mid-90's many theorists gave up hope of finding an exact gauge/string duality. One notable exception is Polyakov who already in 1981 proposed that the string theory dual to a 4-d gauge theory should have a 5-th hidden dimension[20]. In later work[21] he refined this proposal, suggesting that the 5-d metric must be "warped."

4 The Geometry of Dirichlet Branes

In the mid-nineties the Dirichlet branes, or D-branes for short, brought string theory back to gauge theory. The D-branes are soliton-like "membranes" of various internal dimensionalities contained in theories of closed superstrings[2]. A Dirichlet p-brane (or Dp-brane) is a $p+1$ dimensional hyperplane in $9+1$ dimensional space-time where strings are allowed to end. A D-brane is much like a topological defect: upon touching a D-brane, a closed string can open up and turn into an open string whose ends are free to move along the D-brane. For the end-points of such a

string the $p+1$ longitudinal coordinates satisfy the conventional free (Neumann) boundary conditions, while the $9-p$ coordinates transverse to the Dp-brane have the fixed (Dirichlet) boundary conditions; hence the origin of the term "Dirichlet brane." In a seminal paper[22] Polchinski showed that a Dp-brane preserves $1/2$ of the bulk supersymmetries and carries an elementary unit of charge with respect to the $p+1$ form gauge potential from the Ramond-Ramond sector of type II superstring.

For our purposes, the most important property of D-branes is that they realize gauge theories on their world volume. The massless spectrum of open strings living on a Dp-brane is that of a maximally supersymmetric $U(1)$ gauge theory in $p+1$ dimensions. The $9-p$ massless scalar fields present in this supermultiplet are the expected Goldstone modes associated with the transverse oscillations of the Dp-brane, while the photons and fermions provide the unique supersymmetric completion. If we consider N parallel D-branes, then there are N^2 different species of open strings because they can begin and end on any of the D-branes. N^2 is the dimension of the adjoint representation of $U(N)$, and indeed we find the maximally supersymmetric $U(N)$ gauge theory in this setting.

The relative separations of the Dp-branes in the $9-p$ transverse dimensions are determined by the expectation values of the scalar fields. We will be interested in the case where all scalar expectation values vanish, so that the N Dp-branes are stacked on top of each other. If N is large, then this stack is a heavy object embedded into a theory of closed strings which contains gravity. Naturally, this macroscopic object will curve space: it may be described by some classical metric and other background fields. Thus, we have two very different descriptions of the stack of Dp-branes: one in terms of the $U(N)$ supersymmetric gauge theory on its world volume, and the other in terms of the classical charged p-brane background of the type II closed superstring theory. The relation between these two descriptions is at the heart of the connections between gauge fields and strings that are the subject of this talk.

4.1 Coincident D3-branes

Parallel D3-branes realize a $3+1$ dimensional $U(N)$ gauge theory, which is a maximally supersymmetric "cousin" of QCD. Let us compare a stack of D3-branes with the Ramond-Ramond charged black 3-brane classical solution whose metric assumes the form[23]:

$$ds^2 = H^{-1/2}(r) \left[-f(r)(dx^0)^2 + (dx^i)^2 \right]$$
$$+ H^{1/2}(r) \left[f^{-1}(r)dr^2 + r^2 d\Omega_5^2 \right] , \quad (6)$$

where $i = 1, 2, 3$ and

$$H(r) = 1 + \frac{L^4}{r^4} , \qquad f(r) = 1 - \frac{r_0^4}{r^4} .$$

Here $d\Omega_5^2$ is the metric of a unit 5 dimensional sphere, $\mathbf{S}^5$.

In general, a d-dimensional sphere of radius L may be defined by a constraint

$$\sum_{i=1}^{d+1} (X^i)^2 = L^2 \qquad (7)$$

on $d+1$ real coordinates X^i. It is a positively curved maximally symmetric space with symmetry group $SO(d+1)$. Similarly, the d-dimensional Anti-de Sitter space, AdS_d, is defined by a constraint

$$(X^0)^2 + (X^d)^2 - \sum_{i=1}^{d-1} (X^i)^2 = L^2 , \qquad (8)$$

where L is its curvature radius. AdS_d is a negatively curved maximally symmetric space with symmetry group $SO(2, d-2)$. There exists a subspace of AdS_d called the Poincaré wedge, with the metric

$$ds^2 = \frac{L^2}{z^2} \left(dz^2 - (dx^0)^2 + \sum_{i=1}^{d-2} (dx^i)^2 \right) ,$$
$$(9)$$

where $z \in [0, \infty)$. In these coordinates the boundary of AdS_d is at $z = 0$.

The event horizon of the black 3-brane metric (6) is located at $r = r_0$. In the extremal limit $r_0 \to 0$ the 3-brane metric becomes

$$ds^2 = \left(1 + \frac{L^4}{r^4} \right)^{-1/2} \left(-(dx^0)^2 + (dx^i)^2 \right)$$
$$+ \left(1 + \frac{L^4}{r^4} \right)^{1/2} \left(dr^2 + r^2 d\Omega_5^2 \right) . \quad (10)$$

Just like the stack of parallel, ground state D3-branes, the extremal solution preserves 16 of the 32 supersymmetries present in the type IIB theory. Introducing $z = \frac{L^2}{r}$, one notes that the limiting form of (10) as $r \to 0$ factorizes into the direct product of two smooth spaces, the Poincaré wedge (9) of AdS_5, and $\mathbf{S}^5$, with equal radii of curvature L. The 3-brane geometry may be thus viewed as a semi-infinite throat of radius L which for $r \gg L$ opens up into flat $9 + 1$ dimensional space. Thus, for L much larger than the string length scale, $\sqrt{\alpha'}$, the entire 3-brane geometry has small curvatures everywhere and is appropriately described by the supergravity approximation to type IIB string theory.

The relation between L and $\sqrt{\alpha'}$ may be found by equating the gravitational tension of the extremal 3-brane classical solution to N times the tension of a single D3-brane, and one finds

$$L^4 = g_{\text{YM}}^2 N \alpha'^2 . \qquad (11)$$

Thus, the size of the throat in string units is $\lambda^{1/4}$. This remarkable emergence of the 't Hooft coupling from gravitational considerations is at the heart of the success of the AdS/CFT correspondence. Moreover, the requirement $L \gg \sqrt{\alpha'}$ translates into $\lambda \gg 1$: the gravitational approach is valid when the 't Hooft coupling is very strong and the perturbative field theoretic methods are not applicable.

5 The AdS/CFT Correspondence

Consideration of low-energy processes in the 3-brane background[24] indicates that, in the low-energy limit, the $AdS_5 \times \mathbf{S}^5$ throat region ($r \ll L$) decouples from the asymptotically flat large r region. Similarly, the $\mathcal{N} = 4$ supersymmetric $SU(N)$ gauge theory on the stack of N D3-branes decouples in the low-energy limit from the bulk closed string theory. Such considerations prompted Maldacena[3] to conjecture that type IIB string theory on $AdS_5 \times \mathbf{S}^5$, of radius L given in (11), is dual to the $\mathcal{N} = 4$ SYM theory. The number of colors in the gauge theory, N, is dual to the number of flux units of the 5-form Ramond-Ramond field strength.

It was further conjectured in[4,5] that there exists a one-to-one map between gauge invariant operators in the CFT and fields (or extended objects) in AdS_5. The dimension Δ of an operator is determined by the mass of the dual field in AdS_5. For example, for scalar operators one finds that $\Delta(\Delta-4) = m^2 L^2$. Precise methods for calculating correlation functions of various operators in a CFT using its dual formulation were also formulated[4,5]. They involve calculating the string theory path integral as a function of the boundary conditions in AdS_5, which are imposed near $z = 0$.

If the number of colors N is sent to infinity while $g_{\mathrm{YM}}^2 N$ is held fixed and large, then there are small string scale corrections to the supergravity limit[3,4,5] which proceed in powers of $\frac{\alpha'}{L^2} = (g_{\mathrm{YM}}^2 N)^{-1/2}$. If we wish to study finite N, then there are also string loop corrections in powers of $\frac{\kappa^2}{L^8} \sim N^{-2}$. As expected, taking N to infinity enables us to take the classical limit of the string theory on $AdS_5 \times \mathbf{S}^5$.

Immediate support for the AdS/CFT correspondence comes from symmetry considerations[3]. The isometry group of AdS_5 is $SO(2,4)$, and this is also the conformal group in $3 + 1$ dimensions. In addition we have the isometries of $\mathbf{S}^5$ which form $SU(4) \sim SO(6)$. This group is identical to the R-symmetry of the $\mathcal{N} = 4$ SYM theory. After including the fermionic generators required by supersymmetry, the full isometry supergroup of the $AdS_5 \times \mathbf{S}^5$ background is $SU(2,2|4)$, which is identical to the $\mathcal{N} = 4$ superconformal symmetry.

To formulate an AdS/CFT duality with a reduced amount of supersymmetry, we may place the stack of D3-branes at the tip of a 6-dimensional Ricci flat cone,[25,26,27] whose base is a 5-dimensional compact Einstein space Y_5. The metric of such a cone is $dr^2 + r^2 ds_Y^2$; therefore, the 10-d metric produced by the D3-branes is obtained from (10) by replacing $d\Omega_5^2$, the metric on $\mathbf{S}^5$, by ds_Y^2, the metric on Y_5. In the $r \to 0$ limit we then find the space $AdS_5 \times Y_5$ as the candidate dual of the CFT on the D3-branes placed at the tip of the cone. The isometry group of Y_5 is smaller than $SO(6)$, but AdS_5 is the "universal" factor present in the dual description of any large N CFT, making the $SO(2,4)$ conformal symmetry geometric.

The fact that after the compactification on Y_5 the string theory is 5-dimensional supports earlier ideas on the necessity of the 5-th dimension to describe 4-d gauge theories[20]. The z-direction is dual to the energy scale of the gauge theory: small z corresponds to the UV domain of the gauge theory, while large z to the IR.

In the AdS/CFT duality, type IIB strings are dual to the chromo-electric flux lines in the gauge theory, providing a string theoretic set-up for calculating the quark antiquark potential[28]. The quark and anti-quark are placed near the boundary of Anti-de Sitter space ($z = 0$), and the fundamental string connecting them is required to obey the equations of motion following from the Nambu action. The string bends into the interior ($z > 0$), and the maximum value of the z-coordinate increases with the separation r between quarks. An explicit calculation of the string action gives an attractive $q\bar{q}$ potential[28]:

$$V(r) = -\frac{4\pi^2\sqrt{\lambda}}{\Gamma\left(\frac{1}{4}\right)^4 r} \; . \tag{12}$$

Its Coulombic $1/r$ dependence is required by the conformal invariance of the theory. Historically, a dual string description was hoped for mainly in the cases of confining gauge theories, where long confining flux tubes have string-like properties. In a pleasant surprise, we have seen that a string description can be applicable to non-confining theories too, due to the presence of extra dimensions in the string theory.

5.1 Spinning Strings vs. Long Operators

A few years ago it was noted that the AdS/CFT duality becomes particularly powerful when applied to operators with large quantum numbers. One class of such single-trace "long operators" are the BMN operators[29] that carry a large R-charge in the SYM theory and contain a finite number of impurity insertions. The R-charge is dual to a string angular momentum on the compact space Y_5. So, in the BMN limit the relevant states are short closed strings with a large angular momentum, and a small amount of vibrational excitation. Furthermore, by increasing the number of impurities the string can be turned into a large semi-classical object moving in $AdS_5 \times Y_5$. Comparing such objects with their dual long operators has become a very fruitful area of research[30]. Work in this direction has also produced a great deal of evidence that the $\mathcal{N} = 4$ SYM theory is exactly integrable (see[31,32] for recent reviews).

A familiar example of a gauge theory operator with a large quantum number is a twist-2 operator carrying a large spin J, Tr $F_{+\mu}D_+^{J-2}F_+{}^\mu$. In QCD, such operators

play an important role in studies of deep inelastic scattering[33]. In the $\mathcal{N} = 4$ SYM theory, the dual of such a high-spin operator is a folded string spinning around the center of AdS_5.[34] In general, for a high spin, the anomalous dimension of such an operator is[35]

$$\Delta - (J + 2) \to f(\lambda)\ln J \; . \tag{13}$$

Calculating the energy of the spinning folded string, we find that the AdS/CFT prediction is[34]

$$f(\lambda) \to \frac{\sqrt{\lambda}}{\pi} \; , \tag{14}$$

in the limit of large 't Hooft coupling. For small λ, perturbative calculations in the large $\mathcal{N} = 4$ SYM theory up to 3-loop order give[36]

$$f(\lambda) = \frac{1}{2\pi^2}\left(\lambda - \frac{\lambda^2}{48} + \frac{11\lambda^3}{11520} + O(\lambda^4)\right) \tag{15}$$

An approximate extrapolation formula, suggested in[36] works with about 10% accuracy:

$$\begin{aligned}\tilde{f}(\lambda) &= \frac{12}{\pi^2}\left(-1 + \sqrt{1+\lambda/12}\right)\\ &= \frac{1}{2\pi^2}\left(\lambda - \frac{\lambda^2}{48} + \frac{\lambda^3}{1152} + O(\lambda^4)\right)\end{aligned} \tag{16}$$

Note that $\tilde{f}$ has a branch cut running from $-\infty$ to -12. Thus, the series has a finite radius of convergence, in accord with general arguments about planar gauge theory given by 't Hooft. The fact that the branch point is at a negative λ suggests that in the $\mathcal{N} = 4$ SYM theory the perturbative series is alternating, and that there is no problem in extrapolating from small to large λ along the positive real axis. It is, of course, highly desirable to find an exact formula for $f(\lambda)$. Recent work[37] raises hopes that a solution of this problem is within reach.

6 Thermal Gauge Theory from Near-extremal D3-branes

6.1 Entropy

An important black hole observable is the Bekenstein-Hawking (BH) entropy, which is proportional to the area of the event horizon, $S_{BH} = A_h/(4G)$. For the 3-brane solution (6), the horizon is located at $r = r_0$. For $r_0 > 0$ the 3-brane carries some excess energy E above its extremal value, and the BH entropy is also non-vanishing. The Hawking temperature is then defined by $T^{-1} = \partial S_{BH}/\partial E$.

Setting $r_0 \ll L$ in (10), we obtain a near-extremal 3-brane geometry, whose Hawking temperature is found to be $T = r_0/(\pi L^2)$. The small r limit of this geometry is $\mathbf{S}^5$ times a certain black hole in AdS_5. The 8-dimensional "area" of the event horizon is $A_h = \pi^6 L^8 T^3 V_3$, where V_3 is the spatial volume of the D3-brane (i.e. the volume of the x^1, x^2, x^3 coordinates). Therefore, the BH entropy is[38]

$$S_{BH} = \frac{\pi^2}{2} N^2 V_3 T^3 . \qquad (17)$$

This gravitational entropy of a near-extremal 3-brane of Hawking temperature T is to be identified with the entropy of $\mathcal{N} = 4$ supersymmetric $U(N)$ gauge theory (which lives on N coincident D3-branes) heated up to the same temperature.

The entropy of a free $U(N)$ $\mathcal{N} = 4$ supermultiplet, which consists of the gauge field, $6N^2$ massless scalars and $4N^2$ Weyl fermions, can be calculated using the standard statistical mechanics of a massless gas (the black body problem), and the answer is

$$S_0 = \frac{2\pi^2}{3} N^2 V_3 T^3 . \qquad (18)$$

It is remarkable that the 3-brane geometry captures the T^3 scaling characteristic of a conformal field theory (in a CFT this scaling is guaranteed by the extensivity of the entropy and the absence of dimensionful parameters). Also, the N^2 scaling indicates the presence of $O(N^2)$ unconfined degrees of freedom, which is exactly what we expect in the $\mathcal{N} = 4$ supersymmetric $U(N)$ gauge theory. But what is the explanation of the relative factor of $3/4$ between S_{BH} and S_0? In fact, this factor is not a contradiction but rather a

prediction about the strongly coupled $\mathcal{N} = 4$ SYM theory at finite temperature. As we argued above, the supergravity calculation of the BH entropy, (17), is relevant to the $\lambda \to \infty$ limit of the $\mathcal{N} = 4$ $SU(N)$ gauge theory, while the free field calculation, (18), applies to the $\lambda \to 0$ limit. Thus, the relative factor of $3/4$ is not a discrepancy: it relates two different limits of the theory. Indeed, on general field theoretic grounds, in the 't Hooft large N limit the entropy is given by[39]

$$S = \frac{2\pi^2}{3} N^2 f(\lambda) V_3 T^3 . \qquad (19)$$

The function f is certainly not constant: Feynman graph calculations valid for small $\lambda = g_{\mathrm{YM}}^2 N$ give[40]

$$f(\lambda) = 1 - \frac{3}{2\pi^2}\lambda + \frac{3 + \sqrt{2}}{\pi^3}\lambda^{3/2} + \ldots \qquad (20)$$

The BH entropy in supergravity, (17), is translated into the prediction that

$$\lim_{\lambda \to \infty} f(\lambda) = \frac{3}{4} . \qquad (21)$$

A string theoretic calculation of the leading correction at large λ gives[39]

$$f(\lambda) = \frac{3}{4} + \frac{45}{32}\zeta(3)\lambda^{-3/2} + \ldots \qquad (22)$$

These results are consistent with a monotonic function $f(\lambda)$ which decreases from 1 to $3/4$ as λ is increased from 0 to ∞. The $1/4$ deficit compared to the free field value is a strong coupling effect predicted by the AdS/CFT correspondence.

It is interesting that similar deficits have been observed in lattice simulations of deconfined non-supersymmetric gauge theories[41,42,43]. The ratio of entropy to its free field value, calculated as a function of the temperature, is found to level off at values around 0.8 for T beyond 3 times the deconfinement temperature T_c. This is often interpreted as the effect of a sizable coupling. Indeed, for $T = 3T_c$, the lattice estimates indicate that $g_{\mathrm{YM}}^2 N \approx 7$.[42] This challenges an old prejudice that the QGP is inherently very weakly coupled. We now turn to calculations of the shear viscosity where strong coupling effects are even more pronounced.

6.2 Shear Viscosity

The shear viscosity η may be read off from the form of the stress-energy tensor in the local rest frame of the fluid where $T_{0i} = 0$:

$$T_{ij} = p\delta_{ij} - \eta(\partial_i u_j + \partial_j u_i - \frac{2}{3}\delta_{ij}\partial_k u_k) , \quad (23)$$

where u_i is the 3-velocity field. The viscosity can be also determined[44] through the Kubo formula

$$\eta = \lim_{\omega \to 0} \frac{1}{2\omega} \int dt d^3x e^{i\omega t} \langle [T_{xy}(t,\vec{x}), T_{xy}(0,0)] \rangle \quad (24)$$

For the $\mathcal{N} = 4$ supersymmetric YM theory this 2-point function may be computed from absorption of a low-energy graviton h_{xy} by the 3-brane metric[24]. Using this method, it was found[44] that at very strong coupling

$$\eta = \frac{\pi}{8}N^2 T^3 , \quad (25)$$

which implies

$$\frac{\eta}{s} = \frac{\hbar}{4\pi} \quad (26)$$

after $\hbar$ is restored in the calculation (here $s = S/V_3$ is the entropy density). It has been proposed[45] that this value is the universal lower bound on η/s. Indeed, at weak coupling η/s is very large, $\sim \frac{1}{\lambda^2 \ln(1/\lambda)}$, and there is evidence that it decreases monotonically as the coupling is increased[46].

The appearance of $\hbar$ in (26) is reasonable on general physical grounds[45]. The shear viscosity η is of order the energy density times quasi-particle mean free time τ. So, η/s is of order of the energy of a quasi-particle times its mean free time, which is bounded from below by the uncertainty principle to be some constant times $\hbar$. The AdS/CFT correspondence fixes this constant to be $1/(4\pi)$, which is not far from some earlier estimates[47].

For known fluids (e.g. helium, nitrogen, water) η/s is considerably higher than the proposed lower bound[45]. On the other hand, the Quark-Gluon Plasma produced at RHIC is believed to have a very low η/s, within a factor of 2 of the bound (26)[48,47]. This suggests that it is rather strongly coupled. Recently a new term, sQGP, which stands for "strongly coupled Quark-Gluon Plasma," has been coined to describe the deconfined state observed at RHIC[49,50] (a somewhat different term, "Non-perturbative Quark-Gluon Plasma," was proposed in [51]). As we have reviewed, the AdS/CFT correspondence is a theoretical laboratory which allows one to study analytically an extreme example of such a new state of matter: the thermal $\mathcal{N} = 4$ SYM theory at very strong 't Hooft coupling.

In a CFT, the pressure is related to the energy density by $p = 3e$. Hence, the speed of sound satisfies $c_s^2 = dp/de = \frac{1}{3}$. Recent lattice QCD calculations indicate that, while c_s^2 is much lower for temperatures slightly above T_c, it gets close to $1/3$ for $T \geq 2T_c$.[42] Thus, for some range of temperatures starting around $2T_c$, QCD may perhaps be treated as an approximately conformal, yet nonperturbative, gauge theory. This suggests that AdS/CFT methods could indeed be useful in studying the physics of sQGP, and certainly gives strong motivation for continued experimental and lattice research.

Lattice calculations indicate that the deconfinement temperature T_c is around 175 MeV, and the energy density is ≈ 0.7 GeV/fm^3, around 6 times the nuclear energy density. RHIC has reached energy densities around 14 GeV/fm^3, corresponding to $T \approx 2T_c$. Furthermore, in a few years, heavy ion collisions at the LHC are expected to reach temperatures up to $5T_c$. Thus, RHIC and LHC should provide a great deal of useful information about the conjectured quasiconformal temperature range of QCD.

7 String Duals of Confining Theories

It is possible to generalize the AdS/CFT correspondence in such a way that the quark anti-quark potential is linear at large distance. In an effective 5-dimensional approach[21] the necessary metric is

$$ds^2 = \frac{dz^2}{z^2} + a^2(z)\big(-(dx^0)^2 + (dx^i)^2\big) \quad (27)$$

and the space must end at a maximum value of z where the "warp factor" $a^2(z_{\max})$ is finite.[b] Placing widely separated probe quark and anti-quark near $z = 0$, we find that the string connecting them bends toward larger z until it stabilizes at $z_{\max}$ where its tension is minimized at the value $\frac{a^2(z_{\max})}{2\pi\alpha'}$. Thus, the confining flux tube is described by a fundamental string placed at $z = z_{\max}$ parallel to one of the x^i-directions. This establishes a duality between "emergent" chromo-electric flux tubes and fundamental strings in certain curved string theory backgrounds.

Several 10-dimensional supergravity backgrounds dual to confining gauge theories are now known, but they are somewhat more complicated than (27) in that the compact directions are "mixed" with the 5-d (x^μ, z) space. Witten[56] constructed a background in the universality class of non-supersymmetric pure glue gauge theory. While in this background there is no asymptotic freedom in the UV, hence no dimensional transmutation, the background has served as a simple model of confinement where many infrared observables have been calculated using the classical supergravity. For example, the lightest glueball masses have been found from normalizable fluctuations around the supergravity solution[57]. Their spectrum is discrete, and resembles qualitatively the results of lattice simulations in the pure glue theory.

Introduction of a minimal ($\mathcal{N} = 1$) supersymmetry facilitates construction of gauge/string dualities. As discussed earlier, a useful method is to place a stack of D-branes at the tip of a six-dimensional cone, whose base is Y_5. For N D3-branes, one finds the background $AdS_5 \times Y_5$ dual to a superconformal gauge theory. Furthermore, there exists an interesting way of breaking the conformal invariance for spaces Y_5 whose topology includes an $\mathbf{S}^2$ factor. At the tip of the cone over Y one may add M wrapped D5-branes to the N D3-branes. The gauge theory on such a combined stack is no longer conformal; it exhibits a novel pattern of quasi-periodic renormalization group flow, called a duality cascade[58,59] (for reviews, see [60,61]).

To date, the most extensive study of a theory of this type has been carried out for a simple 6-d cone called the conifold, where one finds a $\mathcal{N} = 1$ supersymmetric $SU(N) \times SU(N+M)$ theory coupled to chiral superfields A_1, A_2 in the $(\mathbf{N}, \overline{\mathbf{N} + \mathbf{M}})$ representation, and B_1, B_2 in the $(\overline{\mathbf{N}}, \mathbf{N} + \mathbf{M})$ representation. In type IIB string theory, D5-branes source the 7-form field strength from the Ramond-Ramond sector, which is Hodge dual to the 3-form field strength. Therefore, the M wrapped D5-branes create M flux units of this field strength through the 3-cycle in the conifold; this number is dual to the difference between the numbers of colors in the two gauge groups. An exact non-singular supergravity solution dual to this gauge theory, incorporating the 3-form and the 5-form R-R field strengths, and their back-reaction on the geometry, has been found[59]. This back-reaction creates a "geometric transition" to the deformed conifold

$$\sum_{a=1}^{4} z_a^2 = \epsilon^2 , \quad (28)$$

and introduces a "warp factor" so that the full 10-d geometry has the form

$$ds^2 = h^{-1/2}(\tau)\left(-(dx^0)^2 + (dx^i)^2\right)$$
$$+ h^{1/2}(\tau) d\tilde{s}_6^2 , \quad (29)$$

---

[b]A simple model of confinement[52] is obtained for $a(z) = 1/z$ in (27), i.e. the metric is a slice of AdS_5 cut off at $z_{\max}$. Hadron spectra in models of this type were studied in [53,54,55].

where $d\tilde{s}_6^2$ is the Calabi-Yau metric of the deformed conifold, which is known explicitly.

The field theoretic interpretation of this solution is unconventional. After a finite amount of RG flow, the $SU(N + M)$ group undergoes a Seiberg duality transformation[62]. After this transformation, and an interchange of the two gauge groups, the new gauge theory is $SU(\tilde{N}) \times SU(\tilde{N} + M)$ with the same matter and superpotential, and with $\tilde{N} = N - M$. The self-similar structure of the gauge theory under the Seiberg duality is the crucial fact that allows this pattern to repeat many times. If $N = (k + 1)M$, where k is an integer, then the duality cascade stops after k steps, and we find a $SU(M) \times SU(2M)$ gauge theory. This IR gauge theory exhibits a multitude of interesting effects visible in the dual supergravity background. One of them is confinement, which follows from the fact that the warp factor h is finite and non-vanishing at the smallest radial coordinate, $\tau = 0$, which roughly corresponds to $z = z_{\max}$ in an effective 5-d approach (27). This implies that the quark anti-quark potential grows linearly at large distances. Other notable IR effects are chiral symmetry breaking, and the Goldstone mechanism[63]. Particularly interesting is the appearance of an entire "baryonic branch" of the moduli space in the gauge theory, whose existence has been recently demonstrated also in the dual supergravity language[64].

Besides providing various new insights into the IR physics of confining gauge theories, the availability of their string duals enables one to study Deep-Inelastic and hadron-hadron scattering in this new language[52].

8 Summary

Throughout its history, string theory has been intertwined with the theory of strong interactions. The AdS/CFT correspondence[3,4,5] succeeded in making precise connections between conformal 4-dimensional gauge theories and superstring theories in 10 dimensions. This duality leads to a multitude of dynamical predictions about strongly coupled gauge theories. When extended to theories at finite temperature, it serves as a theoretical laboratory for studying a novel state of matter: a gluonic plasma at very strong coupling. This appears to have surprising connections to the new state of matter, sQGP, observed at RHIC[6].

Extensions of the AdS/CFT correspondence to confining gauge theories provide new geometrical viewpoints on such important phenomena as chiral symmetry breaking and dimensional transmutation. They allow for studying meson and glueball spectra, and high-energy scattering, in model gauge theories.

This recent progress offers new tantalizing hopes that an analytic approximation to QCD will be achieved along this route, at least for a large number of colors. But there is much work that remains to be done if this hope is to become reality: understanding the string duals of weakly coupled gauge theories remains an important open problem.

Acknowledgments

I am grateful to the organizers of the 2005 Lepton-Photon Symposium in Uppsala for giving me an opportunity to present this talk in a pleasant and stimulating environment. I also thank Chris Herzog for his very useful input. This research is supported in part by the National Science Foundation Grant No. PHY-0243680. Any opinions, findings, and conclusions or recommendations expressed in this material are those of the authors and do not necessarily reflect the views of the National Science Foundation.

References

1. M. B. Green, J. H. Schwarz and E. Witten, "Superstring Theory. Vol. 1 and 2", Cam-

bridge Univ. Press.

2. J. Polchinski, "String theory. Vol. 1 and 2," Cambridge Univ. Press.

3. J. M. Maldacena, *Adv. Theor. Math. Phys.* **2**, 231 (1998) [arXiv:hep-th/9711200].

4. S. S. Gubser, I. R. Klebanov and A. M. Polyakov, *Phys. Lett.* B **428**, 105 (1998) [arXiv:hep-th/9802109].

5. E. Witten, *Adv. Theor. Math. Phys.* **2**, 253 (1998) [arXiv:hep-th/9802150].

6. K. Adcox *et al.* [PHENIX Collaboration], *Nucl. Phys.* A **757**, 184 (2005) [arXiv:nucl-ex/0410003];
   I. Arsene *et al.* [BRAHMS Collaboration], *Nucl. Phys.* A **757**, 1 (2005) [arXiv:nucl-ex/0410020];
   B. B. Back *et al.*, *Nucl. Phys.* A **757**, 28 (2005) [arXiv:nucl-ex/0410022];
   J. Adams *et al.* [STAR Collaboration], *Nucl. Phys.* A **757**, 102 (2005) [arXiv:nucl-ex/0501009].

7. R. Dolen, D. Horn and C. Schmid, *Phys. Rev.* **166**, 1768 (1968).

8. G. Veneziano, *Nuovo Cim.* A **57**, 190 (1968).

9. Y. Nambu, in *Symmetries and Quark Models*, ed. R. Chand, Gordon and Breach (1970).

10. H. B. Nielsen, submitted to the 15th International Conference on High Energy Physics, Kiev (1970).

11. L. Susskind, *Nuovo Cim.* **69A** (1970) 457.

12. D. J. Gross and F. Wilczek, *Phys. Rev. Lett.* **30**, 1343 (1973); H. D. Politzer, *Phys. Rev. Lett.* **30**, 1346 (1973).

13. J. Scherk and J. Schwarz, "Dual models for non-hadrons," *Nucl. Phys.* **B81** (1974) 118.

14. K. G. Wilson, *Phys. Rev.* D **10**, 2445 (1974).

15. Y. Nambu, *Phys. Lett.* B **80**, 372 (1979).

16. B. Andersson, G. Gustafson, G. Ingelman and T. Sjostrand, *Phys. Rept.* **97**, 31 (1983).

17. M. Luscher, K. Symanzik and P. Weisz, *Nucl. Phys.* B **173**, 365 (1980).

18. M. Luscher and P. Weisz, *JHEP* **0207**, 049 (2002) [arXiv:hep-lat/0207003].

19. G. 't Hooft, *Nucl. Phys.* B **72**, 461 (1974).

20. A. M. Polyakov, *Phys. Lett.* B **103**, 207 (1981);

21. A. M. Polyakov, *Nucl. Phys. Proc. Suppl.* **68**, 1 (1998) [arXiv:hep-th/9711002].

22. J. Polchinski, *Phys. Rev. Lett.* **75**, 4724 (1995) [arXiv:hep-th/9510017].

23. G. T. Horowitz and A. Strominger, *Nucl. Phys.* B **360**, 197 (1991).

24. I. R. Klebanov, *Nucl. Phys.* B **496**, 231 (1997) [arXiv:hep-th/9702076]; S. S. Gubser, I. R. Klebanov and A. A. Tseytlin, *Nucl. Phys.* B **499**, 217 (1997) [arXiv:hep-th/9703040]; S. S. Gubser and I. R. Klebanov, *Phys. Lett.* B **413**, 41 (1997) [arXiv:hep-th/9708005].

25. S. Kachru and E. Silverstein, *Phys. Rev. Lett.* **80**, 4855 (1998) [arXiv:hep-th/9802183];
   A. E. Lawrence, N. Nekrasov and C. Vafa, "On conformal field theories in four dimensions," *Nucl. Phys.* B **533**, 199 (1998) [arXiv:hep-th/9803015].

26. A. Kehagias, *Phys. Lett.* B **435**, 337 (1998) [arXiv:hep-th/9805131].

27. I. R. Klebanov and E. Witten, *Nucl. Phys.* B **536**, 199 (1998) [arXiv:hep-th/9807080].

28. J. M. Maldacena, *Phys. Rev. Lett.* **80**, 4859 (1998) [arXiv:hep-th/9803002];
   S. J. Rey and J. T. Yee, *Eur. Phys. J.* C **22**, 379 (2001) [arXiv:hep-th/9803001].

29. D. Berenstein, J. M. Maldacena and H. Nastase, *JHEP* **0204**, 013 (2002) [arXiv:hep-th/0202021].

30. A. A. Tseytlin, arXiv:hep-th/0409296.

31. N. Beisert, *Phys. Rept.* **405**, 1 (2005) [arXiv:hep-th/0407277].

32. A. V. Belitsky, V. M. Braun, A. S. Gorsky and G. P. Korchemsky, Int. J. Mod. Phys. A **19**, 4715 (2004) [arXiv:hep-th/0407232].

33. D. J. Gross and F. Wilczek, *Phys. Rev.* D **9**, 980 (1974); H. Georgi and H. D. Politzer, *Phys. Rev.* D **9**, 416 (1974).

34. S. S. Gubser, I. R. Klebanov and A. M. Polyakov, *Nucl. Phys.* B **636**, 99 (2002) [arXiv:hep-th/0204051].

35. G. P. Korchemsky, *Mod. Phys. Lett.* A **4**, 1257 (1989); G. P. Korchemsky and G. Marchesini, *Nucl. Phys.* B **406**, 225 (1993) [arXiv:hep-ph/9210281].

36. A. V. Kotikov, L. N. Lipatov, A. I. Onishchenko and V. N. Velizhanin, *Phys. Lett.* B **595**, 521 (2004) [arXiv:hep-th/0404092].

37. Z. Bern, L. J. Dixon and V. A. Smirnov, arXiv:hep-th/0505205.

38. S. S. Gubser, I. R. Klebanov and A. W. Peet, *Phys. Rev.* D **54**, 3915 (1996) [arXiv:hep-th/9602135]; I. R. Klebanov and A. A. Tseytlin, *Nucl. Phys.* B **475**, 164 (1996) [arXiv:hep-th/9604089].

39. S. S. Gubser, I. R. Klebanov and A. A. Tseytlin, *Nucl. Phys.* B **534**, 202 (1998) [arXiv:hep-th/9805156].

40. A. Fotopoulos and T. R. Taylor, *Phys. Rev.* D **59**, 061701 (1999) [arXiv:hep-th/9811224]; M. A. Vazquez-Mozo, *Phys. Rev.* D **60**, 106010 (1999) [arXiv:hep-th/9905030]; C. J. Kim and S. J. Rey, *Nucl. Phys.* B **564**, 430 (2000) [arXiv:hep-th/9905205].

41. F. Karsch, *Lect. Notes Phys.* **583**, 209 (2002) [arXiv:hep-lat/0106019].

42. R. V. Gavai, S. Gupta and S. Mukherjee, arXiv:hep-lat/0506015.

43. B. Bringoltz and M. Teper, arXiv:hep-lat/0506034.

44. G. Policastro, D. T. Son and A. O. Starinets, *Phys. Rev. Lett.* **87**, 081601 (2001) [arXiv:hep-th/0104066].

45. P. Kovtun, D. T. Son and A. O. Starinets, *JHEP* **0310**, 064 (2003) [arXiv:hep-th/0309213]; *Phys. Rev. Lett.* **94**, 111601 (2005) [arXiv:hep-th/0405231].

46. A. Buchel, J. T. Liu and A. O. Starinets, *Nucl. Phys.* B **707**, 56 (2005) [arXiv:hep-th/0406264].

47. T. Hirano and M. Gyulassy, arXiv:nucl-th/0506049.

48. D. Teaney, *Phys. Rev.* C **68**, 034913 (2003).

49. M. Gyulassy and L. McLerran, *Nucl. Phys.* A **750**, 30 (2005) [arXiv:nucl-th/0405013].

50. E. V. Shuryak, *Nucl. Phys.* A **750**, 64 (2005) [arXiv:hep-ph/0405066].

51. R. Pisarsky, talk available at `http://quark.phy.bnl.gov/~pisarski/talks/unicorn.pdf`.

52. J. Polchinski and M. J. Strassler, *Phys. Rev. Lett.* **88**, 031601 (2002) [arXiv:hep-th/0109174]; *JHEP* **0305**, 012 (2003) [arXiv:hep-th/0209211].

53. H. Boschi-Filho and N. R. F. Braga, *Eur. Phys. J.* C **32**, 529 (2004) [arXiv:hep-th/0209080].

54. J. Erlich, E. Katz, D. T. Son and M. A. Stephanov, "QCD and a holographic model of hadrons," arXiv:hep-ph/0501128; L. Da Rold and A. Pomarol, "Chiral symmetry breaking from five dimensional spaces," *Nucl. Phys.* B **721**, 79 (2005) [arXiv:hep-ph/0501218].

55. G. F. de Teramond and S. J. Brodsky, *Phys. Rev. Lett.* **94**, 201601 (2005) [arXiv:hep-th/0501022].

56. E. Witten, *Adv. Theor. Math. Phys.* **2**, 505 (1998) [arXiv:hep-th/9803131].

57. C. Csaki, H. Ooguri, Y. Oz and J. Terning, *JHEP* **9901**, 017 (1999) [arXiv:hep-th/9806021]; R. C. Brower, S. D. Mathur and C. I. Tan, *Nucl. Phys.* B **587**, 249 (2000) [arXiv:hep-th/0003115].

58. I. R. Klebanov and A. A. Tseytlin, *Nucl. Phys.* B **578**, 123 (2000) [arXiv:hep-th/0002159].

59. I. R. Klebanov and M. J. Strassler, *JHEP* **0008**, 052 (2000) [arXiv:hep-th/0007191].

60. C. P. Herzog, I. R. Klebanov and P. Ouyang, "D-branes on the conifold and N = 1 gauge / gravity dualities," arXiv:hep-th/0205100.

61. M. J. Strassler, arXiv:hep-th/0505153.

62. N. Seiberg, *Nucl. Phys.* B **435**, 129 (1995) [arXiv:hep-th/9411149].

63. S. S. Gubser, C. P. Herzog and I. R. Klebanov, *JHEP* **0409**, 036 (2004) [arXiv:hep-th/0405282].

64. A. Butti, M. Grana, R. Minasian, M. Petrini and A. Zaffaroni, *JHEP* **0503**, 069 (2005) [arXiv:hep-th/0412187].

DISCUSSION

Lorenzo Magnea (University of Torino):
Much of what you said applies to $\mathcal{N} = 4$ supersymmetric Yang-Mills, although many important results have been obtained also for $\mathcal{N} = 1$. Could you comment on the possibility to extend these techniques to the case of non-supersymmetric QCD?

Igor Klebanov: Yes, as I stressed in my talk, the presence of at least $\mathcal{N} = 1$ supersymmetry has so far been very useful in constructing gauge/gravity dualities. But there is no deep reason why dualities of this sort won't work for non-supersymmetric backgrounds. In fact, we already know that they do. One example of this is provided by thermal SYM theories where the temperature breaks all the supersymmetry. A related example is Witten's construction of a non-supersymmetric confining background which, although it does not incorporate asymptotic freedom, has qualitative features in common with QCD. For example, calculations of the low-lying glueball and meson spectra have produced results in reasonable agreement with lattice gauge theory. The string dual of large N non-supersymmetric QCD has not been constructed: since the coupling is weak in the UV, the string dual cannot be approximated by a weakly curved supergravity solution. Therefore, a full stringy solution is required. This is a hard problem, but it is now rather well-posed, and my feeling is that it is solvable.

NEUTRINO PHYSICS

Previous page:

The Gamla Uppsala burial mounds (above). Gamla Uppsala is the site of the first settlement in the area, dating back to the Iron Age. Photo: Uppsala kommun

The Wiks Castle (below), is one of Sweden's best preserved medieval castles. Built during the latter half of the 15th century. (Photo: Donald Griffiths)

ATMOSPHERIC AND ACCELERATOR NEUTRINOS

YOICHIRO SUZUKI

Kamioka Observatory, Institute for Cosmic Ray Research, University of Tokyo
Higashi-Mozumi, Kamioka, Hida-City,
Gifu 506-1205, Japan
E-mail: suzuki@suketto.icrr.u-tokyo.ac.jp

Recent results from the atmospheric neutrino measurements are discussed. The best constraint oscillation parameters of $\Delta m^2 = 2.5 \times 10^{-3} eV^2$ and $\sin^2 2\theta = 1.0$ was obtained by a new treatment of the atmospheric neutrino data. An evidence for the ν_τ appearance in the atmospheric neutrino events was shown by statistical methods. The long baseline oscillation experiment using man-made neutrinos has confirmed the atmospheric neutrino oscillation and obtained consistent parameter regions. The prospects for future accelerator experiments are presented.

1 Introduction

By the last several years of studies on atmospheric[1,2,3], solar[4,5,6,7,8], accelerator[9] and reactor[10] neutrinos, we have obtained the knowledge that $\Delta m_{23} \sim 2.5 \times 10^{-3} eV^2$ is significantly larger than $\Delta m_{12} \sim 8.0 \times 10^{-5} eV^2$, and θ_{23} is nearly maximal and θ_{12} is also large ($\sin^2 2\theta \sim 0.7$). The value of θ_{13} has not yet been determined and only the upper bound of $\sin^2 2\theta_{13} < 0.2$ at $\Delta m_{13}^2 = 2 \times 10^{-3} eV^2$ has been obtained[11].

Due to the mass hierarchy and the smallness of the θ_{13}, the atmospheric and the solar neutrino oscillations are nearly decoupled. However it is an interesting subject to look for sub-dominant effects, for example, the contribution of the solar term on the atmospheric neutrino oscillation, which may reveal a hint of a deviation of θ_{23} from the maximal mixing.

In the frame work of the two flavor oscillation, the oscillation probability is shown by $P(\nu_\mu \to \nu_\tau) = \sin^2 2\theta_{23} \sin^2(1.27 \Delta m_{23}^2 L/E)$. The wavelength of the oscillation, $\lambda = 4\pi E/\Delta m^2$, is proportional to the energy, E, of the neutrinos. Therefore, the oscillatory behavior may be seen in the L/E plot. The mixing angle behaves as a strength of the oscillation. The typical half wave length, $\lambda/2$, for the atmospheric neutrinos with the energy of 1GeV and for $\Delta m^2 = 2.5 \times 10^3$ eV^2 is ~500 km.

2 Atmospheric Neutrinos

The atmospheric neutrinos are produced through the interactions of the primary cosmic rays in the atmosphere. The atmospheric neutrinos consist of the mixture of ν_μ, $\bar{\nu}_\mu$, ν_e, $\bar{\nu}_e$. In the recent development, the secondary particles produced by the primary interactions are treated in three dimensions[12,13]. The error of the absolute neutrino flux is estimated to be about 15% for the low energy neutrinos below 10 GeV and the uncertainty of 0.05 for the primary CR spectrum index above 100 GeV is assigned.

The flux ratio, $(\nu_\mu + \bar{\nu}_\mu))/(\nu_e + \bar{\nu}_e)$, is 2 in low energy limit where all the muons decay before reaching on the surface of the earth, and increases as energy goes up. The uncertainty of this flux ratio is greatly reduced to 3 % in the low energy region. The systematic errors at high energy mostly come from the uncertainty of the production ratio of π/K, and 15% uncertainty is assigned at 100GeV based on the difference of the flux values from the three independent flux calculations[12,13].

The zenith angle distribution is a key to the oscillation analysis. The uniformity of the primary cosmic rays beyond the energy above the geomagnetic cut off indicates the up-down symmetric distribution of the neu-

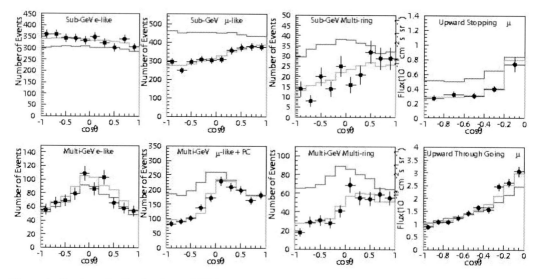

Figure 1. Zenith angle distributions of SK-I atmospheric neutrino data for different event categories. The Points show data and sysmetric solid lines are the expected event distributions. The lines following the data points are the expected from the best fit neutrino oscillation parameters. ν_μ-like events have dificits for the upward going events while the e-like events show no indications.

trino flux. The uncertainty of the ratio is 1∼2% for the entire energy range and the uncertainty of the horizontal and vertical ratio, which is used for the upward going muon analyses, is about 2% which mostly come from the uncertainty of the K productions in the hadronic interactions.

The atmospheric neutrino events in Super-Kamiokande (SK) cover the wide range of the path-length, 10 ∼ 13,000km and cover also the wide range of energy from 0.1 ∼ 10,000 GeV, five orders of magnitudes.

The events are classified in the following category depending on their topology. The contained events have their event vertices in the detector fiducial volume. Those which all the tracks produced are stopped in the detector are called fully contained (FC), and those which some tracks escaped the inner detector are called partially contained (PC) events. The averaged incident neutrino energy is 1 GeV for the FC events and 10 GeV for the PC events. Upward going muons are produced beneath the detector. The mean incident energy of through–going muons is

about 100 GeV and that of stopped in the detector is about 10 GeV.

SK-I, took data between May-1996 and July-2001, was equipped with 11,146 photo-multiplier tubes (PMTs) of 50cm in diameter providing 40% photo-cathode coverage of the inner surface of the detector. SK-II has started in December 2002 and is running as of July 2005 with a reduced photo-cathode coverage of roughly 20% due to the tragic accident happened in Nov-12 in 2001 by which nearly 6,777 PMTs were lost in a couple of seconds. We expect to finish the full restoration work in October 2005, and the so called SK-III with all the PMTs back will be taking data by the summer 2006.

2.1 Two flavor analysis

SK-I has ∼15,000 atmospheric neutrino events taken for effective 1489 days (contained events) and for effective 1678 events (upward going muons). A signature of the neutrino oscillation can be obtained by the double ratio, $R = (\mu/e)_{data}/(\mu/e)_{MC}$, since the uncertainty of the ratio is reduced com-

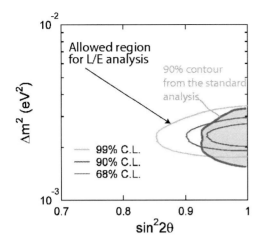

Figure 2. The allowed parameter region for the SK-I data. The solid painted region shows the 90% allowed region from the standard 2 flavor analysis ans the narrow three lines for the allowed regin from the L/E analysis. The lines from inward to outward show 68%, 90% and 99% C.L.. The result from the L/E analysis gives narrower allowed region for Δm^2

2.2 L/E analysis and parameter determination

Since the oscillation wave length is proportional to the neutrino energy, E, one can expect to observe a sinusoidal behavior in the L/E plot. This oscillatory pattern also distinguishes other exotic hypotheses. On the other hand, this analysis gives strong constraint on the determination of Δm^2, since the position of the dip corresponds directly to $\Delta m^2 (\lambda/\text{E}=4\pi/\Delta m^2)$. However, the L/E plot for all the data decreases monotonously and does not reveal any oscillatory patterns at all. We need to select the events with good L/E resolution in order to observe the expected pattern.

We therefore have selected those events with $\delta(\text{E/L})< 70\%$. The selection basically has removed horizontally going events and low energy events. The rejected events poorly determine L. 2726 events, which are about 1/5 of the total 15726 events, have remained after the cuts.

The results[15] are shown in Fig. 3. The dip was observed at around 500 km/GeV, which provides strong confirmation of the neutrino oscillation. This first dip observed cannot be explained by other hypotheses, and we have rejected those hypotheses with significance of 3.4 σ for decay and 3.8 σ for decoherence. Even if we have altered the resolution cut from 50% to 90% at every 10% step, the obtained $\Delta\chi^2$ for those hypotheses does not change. The results, therefore, are very robust.

The best fitted oscillation parameters of $(\Delta m^2, \sin^2 2\theta)=(2.4\times10^{-3}$ eV^2,1.00) for the physical region were obtained. The allowed parameter regions are shown in Fig. 2 as three different lines for the different confidence levels at 60%, 90% and 99% from the inner line to the outer lines. The obtained regions are consistent with that from the standard analysis and give much stronger constraint on Δm^2 even with fewer events.

paring to the uncertainty in the absolute flux normalization. The amount of deviation of R from unity,

$$R_{sub-GeV} = 0.658\pm0.016(stat)\pm0.035(syst)$$

$$R_{multi-GeV} = 0.702^{+0.032}_{-0.030}(stat)\pm0.101(syst),$$

indicates that the mass difference ranges between $10^{-3\sim2}eV^2$. The first indication of the neutrino oscillation was indeed given by this ratio[14].

The zenith angle distribution is shown in Fig. 1. Total 180 momentum and zenith angle bins were used for the fit including the 39 systematic error parameters. The overall normalization was not constrained. The best fit parameter obtained are $\sin^2 2\theta$=1.00 and Δm^2=2.1x10$^{-3}eV^2$ if they are constrained in the physical region. The χ^2 difference to the non oscillation is 303.9. The allowed parameter region is shown in Fig. 2.

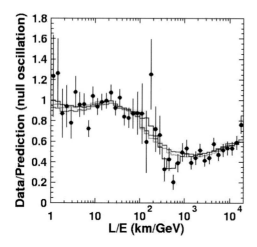

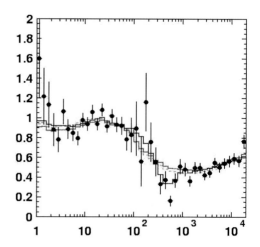

Figure 3. The L/E plots for the selected data about 1/5th of the total data sample. The dip around 500 km/GeV is clearly seen. THe line following the data is obtained from the best fit oscillation parameters and two lines which monotonously decrease, are for the phyposese of neutrino decay and decoherence. The observed oscillatory pattern rejects othere possibilities and only the neutrino oscillation explains the data.

Figure 4. The L/E plots for the combined data of SK-I and SK-II. The dip around 500 km/GeV has become stronger and clearer.

The analysis including the data from SK-II is under study, but the preliminary plot for the combined data set is shown in Fig. 4. The fist dip has become much stronger.

2.3 Two flavor analysis with finner binning

The standard zenith angle analysis gives a good constrain for the mixing angle, and the L/E analysis for the mass square difference. What is the best analysis by taking into account of the advantages of each analysis. The finner binning data enable us to follow the oscillation behavior much better. Especially, high energy data are sensitive to the oscillation dip where we use a single combined energy bin for the standard zenith angle analysis.

A new analysis aiming to get the best parameters is done by using the finner binnning data. The energy of the multi-GeV data are divided into 3 to 5 energy bins and the PC data are sub-divided into the PC-stop and PC-though data like one adopted in the L/E analysis. Furthermore, we have added the multi-ring e-like data sample. A total of 370 bins, 37 momentum bins × 10 zenith bins, are used.

The best fit parameters obtained in the physical region are

$$(\Delta m^2, sin^2 2\theta) = (2.5 \times 10^{-3} eV^2, 1.00).$$

The χ^2 distribution sliced at $sin^2 2\theta = 1$ for the standard analysis, the L/E analysis and the 370 binned analysis have been compared in Fig. 5. The new analysis gives a sharp minimum value at $2.5 \times 10^{-3} eV^2$.

2.4 Tau appearance

The tau events cannot be identified in an event-by-event basis since many hadrons are produced and the incident neutrino directions are not known. By making use of the characteristic of the tau production and by using the fact that tau events can only be seen as upward going events, we can apply a statistical analysis to enhance the tau events.

However, it is not an easy task, because the energy threshold of the tau production is 3.5 GeV and the expected rate for the tau

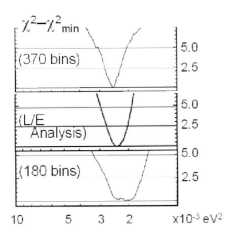

Figure 5. The χ^2 distributions sliced at $\sin^2 2\theta = 1.0$ for the three different analyses, the standard (180 bins) analysis, the L/E analyis and the finner binned (370 bins) analysis. The new finner binned analysis shows sharp minimum at $2.5 \times 10^{-3} \mathrm{eV}^2$ while the 180 bin analysis has a broad minimum covering from 2.0 to $2.5 \times 10^{-3} \mathrm{eV}^2$.

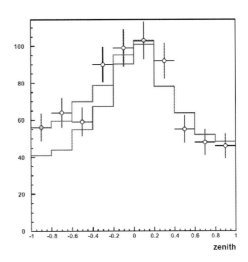

Figure 6. The zenith angle distribution of the event sample enhanced for the tau events. The excess consistent with the expected tau production is seen for the uupward going events. The number of events corresponding to the difference between two lines seen in the upward direction are the calculated tau production in the detector.

production–1 FC charged current ν_τ events per kton per year–is very small compared to the event rate of the usual ν_μ, ν_e interactions, which is about 130 events per kton per year.

We have selected multi-GeV and multi-ring events and constrained the event vertices within the fiducial volume, 2m from the ID PMT surface. Those events for which the most energetic ring is e-like, are selected as initial data. The six distributions, for example, the visible energy, the distance of the event vertex to the μ-e decay electron vertex, number of rings and so on, were used to statistically differentiate tau events from other BG events. Two independent methods, a likelihood analysis and a neural network program, were applied for those distributions.

For the likelihood analysis, the events with higher tau likelihood are selected. The efficiency to select tau events are estimated to be 42% and the contamination from the ν_μ, ν_e interaction events are 3.4%. The zenith angle distribution for the finally selected events is shown in Fig. 6. The clear enhancement due

to the tau production is seen for the upward going events.

The number of the fitted tau events is $145 \pm 48 (\text{stat.})^{+9}_{-36} (\text{syst.})$ while the expected number of tau events is $79 \pm 31 (\text{syst.})$. Another method, the neural network, gives similar results: The number of the fitted tau events is $152 \pm 47 (\text{stat.})^{+12}_{-27} (\text{syst.})$ while the expected number of tau events is $79 \pm 31 (\text{syst.})$. Those numbers are consistent with the expected excess from the oscillated ν_τ events.

3 Long Baseline Accelerator Oscillation Experiments

3.1 K2K experiment

The first long baseline neutrino oscillation experiment using man-made neutrinos is the K2K (KEK to Kamioka) experiment[9] which has started in 1999, one year after the announcement of the discovery of the atmospheric neutrino oscillation.

For $\Delta m^2 = 2.5 \times 10^{-3} \mathrm{eV}^2$ and full mixing,

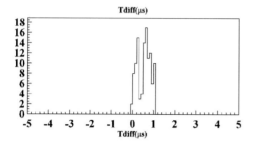

Figure 7. The time distribution of the clustered events within 1.2 μ sec observed in Super-Kamiokande, which are supposed to be created by man-made neutrinos from KEK.

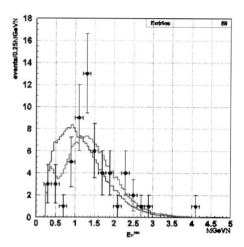

Figure 8. The reconstructed energy distribution of the 58 single ring μ-like events. Two lines, each normalized to the number of observed events, show the expected ones for no oscillation and for the best fit oscillation parameters (distorted spectrum).

by putting the distance of 250km between the neutrino source and the detector and the averaged energy of 1.3 GeV, the expected survival probability is expected to be about 70%.

The total number of protons delivered on the production target (POT) is 0.561×10^{20} for SK-I and 0.488×10^{20} for SK-II. The total POT used for the analysis is 0.992×10^{20} POT. The direction of the beam is controlled well within 1mrad and monitored by the muon distributions measured by the counters placed downstream of the end plug of the decay volume.

The neutrino beam at KEK-PS is produced every 2.2 seconds with the duration of 1.2 μs. The clocks at KEK and Kamioka have been synchronized to the accuracy less than 100ns by using a GPS. The events produced by interactions of man-made neutrinos were selected by using the expected beam arrival time at Super-Kamiokande.

The criteria like ones used for the atmospheric neutrino analysis, for expamle, requirement of fully contained, Evis>30 MeV, fiducial volume of 22.5kt, and so on, were also applied. A total of 112 events were found in the time cluster of 1.2μs. Outside of the cluster, within $\pm 5\mu$s, non of the neutrino events were found as shown in Fig. 7. Those observed events are classified into 1-ring (67 events) and multi-ring (45 events). In the 1-ring event sample, there are 58 μ-like events and 9 e-like events. The systematic uncer-

tainty for the total number of events is 3%, of which the fiducial volume error of 2% is the largest.

3.2 $\nu_\mu \to \nu_x$

The number of expected events at Super-K and the spectrum shape before the oscillation were obtained by using the measured neutrino events in the front detectors located at KEK and the MC simulations. The total number of observed neutrino events in 1kt water Cherenkov detector with a detection efficiency of 74.9% was used to obtain the overall normalization factor. The combined spectrum fit for all the front detectors, 1kt water Cherenkov detector, Muon Range Detector, Scifi and SciBar, was used to obtain the neutrino beam spectrum at KEK site. Then the MC calculation was used to estimate the spectrum at SK site (Far/Near ratio). Finally the expected number of events at SK, $N_{SK}^{pred.} = 155.9 \pm 0.3^{+13.6}_{-15.6}$, was obtained. The 4.1% and $^{+5.6}_{-7.3}$% uncertainties come from the 1kt fiducial volume error and the Far/Near

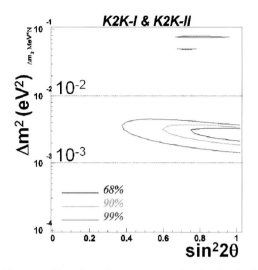

Figure 9. The allowed parameter region for the final data from K2K. The three lines corresponding to 68, 90 and 99% C.L.

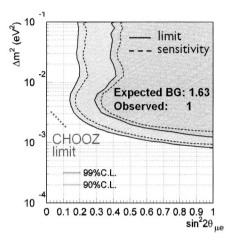

Figure 10. The excluded parameter ragion in Δm^2 and $\sin^2 2\theta_{\mu e}$ for the K2K electron appearance data. Two regions corresponding to the 90 and 99% C.L.. and solid lines show the limit and the dasshed lines show the sensitivity in the Feldman-Cousin Analysis. The CHOOZ limt is also shown.

ratio, respectively.

The energy spectrum of 1-ring μ-like events is shown in Fig. 8. The maximum likelihood method using a convolution of the number of events, the spectrum and the systematic error parameters, are adopted to obtain the oscillation parameters.

The best fit value for the physical region is $(\Delta m^2, \sin^2 2\theta) = (2.76 \times 10^{-3} eV^2, 1.0)$, shown in Fig. 9, which is consistent with the atmospheric neutrino oscillation. Thus, confirmation of the neutrino oscillation was made by man-made neutrinos.

3.3 Electron appearance

In the final sample, we have 9 electron–like events. The cuts applied for those selections were optimized to choose μ-like events, not e-like events. Therefore we have many electron-like candidates in the final sample. In order to look for the electron appearance we need to apply further selection criteria on those 9 candidates to increase the purity of the electron final sample. The tight electron identification algorithm taking into account the opening angle information, the visible energy

cut of 100 MeV and rejection of events with $\mu \to$e decay electron were applied. Then 5 events remained.

We further applied so called π^0 cut which aims to remove π^0 contamination. This cut is effective to remove those events where the energy of one of the γ-rays from π^0 decay is dim, or escaped detection. The cut forced to look for a second ring and the π^0 mass is reconstructed. Those events with the reconstructed mass consistent with π^0 mass were removed. After applying this cut, 1 electron candidate remained. The total efficiency for accepting electrons is 35.7% and 40.9% for K2K-I and K2K-II, respectively. The estimated remaining backgrounds are 1.63 events, 1.25 events from ν_μ interactions and 0.38 events from the beam ν_e interactions, which is consistent with one observed candidate.

The excluded parameter regions are shown in Fig. 10, and the 90% C.L. limit, $\sin^2 2\theta_{\mu e} < 0.18$ at $2.8 \times 10^{-3} eV^2$, was obtained.

4 Future Long Baseline Experiments

The purpose of first generation experiments is to confirm the neutrino oscillation observed in the atmospheric neutrinos. The future accelerator oscillation experiment will explore the region where the atmospheric neutrino study can hardly reach. A search for a definitive θ_{13} value through the electron appearance experiment can reach to the sensitivity of $\sin^2 2\theta_{13} \leq 0.01$ with the combination of a Megawatt neutrino beam and a SK-scale neutrino detector and could be realized around 2010. If the definitive θ_{13} has been determined and if the value is relatively large, then experiments with the conventional technology making use of multi-Megawatt neutrino beams and Megaton neutrino detectors will be able to make detailed studies on the neutrino sector like CP violation, mass hierarchy and so on.

4.1 MINOS

The MINOS[16] 5.4 kt far detector, consisting of interleaved planes of 2.54 cm thick steel plane and 1 cm thick scintillator planes, and a 1.5 T toroidal magnet, is placed in the underground Soudan mine, 735 km from the neutrino source at Fermilab. Protons accelerated to 120 GeV with an intensity of $1.5 \sim 2.5 \times 10^{13}$ppp will produce three different neutrino beams, LE(low energy), ME, HE according to the different configurations of the beam line. The LE beam with the peak energy around 3 GeV with relatively large high energy tail will produce 1300 ν_μ charged current events in the far detector for 2.5×10^{20}POT/yr.

For about 5 years of data taking with 16×10^{20} POT, MINOS will reach the accuracy of $\Delta m^2 < 10\%$. If θ_{13} is close to the CHOOZ limit, then MINOS will see $> 3\sigma$ effect in ~ 3 years of running.

MINOS has started running at the end of 2004. The experiment has already observed beam neutrino interaction in the far detector and also atmospheric neutrino interactions have been observed. The results from the experiment are expected soon.

4.2 CNGS

The high energy ν_μ beam from CERN is optimized for the ν_τ appearance ($< E_\nu >= 17 GeV$) in the detectors at the Gran Sasso Laboratory, 732 km away (CNGS[17]).

OPERA is a emulsion-counter hybrid experiment with the total mass of 1700 tons. For 5 years running with the yearly accumulation of the beam of 4.5×10^{19}POT/yr, OPERA expected to observe 12.4 ν_τ appearance (with 0.8 BG) for $2.4 \times 10^{-3} eV^2$. Preparation of the emulsion films is going on and the first delivery to Gran Sasso was done at January, 2005. The experiment is expected to start in June 2006 with 850 tons of emulsion films.

ICARUS is a liquid Ar detector. The 476.5ton LAr detector (T600) has been sent to Hall C of Gran Sasso and being installed. By summer 2006, T600 will be ready. There is a plan to increase the volume to T1800. With this plan for 5 years of operation, ICARUS expected to observe 6.5 ν_τ appearance with 0.3 BG for $2.5 \times 10^{-3} eV^2$.

4.3 T2K

The long baseline neutrino oscillation experiment from Tokai to Kamioka (T2K) will be expected to start taking data in 2009 with 100 times higher power than K2K[18]. Neutrinos are produced at the 40 GeV proton accelerator at J-PARC in Tokai Village, Japan. The construction of the machine has started in 2001 and will be completed in 2007. The construction of the neutrino beam line has started in 2004 and will be completed in 2008. The baseline to Kamioka is 295km and off-axis beams will be used for the experiment. The accelerator power for phase I is 0.75 MW and Super-Kamiokande will be used. The experiment aims to look for the finite θ_{13} effect through the appearance of electron neutrinos. For a future option, Phase-II to study CP violation, matter effect, mass hierarchy and so on, can be done by increasing the machine power to 4 MW and building a Megaton detector.

The off–axis beam is quasi monochromatic and 2∼3 times more intense than narrow band neutrino beams. The beam energy can be tuned for the oscillation maximum by selecting the off-axis angle.

By choosing the off-axis angle at 2.5 degree, we can make the peak neutrino energy to match the oscillation maximum. T2K expected to see 11,000 total ν_μ and 8,000 charged current interactions for 5 yeas of running. The beam ν_e contamination is 0.4% at ν_μ peak energy. After 5 yrs of running T2K reaches to the sensitivity of $\delta(\Delta m_{23}^2) < 1 \times 10^{-4} eV^2$, $\delta(\sin^2 2\theta_{23}) \sim 0.01$ and $\sin^2 2\theta_{13} \sim 0.008$, about 1/20 of CHOOZ limit.

4.4 NOνA

The ∼1 degree off-axis neutrino beam with the peak energy about 1∼2 GeV will be used for NOνA[19]. The 30kt scintillator detector is placed 819km from Fermilab. The experimental sensitivity is $\delta(\Delta m_{23}^2) < 5\times10^{-5}eV^2$, $\delta(\sin^2 2\theta_{23}) \sim 0.004$ and $\sin^2 2\theta_{13} \sim 0.0044 \sim 0.005$. The experiment is expected to start at around 2010.

5 Summary

The atmospheric L/E analysis gives tighter Δm^2 region and has confirmed the oscillatory behavior. New finner binning analysis with total 370 bins gives the most constraint on the oscillation parameters. The best fit parameters in the physical region are $\Delta m^2 = 2.5 \times 10^{-3} eV^2$ and $\sin^2 2\theta_{23}$=1.0, and the 90% C.L. regions are $2.0 < \Delta m^2 < 3.0\times10^{-3}eV^2$, $\sin^2 2\theta > 0.93$.

The evidence for the ν_τ appearance in atmospheric neutrinos was obtained.

K2K has confirmed the atmospheric neutrino oscillation and the observed energy distortion is consistent with the neutrino oscillation.

The accelerator experiments in future will be expected to bring fruitful outcomes.

References

1. Y. Ashie et al., *Phys. Rev.* **D71**, 112005, (2005); Y. Fukuda et al., *Phys. Rev. Lett.* **82**, 2644, (1999); Y. Fukuda et al., *Phys.Rev.Lett.* **81**, 1562, (1998).
2. G. Giacomelli and A. Margiotta, *Eur. Phys. J.* **C33**, S826, (2004).
3. Sanchez et al., *Phys. Rev.* **D68**, 113004, (2003); W. W. M.Allison et al., *Phys. Lett.* **B449**, 137, (1999).
4. B. T.Cleveland et al., *Astrophys J.* **496**, 505, (1998)
5. J. N.A bdurashitov et al., *Phys. Rev.* **C60**, 055801,(1999)
6. M. Altmann et al., *Phys. Lett.* **B490**, 16, (2000)
7. M. B. Smy et al., *Phys. Rev.* **D69**, 01104, (2004); S. Fukuda et al., *Phys. Lett.* **B539**, 179, (2002); Y. Suzuki, in the Proceedings of Neutrino2000, Sudbury, CANADA, June, 2002; S. Fukuda et al., *Phys. Rev. Lett.* **86**, 5651, (2001); S. Fukuda et al., *Phys. Rev. lett.* **86**, 5656, (2001).
8. Q. R.Ahmad et al., *Phys. Rev. Lett.* **87**, 071301, (2001)
9. E. Aliu et al., *Phys. Rev. Lett.* **94**, 081802, (2005); M. H. Ahn et al.,*Phys. Rev. Lett.* **90**, 041801 ,(2003); S. H. Ahn et al.,*Phys. Lett.* B511, **178**,(2001).
10. T. Araki et al., *Phys. Rev. Lett.* **94**, 081801,(2005).
11. M. Apollonio et al., *Phys. Lett.* **B466**, 415,(1999).
12. M. Honda, T.Kajita, K.Kasahara, and S.Midorikawa (2004); astro-ph/0404457.
13. G. D.Barr, T. K. Gaisser, P. Lipari, S. Robbins, and T. Stanev (2004); astro-

ph/0305208; G. Battistone, A. Ferrari, T. Montauli, and P. R. Sala (2003); hep-ph/0305208.

14. Y. Fukuda., *Phys. Lett.* **B335**, 237,(1994); K. S. Hirata et al., *Phys. Lett.* **B208**, 146,(1992); K. S. Hirata et al., *Phys. Lett.* **B205**, 416,(1988).

15. Y. Ashie et al., *Phys. Rev. Lett.* **93**, 101801,(2004).

16. M. Kordosky et al., Talk given at the 5th International Conference on Non-Accelerator New Physics, Dubna, June 2005; S. Wojcicki et al., Talk given at the XI international Workshops on Neutrino Telescopes, Venice, February 2005.

17. M. Guler et al., CERN/SPSC 2000-028; P. Aprili et al., ICARUS Collaboration, CERN/SPSC-2003-030.

18. Letter of Intent, Neutrino Oscillation Experiment at JHF, Y.Hayato et al., `http://neutrino.kek.jp/jhfnu/` `loi/loi.v2.030528.pdf`

19. NOνA Proposal, NOνA Collaboration, D. Ayres, et al.,hep-ex/0503053.

DISCUSSION

Yee Bob Hsiong (National Taiwan University):

Can you also do the ν_τ anppearance analysis for K2K data similar to atmospheric data? and why?

Yoichiro Suzuki: No, we cannot. The mean energy of the neutrino beam of K2K is 1.4 GeV, which is too low comparing to the tau production threshold energy of 3.5 GeV.

Peter Chleper (University of Hamburg):

For the ν_τ appearance analysis, the error on the Mone Carlo expectation is really large. What is the reason for this?

Yoichiro Suzuki: The most of the errors come from the uncertainty of the cross section near the energy threshold. The oscillation parameter errors, another large uncertainties, are included in the systematic errors of the number of observed events since the oscillation parameters must be included in the process for the signal extraction fits.

REVIEW OF SOLAR AND REACTOR NEUTRINOS

ALAN W. P. POON

Institute for Nuclear and Particle Astrophysics, Nuclear Science Divsion
Lawrence Berkeley National Laboratory
1 Cyclotron Road, Berkeley, CA 94720, USA
E-mail: awpoon@lbl.gov

Over the last several years, experiments have conclusively demonstrated that neutrinos are massive and that they mix. There is now direct evidence for ν_es from the Sun transforming into other active flavors while en route to the Earth. The disappearance of reactor $\bar{\nu}_e$s, predicted under the assumption of neutrino oscillation, has also been observed. In this paper, recent results from solar and reactor neutrino experiments and their implications are reviewed. In addition, some of the future experimental endeavors in solar and reactor neutrinos are presented.

1 Introduction

From the 1960s to just a few years ago, solar neutrino experiments had been observing fewer neutrinos than what were predicted by detail models of the Sun[1,2,3,4,5,6]. The radiochemical experiments, which used ^{37}Cl[7] and ^{71}Ga[8,9,10] as targets, were sensitive exclusively to ν_e. The real-time water Cherenkov detector Super-Kamiokande[11,12,13,14] (and its predecessor Kamiokande[15]) observes solar neutrinos by ν-e elastic scattering, and has sensitivity to all active neutrino flavors. However, its sensitivity to ν_μ and ν_τ is only 1/6 of that for ν_e, and the flavor content of the observed solar neutrino events cannot be determined.

As these terrestrial detectors have different kinematic thresholds, they probed different parts of the solar neutrino energy spectrum. The measured solar neutrino flux exhibited an energy dependence. These observations of an energy dependent flux deficit can be explained only if the solar models are incomplete or neutrinos undergo flavor transformation while in transit to the Earth. Table 1 shows a comparison of the predicted and the observed solar neutrino fluxes for these experiments.

Since 2001, significant advances have been made in solar neutrino physics. The Sudbury Neutrino Observatory

(SNO)[16,17,18,19,20,21] has conclusively demonstrated that a significant fraction of ν_es that are produced in the solar core transforms into other active flavors. One of the most favored explanation for this flavor transformation is matter-enhanced neutrino oscillation, or the Mikheyev-Smirnov-Wolfenstein (MSW) effect[25]. The KamLAND experiment[22,23,24] observes the disappearance of reactor $\bar{\nu}_e$s that is predicted from the neutrino mixing parameters derived from global MSW analyses of solar neutrino results. This provides very strong evidence that MSW oscillation is the underlying mechanism in solar neutrino flavor transformation. In this paper, these advances in solar and reactor neutrino experiments and their physical implications are discussed. A brief overview of the future program in solar neutrinos and reactor antineutrinos will also be presented.

2 Solar Neutrino Flux Measurements at Super - Kamiokande

The Super-Kamiokande (SK) detector is a 50000-ton water Cherenkov detector located in the Kamioka mine, Gifu prefecture, Japan. During the first phase of the experiment SK-I (April 1996 to July 2001), approximately 11200 20-inch-diameter photomulti-

Table 1. Summary of solar neutrino observations at different terrestrial detectors before 2002. The Bahcall-Pinsonneault (BP2001) model predictions of the solar neutrino flux are presented in this table. The experimental values are shown with statistical uncertainties listed first, followed by systematic uncertainties. The Solar Neutrino Unit (SNU) is a measure of solar neutrino interaction rate, and is defined as 1 interaction per 10^{-36} target atom per second. For the Kamiokande and Super-Kamiokande experiments, the predicted and measured neutrino fluxes are listed in units of 10^6 cm^{-2} s^{-1}.

Experiment	Measured Rate/Flux	SSM Prediction (BP2001)[1]
Homestake[7] (^{37}Cl)	$2.56 \pm 0.16 \pm 0.16$ SNU	$7.6\,^{+1.3}_{-1.1}$ SNU
SAGE[8] (^{71}Ga)	$70.8\,^{+5.3}_{-5.2}\,^{+3.7}_{-3.2}$ SNU	
Gallex[9] (^{71}Ga)	$77.5 \pm 6.2\,^{+4.3}_{-4.7}$ SNU	$128\,^{+9}_{-7}$ SNU
GNO[10] (^{71}Ga)	$62.9\,^{+5.5}_{-5.3} \pm 2.5$ SNU	
Kamiokande[15] (νe)	$2.80 \pm 0.19 \pm 0.33$	$5.05\,\left(1^{+0.20}_{-0.16}\right)$
Super-Kamiokande[12] (νe)	$2.32 \pm 0.03\,^{+0.08}_{-0.07}$	

plier tubes (PMTs) were mounted on a cylindrical tank to detect Cherenkov light from neutrino interactions in the inner detector. Since December 2002, the experiment has been operating in its second phase (SK-II) with approximately 5200 PMTs in its inner detector. An additional 1885 8-inch-diameter PMTs are used as a cosmic veto.

2.1 Super-Kamiokande-I

In SK-I and SK-II, neutrinos from the Sun are detected through the elastic scattering process $\nu e \to \nu e$. Because of the strong directionality in this process, the reconstructed direction of the scattered electron is strongly correlated to the direction of the incident neutrinos. The sharp elastic scattering peak in the angular distribution for events with a total electron energy of $5< E <20$ MeV in the SK-I data set is shown in Figure 1. This data set spans 1496 days (May 31, 1996 to July 15, 2001), and the solar neutrino flux is extracted by statistically separating the solar neutrino signal and the backgrounds using this angular distribution. At the analysis threshold of $E=5$ MeV, the primary signal is ν_es from ^{8}B decays in the solar interior. The extracted solar ^{8}B neutrino flux in this SK-I data set ($\phi^{\text{ES}}_{\text{SK}-\text{I}}$) is[26] (in units of 10^6 cm^{-2}s^{-1}):

$$\phi^{\text{ES}}_{\text{SK}-\text{I}} = 2.35 \pm 0.02(\text{stat.}) \pm 0.08(\text{sys.})$$

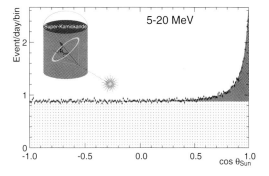

Figure 1. Angular distribution of solar neutrino event candidates in the 1496-day SK-I data set. The shaded area is the solar neutrino elastic scattering peak, and the dotted area represents backgrounds in the candidate data set.

When comparing this measured flux to the BP2001[1] and BP2004[2] model predictions:

$$\frac{\phi^{\text{ES}}_{\text{SK}-\text{I}}}{\phi_{\text{BP2001}}} = 0.465 \pm 0.005(\text{stat.})\,^{+0.016}_{-0.015}(\text{sys.})$$

$$\frac{\phi^{\text{ES}}_{\text{SK}-\text{I}}}{\phi_{\text{BP2004}}} = 0.406 \pm 0.004(\text{stat.})\,^{+0.014}_{-0.013}(\text{sys.}),$$

where the model uncertainties ($\sim 20\%$) have not been included in the systematic uncertainties above.

2.2 Super-Kamiokande-II

With only about half of the photocathode coverage as SK-I, significant improvements have been made to the trigger system in the SK-II detector in order to maintain high

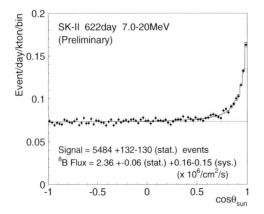

Figure 2. Angular distribution of solar neutrino event candidates in a 622-day data set in SK-II. For the first 159 days of these data, the energy threshold for the analysis was set at $E =8$ MeV, and it was lowered to 7 MeV for the remaining 463 days of data.

trigger efficiency for solar neutrino events. The improved trigger system can trigger with 100% efficiency at $E \sim 6.5$ MeV. Results from a 622-day SK-II data set have recently been released. The solar angular distribution plot is shown in Figure 2. For the first 159 days (Dec. 24, 2002 to July 15, 2003) of these data, the energy threshold for the analysis was set at $E =8$ MeV, and it was lowered to 7 MeV for the remaining 463 days (Jul. 15, 2003 to Mar. 19, 2005). The extracted solar ^{8}B neutrino flux in this 622-day SK-II data set ($\phi^{\mathrm{ES}}_{\mathrm{SK-II}}$) is (in units of 10^6 cm^{-2}s^{-1}):

$$\phi^{\mathrm{ES}}_{\mathrm{SK-II}} = 2.36 \pm 0.06(\mathrm{stat.}) ^{+0.16}_{-0.15}(\mathrm{sys.}),$$

which is consistent with the SK-I results.

3 Sudbury Neutrino Observatory

The Sudbury Neutrino Observatory (SNO) detector is a 1000-tonne heavy water (D$_2$O) Cherenkov detector located near Sudbury, Ontario, Canada. Approximately 9500 8-inch-diameter PMTs are mounted on a spherical geodesic structure to detect Cherenkov light resulting from neutrino interactions. It can make simultaneous measurements of the ν_e flux from ^{8}B decay in the Sun and the flux

of all active neutrino flavors[27] through the following reactions:

$$\begin{aligned} \nu_e + d &\rightarrow p + p + e^- & \text{(CC)} \\ \nu_x + d &\rightarrow p + n + \nu_x & \text{(NC)} \\ \nu_x + e^- &\rightarrow \nu_x + e^- & \text{(ES)} \end{aligned}$$

The charged-current (CC) reaction on the deuteron is sensitive exclusively to ν_e, and the neutral-current (NC) reaction has equal sensitivity to all active neutrino flavors (ν_x, $x = e, \mu, \tau$). Similar to the Super-Kamiokande experiment, elastic scattering (ES) on electron is also sensitive to all active flavors, but with reduced sensitivity to ν_μ and ν_τ. If the measured total ν_x flux (through the NC channel) is greater than the measured ν_e flux (through the CC channel), it would conclusively demonstrate that solar ν_es have undergone flavor transformation since their production in the solar core. Alternatively, this flavor transformation can be demonstrated by comparing the ν_x flux deduced from the ES channel to the ν_e flux.

3.1 Pure D$_2$O phase

The first phase of the SNO experiment (SNO-I) used a pure D$_2$O target. The free neutron from the NC interaction is thermalized, and in 30% of the time, a 6.25-MeV γ ray is emitted following the neutron capture by deuteron. In 2001, the SNO collaboration published a measurement of the ν_e flux, based on a 241-day data set taken from Nov. 2, 1999 to Jan 15, 2001. At an electron kinetic energy T_{eff} threshold of 6.75 MeV [17], the measured ν_e and ν_x fluxes through the CC $\phi^{\mathrm{CC}}_{\mathrm{SNO-I}}$ and ES $\phi^{\mathrm{ES}}_{\mathrm{SNO-I}}$ channels are (in units of 10^6 cm^{-2}s^{-1}):

$$\phi^{\mathrm{CC}}_{\mathrm{SNO-I}} = 1.75 \pm 0.07(\mathrm{stat.}) ^{+0.12}_{-0.11}(\mathrm{sys.})$$
$$\phi^{\mathrm{ES}}_{\mathrm{SNO-I}} = 2.39 \pm 0.34(\mathrm{stat.}) ^{+0.16}_{-0.14}(\mathrm{sys.}).$$

The measured $\phi^{\mathrm{ES}}_{\mathrm{SNO-I}}$ agrees with that from the SK-I detector $\phi^{\mathrm{ES}}_{\mathrm{SK-I}}$. But a comparison of $\phi^{\mathrm{CC}}_{\mathrm{SNO-I}}$ to $\phi^{\mathrm{ES}}_{\mathrm{SK-I}}$, after adjusting for the difference in the energy response of the two detectors, yields (in units of $10^6\,\mathrm{cm}^{-2}\mathrm{s}^{-1}$)

$$\phi^{\mathrm{ES}}_{\mathrm{SK-I}}(\nu_x) - \phi^{\mathrm{ES}}_{\mathrm{SNO-I}}(\nu_e) = 0.57 \pm 0.17,$$

which is 3.3σ away from 0. This measurement not only confirmed previous observations of the solar neutrino deficit from different experiments, it also provided the first indirect evidence, when combined with the SK-I results, that neutrino flavor transformation might be the solution to this long-standing deficit.

In 2002, the SNO collaboration reported a measurement of the total active neutrino flux through the NC channel[18]. This measurement used a T_{eff} threshold of 5 MeV and was based the 306-day data set (Nov. 2, 1999 to May 28, 2001). Under the assumption of an undistorted ^{8}B ν_e spectrum, the non-ν_e component ($\phi^{\mu\tau}_{\mathrm{SNO-I}}$) of the total active neutrino flux is (in units of $10^6\,\mathrm{cm}^{-2}\mathrm{s}^{-1}$)

$$\phi^{\mu\tau}_{\mathrm{SNO-I}} = 3.41 \pm 0.45(\mathrm{stat.})^{+0.48}_{-0.45}(\mathrm{sys.}),$$

which is 5.3σ away from 0. This result was the first direct evidence that demonstrated neutrino flavor transformation. The measured total active neutrino flux confirmed the solar model predictions and provided the definitive solution to the solar neutrino deficit problem.

3.2 Salt Phase

In phase two of the SNO experiment (SNO-II), 2 tonnes of NaCl were added to the D$_2$O target in order to enhance the detection efficiency of the NC channel. The free neutron from the NC channel was thermalized in the D$_2$O and subsequently captured by a ^{35}Cl nucleus, which resulted in the emission of a γ-ray cascade with a total energy of 8.6 MeV. The neutron capture efficiency increased three folds from SNO-I. The CC signal involved a single electron and multiple γs were emitted in the NC channel. This difference in the number of particles in the final state resulted in a difference in the isotropy of the Cherenkov light distribution. The CC

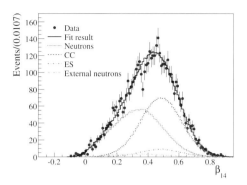

Figure 3. Statistical separation of CC and NC events using Cherenkov light isotropy in SNO-II. The measure of isotropy, β_{14}, is a function with Legendre polynomials of order 1 and 4 as its bases. Because there were multiple γs in the NC signal, the Cherenkov light distribution was more diffuse (smaller β_{14}).

and the NC signals could be statistically separated by this isotropy difference. This separation for events with $T_{\mathrm{eff}} > 5.5$ MeV is shown in Fig. 3 for the 391-day data set (taken from Jul. 26, 2001 to Aug. 28, 2003). This use of light isotropy also removed the need to constrain the ^{8}B ν_e energy spectrum, which can be distorted if the neutrino flavor transformation process is energy dependent, as in SNO-I. The measured energy-unconstrained ν_e and ν_x fluxes through the different channels are[20,21] (in units of $10^6\,\mathrm{cm}^{-2}\mathrm{s}^{-1}$):

$$\phi^{\mathrm{uncon.CC}}_{\mathrm{SNO-II}} = 1.68 \pm 0.06(\mathrm{stat.})^{+0.08}_{-0.09}(\mathrm{sys.})$$
$$\phi^{\mathrm{uncon.ES}}_{\mathrm{SNO-II}} = 2.35 \pm 0.22(\mathrm{stat.})^{+0.15}_{-0.15}(\mathrm{sys.})$$
$$\phi^{\mathrm{uncon.NC}}_{\mathrm{SNO-II}} = 4.94 \pm 0.21(\mathrm{stat.})^{+0.38}_{-0.34}(\mathrm{sys.}).$$

The ratio of the ν_e flux and the total active neutrino flux is of physical significance (which will be discussed later), and is

$$\frac{\phi^{\mathrm{uncon.CC}}_{\mathrm{SNO-II}}}{\phi^{\mathrm{uncon.NC}}_{\mathrm{SNO-II}}} = 0.340 \pm 0.023(\mathrm{stat.})^{+0.029}_{-0.031}(\mathrm{sys.}).$$

4 Search for MSW Signatures in Solar Neutrinos

Recent results from SNO and Super-Kamiokande have conclusively demonstrated that neutrino flavor transformation is the solution to the solar neutrino deficit. The most favored mechanism for this transformation is the Mikheyev-Smirnov-Wolfenstein (MSW) matter-enhanced neutrino oscillation[25]. MSW oscillation can be a resonant effect as opposed to vacuum oscillation, which is simply the projection of the time evolution of eigenstates in free space. A resonant conversion of ν_e to other active flavors is possible in MSW oscillation if the ambient matter density matches the resonant density. Two distinct signatures of the MSW effect are distortion of the neutrino energy spectrum and a day-night asymmetry in the measured neutrino flux. The former signature arises from the energy dependence in neutrino oscillation. When the Sun is below a detector's horizon, some of the oscillated solar neutrinos may revert back to ν_e while traversing the Earth's interior. This ν_e regeneration effect would give an asymmetry in the measured ν_e fluxes during the day and the night.

Both SNO and Super-Kamiokande have done extensive searches for these two signatures in their data. Figure 4 shows the measured electron spectra from SK-I and SK-II, whereas Fig. 5 shows the measured spectra from SNO-I and SNO-II. No statistically significant distortion is seen from either experiment.

The Super-Kamiokande experiment defines the day-night asymmetry ratio $A_{\mathrm{DN}}^{\mathrm{SK}}$ as[14]

$$A_{\mathrm{DN}}^{\mathrm{SK}} = \frac{\Phi_{\mathrm{D}} - \Phi_{\mathrm{N}}}{\frac{1}{2}(\Phi_{\mathrm{D}} + \Phi_{\mathrm{N}})},$$

where Φ_{D} is the measured neutrino flux when the Sun is above the horizon, and Φ_{N} is the corresponding flux when the Sun is below the horizon. The measured day-night asymmetries of the solar neutrino flux by SK-I and SK-II are:

$$A_{\mathrm{DN}}^{\mathrm{SK-I}} = -0.021 \pm 0.020(\text{stat.})^{+0.013}_{-0.012}(\text{sys.})$$
$$A_{\mathrm{DN}}^{\mathrm{SK-II}} = 0.014 \pm 0.049(\text{stat.})^{+0.024}_{-0.025}(\text{sys.}).$$

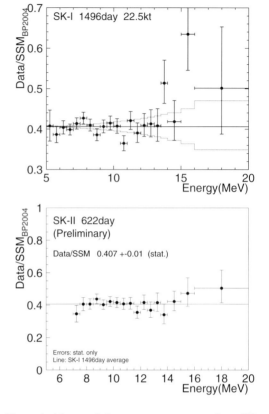

Figure 4. Measured electron energy spectra from SK-I and SK-II. The measured spectra have been normalized to the BP2004 model predictions. The solid lines indicate the mean of the measured ^{8}B neutrino flux. The band in the SK-I spectrum represents the energy-correlated systematic uncertainties in the measurement.

It should be noted that the flux measured by Super-Kamiokande is a mixture of all three active neutrino flavors.

The SNO experiment has also measured the day-night asymmetry of the measured neutrino flux[19,21]. It should be noted that the SNO and the Super-Kamiokande asymmetry ratios are defined differently, such that $A_{\mathrm{DN}}^{\mathrm{SNO}} = -A_{\mathrm{DN}}^{\mathrm{SK}}$. Because SNO can measure the ν_e flux and the total active neutrino flux separately through the CC and the NC channels, it can determine the day-night asymmetry for these fluxes separately. In addition, the asymmetry ratio can be determined with the day-night asymmetry in the NC channel

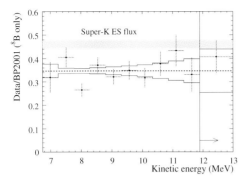

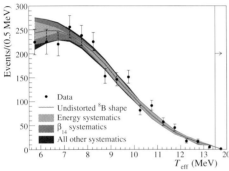

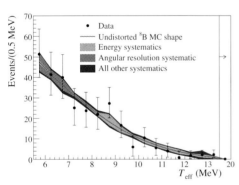

Figure 5. Measured electron energy spectra from SNO-I and SNO-II. *Top*: The ratio of the measured SNO-I CC electron kinetic energy spectrum to the expected undistorted kinetic energy distribution for ^{8}B neutrinos (in BP2001 model) with correlated systematic uncertainties. *Middle*: The measured CC electron kinetic energy spectrum in SNO-II is shown as the data points (with statistical uncertainties only). *Bottom*: The measured ES electron kinetic energy spectrum in SNO-II is shown as the data points (with statistical uncertainties only). In the last two plots, the bands show the accumulated effect of different correlated systematic uncertainties on the electron spectra expected from an undistorted ^{8}B neutrino spectrum.

$A_{NC,DN}^{SNO}$ constrained to 0. With the ^{8}B shape and $A_{NC,DN}^{SNO} = 0$ constraints, the measured day-night asymmetry in the ν_e flux in SNO-I and SNO-II are

$$A_{e,DN}^{SNO-I} = 0.021 \pm 0.049 (\text{stat.})_{-0.012}^{+0.013} (\text{sys.})$$

$$A_{e,DN}^{SNO-II} = -0.015 \pm 0.058 (\text{stat.})_{-0.027}^{+0.027} (\text{sys.}).$$

Because of the presence of ν_μ and ν_τ in the Super-Kamiokande measured flux, its day-night asymmetry is diluted by a factor of 1.55 [21]. Assuming an energy-independent conversion mechanism and only active neutrinos, the SK-I result scales to a ν_e flux asymmetry $A_{e,SK-I} = 0.033 \pm 0.031_{-0.020}^{+0.019}$. Combining the SNO-I and SNO-II values for $A_{e,DN}$ with the equivalent SK-I value $(A_{e,SK-I})$ gives $A_{e,combined} = 0.035 \pm 0.027$. No statistically significant day-night asymmetry has been observed.

5 MSW Interpretation of Solar Neutrino Data

Although no direct evidence for the MSW effect has been observed, the null hypothesis that no MSW oscillation in the solar neutrino results is rejected at 5.6σ [28]. There are two parameters in a two-flavor, active neutrino oscillation model: Δm^2, which is the difference of between the square of the eigenvalue of two neutrino mass states; and $\tan^2 \theta$, which quantifies the mixing strength between the flavor and the mass eigenstates. Each pair of these parameters affects the total solar neutrino spectrum differently, which can give rise to the energy dependence in the ratio between the observed and the predicted neutrino fluxes in different detectors. Using the measured rates in the radiochemical (^{37}Cl and ^{71}Ga) experiments, the solar zenith angle distribution from the Super-Kamiokande experiment, and the day and night energy spectra from the SNO experiment, a global statistical analysis can then be performed to determine the $(\Delta m^2, \tan^2 \theta)$ pair that best describes the data[29]. The best-fit parameters[21]

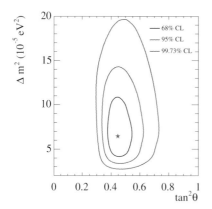

Figure 6. Global neutrino oscillation analysis of solar neutrino data. The solar neutrino data include SNO-I day and night spectra, SNO-II day and night CC spectra and ES and NC fluxes, the SK-I solar zenith spectra, and the rate measurements from the ^{37}Cl (Homestake) and ^{71}Ga (SAGE, GALLEX, GNO) experiments.

are found in the so-called "Large Mixing Angle" (LMA) region:

$$\Delta m^2 = 6.5^{+4.4}_{-2.3} \times 10^{-5} \text{eV}^2$$
$$\tan^2 \theta = 0.45^{+0.09}_{-0.08}.$$

There are two implications to these results. First, maximal mixing (i.e. $\tan^2 \theta = 1$) is ruled out at very high significance. This is in contrary to the atmospheric neutrino sector, where maximal mixing is the preferred scenario. Second, because the "dark-side" ($\tan^2 \theta > 1$) is also ruled out, a mass hierarchy of $m_2 > m_1$ is implied.

6   KamLAND

Previous reactor $\bar{\nu}_e$ oscillation experiments, with reactor-detector distances ("baselines") ranging from ~10 m to ~1 km, did not observe any $\bar{\nu}_e$ disappearance[30]. If CPT is conserved and matter-enhanced neutrino oscillation is the underlying mechanism for the observed flavor transformation in solar neutrinos, one would expect a significant fraction of the reactor $\bar{\nu}_e$s oscillating (primarily through vacuum oscillation) into another fla-

vor at a baseline of 100 to 200 km. The Kamioka Liquid scintillator Anti-Neutrino Detector (KamLAND) experiment[22] has a unique geographic advantage over other previous reactor $\bar{\nu}_e$ experiments; it is surrounded by 53 Japanese power reactors with an average baseline of 180 km.

KamLAND is a 1000-tonne liquid scintillator detector located in the Kamioka mine in Japan. Its scintillator is a mixture of dodecane (80%), pseudocumene (1,2,4-Trimethylbenzene, 20%), and 1.52 g/liter of PPO (2,5-Diphenyloxazole). An array of 1325 17-inch-diameter PMTs and 554 20-inch-diameter PMTs are mounted inside the spherical containment vessel. Outside this vessel, an additional 225 20-inch-diameter PMTs act as a cosmic-ray veto counter.

KamLAND detects $\bar{\nu}_e$s by the inverse β decay process $\bar{\nu}_e + p \rightarrow e^+ + n$, which has a 1.8 MeV kinematic threshold. The prompt signal E_{prompt} in the scintillator, which includes the positron kinetic energy and the annihilation energy, is related to the incident $\bar{\nu}_e$ energy $E_{\bar{\nu}_e}$ and the average neutron recoil energy $\bar{E}_n$ by $E_{\bar{\nu}_e} = E_{\text{prompt}} + \bar{E}_n + 0.8$ MeV. The final state neutron is thermalized and captured by a proton with a mean lifetime of ~200μs. The prompt signal and the 2.2-MeV γ emitted in the delayed neutron capture process form a coincident signature for the $\bar{\nu}_e$ signal.

Table 2 summarizes the two KamLAND reactor $\bar{\nu}_e$ rate measurements[23,24] to-date. The exposure of the two measurements are 162 ton-years and 766 ton-years respectively. The null hypothesis that the observed $\bar{\nu}_e$ rates are statistical downward fluctuations is rejected at 99.95% and 99.998% confidence levels in the two measurements. KamLAND is the first experiment that observes reactor $\bar{\nu}_e$ disappearance.

If neutrino vacuum oscillation is responsible for the disappearance of reactor $\bar{\nu}_e$s in KamLAND, then one might expect a distortion of the E_{prompt} spectrum. The measured

Table 2. Summary of KamLAND $\bar{\nu}_e$ rate measurements.

	First result[23]	Second result[24]
	Data Sets	
Data span	Mar. 4, 2002 to Oct. 6, 2002	Mar. 9, 2002 to Jan. 11, 2004
Live days	145.1 live days	515.1 live days
Exposure	162 ton-year	766.3 ton-year
	Results	
Expected signal	86.8 ± 5.6 counts	365.2 ± 23.7 counts
Background	1 ± 1 count	17.8 ± 7.3 counts
Observed signal	54 counts	258 counts
Systematic uncertainties	6.4%	6.5%
$\bar{\nu}_e$ disappearance C.L.	99.95%	99.998%

E_{prompt} spectrum is shown in Fig. 7. A fit of the observed E_{prompt} spectrum to a simple re-scaled, undistorted energy spectrum is excluded at 99.6%. Also shown in the figure is the best-fit spectrum under the assumption of neutrino oscillations. The allowed oscillation parameter space is shown in Fig. 8, and the best-fit parameters of this KamLAND-only analysis are[24]

$$\Delta m^2 = 7.9^{+0.6}_{-0.5} \times 10^{-5} \mathrm{eV}^2$$
$$\tan^2 \theta = 0.46.$$

Assuming CPT invariance, a global neutrino oscillation analysis can be performed on the solar neutrino results and the KamLAND results. The allowed parameter space is shown in Fig. 9. The best-fit parameters in this global analysis are[21]

$$\Delta m^2 = 8.0^{+0.6}_{-0.4} \times 10^{-5} \mathrm{eV}^2$$
$$\tan^2 \theta = 0.45^{+0.09}_{-0.07}.$$

One immediately notices the complementarity of reactor anti-neutrino and solar neutrino experiments. The former restricts Δm^2, whereas the latter restricts $\tan^2 \theta$ in an orthogonal manner.

7 Future Experimental Solar Neutrino Program

In the next several years, both running solar neutrino experiments SNO and Super-Kamiokande have an ambitious physics program.

The three-flavor mixing matrix element U_{e2} can be written[31] as $\cos \theta_{13} \sin \theta_{12}$, which approximately equals $\sin \theta$ for two-flavor solar neutrino oscillation when θ_{13} is small and when Δm^2 from the solar neutrino sector is much less than that from the atmospheric neutrino sector. For oscillation parameters in the LMA region, the MSW effect can result in ^{8}B neutrinos emerging from the Sun essentially as a pure ν_2 state. The SNO ϕ^{CC}/ϕ^{NC} ratio, a direct measure of ν_e survival probability, is also a direct measure of $|U_{e2}|^2$ ($\sim \sin^2 \theta$). Therefore, one of the primary goals of the SNO experimental program is to make a precision measurement of this fundamental parameter by improving on the CC and NC measurements.

The SNO experiment has entered the third phase (SNO-III) of its physics program. Thirty six strings of ^{3}He and 4 strings of ^{4}He proportional counters have been deployed on a 1-m square grid in the D_2O volume. In SNO-I and SNO-II, the extracted CC and NC fluxes are strongly anti-correlated (the correlation coefficient is -0.53 in SNO-II). This anticorrelation is a significant fraction of the total flux uncertainties. By introduction this array of ^{3}He proportional counters, NC neutrons are detected by $n + {}^3\mathrm{He} \rightarrow p + {}^3\mathrm{H}$;

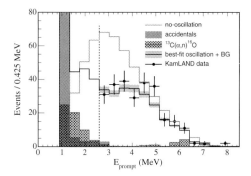

Figure 7. Prompt energy spectrum of $\bar{\nu}_e$ candidate events in the 766-ton-year KamLAND analysis. Also shown are the no-oscillation spectrum and the best-fit spectrum under the assumption of neutrino oscillation.

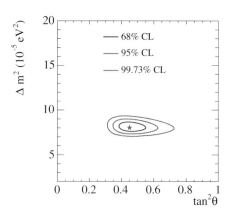

Figure 9. Global neutrino oscillation analysis of solar neutrino experiments and KamLAND. The best fit point is at $(0.45^{+0.09}_{-0.07}, 8.0^{+0.6}_{-0.4} \times 10^{-5} \text{eV}^2)$.

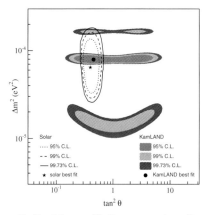

Figure 8. Neutrino oscillation parameter allowed region from the 766-ton-year KamLAND reactor $\bar{\nu}_e$ results (shaded regions) and the LMA region derived from solar neutrino experiments (lines).

whereas the Cherenkov light from the CC electrons are recorded by the PMT array. This physical separation, as opposed to a statistical separation of the CC and NC signals in SNO-I and SNO-II, will allow a significant improvement in the precision of the CC and the NC fluxes. Table 3 summarizes the uncertainties in the CC and NC flux measurements in SNO-I and SNO-II, and the projected uncertainties for the corresponding measurements in SNO-III.

Because the ^{3}He proportional counter array "removes" Cherenkov light signals from

NC interactions, it allows for a search of CC electron spectral distortion at an analysis threshold of $T_{\text{eff}} =4$ to 4.5 MeV. The distortion effects are enhanced at this threshold when compared to those analyses at higher thresholds in SNO-I and SNO-II.

The Super-Kamiokande is scheduled for a detector upgrade from October 2005 to March 2006. After this upgrade, the detector will return to the same photocathode coverage in SK-I. The primary physics goal of the SK-III solar neutrino program is to search for direct evidence of the MSW effect. With the improvements made to the trigger system in SK-II and the anticipated increase in photocathode coverage, the SK-III detector will be able to push the analysis threshold down to $E \sim 4$ MeV.

Figure 10 shows the projected SK-III sensitivity to observing spectral distortion in its solar neutrino signal. In this figure, the sensitivities of several combinations of (Δm^2, $\sin^2\theta$) in the LMA region allowed by solar neutrino measurements (c.f. Fig. 6) are shown. It is possible to discover MSW-induced spectral distortion at $> 3\sigma$ after several years of counting.

Future solar neutrino experiments focus on detecting the low energy pp neutrinos ($E_\nu < 0.42$ MeV), ^{7}Be neutrinos ($E_\nu =$

Table 3. Uncertainties in the CC and NC fluxes in SNO-I, SNO-II and SNO-III. The SNO-III entries are projected uncertainties for 1 live year of data. The total uncertainties are the quadratic sum of the statistical and systematic uncertainties.

	SNO-I		SNO-II		SNO-III	
	$\Delta\phi^{CC}/\phi^{CC}$	$\Delta\phi^{NC}/\phi^{NC}$	$\Delta\phi^{CC}/\phi^{CC}$	$\Delta\phi^{NC}/\phi^{NC}$	$\Delta\phi^{CC}/\phi^{CC}$	$\Delta\phi^{NC}/\phi^{NC}$
Syst.	5.3	9.0	4.9	7.3	3.3	5.2
Stat.	3.4	8.6	3.7	4.2	2.2	3.8
Total	6.3	12.4	6.1	8.4	4.0	6.4

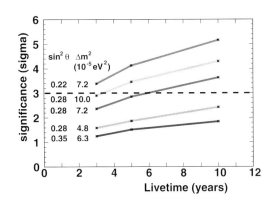

Figure 10. Projected sensitivity to spectrum distortion for SK-III. Each curve represents the significance for observing spectral distortion with the labeled mixing parameters.

0.86 MeV, BR=90%), pep neutrinos ($E_\nu = 1.44$ MeV), or neutrinos from the CNO cycle. These experiments are subdivided into two broad classes: ν_e-only detection mechanism through ν_e charged-current interaction with the target nucleus; and νe elastic scattering which measures an admixture of ν_e, ν_μ and ν_τ. The pp neutrino experiments seek to make high precision measurement of the pp neutrino flux and to constrain the flux of sterile neutrinos. The ^{7}Be and the pep neutrino experiments seek to map out the ν_e survival probability in the vacuum-matter transition region. Figure 11 shows this transition region[32]. Although certain non-standard interaction (NSI) models[33] can match the survival probability in the pp and the ^{8}B energy regimes, they differ substantially from the MSW prediction in the ^{7}Be and pep energy regimes ($E_\nu \sim 1-2$ MeV).

Table 4 is a tabulation of future solar neutrino experiments (adopted from Nakahata[34]). Most of these experiments are in proposal stage; but two liquid scintillator experiments, Borexino and KamLAND, will come online for ^{7}Be solar neutrino measurements in the next year. The construction of Borexino is complete, and it is waiting for authorization to fill the detector with liquid scintillator. The purification system of the KamLAND experiment is being upgraded in order to achieve an ultra-pure liquid scintillator for the ^{7}Be neutrino measurement.

For other future solar experiments, a summary of their current status can be found in the supplemental slides of this conference talk[35].

8 Future Experimental Reactor Anti-neutrino Program

The future reactor anti-neutrino program is focused on determining the neutrino mixing angle θ_{13}. This is the only unknown angle in the neutrino mixing matrix, and its current upper limit is $\sin^2(2\theta_{13}) < 0.2$ (90% C.L.)[36].

Although there are ongoing efforts in developing accelerator-based θ_{13} measurements by searching for $\nu_\mu \to \nu_e$ appearance, such long baseline measurements can be affected by matter effects. The determination of θ_{13} in these appearance experiments is complicated by the degeneracy of mixing parameters (e.g. θ_{23}). A reactor-based θ_{13} measurement can complement accelerator-based experiments by removing these intrinsic ambi-

288

Table 4. Future solar neutrino experiments. Those experiments identified with an asterisk have been funded for the full scale detector.

Experiment	ν source	Reaction	Detector
Charged-Current Detectors			
LENS	pp	$\nu_e{}^{115}\mathrm{In} \to e\,{}^{115}\mathrm{Sn}, e, \gamma$	15 t of In in 200 t liquid scintillator
Lithium	pep, CNO	$\nu_e{}^7\mathrm{Li} \to e\,{}^7\mathrm{Be}$	Radiochemical, 10 t lithium
MOON	pp	$\nu_e{}^{100}\mathrm{Mo} \to e\,{}^{100}\mathrm{Tc}(\beta)$	3.3 t ^{100}Mo foil + plastic scintillator
νe Elastic Scattering Detectors			
Borexino*	^{7}Be	$\nu e \to \nu e$	100 t liquid scintillator
CLEAN	pp, ^{7}Be	$\nu e \to \nu e$	10 t liquid Ne
HERON	pp, ^{7}Be	$\nu e \to \nu e$	10 t liquid He
KamLAND*	^{7}Be	$\nu e \to \nu e$	1000 t liquid scintillator
SNO+	pep, CNO	$\nu e \to \nu e$	1000 t liquid scintillator
TPC-type	pp, ^{7}Be	$\nu e \to \nu e$	Tracking electron in gas target
XMASS	pp, ^{7}Be	$\nu e \to \nu e$	10 t liquid Xe

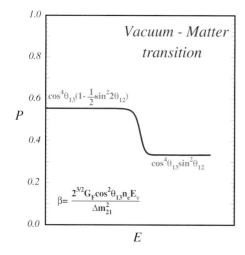

Figure 11. Vacuum-matter transition of ν_e survival probability. In the pp neutrino regime, the survival probability is approximately $(1 - \frac{1}{2}\sin^2 2\theta_{12})$ (for $\theta_{13} \ll 1$). In the energy regime where ^{8}B neutrinos are observed, the survival probability is approximately $\sin^2 \theta_{12}$. Non-standard interactions (NSI) predict a substantially different energy dependence in the transition region, and future ^{7}Be and the pep neutrinos experiments can be used to discriminate these NSI models from the MSW scenario.

guities. In a reactor-based measurement, one searches for $\bar\nu_e$ flux suppression and spectral distortion of the prompt positron signal at different baselines.

The general configuration of such a reactor-based experiment consists of at least two or more detectors: one is placed at a distance of <0.5 km, and the other at 1-2 km. The *near* detector is used to normalize the reactor $\bar\nu_e$ flux, while the *far* detector is used to search for rate suppression and spectral distortion. The baseline of the far detector, $\sim$2 km, is the distance where the survival probability reaches its first minimum. In order to shield the detectors from muon spallation backgrounds, these detectors require overhead shielding. For these experiments, a $<\sim$1% error budget is required in order to reach a $\sin^2 2\theta_{13}$ sensitivity of $<\sim 0.01$.

In addition to θ_{13} measurements, these experiments can also make contributions in measuring the Weinberg angle $\sin\theta_W$ (at $Q^2 =0$), and in investigating the nature of neutrino neutral-current weak coupling. The precision in θ_{12} can also be improved by a reactor anti-neutrino experiment with a baseline of 50 to 70 km [37].

Table 5 is a tabulation of the proposed

Table 5. Proposed reactor anti-neutrino experiments for measuring θ_{13}.

Experiment	Location	Baseline (km)		Overburden (m.w.e.)		Detector size (tons)		$\sin^2(2\theta_{13})$ sensitivity (90% C.L.)
		Near	Far	Near	Far	Near	Far	
Angra dos Reis	Brazil	0.3	1.5	200	1700	50	500	$<\sim 0.01$
Braidwood	USA	0.27	1.51	450	450	65×2	65×2	$<\sim 0.01$
Double Chooz	France	0.2	1.05	50	300	10	10	$<\sim 0.03$
Daya Bay	China	0.3	1.8-2.2	300	1100	50	100	$<\sim 0.01$
Diablo Canyon	USA	0.4	1.7	150	750	50	100	$<\sim 0.01$
KASKA	Japan	0.4	1.8	100	500	8	8	$<\sim 0.02$
Kr2Det	Russia	0.1	1.0	600	600	50	50	$<\sim 0.03$

reactor θ_{13} experiments[38]. More details on the status of these experiments can be found in the supplemental slides of this talk[35].

9 Conclusions

After nearly four decades, the solar neutrino deficit problem is finally resolved. There are now strong evidences for neutrino oscillation from solar neutrino and reactor anti-neutrino experiments. SNO, Super-Kamiokande and KamLAND will improve the precision of the neutrino oscillation parameters. Future solar neutrino experiments will provide high precision tests of the solar model calculations, and will probe the energy dependence of neutrino oscillation. Future reactor anti-neutrino experiments will attempt to measure the unknown mixing angle θ_{13}. Once the magnitude of this angle is known, a new arena of CP studies may be opened up in the neutrino sector in the future.

Acknowledgments

The author thanks the conference organizers for the invitation to speak; and E. Blucher, M. Chen, P. Decowski, H. Ejiri, R. Hazama, K. Ishii, R. Lanou, T. Lasserre, K.-B. Luk, D. McKinsey, M. Nakahata, R. Raghavan, F. Suekane, R. Svoboda, and Y. Takeuchi for providing much of the background information in this talk and proceedings.

References

1. J. N. Bahcall, M. Pinsonneault, S. Basu, *Astrophys. J.* **555**, 999 (2001).
2. J. N. Bahcall and M. H. Pinsonneault, *Phys. Rev. Lett.* **92**, 121301 (2004).
3. J. N. Bahcall, A. M. Serenelli, and S. Basu, astro-ph/0412440.
4. A. S. Brun, S. Turck-Chièze, and J. P. Zahn, *Astrophys. J.* **525**, 1032 (1999)
5. S. Turck-Chièze *et al.*, *Ap. J. Lett.* **555** (2001).
6. S. Turck-Chièze *et al.*, *Phys. Rev. Lett.* **93**, 221102 (2004).
7. B. T. Cleveland *et al.* [Homestake Collaboration], *Astrophys. J.* **496**, 505 (1998).
8. J. N. Abdurashitov *et al.* [SAGE Collaboration], *J. Exp. Theor. Phys.* **95**, 181 (2002).
9. W. Hampel *et al.* [GALLEX Collaboration], *Phys. Lett. B* **447**, 127 (1999).
10. M. Altmann *et al.* [GNO COLLABORATION Collaboration], *Phys. Lett. B* **616**, 174 (2005).
11. S. Fukuda *et al.* [Super-Kamiokande Collaboration], *Nucl. Instr. Meth. A* **501**, 418 (2003).
12. S. Fukuda *et al.* [Super-Kamiokande Collaboration], *Phys. Rev. Lett.* **86**, 5651 (2001).
13. S. Fukuda *et al.* [Super-Kamiokande

Collaboration], *Phys. Rev. Lett.* **86**, 5656 (2001).

14. M. B. Smy *et al.* [Super-Kamiokande], *Phys. Rev. D* **69** 011104 (2004).

15. Y. Fukuda *et al.* [Kamiokande Collaboration], *Phys. Rev. Lett.* **77**, 1683 (1996).

16. J. Boger *et al.* [SNO Collaboration], *Nucl. Instr. and Meth. A* **449**, 172 (2000).

17. Q. R. Ahmad *et al.* [SNO Collaboration], *Phys. Rev. Lett.* **87**, 071301 (2001).

18. Q. R. Ahmad *et al.* [SNO Collaboration], *Phys. Rev. Lett.* **89**, 011301 (2002).

19. Q. R. Ahmad *et al.* [SNO Collaboration], *Phys. Rev. Lett.* **89**, 011302 (2002).

20. S. N. Ahmed *et al.* [SNO Collaboration], *Phys. Rev. Lett.* **92**, 181301 (2004).

21. B. Aharmim *et al.* [SNO Collaboration], arXiv:nucl-ex/0502021.

22. A. Suzuki, talk at the XVIII International Conference on Neutrino Physics and Astrophysics, Takayama, Japan, (1998).

23. K. Eguchi *et al.* [KamLAND Collaboration], *Phys. Rev. Lett.* **90**, 021802 (2003).

24. T. Araki *et al.* [KamLAND Collaboration], *Phys. Rev. Lett.* **94**, 081801 (2005).

25. S. P. Mikheyev and A. Yu. Smirnov, *Sov. J. Nucl. Phys.* **42** 913 (1985); L. Wolfenstein, Phys. Rev. **D17** 2369 (1978).

26. Super-Kamiokande results presented in this paper are preliminary results provided by the collaboration. Since this conference, the Super-Kamiokande collaboration has published a full summary of its SK-I results in J. Hosaka *et al.*, arXiv:hep-ex/0508053.

27. H. H. Chen, *Phys. Rev. Lett.* **55**, 1534 (1985).

28. G.L. Fogli, E. Lisi, A. Marrone, and A.

Palazzo, *Phys. Lett. B* **583** 149 (2004).

29. Examples of such global analyses include J.N. Bahcall, M.C. Gonzalez-Garcia and C. Peña-Garay, *JHEP* **08**, 016 (2004); A. B. Balantekin and H. Yüksel, *Phys. Rev. D* **68**, 113002 (2003); A. Bandyopadhyay, S. Choubey, S. Goswami, S. T. Petcov, and D. P. Roy, *Phys. Lett. B* **583**, 134 (2004); G. L. Fogli *et al.* *Phys. Rev. D* **66**, 053010 (2002); P. C. de Holanda and A. Yu. Smirnov, *Astropart. Phys.* **21**, 287 (2004).

30. A comprehensive review of reactor neutrino oscillation experiments, prior to the KamLAND experiment, can be found in C. Bemporad, G. Gratta, and P. Vogel, *Rev. Mod. Phys.* **74**, 297 (2002).

31. Z. Maki, N. Nakagawa, and S. Sakata, *Prog. Theor. Phys.* **28**, 870 (1962); V. Gribov and B. Pontecorvo, *Phys. Lett. B* **28**, 493 (1969).

32. J.N. Bahcall and C. Peña-Garay, *JHEP* **11**, 004 (2003)

33. O. G. Miranda, M. A. Tortola and J. W. F. Valle, *Nucl. Phys. Proc. Suppl.* **145**, 61 (2005); A. Friedland, C. Lunardini and C. Pena-Garay, *Phys. Lett. B* **594**, 347 (2004).

34. M. Nakahata, *Nucl. Phys. Proc. Suppl.* **145**, 23 (2005).

35. The conference website is http://www.uu.se/LP2005.

36. M. Apollonio *et al.* [CHOOZ Collaboration], *Eur. Phy. J. C* **27**, 331 (2003).

37. A. Bandyopadhyay, S. Choubey, S. Goswami and S. T. Petcov, *Phys. Rev.* **D72**, 033013 (2005)

38. E. Abouzaid *et al.*, *Report of the APS Neutrino Study Reactor Working Group*, (2004) http://www.aps.org/neutrino

DISCUSSION

Jonathan Rosner (U. of Chicago, USA): Can you say more about the proposal to put liquid scintillator in SNO?

Alan Poon: The main objectives of this proposed experiment are to measure the *pep* neutrino flux and to search for geo-neutrinos. The *pep* measurement will probe the vacuum-matter transition that I discussed in the talk.

The project is still in an early proposal stage. There are a number of technical problems that need to be resolved before its realisation. For example, the compatibility of the liquid scintillator and the acrylic vessel has to be established; the longevity of the current photomultiplier tube mounting structure, which was designed to have a life span of 10 years in ultra-pure water, has to be evaluated; and the optical response of the liquid scintillator has to be established.

Peter Rosen (DOE, USA): A comment about MSW. If you restrict your analysis to 2 flavors, then the fact that the solar ν_e survival probability is less than $1/2$ is evidence for MSW. Pure in-vacuum oscillation give a survival probability greater than or equal to $1/2$. Secondly, the best chance to see a day/night effect is to have a detector at the equator. This maximizes tha path of the neutrino through matter, and through the densest part of the Earth.

Alan Poon: Thank you for your comment. Because of time constraint, I did not have enough time to discuss this point further in my talk. It is true that even though we have not directly observed MSW effect, there are very strong indirect evidence that it is the underlying mechanism for neutrino flavor transformation in solar neutrinos. Fogli *et al.* showed that the null hypothesis of no MSW effect in the solar neutrino results is rejected at more than 5σ.

NEUTRINO PHYSICS AT SHORT BASELINE

ERIC D. ZIMMERMAN

Department of Physics, University of Colorado, Boulder, Colo. 80309, USA
E-mail: edz@colorado.edu

Neutrino oscillation searches at short baseline (defined as $\lesssim 1$ km) have investigated oscillations with $\Delta m^2 \gtrsim 0.1 \text{eV}^2$. One positive signal, from the LSND collaboration, exists and is being tested by the MiniBooNE experiment. Neutrino cross-section measurements are being made, which will be important for reducing systematic errors in present and future oscillation measurements.

1 Overview

Acclerator-based neutrino experiments at "short" baseline (defined here as $\lesssim 1$ km) probe high-Δm^2 regions of oscillation parameter space, using beams with energies ranging from stopped muon decay (< 53 MeV) to several hundred GeV. In general, these experiments are sensitive to oscillations with $\Delta m^2 \gtrsim 10^{-2}$ eV^2, making them insensitive to the solar and atmospheric mass scales.

These neutrinos are also used as probes of electroweak physics and nucleon structure. Finally, accelerator neutrinos are being used at short baseline to make neutrino interaction cross-section measurements necessary for analyzing long- and short-baseline oscillation data.

2 Oscillations at high Δm^2

The highest energy neutrino beams have been used in recent years to investigate the possibility of neutrino oscillations at high Δm^2. These have probed primarily $\nu_\mu \to \nu_e$ and $\bar\nu_\mu \to \bar\nu_e$ oscillations at $\Delta m^2 > 10$ eV^2, with sensitivities as low as $\sin^2 2\theta \sim 10^{-3}$. The tightest limits on $\nu_\mu \to \nu_e$, when measured separately from antineutrinos, come from NuTeV[1] (Fermilab E815) above $\Delta m^2 \sim 30$ eV^2, and from BNL E734[2] and E776[3] at lower Δm^2. The most stringent limits on $\nu_\mu \to \nu_\tau$ appearance at high Δm^2 are from the NOMAD[4] and CHORUS[5] detectors at CERN.

3 The LSND signal

The LSND collaboration has published strong evidence for $\bar\nu_\mu \to \bar\nu_e$ oscillations using neutrinos from stopped muon decay[6].

3.1 LSND

LSND used a beam-stop neutrino source at the 800 MeV LAMPF proton accelerator at Los Alamos National Laboratory. The primary source of neutrinos was π^+ and μ^+ decays at rest (DAR) in the target, which yielded ν_μ, $\bar\nu_\mu$, and ν_e with energies below 53 MeV. In addition, π^+ and π^- decays in flight (DIF) provided a small flux of higher-energy ν_μ and $\bar\nu_\mu$. The $\bar\nu_e$ flux was below 10^{-3} of the total DAR rate. The LSND data were collected between 1993 and 1998. The first data set, collected 1993-1995, used a water target that stopped all hadrons and provided 59% of the DAR data set; the remainder of the data came from a heavy metal target composed mostly of tungsten. The collaboration searched for $\bar\nu_e$ appearance using the reaction $\bar\nu_e + p \to e^+ + n$ in a 167-ton scintillator-doped mineral oil (CH$_2$) target/detector. The detector sat 30 m from the target, providing an oscillation scale $L/E \sim 0.6 - 1$ m/MeV. The detector, which was instrumented with 1220 8-inch photomultiplier tubes (PMTs), observed a Cherenkov ring and scintillation light from the positron emitted in the neutrino interaction. An additional handle was the detection of the 2.2 MeV neutron-capture gamma ray from the reac-

parameter space.

3.2 KARMEN and the joint analysis

Another experiment of similar design, the Karlsruhe-Rutherford Medium Energy Neutrino (KARMEN) experiment at the ISIS facility of the Rutherford Laboratory, also searched for $\bar{\nu}_\mu \to \bar{\nu}_e$ oscillations. KARMEN used a similar beam-stop neutrino source, but with a segmented smaller neutrino target (56 tons). KARMEN's sensitivity was enhanced because the lower beam duty factor (10^{-5}) allowed beam-unrelated events to be removed more effectively with a timing cut. In addition, KARMEN had higher flux because it was closer to the target (18 m versus 30 m). This did, however, reduce KARMEN's sensitivity to low-Δm^2 oscillations compared to LSND. KARMEN's final published result[7], using data collected from 1997 to 2001, reported 15 $\bar{\nu}_e$ oscillation candidates with an expected background of 15.8 ± 0.5 events. This result does not provide evidence for oscillations, and indeed can be used to rule out most of the high-Δm^2 portions of the LSND allowed region. However, an analysis of the combined LSND and KARMEN data sets has found regions of oscillation parameter space (Fig. 2) that fit both experiments' data well[8].

3.3 Physics scenarios including LSND

The LSND data indicate a much larger Δm^2 than atmospheric or solar experiments: $\Delta m^2 \sim 0.1 - 10$ eV^2. This led to the paradox of three Δm^2 values all of different orders of magnitude; this is impossible if there are only three neutrino masses. The more common way to account for all the existing oscillation data is to introduce one or more "sterile" neutrino flavors[9]. Other recently proposed exotic physics scenarios include mass-varying neutrinos[10] and a decaying sterile neutrino[11].

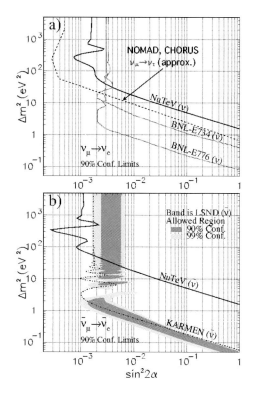

Figure 1. Limits on oscillations of ν_μ (a) and $\bar{\nu}_m u$ (b). All curves represent $\nu_e(\bar{\nu}_e)$ appearance except for NOMAD/CHORUS curve, which is $\nu_\mu \to \nu_\tau$.

tion $n + p \to d + \gamma$. The appropriate delayed coincidence (the neutron capture lifetime in oil is 186 μs) and spatial correlation between the e^+ and γ were studied for DAR $\bar{\nu}_e$ candidates.

In 2001, LSND presented the final oscillation search results, which gave a total $\bar{\nu}_e$ excess above background of $87.9 \pm 22.4 \pm 6.0$ events in the DAR energy range. The dominant background was beam-unrelated events, primarily from cosmic rays. These backgrounds were measured using the 94% of detector livetime when the beam was not on. No significant signal was observed in DIF events; the total $\nu_e/\bar{\nu}_e$ excess above background was $8.1 \pm 12.2 \pm 1.7$ events, consistent with the DAR result. The total events and energy distributions of the DAR and DIF events were used to constrain the oscillation

294

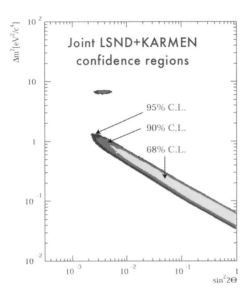
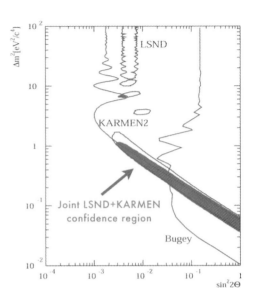

Figure 2. Left: Confidence regions from the joint LSND-KARMEN analysis. Right: LSND-KARMEN region superimposed over confidence region from LSND decay-at-rest data and 90% confidence exclusion regions from KARMEN2 and Bugey.

3.4 MiniBooNE

MiniBooNE[12] (Experiment 898 at Fermilab) is a short-baseline neutrino oscillation experiment whose main purpose is to test the LSND result. It uses an 8 GeV proton beam from the Fermilab Booster to produce pions, which then decay in flight to produce a nearly pure ν_μ flux at a mineral oil Cherenkov detector 500 m away. The detector uses Cherenkov ring shape information to distinguish charged-current ν_μ from ν_e interactions, searching for an excess of ν_e which would indicate oscillations. Data collection began in late 2002.

There are several major differences between MiniBooNE and LSND, which should assure that systematic errors are independent. First, MiniBooNE operates at an energy and oscillation baseline over an order of magnitude greater than LSND: $E_\nu \sim 500 - 1000$ MeV, compared to $30 - 53$ MeV at LSND. The baseline $L = 500$ m, versus 30 m at LSND. L/E remains similar, ensuring that the oscillation sensitivity is maxi-

mized in the same region of parameter space as LSND. MiniBooNE uses the quasielastic neutrino scattering reaction $\nu_e\,^{12}C \rightarrow e^- X$ with the leading lepton's Cherenkov ring reconstructed, rather than LSND's antineutrino interaction with a hydrogen nucleus followed by neutron capture. Finally, MiniBooNE's goal is a factor of ten higher statistics than LSND had.

4 Non-oscillation physics

4.1 Deep inelastic scattering

Neutrino deep inelastic scattering has in the recent past been used for studies of electroweak and nucleon structure physics. NuTeV (FNAL E815) is the most recent and likely to be the last of these studies. Recent results from NuTeV include a measurement[13] of the electroweak mixing angle $\sin^2 \theta_W = 0.2277 \pm 0.0013_{\text{stat}} \pm 0.0009_{\text{syst}}$, a value three standard deviations above the standard model prediction. While several nonperturbative QCD effects (in particular the possibility of isospin violation in the nucleon and asymmetry in the strange sea) could affect this result at the $\lesssim 1\sigma$ level, no standard model effect has fully explained the experimental result. Unfortunately, no high-energy neutrino beams are now operating or under development, so it is unlikely that a new experiment will test this result in the foreseeable future. (A collaboration has proposed to perform a neutrino-based measurement of $\sin^2 \theta_W$ at the Braidwood reactor, using a new method[14]. Sensitivity comparable to NuTeV may be achieved, albeit at lower Q^2.)

NuTeV has also recently published precise measurements[15] of the muon neutrino and antineutrino cross-sections in the energy range $30 < E < 340$ GeV, along with fits to the structure functions $F_2(x, Q^2)$ and $xF_3(x, Q^2)$. Full cross-section tables have been made available at Ref.[16].

4.2 *Neutrino interactions at the GeV scale*

Short-baseline neutrino experiments have recently made some significant measurements of neutrino cross-sections on nuclear targets. These measurements are important for testing nuclear models, and are critical for understanding the large data sets being produced in current and future oscillation experiments.

At the MiniBooNE flux (which is very similar to T2K, and overlaps with the somewhat higher-energy K2K flux), the largest cross-section process is charged current quasielastic scattering (CCQE),

$$\nu_\mu n \to \mu^- p.$$

This process, which is the primary detection mode for oscillation searches, represents $\sim 40\%$ of the total interaction rate.

Charged current pion production (labeled $CC\pi^+$) via a nucleon resonance,

$$\nu_\mu N \to \mu^- \Delta \to \mu^- \pi^+ N',$$

represents about a quarter of the total event rate. The recoil nucleons are generally un-

detected, so these events are difficult to distinguish from a similar final state that can be achieved by scattering coherently off a nucleus:

$$\nu_\mu A \to \mu^- \pi^+ A.$$

Neutral pion production can occur in charged-current scattering through a resonance ($CC\pi^0$), or in either resonant or coherent neutral-current scattering ($NC\pi^0$, with no charged lepton in the final state). The $NC\pi^0$ processes, expected to be about 7% of the total event rate at MiniBooNE, are of particular interest to electron neutrino appearance searches because of the potential for the π^0 to be misidentified as an electron.

Coherent and resonant π^+ and π^0 production have been modeled by Rein and Sehgal[17], and these results are used by all the major neutrino collaborations for Monte Carlo modeling of these processes. At energies in the 1 GeV range, these models have only been tested in the past on proton and deuterium targets. K2K and MiniBooNE are now producing the first tests of these models on nuclear targets and, therefore, the first searches for coherent pion production.

K2K has measured the q^2 distribution of $CC\pi^+$ production using the SciBar fine-grained scintillation detector at K2K's near detector site[18]. The q^2 distribution can be used to distinguish statistically the coherent and resonant fractions, and is therefore a test of the Rein-Sehgal model. That model predicted a coherent π^+ cross-section ratio to all charged-current ν_μ for the K2K flux of 2.67%. The K2K data fit (Fig. 3) to the q^2 distribution, however, showed no evidence for coherent events: the final sample contained 113 events with a background estimate of 111. This results in an upper limit of the cross-section ratio of

$$\frac{\sigma(\text{Coherent } \nu_\mu A \to \mu^- \pi^+ A)}{\sigma(\nu_\mu A \to \mu^- X)} < 0.6\%$$

at 90% confidence level, a significant disagreement from the Rein-Sehgal prediction.

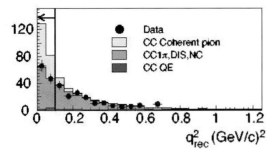

Figure 3. Reconstructed q^2 for charged-current pion candidates in the K2K coherent scattering search sample.

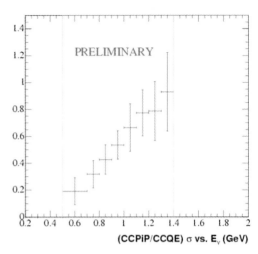

(CCPiP/CCQE) σ vs. E_ν (GeV)

Figure 4. MiniBooNE preliminary measurement of ratio $\sigma(\nu_\mu + {}^{12}\mathrm{C} \to \mu^- + \pi^+ + X)/\sigma(\nu_\mu + {}^{12}\mathrm{C} \to \mu^- + Y)$, where Y represents a final state with no pions. Errors are predominantly systematic.

MiniBooNE has studied the $\mathrm{CC}\pi^+$ process on carbon and, while the results are still preliminary, the angular distribution of leading muons shows a deficit at the extremely forward angles where coherent scattering is expected.

Another preliminary measurement recently released by MiniBooNE is the neutrino energy dependence of the ratio of the $\mathrm{CC}\pi^+$ to CCQE cross-sections. The $\mathrm{CC}\pi^+$ events are readily identified in the MiniBooNE detector by their final state: the pion and muon both leave stopped-μ decay electron signatures after the primary event. The pion is generally below Cherenkov threshold, so the muon ring can be reconstructed cleanly. By using the measured lepton energy and angle relative to the beam direction, a neutrino energy is calculated assuming a missing Δ^{++} mass. This yields $\sim 20\%$ resolution on E_ν for $\mathrm{CC}\pi^+$ events. Normalization quasielastic events are reconstructed similarly, with an assumed proton recoil. The resolution on E_ν is $\sim 10\%$ for the CCQE sample. The measured cross-section ratio is shown in Fig. 4. As expected, the relative contribution of pion production rises from the kinematic threshold, to a level comparable to CCQE for neutrinos above 1 GeV. Systematic errors are dominant at present, and are due primarily to energy scale and photon scattering and extinction models in the mineral oil. MiniBooNE expects these errors to be reduced substantially as modeling of the detector optics is improved.

We can expect future investigations of these processes (and the $\mathrm{CC}\pi^0$ and $\mathrm{NC}\pi^0$ processes) from MiniBooNE in the near future. In addition, a proposal[19] to relocate the SciBar detector to the BooNE beamline is under review by Fermilab at present. In the more distant future, we can expect major improvement in neutrino interaction physics when the MINERνA (E938)[20] experiment at Fermilab and the near detectors at the newly constructed JPARC neutrino beam (T2K)[21] are commissioned later this decade.

Acknowledgments

The author acknowledges members of the LSND, BooNE, and K2K collaborations and support from the U. S. Department of Energy.

References

1. S. Avvakumov *et al.*, *Phys. Rev. Lett.* **89** 011804.

2. L. A. Ahrens *et al.*, *Phys. Rev.* D**31** 2732 (1985).

3. L. Borodovsky *et al.*, *Phys. Rev. Lett.* **68** 274 (1992).

4. P. Astier *et al.*, *Nucl. Phys.* B**611** 3 (2001).

5. E. Eskut *et al.*, *Phys. Lett.* B**497** 8 (2001).

6. A. Aguilar *et al.*, *Phys. Rev.* D**64** 112007 (2001).

7. B. Armbruster *et al.*, *Phys. Rev.* D**65** (2002).

8. E. Church *et al.*, *Phys. Rev.*D **66** 013001 (2002).

9. M. Sorel, J. M. Conrad, and M. Shaevitz, *Phys. Rev.* D**70** 073004 (2004).

10. R. Fardon, A. Nelson, and N. Weiner, *JCAP* **0410** 005 (2004).

11. S. Palomares-Ruiz, S. Pascoli, and T. Schwetz, *JHEP* **0509** 048 (2005).

12. http://www-boone.fnal.gov

13. G. P. Zeller *et al.*, *Phys. Rev. Lett.* **88** 091802 (2002).

14. J. M. Conrad, J. Link, and M. Shaevitz, *Phys. Rev.* D**71** 073013 (2005).

15. M. Tzanov *et al.*, e-print hep-ex/0509010 (2005).

16. http://www-nutev.phyast.pitt.edu/ results_2005/nutev_sf.html

17. D. Rein and L. M. Sehgal, *Nucl. Phys.* **B223** 29 (1983).

18. M. Hasegawa *et al.*, e-print hep-ex/0506008 (2005).

19. http://home.fnal.gov/~wascko/ scibar.pdf

20. http://minerva.fnal.gov

21. http://neutrino.kek.jp/jhfnu

DIRECT NEUTRINO MASS MEASUREMENTS

CHRISTIAN WEINHEIMER

Institut für Kernphysik, Wilhelm-Klemm-Str. 9
Westfälische Wilhelms-Universität Münster
D-48149 Münster, Germany
E-mail: weinheimer@uni-muenster.de

The discovery by many different experiments of the flavour oscillation of neutrinos from different sources proved clearly that neutrinos have non-vanishing masses in contrast to their current description within the Standard Model of particle physics. However, the neutrino mass scale, which is – in addition to particle physics – very important for cosmology and astrophysics, cannot be resolved by oscillation experiments. Although there are a few ways to determine the absolute neutrino mass scale, the only model-independent method is the investigation of the electron energy spectrum of a β decay near its endpoint. The tritium β decay experiments at Mainz and Troitsk using tritium have recently been finished and have given upper limits on the neutrino mass scale of about 2 eV/c^2. The bolometric experiments using ^{187}Re have finished the first round of prototype experiments yielding a sensitivity on the neutrino mass of 15 eV/c^2. The new Karlsruhe Tritium Neutrino Experiment (KATRIN) will enhance the sensitivity on the neutrino mass by an ultra-precise measurement of the tritium β decay spectrum by another order of magnitude down to 0.2 eV/c^2 by using a very strong windowless gaseous molecular tritium source and a huge ultra-high resolution electrostatic spectrometer of MAC-E-Filter type. The recent achievements in test experiments show, that this very challenging experiment is feasible.

1 Introduction

The recent discovery of neutrino oscillation by experiments with atmospheric, solar, reactor and accelerator neutrinos[1] proved that neutrinos mix and that they have non-zero masses in contrast to their current description in the Standard Model of particle physics. Unfortunately, these oscillation experiments are sensitive to the differences of squared neutrino mass states $|\Delta m_{ij}^2| = |m^2(\nu_i) - m^2(\nu_j)|$ *, but not directly to the neutrino masses $m(\nu_i)$ themselves. On the other hand, if one neutrino mass is measured absolutely, the whole neutrino mass spectrum can be calculated using the values Δm_{ij}^2 from the oscillation experiments.

Theories beyond the Standard Model try to explain the smallness of neutrino masses in comparison with the much heavier charged fermions[2]. One prominent explanation is the Seesaw type I mechanism using heavy Majorana neutrinos yielding a hierarchical pattern of neutrino masses. Alternatively, Seesaw type II models usually produce a scenario of quasi-degenerate neutrino masses with the help of a Higgs triplet. Here all masses are 0.1 eV/c^2 or heavier exhibiting small mass differences between each other to explain the oscillations. In this quasi-degenerate case – due to huge abundance of relic neutrinos in the universe left over by the big bang – neutrinos would make up not the major, but a significant contribution to the dark matter. Therefore the open question of the value of the neutrino masses is not only crucial for particle physics to decide between different theories beyond the Standard Model but it is also very important for astrophysics and cosmology.

There are different ways to determine the neutrino mass scale:

- Cosmology

Information on the absolute scale of the neutrino mass can be obtained from astrophysical observations like the power spectrum of matter and the energy distribution in the

---

*In the case of matter effects involved – like for solar neutrinos – the sign of Δm_{ij}^2 can be resolved.

universe at different scales. Usually these analyses use the combination of Cosmic Microwave Background data (*e.g.* from the WMAP satellite), the distribution of the galaxies in our universe, the so-called "Large Scale Structure", and information from the so-called "Lyman α-Forest" or X-ray clusters to describe the distribution at large, medium and small scales, respectively. In most cases they give upper limits on the mass of the neutrinos on the order of several 0.1 eV/c² [4]. In some cases non-zero neutrino masses are found[5], illustrating the dependence on the assumptions and on the data used to obtain the cosmological limits. One should not forget that these models describe 95 % of the matter and energy distribution of the universe by yet non-understood quantities like the cosmological constant[6] and Cold Dark Matter. And the limits on the neutrino mass rely on the existence of the yet not observed relic neutrinos[7].

- Neutrinoless double β decay

One laboratory way to access the neutrino mass scale is the search for the neutrinoless double β decay[3]. This process is a conversion of two neutrons (protons) into two protons (neutrons) within a nucleus at the same time. Usually two electrons (positrons) and two neutrinos (antineutrinos) are emitted. In the case that the neutrino is a Majorana particle (particle is equal to its antiparticle), the double β decay could occur without emission of any neutrinos. This transition is directly proportional to the neutrino mass (in the absence of right-handed weak charged currents or the exchange of other new particles). The observable of double β decay is the so-called effective neutrino mass

$$m_{\text{ee}} = \sum_i |U_{\text{ei}}^2 \cdot m(\nu_{\text{i}})| \qquad (1)$$

which is a coherent sum over all neutrino mass eigenstates $m(\nu_i)$ contributing to the electron neutrino with their (complex) mixing matrix elements U_{ei}. A subgroup of the

most sensitive experiment, the Heidelberg-Moscow experiment using 5 low-background, highly-enriched and high-resolution ^{76}Ge detectors in the Gran Sasso underground lab, has claimed evidence for having observed neutrinoless double β decay. Recently new data and a re-analysis of the old data have been presented[9] showing a line at the position expected for neutrinoless double β decay with 4 σ significance. Due to the uncertainties of the nuclear matrix element this signal translates into $0.1 \text{ eV}/c^2 \leq m_{ee} \leq 0.9 \text{ eV}/c^2$. Clearly, this yet unconfirmed result requires further checks, which are under way by several experiments[3].

- Direct neutrino mass determination

In contrast to the other methods, the direct method does not require further assumptions. The neutrino mass is determined using the relativistic energy-momentum relationship. Therefore $m^2(\nu)$ is the observable in most cases.

The non-observation of a dependence of the arrival time on energy of supernova neutrinos from SN1987a gave a generally accepted upper limit on the neutrino mass of 5.7 eV/c² [10]. Unfortunately nearby supernova explosions are too rare and too less understood to allow a further improvement to a sub-eV sensitivity on the neutrino mass.

Therefore, the investigation of the kinematics of weak decays – and with regards to eV and sub-eV sensitivities – the electron energy spectrum of a β decay is still the most sensitive model-independent and direct method to determine the neutrino mass. The β spectrum exhibits the value of a non-zero neutrino mass when the neutrino is emitted non-relativistically. This is the case in the vicinity of the endpoint E_0 of the β spectrum where nearly all decay energy is given to the β electron. Therefore, the mass of the electron neutrino is determined by investigating precisely the shape of the β spectrum near its endpoint E_0 (see fig. 1). From fig. 1 it

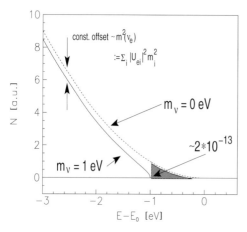

Figure 1. Expanded β spectrum around its endpoint E_0 for $m(\nu_e) = 0$ (dashed line) and for a arbitrarily chosen neutrino mass of 1 eV/c^2 (solid line) The offset between the two curves explains what the "$m(\nu_e)$" is: the average over all neutrino mass states with their contribution according to the neutrino mixing matrix U. In the case of tritium, the gray shaded area corresponds to a fraction of $2 \cdot 10^{-13}$ of all tritium β decays

is clearly visible that the main requirement for such an experiment is to cope with the vanishing count rate near the endpoint by providing the strongest possible signal rate at lowest background rate. Additionally, to become sensitive to the neutrino mass dependent shape of the β spectrum a high energy resolution on the order of eV is required.

Tritium is the standard isotope for this study due to its low endpoint of 18.6 keV, its rather short half-life of 12.3 y, its superallowed shape of the β spectrum, and its simple electronic structure. Tritium β decay experiments have been performed in search for the neutrino mass for more than 50 years. ^{187}Re is a second isotope suited to determine the neutrino mass. The disadvantage of its primordial half life can be compensated by using large arrays of cryogenic bolometers.

For each neutrino mass state $m(\nu_i)$ contributing to the electron neutrino a, a kink at $E_0 - m(\nu_i)c^2$ with a size proportional to $|U_{ei}^2|$ will occur. However, due to the

smallness of differences of squared neutrino masses, Δm_{ij}^2, observed in oscillation experiments and as a consequence of the limited sensitivity of present and upcoming direct neutrino mass experiments, only an incoherent sum or an average neutrino mass can be obtained[8], which can be defined as the *electron neutrino mass* $m(\nu_e)$ by

$$m^2(\nu_e) = \sum_i |U_{ei}|^2 \cdot m^2(\nu_i) \qquad (2)$$

Comparing equations (1) and (2) it is obvious that the neutrinoless double β decay and the investigation of the β decay spectrum yield complementary information. In the former case complex phases of the neutrino Majorana mixing matrix U can lead to a partial cancellation. Also the still large uncertainties of the nuclear matrix element and the possibility, that the exchange of right-handed currents or more exotic particles than the neutrino can add to the neutrinoless double β decay amplitude, disfavours double β decay for a precise neutrino mass determination. On the other hand the high sensitivity of the next generation of double β decay experiments and their unique possibility to prove the Majorana nature of neutrinos underline the very high importance of neutrinoless double β decay experiments.

This paper is organised as following: In section 2 the recent direct neutrino mass experiments investigating the β decay spectra of tritium are presented, whereas the ^{187}Re experiments are shortly reviewed in section 3. Section 4 describes the new KATRIN experiment aiming for a 0.2 eV/c^2 neutrino mass sensitivity. The conclusions are given in section 5.

2   Neutrino mass experiments from tritium β decay

A major break-through in tritium β decay experiments was achieved in the nineties by a new type of spectrometer, the socalled MAC-E-Filter (Magnetic Adiabatic

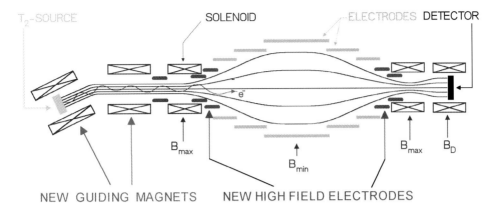

Figure 2. The upgraded Mainz setup shown schematically. The outer diameter amounts to 1 m, the distance from source to detector is 6 m.

Collimation followed by an Electrostatic Filter), which was developed independently at Mainz, Germany and at Troitsk, Russia[11,12]. This integrating spectrometer provides high luminosity and low background combined with a large energy resolution.

The two recent tritium β decay experiments at Mainz and at Troitsk use similar MAC-E-Filters with an energy resolution of 4.8 eV (3.5 eV) at Mainz (Troitsk). The spectrometers differ slightly in size: The diameter and length of the Mainz (Troitsk) spectrometer are 1 m (1.5 m) and 4 m (7 m). The major differences between the two setups are the tritium sources: Mainz uses a thin film of molecular tritium quench-condensed on a cold graphite substrate as tritium source, whereas Troitsk has chosen a windowless gaseous molecular tritium source. After the upgrade of the Mainz experiment in 1995-1997 both experiments run with similar signal and similar background rates.

2.1 The Mainz Neutrino Mass Experiment

Fig. 2 shows the Mainz setup after its upgrade in 1995-1997, which included the installation of a new tilted pair of superconducting solenoids between the tritium source and spectrometer and the use of a new cryo-stat providing temperatures of the tritium film below 2 K. The first measure eliminated source correlated background and allowed the source strength to be increased significantly. The second measure avoids the roughening transition of the homogeneously condensed tritium films with time[13]. The upgrade of the Mainz setup was completed by the application of HF pulses on one of the electrodes in between measurements every 20 s, and a full automation of the apparatus and remote control. This former improvement lowers and stabilises the background, the latter one allows long–term measurements.

Figure 3 shows the endpoint region of the Mainz 1998, 1999 and 2001 data in comparison with the former Mainz 1994 data. An improvement of the signal-to-background ratio by a factor 10 by the upgrade of the Mainz experiment as well as a significant enhancement of the statistical quality of the data by longterm measurements are clearly visible. The main systematic uncertainties of the Mainz experiment are the inelastic scattering of β electrons within the tritium film, the excitation of neighbour molecules due to the β decay, and the self-charging of the tritium film by its radioactivity. As a result of detailed investigations in Mainz [14,15,16] – mostly by dedicated experiments – the systematic corrections became much better understood and their uncertainties were reduced significantly.

The high-statistics Mainz data from 1998-2001 allowed the first determination of the probability of the so-called neighbour excitation[†] to occur in $(5 \pm 1.6 \pm 2.2)$ % of all β decays[16] in good agreement with the theoretical expectation[17].

The most sensitive analysis on the neutrino mass, in which only the last $70\,\mathrm{eV}$ of the β spectrum below the endpoint are used, resulted in the following for Mainz 1998, 1999 and 2001 data[16]

$$m^2(\nu_e) = (-0.6 \pm 2.2 \pm 2.1)\ \mathrm{eV}^2/\mathrm{c}^4 \quad (3)$$

which corresponds to an upper limit of

$$m(\nu_e) < 2.3\ \mathrm{eV}/\mathrm{c}^2 \quad (95\ \%\ \mathrm{C.L.}) \quad (4)$$

This is the lowest model-independent upper limit of the neutrino mass obtained thus far.

2.2 The Troitsk Neutrino Mass Experiment

The windowless gaseous tritium source of the Troitsk experiment [18] is essentially a tube of 5 cm diameter filled with T_2 resulting in a column density of 10^{17} molecules/cm^2. The source is connected to the ultrahigh vacuum of the spectrometer by a series a differential pumping stations.

From its first data taking in 1994 the Troitsk experiment reports an anomalous excess in the experimental β spectrum as a sharp step of the count rate at a varying position of a few eV below the endpoint of the β spectrum E_0 [18], which seems to be an experimental artefact appearing with varying intensity at the Troitsk setup. The Troitsk experiment is correcting for this anomaly by fitting an additional line to the β spectrum run-by-run.

[†]The sudden change of the nuclear charge during β decay results to a certain percentage in the excitation of neighbouring molecules.

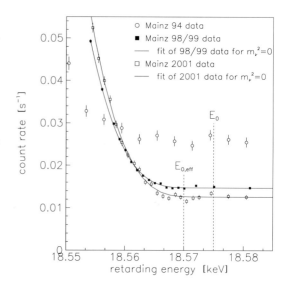

Figure 3. Averaged count rate of the 1998/1999 data filled squares) with fit (line) and of the 2001 data (open squares) with fit (line) in comparison with previous Mainz data from 1994 (open circles) as a function of the retarding energy near the endpoint E_0 and effective endpoint $E_{0,eff}$ (accounting for the width of response function of the setup and the mean rotation-vibration excitation energy of the electronic ground state of the ^{3}HeT$^+$ daughter molecule)[16].

Combining the 2001 results with the previous ones since 1994 gives[19]

$$m^2(\nu_e) = (-2.3 \pm 2.5 \pm 2.0)\ \mathrm{eV}^2/\mathrm{c}^4 \quad (5)$$

from which the Troitsk group deduce an upper limit

$$m(\nu_e) < 2.05\ \mathrm{eV}/\mathrm{c}^2 \quad (95\ \%\ \mathrm{C.L.}) \quad (6)$$

The values of eq. (5) and (6) do not include the systematic uncertainty which is needed to account for, when the timely-varying anomalous excess count rate at Troitsk is described run-by-run with an additional line.

3 Rhenium β decay experiments

Due to the complicated electronic structure of ^{187}Re and its primordial half life the advantage of the 7 times lower endpoint energy

E_0 of ^{187}Re with respect to tritium can only be exploited if the β spectrometer measures the entire released energy, except that of the neutrino. This situation can be realized by using a cryogenic bolometer as the β spectrometer, which at the same time contains the β emitter ^{187}Re.

One disadvantage connected to this method is the fact that one measures always the entire β spectrum. Even for the case of the very low endpoint energy of ^{187}Re, the relative fraction of events in the last eV below E_0 is of order 10^{-10} only (compare to figure 1). Considering the long time constant of the signal of a cryogenic bolometer (typically several hundred μs) only large arrays of cryogenic bolometers could deliver the signal rate needed.

Two groups are working on ^{187}Re β decay experiments at Milan (MiBeta) and Genoa (MANU2) using $AgReO_4$ and metallic rhenium, respectively. Although cryogenic bolometers with an energy resolution of 5 eV have been produced with other absorbers, this resolution has yet not been achieved with rhenium. The lowest neutrino mass limit of $m(\nu_e) < 15$ eV/c^2 comes from MiBeta[20]. Further improvements in the energy resolution and the number of crystals are envisaged aiming for a sensitivity of a few eV/c^2.

4 The KATRIN experiment

The very important tasks presented in the introduction – to distinguish hierarchical from quasi-degenerate neutrino mass scenarios and to check the cosmological relevance of neutrino dark matter for the evolution of the universe – require the improvement of the direct neutrino mass search by one order of magnitude at least.

The KATRIN collaboration has taken this challenge and is currently setting up an ultra-sensitive tritium β decay experiment based on the successful MAC-E-Filter spectrometer technique and a very

strong Windowless Gaseous Tritium Source (WGTS)[21,22] at the Forschungszentrum Karlsruhe, Germany. The international KATRIN collaboration consists of groups from the Czech Republic, Germany, Russia, UK and US, which combines the worldwide expertise in tritium β decay, groups providing special knowledge, with the strength and the possibilities of a big national laboratory including Europeans biggest tritium laboratory. Figure 4 shows a schematic view of the proposed experimental configuration.

The WGTS (see fig. 5) allows for the measurement of the endpoint region of the tritium β decay and consequently the determination of the neutrino mass with a maximum of signal strength combined with a minimum of systematic uncertainties from the tritium source. The WGTS consists of a 10 m long cylindrical tube of 90 mm diameter filled with molecular tritium gas of high isotopic purity (> 95 %). The tritium gas will be continuously injected by a capillary at the middle with a rate of about 4.7 Ci/s and pumped out by a series of differential turbo molecular pump stations at both ends giving rise to a density profile over the source length of nearly triangular shape with a total column density of $5 \cdot 10^{17}$/cm^2 providing a count rate about a factor 100 larger than in Mainz and Troitsk. The β electrons are leaving the WGTS directly to both ends following the magnetic field lines within a solenoidal magnetic field of 3.6 T whereas the pumped-out tritium gas is then purified by a palladium membrane filter and re-circulated. About one percent of the pumped-out gas is given to a tritium recovery and isotopic separation system. Of special importance is the control of the column density on the 1 per mil level by regulating the pressure in the tritium supply buffer vessel, the temperature of the WGTS tube and the isotopic composition of the gas. With the "Test of Inner LOop" setup TILO the possibilities to achieve this stability goal are being checked

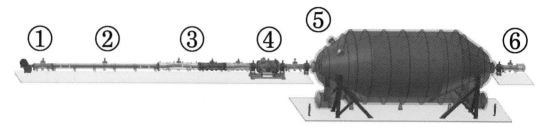

Figure 4. Schematic view of the KATRIN experiment with the rear monitoring and calibration system (1), the windowless gaseous tritium source (WGTS) (2), the differential and cryopumping electron transport section (3), the pre spectrometer (4), the main spectrometer (5) and the electron detector array (6). The main spectrometer has a length of 24 m, a diameter of 10 m, the overall length over the experimental setup amounts to about 70 m. Not shown is the monitor spectrometer.

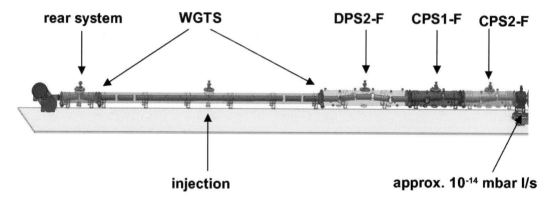

Figure 5. Schematic view of the KATRIN windowless gaseous tritium source (WGTS) and the differential (DPS2-F) and cryopumping (CPS1-F, CPS2-F) electron transport section. The calibration and the monitoring system (read system) is located further upstream of the WGTS. The maximum tritium leakage to the pre spectrometer amounts to 10^{-14} mbarl.

at the Forschungszentrum Karlsruhe, yielding the required sub-permil stability over many hours. To allow a very stable and low WGTS temperature at about 27 K the WGTS tube will be placed inside a pressure-stabilised LNe cryostat. The isotopic composition will be continuously monitored with the help of Laser-Raman spectroscopy provided by the Swansea group. Calculations and simulations by KATRIN and experiments at Troitsk investigate the possible electrical potentials within the WGTS due to the plasma. It seems that this effect could be avoided if the magnetic flux tube is sufficiently grounded at the rear side. The WGTS setup is under construction at a company.

The electron transport system adiabatically guides β decay electrons from the tritium source to the spectrometer by a system of superconducting solenoids at a magnetic field of 5.6 T. At the same time it is eliminating any tritium flow towards the spectrometer by a differential pumping system consisting of 1 m long tubes inside the magnets alternated by pump ports with turbo molecular pumps yielding a tritium reduction factor of about a factor of 10^7 according to simulations by the ASTeC vacuum group. This first part of the tritium eliminating system is currently being fabricated by a company. To reduce the molecular beaming effect, the direct line-of-sight is blocked by 20 degree bents between each pair of superconducting magnets of 1 m length (see fig. 5). In the second part the surfaces of the liquid helium cold vacuum tube act as a cryotrapping section to suppress the tritium partial pressure further to an insignificant level. With the dedicated TRAP experiment at Forschungszentrum Karlsruhe the cryosorption of tritium molecules at LHe cold surfaces which are covered by Argon frost is being checked. A mockup of two tubes of the cryopumping section tritium was build yielding no measurable penetration of D_2 molecules. After the Lepton Photon 2005 conference the TRAP experiment was repeated with tritium gas. Again, no tritium gas was detected at the other end of the system even after many days. This is an important result, as it was not clear whether the radioactivity of tritium and the corresponding local heating due to a decay would cause a slow migration of tritium through the system.

Between the tritium source and the main spectrometer a pre-spectrometer of MAC-E-Filter type will be installed (see fig. 6). It acts as an electron pre-filter running at a retarding energy about 200 eV below the endpoint of the β spectrum to reject all β electrons except the very high energetic ones in the region of interest close to the endpoint E_0. This minimises the chances that β electrons cause background in the main spectrometer by ionisation of residual gas.

A key component of the new experiment will be the large electrostatic main spectrometer with a diameter of 10 m and an overall length of about 23 m. This high-resolution MAC-E-Filter will allow to scan the tritium β decay endpoint at a resolution of $\Delta E = 0.93$ eV, which is – at a much higher luminosity – a factor of 4-5 better than the MAC-E-Filters in Mainz and Troitsk.

Although limiting the electron input rate by the pre-spectrometer, stringent vacuum conditions have to be fulfilled to suppress background from the main spectrometer. Special selection of materials as well as surface cleaning and out-baking at 350 °C will allow to reach a residual gas pressure of better than 10^{-11} mbar. To reduce the size of

Figure 6. Schematic view of the KATRIN pre spectrometer consisting of spectrometer vessel and two superconduction solenoids. The spectrometer vessel is set on high voltage and acts as retarding electrode. The inner electrode system is built in the central part as a nearly massless wire structure put on a little bit more negative potential than the vacuum vessel to reject secondary electrons from the walls (see later in the text).

surfaces inside the vacuum chamber, not a complex solid electrode system will be installed, but the vacuum vessel itself will be put on high voltage and thus will create the electric retarding potential.

A new idea of strong background suppression has be developed and successfully tested at the Mainz spectrometer. It will be applied also to the KATRIN spectrometers: the vessel walls at high potential will be covered by a system of nearly massless wire electrodes, which are put to a slightly more negative potential. Secondary electrons ejected by cosmic rays or environmental radioactivity from the vessel wall, will thus be repelled and prohibited from entering the magnetic flux tube which is connected to the detector. This new method of strong background reduction resulted at the Mainz spectrometer in a factor 10 reduced background rate. To achieve an even higher suppression factor the KATRIN main spectrometer will be instrumented by a two-layer wire electrode system being under construction at Münster. Another advantage of such an inner wire electrode is, that the retarding voltage can be stabilised more easily.

These new ideas and other new technical solutions will be applied also to the KATRIN pre-spectrometer, which has already being set up at the Forschungszentrum Karlsruhe. The vacuum tests with the pre-spectrometer have been successfully finished yielding at a temperature of -20 °C a final pressure of less than 10^{-11} mbar and an outgasing rate of less than 10^{-13} mbar l/s cm^2. Both values are better than the KATRIN requirements showing that the material, the pumping scheme and the surface treatments are right. A wire electrode system for background reduction – build at Seattle – has been installed in the KATRIN pre-spectrometer. Becoming instrumented with an scanning electron gun and a 64-pixel silicon PIN-detector the electromagnetic and background properties of the KATRIN pre-spectrometer will be investigated soon.

The final KATRIN detector requires high efficiency for electrons at $E_0 = 18.6$ keV and low γ background. A high energy resolution of $\Delta E < 600$ eV for 18.6 keV electrons should suppress background events at different energies. The present concept of the detector developed at Seattle and Karlsruhe is based on a large array of about 400 PIN photodiodes surrounded by low-level passive shielding and an active veto counter to reduce background. A possible post-acceleration of the β electrons to about 50 keV will reduce the background rate around the signal line further

After publishing the Letter of Intent[21] the KATRIN collaboration has done significant work to increase the sensitivity of the experiment. The major improvements of the setup to increase the statistics are the design of a tritium re-circulating and purification system providing a near to maximum tritium purity of > 95 %, the increase of the diameter of the windowless tritium source from 75 mm to 90 mm and, correspondingly, of the diameter of the main spectrometer from 7 m to 10 m. Additionally an optimisation of the measurement point distribution around the endpoint has been performed.

To reduce the systematic uncertainties the instrumental improvements have been developed as well as plans for dedicated experiments and their analysis in order to determine systematic corrections have been worked out. The main systematic uncertainties comprise the inelastic scattering within the tritium source and the stability of the retarding voltage of the main spectrometer. The former will be determined and repeatedly monitored with the help of a high-precision electron gun injecting electrons from the rear system. For the latter, a dedicated high-precision high voltage divider with a precision in the ppm range has been developed with the support of the Physikalisch Technische Bundesanstalt at Braunschweig, Germany (see fig. 7) and it is being tested.

For redundancy, the retarding high voltage of the main spectrometer is applied in parallel to a third spectrometer, the monitor spectrometer[‡], which continuously measures a sharp electron line. Different energetically well-defined sources are in preparation by the Rez, Münster and Bonn groups, e.g. a photoelectron source consisting of a cobalt foil irradiated by γs from ^{241}Am, or a condensed ^{83m}Kr conversion electron source. Further systematic uncertainties are the electrical potential distribution within the WGTS, which will be checked by running the WGTS in a second "high temperature regime" of 120-150 K with the conversion electron emitter ^{83m}Kr added to the gaseous molecular tritium and the source contamination by other hydrogen isotopes than tritium, which will be monitored with the help of laser Raman spectroscopy.

The detailed simulations of the KATRIN

[‡]The Mainz spectrometer will be modified for this purpose into a high-resolution spectrometer with $\Delta E \approx 1$ eV.

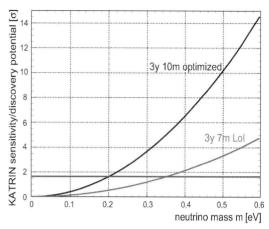

Figure 8. KATRIN's discovery potential or sensitivity in units of the total uncertainty σ for a 3 years measurement as function of the neutrino mass for the improved KATRIN setup with the 10 m spectrometer (black) [22] and the previous KATRIN design with the 7 m spectrometer from the KATRIN letter of intent [21] (gray). For a given neutrino mass the y-axis shows the difference to zero of the measured neutrino mass value in units of the total standard deviation σ, The horizontal line shows the upper limit with 90 % C.L. in case that no neutrino mass is found.

Figure 7. Precision high-voltage divider developed at Münster with the help of the Physikalisch Technische Bundesanstalt Braunschweig/Germany. Shown is the maintenance position. A helix of 100 high precision resistors form the primary divider. A secondary divider gives the electrical potential to the field-shaping copper electrodes, whereas a third capacitive divider protects the divider in case of voltage break-downs. In operational mode, the whole setup up is placed in a temperature controlled vessel filled with dry nitrogen.

experiment yield the following (see fig. 8): A sensitivity of 0.20 eV/c^2 will be achieved with the KATRIN experiment after 3 years of pure data taking. Statistical and systematic uncertainties contribute about equally. This value of 0.20 eV/c^2 corresponds to an upper limit with 90 % C.L. in the case that no neutrino mass will be observed. To the contrary, a non-zero neutrino mass of 0.30 eV/c^2 would be detected with 3 σ significance, a mass of 0.35 eV/c^2 even with 5 σ.

The design of the experiment is nearly finished and a detailed description was documented in[22]. Some parts (e.g. the prespectrometer) have been setup already. Four major components have been ordered, the windowless gaseous tritium source, the differential pumping system, the main spectrometer vessel and the helium liquefier. Many dedicated test experiments are being performed at different places to investigate the inner tritium loop, the cryotrapping, methods to improve the vacuum conditions, new background suppression methods, calibration sources, detector and data acquisition, etc. The ground-breaking of the new KATRIN halls at the Forschungszentrum Karlsruhe has been celebrated in late summer 2005. The full setup of the KATRIN experiment will be finished and commissioned in 2009.

5 Conclusions

Neutrino oscillation experiments have already pointed to new physics beyond the Standard Model by proving that neutrinos mix and that they have non-zero mass. The next goal is to determine the absolute scale of the neutrino mass due to its high importance for particle physics, astrophysics and cosmology.

Among various ways to address the absolute neutrino mass scale the investigation of the shape of β decay spectra around the endpoint is the only model-independent method. Secondly this direct method is complementary to the search for the neutrinoless double β decay and to the information from astrophysics and cosmology.

The investigation of the endpoint spectrum of the tritium β decay is still the most sensitive direct method. The tritium β decay experiments at Mainz and Troitsk have been finished yielding upper limits of about 2 eV/c^2. The new KATRIN experiment is being set up at the Forschungszentrum Karlsruhe by an international collaboration. KATRIN will enhance the sensitivity on the neutrino mass further by one order of magnitude down to 0.2 eV/c^2.

Acknowledgments

The work by the author for the KATRIN experiment is supported by the German Bundesministerium für Bildung und Forschung and within the virtual institute VIDMAN by the Helmholtz Gemeinschaft.

References

1. Y. Suzuki *these proceedings*, A. Poon *these proceedings*
2. *e.g.* R. N. Mohapatra *et al.*, arXiv:hep-ph/0510213
3. O. Cremonesi, *these proceedings*
4. M. Tegmark, arXiv:hep-ph/0503257
5. S. W. Allen *et al.*, arXiv:astro-ph/0303076
6. S. Hannestad, arXiv:astro-ph/0505551
7. J. F. Beacom *et al.*, arXiv:astro-ph/0404585
8. Ch. Weinheimer, ch, 2 of "Massive Neutrinos", ed. G. Altarelli and K. Winter, *Springer Tracts in Modern Physics*, Springer, p25-52 (2003)
9. K V. Klapdor-Kleingrothaus *et al.*, *Phys. Lett. B* **586**, 198 (2004)
10. S. Eidelman *et al.* (Particle Data Group), *Phys. Lett. B* **592**, 1 (2004)
11. A. Picard *et al.*, *Nucl. Instr. Meth. B* **63**, 345 (1992)
12. V M. Lobashev *et al.*, *Nucl. Inst. and Meth.* **A240**, 305 (1985) V. Lobashev *et al.*,
13. L. Fleischmann *et al.*, *Eur. Phys. J. B* **16**, 521 (2000)
14. V. N. Aseev *et al.*, *Eur. Phys. J. D* **10**, 39 (2000)
15. B. Bornschein *et al.*, *J. Low Temp. Phys.* **131**, 69 (2003)
16. C. Kraus *et al.*, *Eur. Phys. J. C* **40**, 447 (2005)
17. W. Kolos *et al.*, *Phys. Rev. A* **37**, 2297 (1988)
18. V. M. Lobashev *et al.*, *Phys. Lett. B* **460**, 227 (1999)
19. V. M. Lobashev , *Nucl. Phys. A* **719**, 153c (2003)
20. M. Sisti *et al.*, *Nucl. Instr. Meth. A* **520**, 125 (2004)
21. A. Osipowicz *et al.*, arXiv:hep-ex/0109033
22. J. Angrik *et al.*, FZK Scientific Report 7090

DISCUSSION

Louis Lyon (Oxford University):

Is it known why many previous experiments gave a negative mass squared for the electron neutrino? Was it just due to omitting some systematic effect on the resolution?

Christian Weinheimer: Today we understand the problems of the Mainz and of the Troitsk experiment in the early nineties: Both experiments observed a trend towards negative values of $m^2(\nu_e)$ for larger fitting intervals (*e.g.* 100 - 500 eV below the endpoint) due to additional energy losses which were not taken into account in the data analysis. At Mainz it was the roughening transition of the tritium films, which was avoided by lower tritium film temperatures since 1997 as this effect follows an Arrhenius law. At Troitsk electrons, which were trapped in the magnetic bottle-like configuration of the tritium source and which escaped through large-angle scattering were neglected in the first analysis. Since any underestimated or not-accounted broadening of the measured β spectrum with Gaussian width σ results in a negative shift of $m^2(\nu_e)$ according to $\Delta m^2 = -2\sigma^2$ (R.G.H. Robertson, D.A. Knapp, Ann. Rev. Nucl. Sci. 38, 185 c(1988)), it is rather likely that the other experiments observing negative values of $m^2(\nu_e)$ has overlooked an energy loss-like process.

DOUBLE BETA DECAY: EXPERIMENT AND THEORY

OLIVIERO CREMONESI

INFN Sezione di Milano, Via Celoria 16, I-20133 Milano
and Universitá di Milano-Bicocca, Piazza della Scienza 3, I-20126 Milano
E-mail: oliviero.cremonesi@mib.infn.it

The present status of experiments searching for neutrinoless double-beta decay ($\beta\beta(0\nu)$) is reviewed and the results of the most sensitive experiments discussed. Phenomenological aspects of $\beta\beta(0\nu)$ are introduced and most relevant aspects for neutrino physics discussed. Given the observation of neutrino oscillations and the present knowledge of neutrino masses and mixing parameters, a possibility to observe $\beta\beta(0\nu)$ at a neutrino mass scale $m_\nu \approx$10-50 meV could actually exist. The achievement of the required experimental sensitivity is a real challenge faced by a number of new proposed projects. A review of the most relevant of them is given.

1 Introduction

The existence of neutrino oscillations, well proved during the last years by the results of the neutrino oscillation experiments, has convincingly shown that neutrinos have a finite mass and mixing. While a number of new oscillation experiments aiming at improving our present knowledge of neutrino mass and mixing paramenters has been proposed and are currently under preparation, some relevant neutrino properties are inaccessible to them. In fact, oscillation experiments can only measure the differences of the neutrino masses squared ($|\Delta m_{ij}^2|$) and, while Δm_{ij}^2 sign could be in principle measured through oscillation matter effects, only experiments sensitive to a linear combination of the neutrino masses have a reasonable probability to measure the absolute neutrino mass scale in a near future. This is the case for kinematic measurements of the β spectrum end-point, $\beta\beta(0\nu)$ and cosmological measurements which are respectively sensitive to

$$\langle m_\beta \rangle \equiv \sqrt{\sum_{k=1}^{3} |U_{ek}^L|^2 m_k^2} \qquad (1)$$

$$\langle m_{\beta\beta} \rangle \equiv \sum_{k=1}^{3} |U_{ek}^L|^2 m_k e^{i\phi_k} \qquad (2)$$

$$\langle m_{COSM} \rangle \equiv \sum_{k=1}^{3} m_k \qquad (3)$$

where m$_k$ are the neutrino mass eigenvalues and U_{ek}^L are the elements of the first line of the neutrino mixing matrix ($c_{ij} \equiv \cos(\theta_{ij})$, $s_{ij} \equiv \sin(\theta_{ij})$)

$$U^L = U_{atm} \times U_{cross} \times U_{solar} \times U_{Maj} =$$
$$\begin{pmatrix} 1 & 0 & 0 \\ 0 & c_{23} & s_{23} \\ 0 & -s_{23} & c_{23} \end{pmatrix} \times \begin{pmatrix} c_{13} & 0 & s_{13}e^{i\delta} \\ 0 & 1 & 0 \\ -s_{13}e^{i\delta} & 0 & c_{13} \end{pmatrix} \times$$
$$\begin{pmatrix} c_{12} & s_{12} & 0 \\ -s_{12} & c_{12} & 0 \\ 0 & 0 & 1 \end{pmatrix} \times \begin{pmatrix} e^{i\alpha_1/2} & 0 & 0 \\ 0 & e^{i\alpha_2/2} & 0 \\ 0 & 0 & 1 \end{pmatrix} (4)$$

Two possible hierarchies are then implied by current available data: the normal (m$_1 \approx$m$_2$ <<m$_3$) and the inverted hierarchy (m$_3$ <<m$_1 \approx$m$_2$) while nothing can be yet inferred about the lowest mass eigenvalue.

A similar ignorance holds also for the neutrino nature (Dirac/Majorana). In the SM neutrinos are Dirac particles by construction (i.e. in order to conserve lepton numeber L). In the limit of vanishing masses however lepton number conservation can be equivalently stated in terms of neutrino helicity properties (neutrinos of a given helicity are always coupled to leptons of the same sign) and the Majorana or Dirac descriptions for neutrino are equivalent (i.e. don't change

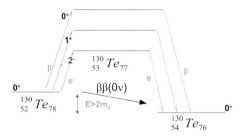

Figure 1. Scheme of a $\beta\beta$ transition.

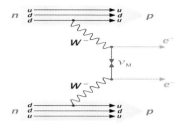

Figure 2. Simplified scheme of the $\beta\beta$ transitions.

the physical content of the theory). For finite neutrino masses however the two descriptions are no more equivalent and can give rise to different physical scenarios (e.g. mass generation mechanisms)[1]. $\beta\beta(0\nu)$ is the most sensitive tool to establish the nature of the neutrino. It can proceed in fact only if neutrinos are Majorana massive particles and can test this hypothesis with extreme accuracy.

The measurement of neutrino masses, mixing angles and phases, as well as the assessment of the Dirac/Majorana character of neutrinos are going to be considered primary goals of the next generation experiments. After the discovery of neutrino oscillations which provided us with the first clear evidence of phenomena beyond the reach of Standard Model (SM), these items represent in fact a unique tool to see what new Physics lies beyond SM predictions. In this framework $\beta\beta(0\nu)$ searches play a unique role giving the possibility to probe the Majorana character of neutrinos while obtaining informations on the neutrino mass hierarchy and Majorana phases with unique sensitivity.

The sum of the masses of the neutrinos of the three flavors is presently constrained to values from 0.7 to 1.7 eV from the Wilkinson microwave anisotropy probe full sky microwave map, together with the survey of the 2dF galaxy redshift [2,3,4,5,6]. A claim for a nonzero value of 0.64 eV has also been proposed[7]. Although these values are more constraining than upper limits of 2.2 eV for m_ν obtained so far in experiments on single-beta decay, they are strongly model dependent and therefore less robust with respect to laboratory measurements. On the other hand the best sensitivity expected for next generation single-beta decay experiments is of the order of ~ 0.2 eV (KATRIN[8] and Re μbolometers[9]). Present, and near future $\beta\beta(0\nu)$ experiments could reach a sensitivity to span the $\langle m_\nu \rangle$ region 0.1-1 eV. Different plans to overcome the challenge implied by the achievement of such a sensitivity, thus probing the inverse mass hierarchy, have been proposed. They will be reviewed in the section devoted to the future projects.

Unfortunately, uncertainties in the transition nuclear matrix elements still affect the interpretation of the experimental results and new efforts to overcome such a problem are strongly required (new theoretical calculations and experimental analyses of different $\beta\beta(0\nu)$ active isotopes). Complementary informations from beta experiments and astrophysics will be also crucial to solve this problem.

Several review articles covering both the experimental and the theoretical aspects and implications of $\beta\beta$ have been recently issued [10,11,12,13,14,15,16]. We will refer to them for more details on the subject.

2 Double Beta Decay

Double Beta Decay (DBD) is a rare spontaneous nuclear transition in which the charge of two isobaric nuclei changes by two units with the simultaneous emission of two elec-

312

trons. The parent nucleus must be less bound than the daughter one, while it is generally required that both be more bound than the intermediate one, in order to avoid (or at least inhibit) the occurrence of the sequence of two single beta decays (Fig.1). These conditions are fulfilled in nature for a number of even-even nuclei. The decay can then proceed both to the ground state or to the first excited states of the daughter nucleus. Nuclear transitions accompanied by positron emission or electron capture processes are also possible. They are however characterized by poorer experimental sensitivities and will not be discussed in the following. Several $\beta\beta$ modes are possible. The most popular are the 2ν mode

$$\,^A_Z X \to\,^A_{Z+2} X + 2e^- + 2\overline{\nu} \qquad (5)$$

which observes the lepton number conservation and it is allowed in the framework of the Standard Model (SM) of electro-weak interactions, and the 0ν mode

$$\,^A_Z X \to\,^A_{Z+2} X + 2e^- \qquad (6)$$

which violates the lepton number and has been recognized since a long time as a powerful tool to test neutrino properties[17]. A third decay mode in which one or more light neutral bosons χ (Majorons)) are also emitted

$$\,^A_Z X \to\,^A_{Z+2} X + 2e^- + N\chi \qquad (7)$$

is often considered. $\beta\beta(0\nu, \chi)$ requires the existence of a Majoron, a massless Goldstone boson that arises upon a global breakdown of B–L symmetry, where B and L are the baryon and the lepton number respectively. Interest in the Majoron is due to its possible signicant role in the history of the early Universe and in the evolution of stars. The model of a triplet Majoron[18] was disproved in 1989 by the LEP results on the decay width of the Z0 boson. Some new models were however recently proposed[19,20], where $\beta\beta(0\nu, \chi)$ decay is possible withouth any contradictions with the LEP data. DBD models involving the emission of two Majorons were also

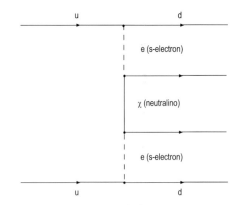

Figure 3. Example of a $\beta\beta(0\nu)$ suspersymmetric diagram involving the exchange of a neutralino.

proposed[21,22]. Due to the variety of proposed models with the term *Majoron*, one generally means today a massless or light boson associated with neutrino.

In the following, we will focus our attention only to the neutrinoless DBD mode. In all envisaged modes, $\beta\beta(0\nu)$ is a very rare transition, hence characterized by very long lifetimes. Besides the exchange of light or heavy Majorana neutrinos (fig. 2), $\beta\beta(0\nu)$ can be mediated by the exchange of a variety of unconventional particles (e.g. SUSY partners, fig 3).

Actually $\beta\beta(0\nu)$ can be meant as a sort of *black box* involving the emission of two electrons by a two–nucleon system with just a number of constraints on the process observables (electron sum energy, single electron energies, single electron energies and decay rate decay). Many models beyond SM with lepton number violation can therefore contribute (e.g. left-right symmetric models, supersymmetric models with or withouth R-parity conservation) and $\beta\beta(0\nu)$ translates in a sensitive tool to obtain constraints on the relevant parameters of these models. Present experimental sensitivities on supersymmetric model predictions are actually lower than those of the LHC experiments. This will not be however true for next generation $\beta\beta(0\nu)$ experiments which will therefore represent a powerful tool after the LHC era.

Independently of the actual mechanism mediating the decay, $\beta\beta(0\nu)$ observation would necessarily imply that neutrinos are Majorana massive particles and would moreover allow to fix the absolute neutrino mass scale. Lacking, however, any evidence for $\beta\beta(0\nu)$, experimental lower limits on the decay lifetimes can only translate into independent limits on each of the possible contributions to the decay amplitude. Disregarding more unconventional contributions (SUSY or left-right symmetric models), the $\beta\beta(0\nu)$ rate is usually expressed as

$$[T_{1/2}^{0\nu}]^{-1} = G^{0\nu}|M^{0\nu}|^2|\langle m_\nu\rangle|^2 \qquad (8)$$

where $G^{0\nu}$ is the (exactly calculable) phase space integral, $|M^{0\nu}|^2$ is the nuclear matrix element and $\langle m_\nu\rangle$ is the already mentioned *effective neutrino mass* parameter (eq. 1) measured in $\beta\beta(0\nu)$ which, in the limit of small neutrino masses becomes

$$\langle m_\nu\rangle = c_{12}^2 c_{13}^2 m_1 + s_{12}^2 c_{13}^2 e^{i\alpha_1} m_2 + s_{13}^2 e^{i\alpha_2} m_3 \qquad (9)$$

The presence of the phases α_k implies that cancellations are possible. Being equivalent to two degenerate Majorana neutrinos with opposite CP phase, such cancellations are complete for a Dirac neutrino. This stresses once more the fact that $\beta\beta(0\nu)$ can occur only through the exchange of Majorana neutrinos.

It should be stressed that $\beta\beta(0\nu)$ represents the only possibility to measure the neutrino Majorana phases. Predictions on $\langle m_\nu\rangle$ based on the most recent neutrino oscillation results have been recently derived by various authors[14,23]. The most striking aspect of such predictions for the $\beta\beta(0\nu)$ community is that, finally, a definite goal (other than verifying L conservation) exists: an experimental sensitivity in the range $\langle m_\nu\rangle \sim$10-50 meV could definitely rule out inverse and quasi-degenerate hierarchies thus assessing a direct neutrino mass hierarchy. $\beta\beta(0\nu)$ observation at larger $\langle m_\nu\rangle$ scales would be equally impor-

tant (expected sensitivities are better than for complementary kinematic and cosmological measurements) but its occurrence is based on more optimistic assumptions (degenerate hierarchy is still considered somehow unnatural).

As it is apparent from eq. (8) the derivation of the crucial parameter $\langle m_\nu\rangle$ from the experimental results on $\beta\beta(0\nu)$ lifetime requires a precise knowledge of the transition Nuclear Matrix Elements (NME). Unfortunately this is not an easy job and a definite knowledge of NME values and uncertainties is still lacking in spite of the large attention attracted by this area of research. Many, often conflicting evaluations are available in the literature and it is unfortunately not so easy to judge their correctness or accuracy. A popular even if doubtful attitude consists in considering the spread in the different evaluations as an estimate of their uncertainties. In such a way one obtains a spread of about one order of magnitude in the calculated half lifetimes (Tab. 1) which corresponds to the factor of $\sim$3 in $\langle m_\nu\rangle$ generally used in the interpretation of $\beta\beta(0\nu)$ experimental results.

Outstanding progress has been achieved over the last years mainly due to the application of the Quasi Random Phase Approximation (QRPA) method and its extensions. In spite of the still observed spread in the calculation results, the QRPA method seems presently the only possibility for heavy nuclei; we will refer to it in the following for the comparison of experimental results. Renewed interest in Shell Model calculations has been on the other hand boosted by the fast development of computer technologies even if the reliability of the method is still debated, especially for high Z nuclei. Alternative approaches (e.g. OEM) have also been pursued. Comparison with experimental $\beta\beta(2\nu)$ rates has often been suggested as a possible way out (direct test of the calculation method). The evaluation methods for the two decay modes show however relevant differences (e.g. the neutrino propagator) and the effectiveness of such a comparison is still controversial. Recently, encouraging results have been obtained in[24], where calculations obtained after fixing QRPA parameters on the $\beta\beta(2\nu)$ experimental results show a surprising stability. Discussions on the interpretation of this result are however still open and experimental $\beta\beta(2\nu)$ results are presently not available for all isotopes of interest for $\beta\beta(0\nu)$ experimental searches. It is clear that a big improvement in the calculation of NME or at least in the estimate of their uncertainties would be welcome. New calculation methods should be pursued while insisting on the comparison with dedicated measurements coming from various areas of nuclear physics[25]. On the other hand, an experimental effort to investigate as many as possible $\beta\beta$ emitters should be addressed. To this end it should be stressed that the $\beta\beta(0\nu)$ observation of one nucleus is likely to lead to searches and eventually to observation of the decay of other nuclei contributing to definitely solve the NME calculation precision problem[26].

3 Experimental approaches

Two main general approaches have been followed so far to investigate $\beta\beta$: i) indirect or inclusive methods, and ii) direct or counter methods. Inclusive methods are based on the measurement of anomalous concentrations of the daughter nuclei in properly selected samples, characterized by very long accumulation times. They include geochemical and radiochemical methods and, being completely insensitive to different $\beta\beta$ modes, can only give indirect evaluations of the $\beta\beta(0\nu)$ and $\beta\beta(2\nu)$ lifetimes. They have played a crucial role in $\beta\beta$ searches especially in the past.

Counter methods are based instead on the direct observation of the two electrons emitted in the decay. Different experimental parameters (energies, momenta, topology,

etc) can then be registered according to the different capabilities of the employed detectors. These methods are further classified in *passive* (when the observed electrons originate in an external sample) and *active* source experiments (when the source of $\beta\beta$'s serves also as detector). In most cases the various $\beta\beta$ modes are separated by the differences in the recorded electron sum energy: a continuous bell ditribution for $\beta\beta(2\nu)$ and $\beta\beta(0\nu, \chi)$, and a sharp line at the transition energy for $\beta\beta(0\nu)$. Direct counting experiments with very good energy resolution are presently the most attractive approach for $\beta\beta(0\nu)$ searches.

Experimental evidence for several $\beta\beta(2\nu)$ decays as well as improved lower limits on the lifetimes of many $\beta\beta(0\nu)$ emitters (Tab. 2) have been provided using the measured two-electron sum energy spectra, the single electron energy distributions and, in some cases, the event topology.

Various different conventional counters have been used in $\beta\beta$ direct searches: solid state devices (Germanium spectrometers and Silicon detector stacks), gas counters (time projection chambers, ionization and multiwire drift chambers) and scintillators (crystal scintillators and stacks of plastic scintillators). New techniques based on the use of low temperature *true* calorimeters have been on the other hand proposed and developed in order to improve the experimental sensitivity and enlarge the choice of suitable candidates for $\beta\beta$ searches investigable with an *active source* approach. A common feature of all $\beta\beta$ experiments has been the constant fight against backgrounds caused mainly by environmental radioactivity, cosmic radiation and residual radioactive contaminations of the detector setup elements. The further suppression of such backgrounds will be the actual challenge for future projects whose main goal will be to maximize $\beta\beta(0\nu)$ rate while minimizing background contributions.

In order to compare the performance of

Table 1. Theoretically evaluated $\beta\beta(0\nu)$ half-lives (units of 10^{28} years for $\langle m_\nu \rangle$ = 10 meV). The first two columns refer to Shell Model calulations while the other are based on the QRPA method.

Isotope	[27]	[28]	[29]	[30]	[31]	[32]	[33]	[24]
^{48}Ca	3.18	8.83	-	-	-	-	2.5	-
^{76}Ge	1.7	17.7	14.0	2.33	0.26–0.58	3.2	3.6	5.7
^{82}Se	0.58	2.4	5.6	0.6	-	0.8	1.5	1.7
^{100}Mo	-	-	1.0	1.28	-	0.3	3.9	3.4
^{116}Cd	-	-	-	0.48	-	0.78	4.7	2.1
^{130}Te	0.15	5.8	0.7	0.5	0.077–1.1	0.9	0.85	2.2
^{136}Xe	-	12.1	3.3	2.2	-	5.3	1.8	4.6
^{150}Nd	-	-	-	0.025	-	0.05	-	0.23
^{160}Gd	-	-	-	0.85	-	-	-	-

the different $\beta\beta$ experiments an experimental *sensitivity* or detector *factor of merit* is usually introduced. This is defined as the process half-life corresponding to the maximum signal n_B that could be hidden by the background fluctuations at a given statistical C.L. At 1σ level ($n_B = \sqrt{BTM\Delta}$), :

$$F_{0\nu} = \tau_{1/2}^{Back.Fluct.} = \ln 2\, N_{\beta\beta}\epsilon\frac{T}{n_B}$$

$$= \ln 2\frac{x\,\eta\,\epsilon\,N_A}{A}\sqrt{\frac{M\,T}{B\,\Delta}} \quad (68\%\text{C.L.})\,(10)$$

where B is the background level per unit mass and energy, M is the detector mass, T is the measure time, Δ is the FWHM energy resolution, $N_{\beta\beta}$ is the number of $\beta\beta$ decaying nuclei under observation, η their isotopic abundance, N_A the Avogadro number, A the compound molecular mass, x the number of $\beta\beta$ atoms per molecule, and ϵ the detection efficiency.

Despite its simplicity, equation (10) has the unique advantage of emphasizing the role of the essential experimental parameters: mass, measuring time, isotopic abundance, background level and detection efficiency. $F_{0\nu}$ can be thought as the inverse of the minimum rate which can be detected in a period T of measurement. When no counts are recorded in the relevant energy interval over a statistically significant period of time (a condition usually but incorrectly referred to as

"zero background case"), the term n_B in eq. 10 is constant (e.g. 2.3 at 90 % C.L.) and one gets a factor-of-merit which scales linearly with T and the detector mass:

$$F_{0\nu}^0 = \ln 2\, N_{\beta\beta}\frac{T}{2.3} \quad (90\%\text{C.L.})\,. \quad (11)$$

Using eq. 8 one can then easily obtain the experimental sensitivity on $\langle m_\nu \rangle$

$$F_{\langle m_\nu \rangle} = \sqrt{\frac{1}{F_{0\nu}^{Exp}G_{0\nu}|M_{0\nu}|^2}} \quad (12)$$

$$= \left[\frac{A}{x\eta\epsilon N_A G^{0\nu}|M_{0\nu}|^2}\right]^{1/2}\left[\frac{B\Delta}{MT}\right]^{1/4}$$

$$F_{\langle m_\nu \rangle}^0 = \left[\frac{A}{x\eta\epsilon N_A G^{0\nu}|M_{0\nu}|^2}\right]^{1/2}\frac{1}{\sqrt{MT}} \quad (13)$$

It now evident that only the second generation experiments characterized by very large masses (possibly isotopically enriched), good energy resolutions and extremely low background levels will have an actual chance to reach the $\langle m_\nu \rangle$ region below 50 meV. The selection of favourable $\beta\beta$ nuclei and the use of special techniques to suppress background (e.g. topological informations) will of course help in reaching the goal. In particular, the effectiveness in reaching the estimated background levels will be the actual measure of a given experiment's chances. Extreme care will have to be dedicated to all possible background contributions including environmental radioactivity, cosmogenically and artifi-

316

cially induced activity, natural activity of the
setup materials and $\beta\beta(2\nu)$.

4 Present experiments

Impressive progress has been obtained dur-
ing the last years in improving $\beta\beta(0\nu)$ half-
life limits for a number of isotopes and in
systematically cataloging $\beta\beta(2\nu)$ rates (Tab.
2). The effort to cover as many as possi-
ble $\beta\beta$ nuclei thus allowing a diret check for
$\beta\beta(2\nu)$ NME elements is evident.

Optimal $\beta\beta(0\nu)$ sensitivities have been
reached in a series of experiments based on
the calorimetric approach. In particular, the
best limit on $\beta\beta(0\nu)$ still comes from the
Heidelberg-Moscow (HM) experiment[36] on
^{76}Ge, even if a similar sensitivity and result have been obtained also by the IGEX
experiment[38] (Tab. 2). In both cases a
large mass (several kg) of isotopically enriched Germanium diodes (86 %), was installed deep underground under heavy shields
for gamma and neutron environmental radiation. Extremely low background levels were
then achieved thanks to a careful selection
of the setup materials and further improved
by the use of pulse shape discrimination techniques. Taking into account the uncertainties
in the NME calculations, such experiments
indicate an upper limit in the range of 0.3-1
eV for $\langle m_\nu \rangle$.

In spite of the above mentioned limit
on the half-lifetime of ^{76}Ge $\beta\beta(0\nu)$ published by the HM collaboration in the sum-
mer 2001, a small subset of the same collaboration (KDHK in the following) surprised
the $\beta\beta$ community in January 2002 with a
first claim of evidence for $\beta\beta(0\nu)$ [47] with
$T_{1/2}^{0\nu} = 0.8 - 18.3 \times 10^{25}$ y (best value $T_{1/2}^{0\nu} =
1.5 \times 10^{25}$ y) corresponding to a $\langle m_\nu \rangle$ range
of 0.11-0.56 eV (best value 0.39 eV). The re-
sult was based on a re-analysis of the HM
data including: i) an automatic peak detection method; ii) identification of the found
lines; iii) narrowing of the fit interval to ex-

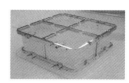

Figure 4. The CUORICINO tower and detail of two
modules.

clude any contribution from recognized lines.
The statistical significance of the result and
the interpretation of the structures observed
in the background spectrum were criticised in
a series of papers following the claim[48,13,49]
and were even contested by other members
of the HM collaboration[50]. A long debate
started[51] until the claim was confirmed 2
years later[52] after a more accurate reanalysis
of the collected data and the inclusion of additional statistics up to the data acquisizion
final date (May 2003). The final reported
result is $T_{1/2}^{0\nu} = 0.69 - 4.18 \times 10^{25}$ y (best
value $T_{1/2}^{0\nu} = 1.19 \times 10^{25}$ y) corresponding
to a $\langle m_\nu \rangle$ range of 0.24-0.58 eV (best value
0.44 eV). The new claim looks actually more
convincing but, due to the relevance of the
result, a more extensive substantiation and
review is needed. In particular the analysis
of the background has still some weak point
(e.g. not all observed lines have been identified) and a definitive validation is expected
by other currently running experiments (e.g.
CUORICINO) or the very sensitive next generation $\beta\beta(0\nu)$ projects.

The most severe limit of the calorimetric approach with conventional detectors is
the small number of $\beta\beta(0\nu)$ possible isotopes that can can be studied (e.g. ^{76}Ge,
^{136}Xe, ^{48}Ca). A solution to this problem,
was suggested[53] and developed[54] by the Mi-

Table 2. Best reported results on $\beta\beta$ processes. $\beta\beta(0\nu)$ limits are at 90% C.L. $\beta\beta(2\nu)$ results are averaged values, taken from ref.[34] except when explicitly noted. The effective neutrino mass ranges are obtained according to the QRPA calculations reported in Table 1.

Isotope	$T_{1/2}^{2\nu}$ (y)	$T_{1/2}^{0\nu}$ (y)	$\langle m_\nu \rangle$ (eV)
^{48}Ca	$4.3^{+2.1}_{-1.0} \times 10^{19}$	$> 6.8 \times 10^{21(35)}$	$< 19 - 36$
^{76}Ge	$1.42^{+0.09}_{-0.07} \times 10^{21}$	$> 1.9 \times 10^{25(36)}$	$< 0.1 - 0.9$
		$> 1.6 \times 10^{25(37,38)}$	$< 0.1 - 0.95$
^{82}Se	$(9.8 \pm 2.0 \pm 1.0) \times 10^{19(39)}$	$> 1.9 \times 10^{23}$ $^{(39)}$	$< 1.8 - 5.4$
^{96}Zr	$2.1^{+0.8}_{-0.4} \times 10^{19}$	$> 1 \times 10^{21(40)}$	
^{100}Mo	$(7.11 \pm 0.02 \pm 0.54) \times 10^{18(39)}$	$> 4.6 \times 10^{23(39)}$	$< 0.8 - 2.9$
^{116}Cd	$3.3^{+0.4}_{-0.3} \times 10^{19}$	$> 7 \times 10^{22(41)}$	$< 2.6 - 8.2$
^{128}Te	$(2.5 \pm 0.4) \times 10^{24}$	$> 7.7 \times 10^{24(42)}$	
^{130}Te	$(0.9 \pm 0.15) \times 10^{21}$	$> 1.8 \times 10^{24(43)}$	$< 0.2 - 1.1$
^{136}Xe	$> 8.1 \times 10^{20(44)}$	$> 1.2 \times 10^{23(45)}$	$< 1.2 - 2.1$
^{150}Nd	$(7.0 \pm 1.7) \times 10^{18}$	$> 1.2 \times 10^{21(46)}$	$< 4.6 - 13.8$

lano group. It is based on the use of low temperature calorimeters (*bolometers*) which, besides providing very good energy resolutions, can in practice avoid any constraint in the choice of the $\beta\beta$ emitter. Due to their very simple concept (a massive absorber in thermal contact with a suitable thermometer measuring the temperature increase following an energy deposition), they are in fact constrained by the only requirement of finding a compound allowing the growth of a diamagnetic and dielectric crystal. Extremely massive detectors can then be built, by assembling large crystal arrays. Thermal detectors have been pioneered by the Milano group for ^{130}Te (chosen, because of its favourable nuclear factor-of-merit and large natural isotopic abundance, within a large number of other successfully tested $\beta\beta$ emitters) in a series of constantly increasing mass experiments carried out at *Laboratori Nazionali del Gran Sasso* (LNGS, 3400 m.w.e.), whose last extension is CUORICINO, an array of 62 TeO$_2$ crystals arranged in a 13 storey tower (fig. 4), totalling a mass of 40.7 kg and operating at LNGS since 2003. Its operating temperature is ~12 mK, with an average energy resolution of ~7 keV at the 2615 line of ^{208}Tl and a very satisfactory detector reproducibil-

ity. Besides being a sensitive experiment on $\beta\beta(0\nu)$ of ^{130}Te (Tab. 3) CUORICINO is the most effective proof of the feasibility of low background large mass arrays of bolometers. No evidence for $\beta\beta(0\nu)$ of ^{130}Te has been reported up to now, with a lower limit on the half-lifetime of 1.8×10^{24} y (90% CL)[43]. This corresponds to a range 0.2–1.1 on $\langle m_\nu \rangle$, partially overlapping the KDHK claim.

The other most sensitive currently running $\beta\beta$ experiment is NEMO3 (fig. 5), installed in the Frejus underground laboratory[55] at a depth of ~ 4800 m.w.e. since 2003. It consists of a 3×3 meters cylindrical tracking (wire chambers filled with an ethyl-alcohol mixture, operated in the Geiger mode) and calorimetric (1940 plastic scintillators) system immersed in a weak magnetic field. 20 azimuthal sections can support a different source foils for a total of ~10 kg of enriched isotopes, currently dominated by 6.9 kg of ^{100}Mo and 0.9 kg of ^{82}Se. The combination of time of flight measurements, magnetic tracking and calorimetry allow precision characterization of double beta decays and rejection of backgrounds. Despite a relatively modest energy resolution, implying a non negligible background contribution from $\beta\beta(2\nu)$, the expected sensitivity on $\langle m_\nu \rangle$ is

318

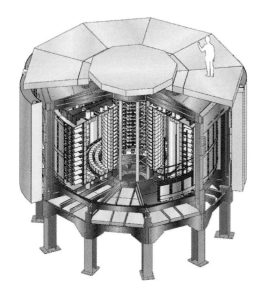

Figure 5. Scheme of the NEMO3 detector.

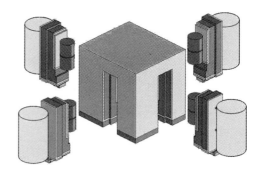

Figure 6. Scheme of four Majorana cryostats and their shielding.

of the order of 0.1 eV (Tab. 3) on the basis of an excellent control of the backgrounds. As for its previous versions, $\beta\beta(2\nu)$ of various $\beta\beta$ isotopes remains a primary goal. Relevant results on $\beta\beta(2\nu)$ and $\beta\beta(0\nu)$ of ^{100}Mo and ^{82}Se have been reported (Tab. 2).

5 Future Experiments

Most of the criteria to be considered when optimizing the design of a new $\beta\beta(0\nu)$ experiment follow directly from eq. 10: i) a well performing detector (e.g. good energy resolution and time stability) giving the maximum number of informations (e.g. electron energies and event topology); ii) a reliable and easy to operate detector technology requiring a minimum level of maintenance (long underground running times); iii) a very large (possibly isotopically enriched) mass, of the order of one ton or larger; iv) an effective background suppression strategy.

Unfortunately, these simple criteria are often conflicting, and simultaneous optimisation is rarely possible. As discussed above, the best results have been so far pursued

exploiting the calorimetric approach which characterizes also most of the future proposed projects.

Actually, a series of new proposals has been boosted by the renewed interest in $\beta\beta(0\nu)$ following neutrino oscillation discovery. They can be classified in three broad classes: i) dedicated experiments using a conventional detector technology with improved background suppression methods (e.g. GERDA, MAJORANA); ii) experiments using unconventional detector (e.g. CUORE) or background suppression (e.g. EXO, SuperNEMO) technologies; iii) experiments based on suitable modifications of an existing setup aiming at a different search (e.g. CAMEO, GEM).

Present and near future experiments expected sensitivities are compared in Tab.3. In some cases technical feasibility tests are requireed, but the crucial issue is still the capability of each project to pursue the expected background suppression. Although all proposed projects show interesting features for a second generation experiment, we will focus in the following only to the few experiments characterized by a reasonable technical feasibility within the next five years.

MAJORANA and GERDA are both phased programs representing large scale extensions of past successful experiments on ^{76}Ge $\beta\beta(0\nu)$. They propose to use large en-

Table 3. Expected 5 y sensitivities of present and near future (funded) experiments.

Experiment	Isotope	$F_{0\nu}^{5y}$ $(10^{26}$ y$)$	$\langle m_\nu \rangle$ (eV)
CUORICINO[56]	^{130}Te	0.09	0.09-0.5
NEMO3[39]	^{100}Mo	0.04	0.3-1
NEMO3[39]	^{82}Se	0.008	0.9-2.7
CUORE[56]	^{130}Te	2.1-6.5	0.01-0.1
GERDA I[57]	^{76}Ge	0.25	0.1-0.75
GERDA II[57]	^{76}Ge	2	0.04-0.26
MAJORANA-180[58]	^{76}Ge	1	0.05-0.37
EXO-200[59]	^{136}Xe	0.15	0.35-0.59
MOON-200[60]	^{100}Mo	0.3	0.1-0.36

riched (86% in ^{76}Ge) germanium crystals as both source and detector.

Evolved from the HM experiment, GERDA[57] aims at implementig the concept of Ge diodes immersed in a LN$_2$ bath[61] for a radical background suppression (mainly due to environmental and setup radioactivity). The goal of the first phase (consisting just of the HM and IGEX detectors) is a validation of the KHDK results, while the second (and possibly the third) phase will aim at a $\langle m_\nu \rangle$ sensitivity well below 100 meV (Tab. 3). GERDA has been approved by the German and Italian Scientific committees. The LN$_2$ cryostat and the water tank are under construction. Phase I data taking is expected for 2007 at LNGS while further 20 kg of isotopically enriched Ge diodes are presently being prepared for the phase II.

The proposed initial configuration of MAJORANA[58], a mainly USA proposal with important Canadian, Japanese, and Russian contributions, would consist of 171 segmented n-type germanium crystals (180 kg), distributed in 3 independent ultra-clean electro-formed conventional cryostats of 57 crystals each. The whole assembly would be enclosed in a low-background passive shield and active veto and be located deep underground (fig. 6). The full MAJORANA-180 design is an evolution of the IGEX exper-

iment and would be operational in seventh years after approval.

The main advantage of the Ge based proposals is the well understood performance of germanium detectors besides an excellent resolution of ~0.16% FWHM, essentially eliminating any contamination of the $\beta\beta(0\nu)$ signal by $\beta\beta(2\nu)$. Other backgrounds are reduced by using ultra-clean materials and techniques together with close packing of the crystals. The use of germanium crystals also allows further background suppression via pulse-shape discrimination and segmentation. The modular approach allows an easy scaling of these experiments to larger size with the only possible limitation of the cost.

CUORE[56] (*Cryogenic Underground Detector for Rare Events*) is a very large extension of the TeO$_2$ bolometric array concept pioneered by the Milano group at the Gran Sasso Laboratory since the eighties. CUORE would consist in a rather compact cylindrical structure of 988 cubic natural TeO$_2$ crystals of 5 cm side (750 g), arranged into 19 separated *towers* (13 *planes* of 4 crystals each) and operated at a temperature of 10 mK (fig. 7). The expected energy resolution is ~5 keV FWHM at the $\beta\beta(0\nu)$ transition energy (2.528 MeV). A background level lower than 0.01 c/keV/kg/y is expected by extrapolating the CUORICINO background results

Figure 7. Scheme of the CUORE detector.

and the dedicated CUORE R&D measurements to the CUORE setup. CUORE technical feasibility (good energy resolution and low background) is demonstrated by the successful operation of CUORICINO which represents just a slightly modified version of one of its 19 towers. The high natural abundance of ^{130}Te results in a relatively low cost for a detector sensitive to the degenerate neutrino mass region (Tab. 3). Thanks to the bolometer's versatility, alternative options with respect to TeO$_2$ could be taken into consideration and moreover the cost of enriched ^{130}Te needed to extend the sensitivity, is lower than for other isotopes. CUORE has been approved by the Italian and LNGS Scientific committes and is under construction at LNGS. It is presently the only second generation experiment with a timely schedule for data taking in 2010.

EXO[59] (Enriched Xenon Observatory) is a challenging project based on a large mass ($\sim$ 1–10 tons) of isotopically enriched (85% in ^{136}Xe) Xenon. An ingenuous tagging of the doubly charged Ba isotope produced in the decay ($^{136}Xe \rightarrow^{136} Ba^{++} + 2e^-$) would allow an excellent background suppression. After reduction to Ba$^+$ ion, the initial 6 $^2S_{1/2}$ state would be excited to the 6 $^2P_{1/2}$ by

means of a first 493 nm laser pulse, to be successively re-excited to the 6 $^2P_{1/2}$ by a second 650 nm laser beam after the decay (30% B.R.) to the metastable 5 $^4D_{3/2}$ state. De-excitation to the original 6 $^2P_{1/2}$ state would then be followed by the emission of a 493 nm photon. The technical feasibility of such an ambitious project aiming at a complete suppression of all the backgrounds requires a hard R&D phase. The unavoidable $\beta\beta(2\nu)$ contribution is however a serious concern due to the poor energy resolution of Xe detectors. In order to improve energy resolution, the energy deposited in the detector is measured by both charge and scintillation while only the ionization signal is used to localize the event vertex for signal identification and background rejection. Two detector concepts have been originally considered: a high pressure gas TPC and a LXe chamber. The second option is however presently preferred. Liquid xenon implies in fact a more compact structure, can be easily purified with commercial systems and the cryogenic system used to keep the detector cold would provide a radioactively clean shielding. In addition, the lifetime of the recoiling ^{136}Ba ion is long enough to allow laser tagging.

A smaller prototype experiment (EXO-200) with a Xe mass of 200 kg (80% ^{136}Xe), is presently under construction to be installed at the WIPP underground laboratory. The prototype has no barium tagging and should reach an energy resolution of $\sim$1.6% at 2.5 MeV. The primary goal is to measure ^{136}Xe $\beta\beta(2\nu)$ and to study $\beta\beta(0\nu)$ with a sensitivity of $\sim 10^{25}$ y in two years of data taking. In parallel, different approaches for barium tagging to be incorporated in the final full-scale experiment are being investigated, the most promising being the one based on the extraction of the ions from the liquid xenon and transfer to an ion trap for laser tagging. The 200 kg prototype is fully funded and should be operational at WIPP in 2008-2009.

The proposed Super-NEMO experiment

is the only based on a *passive source* approach. It is an extension of the NEMO3 concept, properly scaled in order to accommodate ~100 kg of ^{82}Se foils spread among 20 detector modules. The proposed geometry is planar. The energy resolution will be improved from 12% FWHM to 7% FWHM to improve the signal detection efficiency from 8% to 40% and reduce the $\beta\beta(2\nu)$ contribution. The detector modules will have an active water shield to further reduce any cosmic ray backgrounds. The proposed detector dimensions will require a larger hall than is currently available at Frejus. Although an expansion of the facility is possible, other possible locations are being investigated. If funded, Super-NEMO is projected to start operations in 2011. It is proposed by an international collaboration led by French physicists. Although the detector design itself is in principle scalable to 1000 kg due to its modularity, serious limitations could be implied by the radio-purity of the detector systems and by the low volume fraction occupied by the foils.

Based on a passive source approach the MOON project[60] plans to use natural molybdenum (9.63 %) to detect not only $\beta\beta$ but also solar neutrinos. To be installed in the Oto laboratory (Japan), it would consist in a large tracking calorimeter made by thin foils of enriched ^{100}Mo interleaved with specially designed scintillators. The possibility to use bolometeric detectors has also been considered. Background contributions would be substantially reduced by event topology reconstruction, leaving $\beta\beta(2\nu)$ (which for ^{100}Mo is relatively high) as the dominant source. The scintillator detection technique could support also isotopes other than ^{100}Mo. If ^{82}Se would be used the backgrounds would be substantially reduced but at the expense of losing solar neutrino detection capability. Current proposals call for a 200 kg stage, followed by a 1 ton phase.

The CAMEO[62] proposal would use 1 ton

of scintillating ^{116}CdWO$_4$ cristals inside the Borexino detector. CANDLES[63] would be based instead on the use of CaF$_2$ in liquid scintillator. COBRA[64] is based on the use of CdTe or CdZnTe diodes (~10 kg) to investigate Cd and Te $\beta\beta$ isotopes in a calorimetric approach. Promising results of measurements carried out at LNGS with a reduced scale prototype have been reported[65]. Similar to GERDA, GEM[66] would consist in a slight modification of the original GENIUS proposal[67], in which the complex LN huge cryostat has been replaced by a definitely smaller one inserted in a large pure water container (e.g. Borexino). DCBA[63] proposes the use of a modular 3-D tracking (drift chamber) in a uniform magnetic field to study ^{150}Nd $\beta\beta(0\nu)$; the expected sensitivity based on the analysis of the single electron energy distributions seems unfortunately untenable. An interesting $\beta\beta(0\nu)$ sensitivity has been claimed also by the XMASS[68] solar neutrino collaboraion.

6 Conclusions

A renewed interest in $\beta\beta(0\nu)$ searches has been stimulated by recent neutrino oscillation results. Neutrinoless $\beta\beta$ is finally recognized as a unique tool to measure neutrino properties (nature, mass scale, intrinsic phases) unavailable to the successful experiments on neutrino oscillations. Present $\langle m_\nu \rangle$ sensitivities are still outside the range required to test the inverted neutrino mass hierarchy. A phased $\beta\beta(0\nu)$ program based on a number of newly proposed experiments (possibly on different $\beta\beta(0\nu)$ isotopes) is of primary importance and has been supported by APS Multidivisional Neutrino Study[69]. The success of such a program strongly depends on the true capability of the proposed projects to reach the required background levels in the $\beta\beta(0\nu)$ region. An experimental confirmation of the (sometimes optimistic) background predictions of the various projects is

322

therefore worthwhile and the construction of preliminary intermediate scale setups is recommended. Actually, for most of the proposed projects (CUORE, GERDA, MAJORANA, EXO and MOON) a smaller scale detector has already been funded. CUORE is the first full scale project funded and under construction.

The claimed evidence for a $\beta\beta(0\nu)$ signal in the HM data could be soon verified by the presently running experiments and in any case, by the forthcoming next generation experiments.
A strong effort to improve the NME evaluation should be encouraged while stressing the need of experiments addressed to different nuclei.

References

1. B. Kayser in *The Physics of Massive Neutrinos* (World Sci., Singapore, 1989).
2. V. Barger et al., *Phys. Lett.* B **595**, 55 (2004).
3. S. Hannestad et al., *J. Cosmol. Astropart. Phys.* **JCAP05**, 004 (2003).
4. M. Tegmark et al., *Phys. Rev.* D **69**, 103501 (2004)
5. D. N. Spergel et al., *Astrophys. J. Suppl.* **148**, 175 (2003).
6. P. Crotty et al., *Phys. Rev.* D **69**, 123007 (2004).
7. S. W. Allen et al., *Mon. Not. R. Astron. Soc.* **346**, 593 (2003).
8. V. M. Lobashev, *Nucl. Phys.* A **719**, 153 (2003).
9. C. Arnaboldi et al., "MARE: Microcalorimeter Arrays for a Rhenium Experiment", Milano Int. Note, Spring 2005.
10. S. R. Elliott and P. Vogel, *Annu. Rev. Nucl. Part. Sci.* **52**, 115 (2002); hep-ph/0202264.
11. J. D. Vergados, *Phys. Rep.* **361**, 1 (2002).
12. J. Suhonen and O. Civitarese, *Phys. Rep.* **300**, 123 (1998).
13. F. Feruglio, A. Strumia and F. Vissani, *Nuclear Phys.* B **637**, 345 (2002); hep-ph/0201291.
14. A. Strumia and F. Vissani, hep-ph/0503246.
15. V. I. Tretyak and Y. Zdesenko, *Atomic Data and Nuclear Data Tables* **80**, 83 (2002) and

61, 43 (1995).
16. A. S. Barabash, *Phys. of At. Nucl.* **67**, 438 (2004).
17. W. H. Furry, *Phys. Rev.* **56**, 1184 (1939).
18. G. Gelmini and M .Roncadelli, *Phys.Lett.* B **99**, 1411 (1981).
19. R. N. Mohapatra and P. B. Pal in *Massive Neutrinos in Physics and Astrophysics* (World Sci., Singapore, 1991).
20. Z. G. Berezhiani, A. Yu. Smirnov and J. W. F. Valle,, *Phys.Lett.* B **291**, 99 (1992).
21. R. N. Mohapatra and E. Takasugi, *Phys.Lett.* B **211**, 192 (1988).
22. P. Bamert, C. P. Burgess, and R. N. Mohapatra, *Nucl. Phys.* B **449**, 25 (1995).
23. S. Pascoli, S. T. Petcov and T. Schwetz, hep-ph/0505226.
24. V. A. Rodin et al., *Phys. Rev.* C **68**, 044302 (2003) and nucl-th/0503063.
25. H. Ejiri, *Phys. Rep.* **338**, 265 (2000).
26. S. M. Bilenky and S. T. Petcov, hep-ph/0405237.
27. W. C. Haxton and G.J. Stephenson Jr., *Progr. Part. Nucl. Phys.* **12**, 409 (1984).
28. E. Caurier et al., *Nucl. Phys.* A **654**, 973 (1999).
29. J. Engel et al., *Phys. Rev.* C **37**, 731 (1988).
30. A. Staudt et al., *Europhys. Lett* **13**, 31 (1990).
31. A. Staudt, T.T.S. Kuo, H. Klapdor-Kleingrothaus, *Phys. Rev.* C **46**, 871 (1992).
32. A. Faessler and F. Simkovic, *J. Phys.* G **24**, 2139 (1998).
33. G. Pantis et al., *Phys. Rev.* C **53**, 695 (1996).
34. A. S. Barabash, *Czech. J. Phys.* **52**, 567 (2002).
35. You Ke et al., *Phys. Lett.* B **265**, 53 (1991).
36. H. V. Klapdor-Kleingrothaus et al. (HM Coll.), *Eur. Phys. J.* A **12**, 147 (2001) and hep-ph/0103062
37. C. E. Aalseth et al., *Phys. Rev.* C **59**, 2108 (1999).
38. C. E. Aalseth et al. (IGEX Coll.), hep-ex/0202026.
39. Shitov et al. (NEMO3 Coll.), *Proc. of the NANP05 Conference*, Dubna, Russia, June 20–25 2005.
40. R. Arnold et al., *Nucl. Phys.* A **658**, 299 (1999).
41. F. A. Danevich et al., *Phys. Rev.* C **62**, 044501 (2000).

42. T. Bernatowicz et al., *Phys. Rev.* C **47**, 806 (1993).

43. C. Arnaboldi et al., *Phys. Rev. Lett.* **95**, 142501 (2005).

44. J. M. Gavriljuk et al., *Phys. Rev.* C **61**, 035501 (2000).

45. R. Bernabei, *Phys. Lett.* B **546**, 23 (2002).

46. A. De Silva et al., *Phys. Rev.* C **56**, 2451 (1997).

47. H. V. Klapdor-Kleingrothaus, A. Dietz, H.V. Harney and I.V. Krivosheina, *Modern Physics Letters* A **16**, 2409 (2001) and hep-ph/0201231.

48. C. E. Aalseth, et al., hep-ex/0202018.

49. Yu.G Zdesenko et al., *Phys. Lett.* B, **546**, 206 (2002).

50. A. M. Bakayarov et al., hep-ex/0309016.

51. H. V. Klapdor-Kleingrothaus, hep-ph/0205228.

52. H. V. Klapdor-Kleingrothaus et al., *Nucl. Instr. and Meth.* A **522**, 371 (2004) and *Phys. Lett.* B **578**, 54 (2004).

53. E. Fiorini and T. Niinikoski, *Nucl. Instr. and Meth.* A **224**, 83 (1984).

54. A. Alessandrello et al., *Nucl. Instr. and Meth.* A **440**, 397 (1998) and *Phys. Lett.* B **420**, 109 (1998)and *Phys. Lett.* B **433**, 156 (1998).

55. R. Arnold et al., *Nucl. Instr. and Meth.* A **536**, 79 (2005).

56. C. Aarnaboldi et al., hep-ex/0501010.

57. I. Abt et al., hep-ex/0404039.

58. C. E. Aalseth et al., hep-ex/0201021.

59. M. Danilov et al., *Phys. Lett.* B **480**, 12 (2000).

60. H. Ejiri et al., *Phys. Rev. Lett.* **85**, 2917 (2000).

61. H. V. Klapdor-Kleingrothaus HV, hep-ph/0103074.

62. G. Bellini et al., *Eur. Phys. J.* C **19**, 43 (2001).

63. T. Kishimoto et al., Osaka University 2001 Annual Report.

64. K. Zuber, *Phys.Lett.* B **519**, 1 (2001).

65. H.Kiel et al., *Nuclear Physics* A **723**, 499 (2003) and nucl-ex/0301007.

66. Y. G. Zdesenko et al., *J. Phys.* G **27**, 2129 (2001) and phys-ex/0106021.

67. H. V. Klapdor-Kleingrothaus et al., hep-ph/9910205.

68. S. Moriyama et al., *Proc. of XENON01 workshop*, December 2001, Tokyo, Japan and Y. Takeuchi, *Proc. of ICHEP04*, August 2004, Beijing, China.

69. S. Freedman et al., *The Neutrino Matrix*, APS Multidivisional Neutrino Study, October 2004.

NEUTRINO OSCILLATIONS AND MASSES

SRUBABATI GOSWAMI

Harish-Chandra Research Institute, Chhatnag Road, Jhusi, Allahabad -211 019, India
E-mail: sruba@mri.ernet.in

The present status of neutrino oscillation parameters are summarised. The sensitivity of the future planned and proposed experiments are discussed.

1 Parameters of Neutrino mass matrix

Assuming three neutrino flavours there are nine parameters in the light neutrino mass matrix – 3 masses m_1, m_2 and m_3, 3 mixing angles θ_{12}, θ_{13} and θ_{23}, 1 Dirac CP phase and 2 Majorana CP phases. Oscillation experiments are sensitive to the mass squared differences – Δm_{21}^2, $\Delta m_{31}^2 \approx \Delta m_{32}^2$, mixing angles and the Dirac CP phase. Majorana CP phases are observable only in lepton number violating processes. Information on absolute neutrino mass scale comes from tritium beta decay[1], neutrino-less double beta decay[2] and Cosmology[3]. In this article I will discuss the present constraints on the oscillation parameters coming from solar, atmospheric, accelerator and reactor neutrino data. I will also outline how the precision in determination of these parameters can be improved in future experimental facilities.

2 Current Status of Neutrino Oscillation Parameters

2.1 Solar Neutrino Oscillation Parameters

Global solar data from the radiochemical experiments Homestake, Gallex, GNO and Sage and the Čerenkov experiments Superkamiokande (SK) and Sudbury Neutrino Observatory (SNO)[4], constrain the parameters $\Delta m_\odot^2 \equiv \Delta m^2{}_{21}$, $\theta_\odot \equiv \theta_{12}$. Fig. 1 shows the allowed area in the $\Delta m_{21}^2 - \sin^2 \theta_{12}$ plane from global solar data[5]. The left panel shows

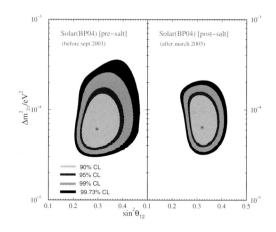

Figure 1. The allowed area in $\Delta m_{21}^2 - \sin^2 \theta_{12}$ parameter space from global solar analysis with and without the SNO salt data.

the allowed area before the salt phase while the right hand panel shows the allowed area including the full salt data (salt-I+salt-II) from SNO[6]. The inclusion of salt data is seen to tighten the constraints on the parameters. The reason for this can be understood if we consider the expressions for charged currents (CC) and neutral current (NC) rates for SNO. For oscillation to active neutrinos $R_{CC} \approx f_B P_{ee} \approx f_B \sin^2 \theta_{12}$ for ^8B neutrinos and $R_{NC} = f_B$. The SNO salt data determines R_{NC} and hence f_B with a greater precision. Therefore regions which could earlier be allowed by adjusting f_B gets disfavoured now.

In fig. 2 we show the impact of KamLAND results[7] on the allowed region in the $\Delta m_{21}^2 - \sin^2 \theta_{12}$ parameter space. The

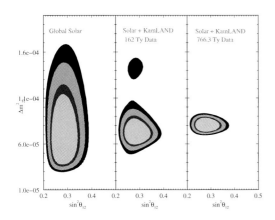

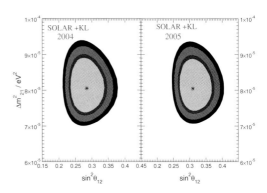

Figure 2. The change in allowed area in $\Delta m_{21}^2 - \sin^2 \theta_{12}$ parameter space with KamLAND data.

Figure 3. The allowed area in $\Delta m_{21}^2 - \sin^2 \theta_{12}$ parameter space from global solar+KamLAND data.

survival probability for KamLAND is :

$$P_{\bar{e}\bar{e}}^{KL} = 1 - \sin^2 2\theta_{12} \sin^2 \left(\frac{1.27 \Delta m_{21}^2 L}{E_\nu} \right)$$

neglecting the small matter effect for lower values of Δm_{21}^2. Assuming CPT conservation i.e same oscillation parameters for neutrinos and antineutrinos KamLAND is sensitive to $\Delta m^2{}_{21}$ in the so called Large-Mixing Angle (LMA) region. The middle panel in fig. 2 shows the allowed region with inclusion of 162 Ty KamLAND data declared in 2002 while the rightmost panel shows the allowed region after inclusion of the 766.3 Ty KamLAND data declared in 2004. The 162 Ty KamLAND data splits the LMA region in two parts called the low and high LMA regions. The high-LMA region gets disfavored at more than 3σ with the 766.3 Ty KamLAND data.

In fig. 3 we show the allowed area in the $\Delta m_{21}^2 - \sin^2 \theta_{12}$ parameter space from combined solar and KamLAND results [5]. The left panel shows the allowed area in 2004 and the right panel shows the allowed area in 2005 after inclusion of the phase 2 SNO salt data declared in March 2005. The best-fit including the latest salt data comes at $\Delta m^2{}_{21} = 8.0 \times 10^{-5} \text{eV}^2$ and $\sin^2 \theta_{12} = 0.31$ The 3σ allowed ranges as obtained from the righthand panel of fig. 2 are $\Delta m^2{}_{21} = (7.0$-

9.3) $\times 10^{-5} \text{eV}^2$ and $\sin^2 \theta_{12} = 0.24$ -0.41. Maximal mixing is ruled out at almost 6σ from the present data. There is strong evidence of MSW matter effect in the present data with no-MSW being ruled out at more than 5σ [8]. Matter effect inside the sun ensures that $\Delta m_{21}^2 > 0$ and disfavours the $\theta > \pi/4$ (Dark-Side) solutions.

In Table 1 we present the 3σ allowed ranges of Δm_{21}^2 and $\sin^2 \theta_{12}$, obtained using different data sets. The only solar and solar + 766.3 Ty KamLAND analysis inlcudes the latest SNO salt results. We also show the uncertainty in the value of the parameters through a quantity "spread" which we define as

$$\text{spread} = \frac{\Delta m^2 (\sin^2 \theta)_{max} - \Delta m^2 (\sin^2 \theta)_{min}}{\Delta m^2 (\sin^2 \theta)_{max} + \Delta m^2 (\sin^2 \theta)_{min}}$$

Table 1 illustrates the remarkable sensitivity of KamLAND in reducing the uncertainty in Δm_{21}^2. But θ_{12} is not constrained much better than the current set of solar experiments. The reason for this is the average energy and distance in KamLAND corresponds to a Survival Probability MAXimum (SP-MAX) i.e $\sin^2 (\Delta m^2{}_{21} L/4E) \approx 0$. This means that the coefficient of the $\sin^2 2\theta_{12}$ term in $P_{\bar{e}\bar{e}}^{KL}$ is relatively small, weakening the sensitivity of KamLAND to θ_{12}. The precision in θ_{12} can be improved by reducing the baseline length such that one gets

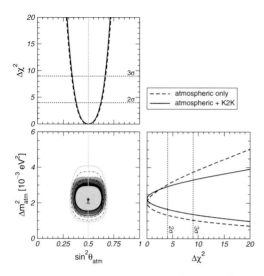

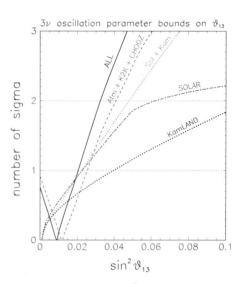

Figure 4. The allowed area in $\Delta m^2_{atm} - \sin^2 \theta_{atm}$ parameter space from atmospheric and K2K data

Figure 5. The bound on $\sin^2 \theta_{13}$ from various data sets

a minimum of the $\bar{\nu}_e$ survival probability (SPMIN) where $\sin^2 (\Delta m^2_{21} L/4E) = 1$. This corresponds to a distance $L = 1.24(E/MeV)(eV^2/\Delta m^2_{12})$ m [9].

2.2 Atmospheric Neutrino Oscillation Parameters

In fig. 4 we show the allowed area in the $\Delta m^2_{atm} - \sin^2 \theta_{atm}$ space from two–generation analysis of the SK atmospheric[10] and K2K data[11]. The figure is taken from[12]. The best-fit values of parameters obtained from combined SK+K2K analysis are: $\Delta m^2_{31} = 2.2 \times 10^{-3}$ eV^2 and $\sin^2 \theta_{23} = 0.5$ [12]. We also show the 1 parameter plots of $\Delta \chi^2$ vs Δm^2_{atm} or $\sin^2 \theta_{atm}$. These plots are marginalised over the undisplayed parameters. We find that Δm^2_{atm} values are constrained further by inclusion of K2K data but the θ_{atm} is not constrained any better. For a two generation $\nu_\mu - \nu_\tau$ analysis the relevant probability is the vacuum oscillation probability which has $\theta_{23} - (\pi/2 - \theta_{23})$ symmetry. The spread in Δm^2_{atm} is $\sim 40\%$ whereas the that in $\sin^2 \theta_{atm}$ is $\sim 33\%$ at 4σ from the 1 parameter plots.

2.3 Status of θ_{13}

In a combined three generation analysis the third angle θ_{13} couples the solar and atmospheric neutrino oscillations. In fig. 5 we show the bounds on $\sin^2 \theta_{13}$ from 1 parameter plots[8]. Combined analysis of solar, atmospheric, reactor and accelerator data give $\sin^2 \theta_{13} \lesssim 0.04$ at 3σ [8,13]. This bound is sensitive to value of Δm^2_{31} [14]. Since this mixing angle is very small the two-generation allowed areas does not change significantly with inclusion of the third generation.

2.4 Accommodating the LSND Signal

The LSND experiment at Los Alamos have declared positive evidence for $\nu_\mu - \nu_e$ oscillation corresponding to $\Delta m^2 \sim$ eV^2 [15]. Thus this experiment introduces another mass scale for oscillations and a three generation explanation of solar, atmospheric and LSND observations cannot be obtained. The way out includes addition of extra sterile neutrinos or assumption of CPT violation. The simplest possibility is to add one additional sterile neutrino. In such a scenario there can be two possibilites the so called 2+2 and 3+1 mass schemes[16]. The 2+2 mass scheme im-

Table 1. 3σ allowed ranges and % spread of Δm^2_{21} and $\sin^2 \theta_{12}$ obtained from 2 parameter plots.

Data set used	Range of Δm^2_{21} eV^2	spread in Δm^2_{21}	Range of $\sin^2 \theta_{12}$	spread in $\sin^2 \theta_{12}$
only sol	3.1 - 21.0	74%	$0.23 - 0.43$	30%
sol+162 Ty KL	4.9 - 10.7	37%	$0.21 - 0.39$	30%
sol+ 766.3 Ty KL	7.0 - 9.3	14%	$0.25 - 0.41$	24%

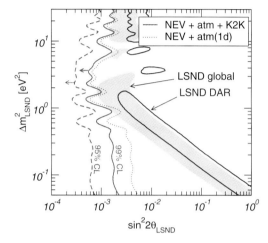

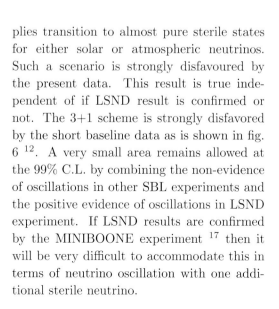

Figure 6. The allowed area in 3+1 mass scheme

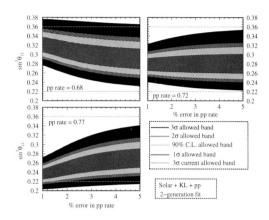

Figure 7. The constraints on $\sin^2 \theta_{12}$ from a generic $pp\ \nu - e$ scattering experiment. The horizontal lines indicate the current 3σ allowed ranges.

plies transition to almost pure sterile states for either solar or atmospheric neutrinos. Such a scenario is strongly disfavoured by the present data. This result is true independent of if LSND result is confirmed or not. The 3+1 scheme is strongly disfavored by the short baseline data as is shown in fig. 6 [12]. A very small area remains allowed at the 99% C.L. by combining the non-evidence of oscillations in other SBL experiments and the positive evidence of oscillations in LSND experiment. If LSND results are confirmed by the MINIBOONE experiment [17] then it will be very difficult to accommodate this in terms of neutrino oscillation with one additional sterile neutrino.

3 Future Precision of Neutrino Oscillation Parameters

3.1 Constraints from Phase III of SNO

In its phase-III the SNO experiment will use ^3He counters to measure the NC events. The projected total error for NC events from this phase is 6% as compared to present error of 8%. The error in CC measurements are also expected to reduce and hence the CC/NC ratio will be better determined. Since $R_{CC}/R_{NC} \approx \sin^2 \theta_{12}$, a reduced error in this ratio is expected to increase the precision in measurement of $\sin^2 \theta_{12}$. A projected analysis including a reduced error for CC and NC for SNO3 gives the 3σ spreads (1 param, $\Delta\chi^2 = 9$) as $\Delta m^2_{21} = 12\%$; and $\sin^2 \theta_{12} = 18\%$.

328

Table 2. 3σ allowed ranges and % spread of Δm^2_{21} and $\sin^2\theta_{12}$ obtained from 1 parameter plots.

Data set used	spread in Δm^2_{21}	spread in $\sin^2\theta_{12}$
sol2005 + 766.3 Ty KL	12%	22%
sol2005 + 766.3 Ty KL + SNO3	12%	18%
sol2005 + 3 kTy KL + SNO3	6%	16%
sol2005 + 3 kTy KL + SNO3 + pp	6%	12%
3 kTy KL	7%	32%
5 Yr SK-Gd	2%	18%
Reactor expt at SPMIN (3kTy)	5%	6%

3.2 Potential of LowNu experiments for θ_{12}

The pp flux is known with 1% accuracy from Standard Solar Models [a]. Since the pp neutrino energy spectrum extends up to 0.42 MeV only, for Δm^2_{21} in the LMA region, the pp neutrino oscillations are practically not affected by matter effects. Thus, to a good approximation, the pp neutrino oscillations are described by the ν_e survival probability for vacuum oscillations, in which the oscillatory term is strongly suppressed by the averaging over the region of neutrino production in the Sun[18]:

$$P^{2\nu}_{ee}(pp) \cong 1 - \frac{1}{2}\sin^2 2\theta_{12}.$$

In fig. 7 we plot the sensitivity of $\sin^2\theta_{12}$ as a function of % error in pp scattering rate for three illustrative values of measured pp rate [19]. We find that the error in $\sin^2\theta_{12}$ is $\sim 14\%(\sim 21\%)$ at 3σ for 1% (3%) error in the measured value of the pp rate. One can also conclude that the error in pp rate should be $\lesssim 4\%$ to cause improvement on the exisiting range of θ_{12} marked by horizontal lines.

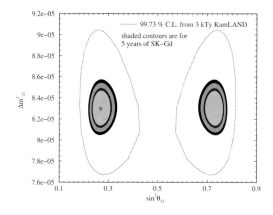

Figure 8. The allowed area from 5 years of simulated SK-Gd data

3.3 Constraints from Future Reactor Experiments

The proposed Super-Kamiokande with 0.1% Gadolinium (SK-Gd) would be a very big reactor anti-neutrino detector with a statistics of about 43 times that expected in Kam-LAND [20] [b]. In fig. 8 we plot the allowed area in the $\Delta m^2_{21} - \sin^2\theta_{12}$ plane for 5 years projected data of Sk-Gd[21]. We also plot the 3σ allowed area from KamLAND on this plot for purposes of comparison. The plot shows that Sk-Gd can further reduce the allowed ranges of Δm^2_{21} and $\sin^2\theta_{12}$. The 99% C.L. spread of Δm^2_{21} obtained from this plot is

[a]In comparison, the uncertainty for ^8B is $\sim 20\%$ and for ^7Be is $\sim 10\%$

[b]In [20] the detector was named GADZOOKS!.

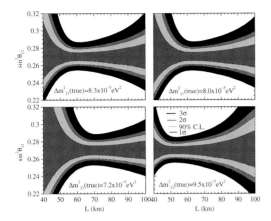

Figure 9. The sensitivity for a reactor experiment at SPMIN

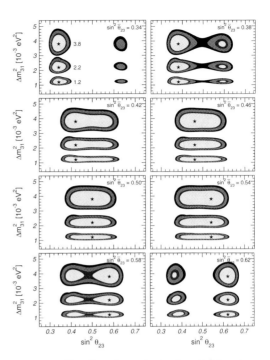

Figure 10. The allowed area in $\Delta m^2_{31} - \sin^2 \theta_{23}$ plane from future SK data

2.4%, demonstrating a remarkable sensitivity.

As we had discussed in section 2.1, θ_{12} sensitivity is best at SPMIN. In fig. 9 we plot the allowed range of $\sin^2 \theta_{12}$ as a function of source-detector distance for a same kind of experiment as KamLAND but at a baseline that corresponds to SPMIN [19]. These plots are presented for four different values of Δm^2_{21} from the current allowed range. Most optimal baseline for the new global best-fit $\Delta m^2_{21} = 8.0 \times 10^{-5}$ eV^2 is about 63 km. For a higher(lower) value of Δm^2_{21} the SPMIN comes at a lower(higher) distance. We can achieve a $\sim 6\%(2\%)$ accuracy at $3\sigma(1\sigma)$ in $\sin^2 \theta_{12}$ in this reactor experiment. In[22] such an experiment in Japan named SADO is proposed. In table 2 we summarise the present and future 3σ % spreads of the parameter Δm^2_{21} and $\sin^2 \theta_{12}$. The table shows that Sk-Gd experiment is best for Δm^2_{21} while a KamLAND like experiment at SPMIN is best for $\sin^2 \theta_{12}$.

3.4 Constraints from future SK atmospheric data

In fig. 10 we show the allowed area including 20 years of SK statistics. The panels are labeled by θ_{23} values for which the SK spectrum is simulated. The Δm^2_{32} values at which the spectrum is simulated for each θ_{23} is indicated in the first panel. The middle contour of the first panel in the third row corresponds to spectrum generated at the present best-fit values. The 3σ spreads as obtained from this contour are $\Delta m^2_{32} = 17\%$ and $\sin^2 \theta_{23} = 24\%$ [23]. Atmospheric neutrino flux measurements in large magnetized iron calorimeter detectors can reduce the uncertainty[24]. Such a detector facilitiy called Indian Neutrino Observatory (INO) is currently planned for location in India [25].

3.5 Potential of long baseline experiments

In Table 3 we list the upcoming terrestrial experiments in the next ten years [26]. The constraints obtained on atmospheric neutrino oscillation parameters using mainly the disappearance channel for Δm^2_{32} and θ_{23} the 3σ spreads are obtained in Table 4. The pre-

Table 3. Upcoming terrestrial experiments in the "next ten years"

Label	L	$\langle E_\nu \rangle$	channel
Conventional beam experiments:			
MINOS	735 km	3 GeV	$\nu_\mu \to \nu_\mu, \nu_e$
ICARUS	732 km	17 GeV	$\nu_\mu \to \nu_\mu, \nu_e, \nu_\tau$
OPERA	732 km	17 GeV	$\nu_\mu \to \nu_\mu, \nu_e, \nu_\tau$
Superbeams:			
T2K	295 km	0.76 GeV	$\nu_\mu \to \nu_\mu, \nu_e$
NOνA	812 km	2.22 GeV	$\nu_\mu \to \nu_\mu, \nu_e$
Reactor experiments:			
CHOOZ II	1.05 km	~ 4 MeV	$\bar{\nu}_e \to \bar{\nu}_e$
Reactor-II	1.70 km	~ 4 MeV	$\bar{\nu}_e \to \bar{\nu}_e$

cision depends on true value of Δm_{31}^2. The Table is obtained for $\Delta m_{31}^2 = 2 \times 10^{-3}$ eV^2, $\sin^2 \theta_{23} = 0.5$. The bounds obtained are sensitive mainly to $\sin^2 2\theta_{23}$ and therefore the sensitivity is not good specially near $\theta_{23} = \pi/4$.

Table 4. 3σ spread in Δm_{31}^2 and $\sin^2 \theta_{23}$ from future experiments

	$\lvert \Delta m_{31}^2 \rvert$	$\sin^2 \theta_{23}$
current	44%	39%
MINOS+CNGS	13%	39%
T2K	6%	23%
NOνA	13%	43%
Combination	4.5%	21%

The figs. 11 [27] and 12 [23] show the potential of the future experiments to measure the deviation of maximality for θ_{23}. The unshaded regions represent the values of θ_{23} and Δm_{32}^2 for which maximal mixing cannot be rejected whereas for the parameters in the shaded regions maximal mixing can be excluded. From the first panel in the second row of fig. 11 one can say that maximality can be tested to $\sim 14\%$ at 3σ for $\Delta m_{32}^2 = 2.5 \times 10^{-3}$ eV^2 after 10 years. The results of fig. 11 are generated for the true value of $\theta_{13} = 0$ and for non-zero values of

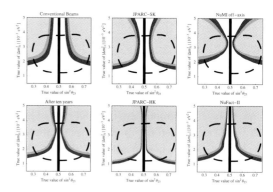

Figure 11. The regions from dark to light shading show the true values of $\sin^2 \theta_{23}$ and Δm_{31}^2 where maximal θ_{23} can be disfavoured at 1, 2 and 3σ.

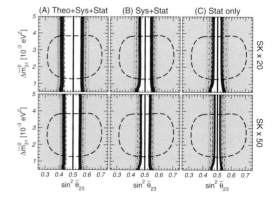

Figure 12. Same as in fig. 11 for atmospheric neutrinos and future SK data

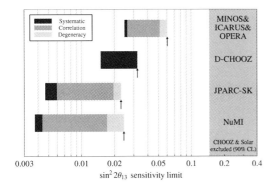

Figure 13. The sensitivity to $\sin^2\theta_{13}$ in the next ten years.

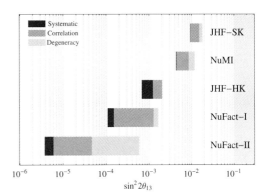

Figure 14. The sensitivity to $\sin^2\theta_{13}$

θ_{13} close to the current bound the results do not change significantly. Fig. 12 is for atmospheric neutrino events in future SK. The sensitivity to deviation of maximality comes from Δm^2_{21} driven oscillations which gives rise to an excess in sub-GeV electron events. At present this electron excess is not at a statistically significant level. But with 20 years of SK statistics maximality can be tested to $\sim 21\%$ at 3σ.

3.6 Future sensitivity of θ_{13}

The appearance channel $\nu_\mu \rightarrow \nu_e$ in beam experiments can be used to probe θ_{13}. The probability for $\nu_\mu - \nu_e$ transition can be approximated as [26]

$$P_{\mu e} \approx \sin^2 2\theta_{13}\sin^2\theta_{23}\sin^2\Delta_{31} + O(\alpha, \alpha^2)$$

where, $\alpha = \Delta_{21}/\Delta_{31}$. The current global data gives best-fit $\alpha = 0.03$ [12] The $O(\alpha)$ terms contain the CP phase δ_{CP}. The sensitivity to θ_{13} in this channel is complicated by the problem of parameter correlations which arises because of our imprecise knowledge of other parameters and degeneracies which are inherent in a three generation analysis due to the presence of non-zero $\delta_{\rm CP}$. The degeneracies are the $(\delta_{\rm CP}, \theta_{13})$ ambiguity, the $\text{sign}(\Delta_{31})$ or mass hierarchy degeneracy and the $(\theta_{23}, \pi/2 - \theta_{23})$ degeneracy, combining to give an overall eight-fold degeneracy [28].

Clean measurement of $\sin^2 2\theta_{13}$ is possible in reactor experiments which measures the survival probability to $P_{\nu_e\nu_e}$ which is free from the CP phase [29]:

$$P(\bar\nu_e \rightarrow \bar\nu_e) \approx 1 - \sin^2 2\theta_{13}\sin^2\Delta_{31} + O(\alpha^2)$$

The determination of $\sin^2\theta_{13}$ in reactor experiments is dominated by systematics.

In figs. 13 and 14 we plot the $\sin^2 2\theta_{13}$ sensitivity for different planned and proposed experiments. In fig. 13 the D-CHOOZ reactor experiment is found to have the highest sensitivity (~ 0.03) to $\sin^2 2\theta_{13}$. The sensitivity for accelerator experiments measuring $P_{\mu e}$ is spoilt by the effect of parameter correlations and degeneracies. Better sensitivity to $\sin^2 2\theta_{13}$ can be obtained in neutrino factories as is seen from fig. 14.

The potential of low and high energy solar neutrino flux measurement for constraining θ_{13} was considered in[13]. The survival probability for low energy (pp) solar neutrinos can be approximated as
• $P_{ee} \approx \cos^4\theta_{13}(1 - 0.5\sin^2 2\theta_{12})$
whereas for high energy ^8B neutrinos the probability is
• $P_{ee} \approx \cos^4\theta_{13}\sin^2\theta_{12}$.
The above equations show that for $\theta_{13} \neq 0$ pp drives θ_{12} to lower values whereas ^8B drives θ_{12} to higher values. Because of these opposing trends combination of low and high energy events can constrain θ_{13} [13]. In fig. 18 we plot the 3σ sensitivity of $\sin^2\theta_{13}$ as

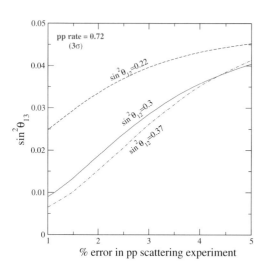

Figure 15. Sensitivity to $\sin^2\theta_{13}$ as a function of % error in pp flux measured in a scattering experiment

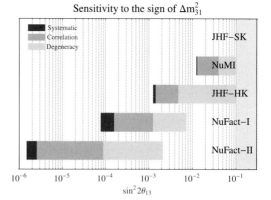

Figure 16. Sensitivity to the sign of Δm^2_{atm}

a function of % error in a typical pp scattering experiment assuming a sample rate of 0.72. The sensitivity to $\sin^2\theta_{13}$ that can be attained is ~ 0.01 for 1% error in pp scattering rate. Since P_{ee} is free of the CP phase this sensitivity is not affected by the $\delta_{CP}-\theta_{13}$ degeneracy. However the sensitivity depends on the value of $\sin^2\theta_{12}$ as is shown in fig. 18.

3.7 Ambiguity in Mass Hierarchy

The sign of the atmospheric mass difference ($\Delta m^2_{31} \approx \Delta m^2_{32}$) is not yet known. Experiments sensitive to matter effects can be used to probe this. In presence of matter the 1-3 mixing angle becomes

$$\tan 2\theta_{13}{}^m = \frac{\Delta m^2_{32}\sin 2\theta_{13}}{\Delta m^2_{32}\cos 2\theta_{13} \pm 2\sqrt{2}G_F n_e E}$$

where the $+$ sign is for antineutrinos and the $-$ sign is for neutrinos. It is evident then that for $\Delta m^2_{\text{atm}} > 0$ matter resonance occurs in neutrinos whereas for $\Delta m^2_{\text{atm}} < 0$ matter resonance occurs in anti neutrinos and this difference can be used to ascertain the mass hierarchy. Also we see that matter effects for Δm^2_{atm} channel depend crucially on

θ_{13} and both parameters get related. In fig. 16 we plot the sensitivity of the future superbeam and neutrino factory experiments to the sign of Δm^2_{atm}. The figure reveals that the sensitivity is limited by the correlation and degeneracies. For the Superbeam experiments JHF-HK or NUMI the hierarchy cannot be ascertained even at the present limit of $\sin^2 2\theta_{13}$. For neutrino factories, even though the problem of correlations and degeneracies is there, hierarchy can be determined for $\sin^2 2\theta_{13} \sim 0.01 - 0.001$. To avoid the problem of degeneracies, synergistic use of two different experiments or two different baselines have been considered by many people[30]. Use of magic baselines, where the terms containing δ_{CP} vanish have also been considered as possible alternatives[31].

3.8 Determining Hierarchy by Atmospheric Neutrinos

The ν_μ survival rate in atmospheric neutrino events can provide a novel and useful method of determining the hierarchy[32,33,34]. For pathlength and energy ranges relevant to atmospheric neutrinos, this rate obtains significant matter sensitive variations not only from resonant matter effects in $P_{\mu e}$ but also from those in $P_{\mu\tau}$. In fig. 17 The muon survival probability $P_{\mu\mu}$ in matter plotted as a function of E (GeV) for two pathlengths, L

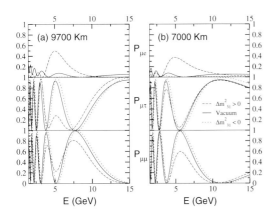

Figure 17. The muon survival and conversion probabilites as a function of energy.

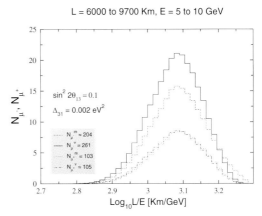

Figure 18. The number of μ^+ and μ^- events as a function of L/E in a magentised iron calorimeter detector for 100 kiloton-year exposure

= 9700 and 7000 km and for the two signs of Δm^2_{atm}. The survival probability in vacuum is also shown for comparison. For negative Δm^2_{atm} (or inverted hierarchy), there is no discernible difference between vacuum and matter survival probabilities. For positive Δm^2_{atm} (or normal hierarchy), the $P_{\mu\mu}$ for 7000 km suffers a sharp decrease in the energy range 5 - 10 GeV. This is due to a corresponding increase in $P_{\mu e}$ since $P_{\mu\mu} = 1 - P_{\mu e} - P_{\mu\tau}$ and the matter effect in $P_{\mu\tau}$ is not appreciable for 7000 km. However for 9700 km there is a drop in $P_{\mu\tau}$ which is as high as 70%, in the energy range 4 - 6 GeV. This drop in $P_{\mu\tau}$ overcomes the rise in $P_{\mu e}$. Thus the net change in $P_{\mu\mu}$ is an increase of the matter value over its vacuum value[35]. It was shown in [34] that problem of δ_{CP} degeneracy is less at these baselines. The matter effect at these baselines and energies for atmospheric neutrinos passing through the earth can be exploited for determining the mass hierarchy using μ^- rates in magnetized iron calorimeter detectors like INO.

In fig. 18 we plot the number of μ^+ and μ^- events in vacuum and matter for L = 6000-9700 km and E = 5 to 10 GeV. In this figure there is a drop in N^-_μ in matter with respect to its vacuum value due to a rise in $P_{\mu e}$

in matter whereas $\text{N}^{\text{mat}}_{\mu^+} \approx \text{N}^{\text{vac}}_{\mu^+}$. This corresponds to the righthand panel of fig. 17. In fig. 19 the matter effect in $P_{\mu\mu}$ in the bin with $\text{Log}_{10} L/E = 3.2 - 3.35$ km/GeV is due both to $P_{\mu e}$ and $P_{\mu\tau}$ corresponding to the probabilites of the lefthand panel of fig. 17 and therefore $\text{N}^{\text{mat}}_{\mu^-}$ is higher than the vacuum value. This difference between matter and vacuum values for μ^- events in chosen L/E bins can give rise to a 3-4σ signal for matter effects in ν_μ and hence for $\Delta m^2_{31} > 0$ or normal hierarchy in a charge discriminating iron calorimeter detector and for $\sin^2 2\theta_{13} = 0.1$ [34]. Water Čerenkov detectors cannot distinguish between the lepton charge and what they observe is $\mu^+ + \mu^-$ events. Therefore the sensitivity of the water Čerenkov detectors to matter effects is less compared to that of charge discriminating detectors. However, the dominance of muon rates over those of anti-muon rates and the resulting higher statistics makes it possible to study matter effects in such detectors also [36].

4   Conclusion and Outlook

• The small observed neutrino masses cannot be explained by Standard Model

334

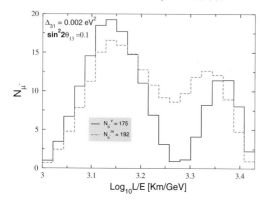

L = 8000 to 10700 Km, E = 4 to 8 GeV

Δ_{31} = 0.002 eV^2
$\sin^2 2\theta_{13}$ =0.1

N_p^ν = 175
N_p^m = 192

Log_{10}L/E [Km/GeV]

Figure 19. Same as in figure 19 but for a different L and E range

• This is so far the only observational signal of physics Beyond the Standard Model
• The theoretical challenges posed by small neutrino Masses are:
— Why $m_{\nu_i} \ll m_l, m_u, m_d$?
—Why two large and one small angle in neutrino sector while all the angles in quark sector are small ?
—Why the hierarchy for neutrino masses is not strong unlike the quarks and leptons ?
• Hierarchy, θ_{13}, deviation from maximality can be good discriminator between various mass models
• Precision neutrino measurements can help in choosing between various alternatives.

References

1. C. Weinheimer, this proceedings.
2. O. Cremonesi, this proceedings.
3. S. Hannested, this proceedings.
4. B. T. Cleveland *et al.*, *Astrophys. J.* **496**, 505 (1998) ;J. N. Abdurashitov *et al.* [SAGE Collaboration], *J. Exp. Theor. Phys.* **95**, 181 (2002) ;W. Hampel *et al.* [GALLEX Collaboration], *Phys. Lett.* **B447**, 127 (1999); C. Cattadori, Talk at Neutrino 2004, Paris, France, http://neutrino2004.in2p3.fr; S. Fukuda *et al.* [Super-Kamiokande Collaboration], *Phys. Lett.* **B539**, 179 (2002). *Phys. Rev. Lett.* **89**, 011301 (2002). Q. R. Ahmad *et al.* [SNO Collaboration], *Phys. Rev. Lett.* **89**, 011302 (2002)
5. This figure is obtained in collaboration with A. Bandyopadhyay and S. Choubey.
6. . N. Ahmed *et al.* [SNO Collaboration], *Phys. Rev. Lett.* **92**, 181301 (2004).
7. K. Eguchi *et al.*, [KamLAND Collaboration], *Phys.Rev.Lett.* **90** (2003) 021802. T. Araki *et al.*, [KamLAND Collaboration], hep-ex/0406035.
8. G. L. Fogli, E. Lisi, A. Marrone and A. Palazzo, arXiv:hep-ph/0506083.
9. A. Bandyopadhyay, S. Choubey, and S. Goswami, *Phys. Rev. D* **67**, 113011 (2003).
10. SUPER-KAMIOKANDE collaboration, Y. Ashie *et al.* *Phys. Rev. Lett.* **93**, 101801 (2004).
11. T. Nakaya *et al.*, talk at Neutrino 2004, Paris, France. http://neutrino2004.in2p3.fr
12. M. Maltoni, T. Schwetz, M. A. Tortola and J. W. F. Valle, arXiv:hep-ph/0405172;
13. S. Goswami and A. Y. Smirnov, arXiv:hep-ph/0411359.
14. S. Goswami, talk at Neutrino 2004, Paris, France http://neutrino2004.in2p3.fr/; S. Goswami, A. Bandyopadhyay and S. Choubey [hep-ph/0409224].
15. C. Athanassopoulos *et al.*, *Phys. Rev. Lett.* **77**, 3082 (1996); *ibid.* **81**, 1774 (1998).
16. S. Goswami, *Phys. Rev.* **D55**, 2931 (1997).
17. S. Brice *et al.*, Talk given at Neutrino 2004, Paris, France. http://neutrino2004.in2p3.fr; E. Zimmerman, this proceedings.
18. S.T. Petcov and J. Rich, *Phys. Lett.* **B224**, 401 (1989).

19. A. Bandyopadhyay, S. Choubey, S. Goswami and S. T. Petcov, *Phys. Rev.* **D72**, 033013 (2005).

20. J. F. Beacom and M. R. Vagins, arXiv:hep-ph/0309300.

21. S. Choubey and S. T. Petcov, *Phys. Lett.* **B594**, 333 (2004).

22. H. Minakata, H. Nunokawa, W. J. C. Teves and R. Zukanovich Funchal, arXiv:hep-ph/0407326.

23. M. C. González-García, M. Maltoni and A. Y. Smirnov, *Phys. Rev.* **D70**, 093005 (2004).

24. N. Y. Agafonova *et al.* [MONO-LITH Collaboration], LNGS-P26-2000, `http://castore.mi.infn.it/~monolith/`

25. See `http://www.imsc.res.in/~ino` and working group reports and talks therein.

26. M. Lindner, arXiv:hep-ph/0503101.

27. S. Antusch *et al.* *Phys. Rev.* **D70**, 097302 (2004).

28. See for e.g. V. Barger, D. Marfatia and K. Whisnant, *Phys. Rev.* **D65**, 073023 (2002) hep-ph/0112119;
H. Minakata, H. Nunokawa and S. J. Parke, *Phys. Rev.* **D66**, 093012 (2002) hep-ph/0208163.

29. P. Huber, M. Lindner, T. Schwetz and W. Winter, *Nucl. Phys.* **B665**, 487 (2003). H. Minakata, H. Sugiyama, O. Yasuda, K. Inoue and F. Suekane, Phys. Rev. D **68**, 033017 (2003) [Erratum-ibid. D **70**, 059901 (2004)].

30. O. Mena Requejo, S. Palomares-Ruiz and S. Pascoli, hep-ph/0504015 and references therein.

31. P. Huber and W. Winter, *Phys. Rev.* **D68**, 037301 (2003), hep-ph/0301257;
S. K. Agarwalla, A. Raychaudhuri and A. Samanta, hep-ph/0505015.

32. S. Palomares-Ruiz and S. T. Petcov, hep-ph/0406096.

33. D. Indumathi and M. V. N. Murthy, *Phys. Rev.* **D71**, 013001 (2005) [hep-ph/0407336].

34. R. Gandhi *et al.*, hep-ph/0411252.

35. R. Gandhi, *et al.* *Phys. Rev. Lett.* **94**, 051801 (2005).

36. R. Gandhi *et al.* arXiv:hep-ph/0506145.

DISCUSSION

Sachio Komamiya (University of Tokyo):
What is the sensitivity in sign of Δm_{31}^2 as a function of θ_{13} of the long baseline experiments?

Srubabati Goswami: The sensitivity of sign of Δ_{31} as a function of $\sin^2 \theta_{13}$ is shown in fig. 16. For the long baseline experiments there is not much sensitivity even at the current upper limit. The reason is the sensitivity to sign of Δm_{31}^2 comes mainly from matter effects and distance traversed by the neutrinos for these experiments are not large enough for matter effects to develop fully. From fig. 16 it is seen that even for superbeam experiments like NuMI and JHF-HK because of correlations and degeneracies sign of Δm_{31}^2 cannot be ascertained even at the present upper limit of $\sin^2 \theta_{13}$.

ASTROPARTICLE PHYSICS AND COSMOLOGY

Previous page:
The Botanical Gardens (above).

A viking rune stone in the university park gardens (below)

ULTRA HIGH ENERGY NEUTRINO TELESCOPES

PER OLOF HULTH

Department of Physics, Stockholm University, AlbaNova University Center
SE-10691 Stockholm, Sweden
E-mail: hulth@physto.se

The Neutrino Telescopes NT-200 in Lake Baikal, Russia and AMANDA at the South Pole, Antarctica have now opened the field of High Energy Neutrino Astronomy. Several other Neutrino telescopes are in the process of being constructed or very near realization. Several thousands of atmospheric neutrinos have been observed with energies up to several 100 TeV but so far no evidence for extraterrestrial neutrinos has been found.

1 Introduction

This paper will mainly discuss Ultra High Energy Neutrino telescopes based on the optical Cherenkov technique. For acoustic and radio based telescopes see Rene Ong's talk "Future facilities in astroparticle physics and cosmology" at this conference.

1.1 Why Neutrino Astronomy?

The Universe is not transparent for ultra high energy photons since these interact with the photons from the Cosmic Microwave Background (CMB) from the Big Bang.

$$\gamma + \gamma_{CMB} \rightarrow e^+ e^- \tag{1}$$

A similar process occurs with infrared background photons. Photons with energies about 10^{15} eV, for example, will only reach us from sources within our own galaxy. Figure 1 shows the observable distance in space as a function of the energy of photons and protons[1]. The maximum observed energy of the photons is about 10^{13} eV limiting the distances to the sources to be within 100 Mpc. Photons are not suitable to transmit information about very high energy processes far out in space.

The cosmic microwave background will also reduce the mean free path in space for protons with energies above $10^{19.5}$ eV via the interaction

$$p + \gamma_{CMB} \rightarrow \Delta^+ \tag{2}$$

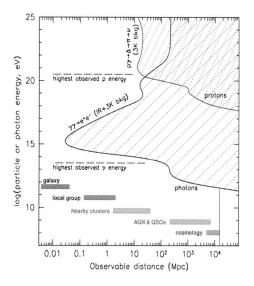

Figure 1. Observable distance in Universe as a function of particle energy[1].

This process is called GZK (after Greisen, Zatsepin, and Kusmin) and limits the distance in the Universe for ultra high energy protons. However, the use of cosmic rays for "astronomy" is not possible because intergalactic magnetic fields can change the trajectories for charged particles with energies below 10^{19} eV. Figure 1 shows the range in space as a function of the energy of the protons. The GZK process will produce UHE neutrinos via the $\pi^+ \rightarrow \mu^+ + \nu_\mu \rightarrow e^+ + \nu_e + \nu_\mu$ from the decay of the Δ^+. The detection of the "GZK neutrinos" and the observation of the expected GZK cut off in the

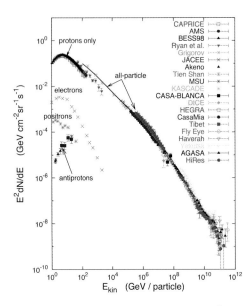

Figure 2. Cosmic ray flux vs. energy[2].

cosmic ray spectrum are two very important tasks in astroparticle physics.

Figure 2 shows the observed cosmic ray flux for different particles compiled by Gaisser[2]. The spectrum is described by power-laws with different spectral index, $\gamma=2.7$ below 10^{15} eV (the so-called "knee"), and $\gamma=3$ for energies above 10^{15} eV. This has been given the interpretation that the particles (mainly protons) are leaking out from our galaxy. The spectrum changes again to a harder spectrum above $10^{18.5}$ eV (the "ankle") which is interpreted to be due to an influx of extragalactic particles. The leading theory to explain the cosmic rays below 10^{15} eV is Fermi acceleration in galactic supernova remnants (SNR) shocks but so far no direct evidence supporting this has been obtained.

The chemical components in the Cosmic rays are identified up to about 10^{14} eV and are dominated by incoming protons. Above 10^{14} eV the fraction of heavier chemical elements seems to increase. The cosmic rays with energies above 10^{19} eV should not be deflected by magnetic fields in space and in principle point back to the source. It is inter-

esting to notice that no special source candidates seem to exist within the GZK range for protons.

The highest observed cosmic ray energies correspond to an accelerator with 10^7 times higher energy than what will be available at the Large Hadron Collider (LHC) at CERN in 2008. The maximum centre-of-mass energy in the cosmic ray interactions (assuming incoming protons) is about 25 times higher than what will be available in LHC. What kinds of sources are able to create particles with such energy?

Using cosmic neutrinos as probes opens up a new window to the Universe. The neutrinos traverse huge amount of matter without being absorbed. They are not deflected by magnetic fields and thus points back to the source. Observing neutrinos might reveal information about hidden processes close to the source. The neutrinos are also not reprocessed at the sources.

Further possibilities to study the universe will come from observing cosmic neutrinos in coincidence with photons, and/or gravitational waves.

So far only two extraterrestrial sources of neutrinos have been seen, the Sun and the Supernova SN1987a in the Large Magellanic Cloud. Both are, however, low energy neutrino sources of a few tens of MeV.

1.2 Possible cosmic High Energy Neutrino sources

Ultra High Energy (UHE) neutrinos are expected to be produced at the cosmic ray accelerators and at possible top-down sources, where neutrinos are produced by decays or annihilation of heavy particles. Candidates for galactic sources for cosmic rays and neutrinos are micro quasars and Supernova Remnants (SNR). Extragalactic sources might be Active Galactic Nuclei (AGN) or gamma ray bursts (GRB). Accelerated protons interact with photons or matter in the vicinity of the source producing pions

$$p+\gamma \rightarrow \Delta^+ \rightarrow p\pi^0 \text{ or } n\pi^+ \qquad (3)$$

The decay of the Δ^+ will produce about the same number of photons from π^0 as number of neutrinos from the π^+ decay.

For Fermi acceleration the expected neutrino flux will follow an E_ν^{-2} spectrum.

The expected flux of cosmic neutrinos can be estimated using the observed flux of high energy cosmic rays and photons assuming different production mechanisms. Many different models exist[3,4]. The expected upper bound for the total diffuse flux of neutrinos $dN/dE_\nu \sim 5 \cdot 10^{-8} E_\nu^{-2}$ GeV^{-1} cm^{-2} s^{-1} sr^{-1}.

The neutrino flavour composition at the source is ν_e:ν_μ:$\nu_\tau = 1$:2:0. Due to vacuum oscillations the flavour composition at the detector will change the to ν_μ:ν_e: $\nu_\tau = 1$:1:1.

1.3   Particle physics and neutrino astronomy

There are several questions in particle physics, which can be studied using Ultra High Energy Neutrino telescopes. Perhaps the most important one is the question about the cold dark matter in the universe, which is about six times more common than the baryonic matter. The most popular hypothesis for the nature of non-baryonic dark matter is that it consists of the lightest Supersymmetric particle (assuming R-parity conservation), the neutralino, left from the Big Bang. The mass of the neutralino is expected to be in the GeV-TeV range. The neutralinos scatter weakly on baryonic matter, loose energy and may be trapped in the gravitational field in heavy objects like the Sun and the Earth. In the centre of these objects the accumulated neutralinos will annihilate and, among other particles, neutrinos will be produced. Observing high energy neutrinos from the Sun or the centre of the Earth could be an evidence for Supersymmetric particles.

Other areas of interest for particle physics are e.g. that a neutrino telescope of the size of one km^3 will reconstruct more than 100 000 atmospheric neutrinos per year with energies up to 10^{15} eV, that the cross section of ultra high energy neutrinos is sensitive to effects of extra dimensions, that there is the possibility to search for magnetic monopoles and that the weak equivalence principle and the Lorentz invariance are possible to test.

1.4   Detection of neutrinos

The expected low flux of cosmic neutrinos and the low cross-section for neutrino interactions demands very large detector mass in order to obtain an expected signal. The probability for a neutrino to interact e.g. in 1 km of water is only $4 \cdot 10^{-7}$ / TeV.

Using optical Cherenkov detectors the choice falls on optical transparent natural media like water or ice.

High energy ν_μ's produce muons with a range in water or ice of several km (about 1 km at 300 GeV) allowing muons created far outside the instrumented detector volume to be detected. The mean angular difference between the incoming neutrino and the outgoing muons falls approximately as $E^{-0.5}$ and is about 1 degree at 1 TeV.

Electron neutrinos, tau neutrinos (at moderate energies) and neutral current interactions will produce "cascades" in which most of the secondary particles will interact and stop within a few tens of metres. The Cherenkov light will, to first order, look like it is coming from a point source inside the large detector volume. For ν_τ's at energies above several PeV the decay length for the tau will be hundreds of meters, allowing detection of the two cascades ("double bang events") from the primary interaction and the decay of the tau.

The cascades from UHE neutrino interactions can also be detected by coherent radio Cherenkov emission from the Askaryan effect[6] in matter and by acoustic waves created by the heated interaction volume. See R. Ong's presentation in this conference. Radio and acoustic detection of neutrino interactions have a higher energy threshold ($> 10^{17}$ eV) than optical Cherenkov telescopes.

Optical Cherenkov telescopes are planned to reach a volume of the order of one km^3. To reach even larger detector volumes, radio and acoustic techniques become more favourable, due to larger attenuation lengths.

The flux of down-going muons from cosmic ray interactions in the atmosphere dominates by many orders of magnitude the flux of neutrino induced muons even at large depth. In order to filter out the atmospheric muons, the Earth is used as filter and only upward going muons are accepted as neutrino induced muons. However, at very large energies the expected cosmic neutrino energy will surpass the energies of the atmospheric muons allowing an acceptance of neutrinos from above the horizon.

The Earth is opaque for ν_μ and ν_e neutrinos with energies above 10^{14} eV, giving the acceptance for PeV neutrinos around the direction of the horizon. For EeV neutrinos the acceptance is even above the horizon. The τ neutrinos will also interact in the Earth but the produced taus will decay before interacting, allowing the ν_τ to continue through the Earth but at a lower energy.

2 Neutrino telescopes

High energy neutrino detectors deep in the ocean for Neutrino Astronomy were proposed already in the 1960s[5]. The first high energy neutrino telescope was designed by the DUMAND collaboration, which aimed to deploy at a depth of 4500 metres in the ocean outside Hawaii. The project was closed in 1995 after leakage problems at the large depth. The experience gained with this project has been important for the development of the field. It took almost 40 years from the first ideas until

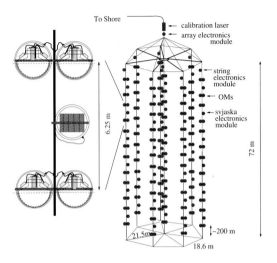

Figure 3. The Baikal NT-200 neutrino telescope.

the first high energy neutrinos (atmospheric ones so far) were detected by this kind of telescopes.

2.1 Running telescopes

Today there are two high energy neutrino telescopes taking data. The NT-200 detector in Lake Baikal, Russia, and the AMANDA telescope at the geographical South Pole, Antarctica. The NT-200 telescope is deployed at a depth of 1100 m and has been taking data since 1998. It consists of eight strings with 192 optical modules arranged in pairs and is 72 m in height, see figure 3. The effective area is >2000 m^2 for muon energies above 1 TeV. The telescope is deployed and the maintenance is done during spring using the lake ice as a platform. The first observed neutrino interaction (ν_μ) was reported by the Baikal collaboration in 1996 using only the first four strings of the telescope. So far about 300 atmospheric neutrinos have been detected. The angular resolution for muons is about 4°. In spring 2005 three new strings were deployed at a radius of 100 m below the

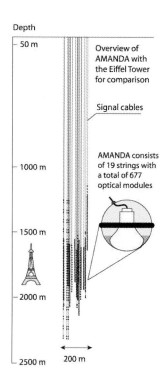

Figure 4. The AMANDA neutrino telescope

AMANDA telescope using the first 10 strings (AMANDA-B10). AMANDA is running in coincidence with the air shower detector, SPASE-II situated on the surface. SPASE-II is observing the electron component of the air shower whilst AMANDA detects the muon component. The muons observed inside AMANDA in coincidence with SPASE are used for calibration of AMANDA as well as for composition studies of the cosmic rays. For the period 2000-2003 AMANDA has reported about 3300 atmospheric neutrinos. The angular resolution for muon tracks in AMANDA is about $2.5°$.

Some recent results from the Baikal and AMANDA detectors are given in section 3.

2.2 Telescopes under constructions

In the Mediterranean there are three projects aiming to build neutrino telescopes deep in the sea. In seawater the amount of radioactive background is much larger than in Antarctic ice due to the ^{40}K decays giving Cherenkov light. A typical noise rate in a photomultiplier deep in the sea is in the order of 50 kHz whilst in ice it is only about 1 kHz. Because of the high rate in sea it is necessary to run two close-by photomultipliers in coincidence. The absorption length is shorter in water but the scattering length is larger than in ice.

Outside the French coast the ANTARES collaboration is preparing the deployment of 12 strings with in total 900 photomultipliers at a depth of 2400 m. Two prototype strings were successfully deployed in April 2005. The first real string is planned to be deployed and connected at the end of 2005 and the final configuration completed during 2007. The strings are deployed from a ship and connected to a Junction Box on the seabed with a Remotely Operated Vehicle (ROV). The final size of ANTARES is comparable with the AMANDA telescope at the South Pole.

The NESTOR collaboration is prepar-

NT200 increasing the sensitivity for cascade events with a factor of four. The new configuration is named NT-200+.

The AMANDA telescope at the South Pole in Antarctica (figure 4) is using the transparent ice as detector medium instead of water. The telescope consists of 19 strings with in total 677 optical modules. The distance between the optical modules in a string varies between 10 m and 20 m. The main sensitive volume is situated between 1500 m and 2000 m below the ice surface. The effective area for AMANDA is about 10^4 m^2 for 1 TeV muons. The optical modules are deployed in holes (about 60 cm in diameter) in the ice, which is melted by a hot water drilling system. The telescope was completed in January 2000 and is continuously taking data except for short interruptions in November and December due to service. Data were already taken by the partially finished

344

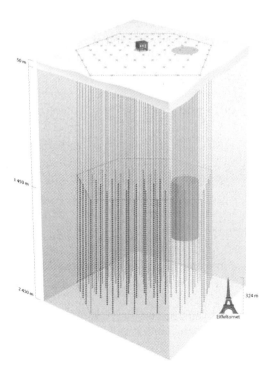

Figure 5. The IceCube Neutrino Observatory with the air shower IceTop at the surface and the IceCube telescope inside the ice. The AMANDA telescope is seen inside IceCube as a darker cylinder.

ing a tower-based telescope outside Pylos, Greece, at a depth of about 4000 m. The telescope will consist of 12 floors with 6 arms, each with one optical module facing upwards and one downwards. The diameter of the floor is 32 m and vertical distance between floors is 30 m. One floor was successfully deployed and operated in 2003 [7].

The NEMO collaboration investigates a site for a possible km^3 telescope 80 km outside the coast of Sicily (Capo Passero) at about 3500 m depth. They are investigating optical properties of the water, optical background, deep-sea currents and other parameters. They are also developing technologies for a large telescope with 5832 PMTs, 81 towers, 18 floors, 20 m bar length, with one PMT facing downwards and one horizon-

tally. An interesting technique has been developed to deploy a tower with optical modules to the bottom of the ocean by folding the floors with bars carrying the optical modules into a very compact unit. When the unit is placed on the bottom the bars are released to float up to their positions in the tower. In order to test technical solutions for a large telescope the collaboration has started a prototyping activity at a site 25 km off the coast of Sicily outside Catania at a depth of 2000 m. The project is named NEMO-Phase 1 and will consist of junction box, underwater cables and two towers, one with four floors and one with 16 floors. An electro-optical cable connected to the shore station was installed in January 2005 together with prototype instrumentation for acoustic detection.

The IceCube Neutrino Observatory (figure 5) at the Amundsen-Scott base at the South Pole, Antarctica, is a funded project, which has started to be installed. It consists of the neutrino telescope, IceCube, deep in the ice and the IceTop air shower array at the surface above IceCube. In the ice, at least 70 strings with 60 Digital Optical Modules (DOM) each will be deployed between 1450 m and 2450 m depth (17 m between optical modules). The distance between the strings is 125 m. The instrumented deep ice part will cover about 1 km^3. The IceTop air shower array will consist of two ice Cherenkov tanks placed close to each IceCube hole. The surface array will be used for calibration and background studies as well as for cosmic ray studies using the combined detector. The AMANDA telescope (inside the volume of IceCube) will be integrated in IceCube and used as a low energy part of IceCube. The modules in IceCube can also be used as veto in order to improve AMANDAs acceptance for neutrinos from neutralino annihilation in the Sun. The first IceCube string and four IceTop stations were successfully deployed in January 2005. Ten more strings are planned to be deployed during the coming season.

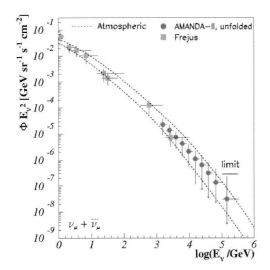

Figure 6. Atmospheric neutrino flux vs. neutrino energy from AMANDA-and the Frejus experiment.

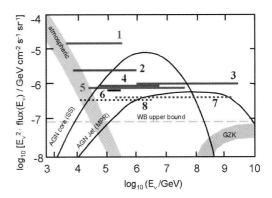

Figure 7. All-flavour neutrino limits and sensitivity on an $E^2 \frac{dN}{dE}$ plot. Neutrino oscillations are taken into account. (1) The MACRO ν_μ analysis for 5.8 years (limit multiplied by three for oscillations)[16]. (2) The AMANDA-B10 ν_μ analysis from1997 (multiplied by three for oscillations)[17]. (3) AMANDA-B10 ultra-high energy neutrinos of all flavours 1997[18]. (4) AMANDA all-flavour cascade limit from 2000[19]. (5) Baikal cascades 1998-2003[20]. (6) The preliminary results of the 2000 AMANDA ν_μ analysis (multiplied by three for oscillations). The limits derived after unfolding the atmospheric neutrino spectrum. (7) The sensitivity for AMANDA ultra-high energy neutrinos of all flavours 2000[21]. (8) 2000 to 2003 AMANDA ν_μ sensitivity (multiplied by three for oscillations)[22].

The last string is expected to be deployed in 2010. The telescope is modular and will add new strings into the data acquisition system as soon as they are deployed and commissioned, giving an increasing sensitivity year by year.

2.3 Future Cherenkov telescopes

The different Mediterranean groups are aiming to collaborate in order to build one full size km^3 neutrino telescope which nicely will complement IceCube in the northern hemisphere allowing a full sky coverage. A European Design Study activity for a km^3 (KM3NeT) telescope has been approved by EU. A completed northern km^3 Mediterranean telescope could be finished around 2012 at the earliest .

2.4 Multi km^3 neutrino detectors

To detect GZK neutrinos, detector volumes even larger than the km^3 scale are planned, relying on radio- or acoustic detection. The ANITA project[8] is a balloon-borne radio experiment using a large fraction of the Antarctic ice sheet as neutrino interaction medium.

A test flight was carried out in 2005 and a full flight is scheduled for end of 2006.

3 Results

3.1 Atmospheric Neutrinos

The cosmic rays interacting in the atmosphere produce a flux of atmospheric neutrinos. These are background when searching for high energy cosmic neutrinos. The flux spectrum for the atmospheric neutrinos is about $E_{\nu_\mu}^{-3.7}$ and differs from the "expected" $E_{\nu_\mu}^{-2}$ for extra-terrestrial neutrinos. The atmospheric neutrinos are very useful to study the efficiency of the detector,

AMANDA has used a regular unfolding method to estimate the atmospheric neutrino energies from the observed upward going muon energies. Figure 6 shows the preliminary atmospheric neutrino flux for

Amanda search for neutrinos
from 33 preselected objects

Candidate	$\delta(°)$	α(h)	$n_{\rm obs}$	$n_{\rm b}$	$\Phi_\nu^{\rm lim}$	Candidate	$\delta(°)$	α(h)	$n_{\rm obs}$	$n_{\rm b}$	$\Phi_\nu^{\rm lim}$
TeV Blazars						*SNR & Pulsars*					
Markarian 421	38.2	11.07	6	5.6	0.68	SGR 1900+14	9.3	19.12	3	4.3	0.35
Markarian 501	39.8	16.90	5	5.0	0.61	Geminga	17.9	6.57	3	5.2	0.29
1ES 1426+428	42.7	14.48	4	4.3	0.54	Crab Nebula	22.0	5.58	10	5.4	1.3
1ES 2344+514	51.7	23.78	3	4.9	0.38	Cassiopeia A	58.8	23.39	4	4.6	0.57
1ES 1959+650	65.1	20.00	5	3.7	1.0						
GeV Blazars						*Microquasars*					
QSO 0528+134	13.4	5.52	4	5.0	0.39	SS433	5.0	19.20	2	4.5	0.21
QSO 0235+164	16.6	2.62	6	5.0	0.70	GRS 1915+105	10.9	19.25	6	4.8	0.71
QSO 1611+343	34.4	16.24	5	5.2	0.56	GRO J0422+32	32.9	4.36	5	5.1	0.59
QSO 1633+382	38.2	16.59	4	5.6	0.37	Cygnus X1	35.2	19.97	4	5.2	0.40
QSO 0219+428	42.9	2.38	4	4.3	0.54	LS I +61 303	61.2	2.68	3	3.7	0.60
QSO 0954+556	55.0	9.87	2	5.2	0.22	Cygnus X3	41.0	20.54	6	5.0	0.77
QSO 0954+556	55.0	9.87	2	5.2	0.22	XTE J1118+480	48.0	11.30	2	5.4	0.20
						CI Cam	56.0	4.33	5	5.1	0.66
Miscellaneous											
3EG J0450+1105	11.4	4.82	6	4.7	0.72	J2032+4131	41.5	20.54	6	5.3	0.74
M 87	12.4	12.51	4	4.9	0.39	NGC 1275	41.5	3.33	4	5.3	0.41
UHE CR Doublet	20.4	1.28	3	5.1	0.30	UHE CR Triplet	56.9	11.32	6	4.7	0.95
AO 0535+26	26.3	5.65	5	5.0	0.57	PSR J0205+6449	64.8	2.09	1	3.7	0.24
PSR 1951+32	32.9	19.88	2	5.1	0.21						

Table 1. Results from the AMANDA search for neutrinos from selected objects. δ is the declination in degrees, α the right ascension in hours, n_{obs} is the number of observed events, and n_b the expected background. $\Phi_\nu^{\rm lim}$ is the 90% CL upper limit in units of 10^{-8}cm^{-2}s^{-1} for a spectral index of 2 and integrated above 10 GeV. These results are preliminary (systematic errors are not included).

AMANDA data taken during 2000[11]. The maximum neutrino energy observed is about 300 TeV (about 1000 times higher than available neutrino beams at FNAL or CERN). Also shown are the lower energy results from Frejus[9]. The dotted curves in figure 6 are the horizontal and vertical fluxes parameterized according to Volkov above 100 GeV and Honda below 100 GeV[10].

3.2 Diffuse fluxes

Neutrinos from different cosmic sources with too few events to be detected as individual point sources will add up as a "diffuse" flux of cosmic neutrinos. The signal for diffuse cosmic neutrinos will be seen as high energy events above the atmospheric neutrino energy distribution due to the expected harder spectrum. In figure 6 a 90 % CL upper limit for an extraterrestrial $E_{\nu_\mu}^{-2}$ flux component is shown between 100 TeV and 300 TeV giving an upper limit of $E^2\,\Phi_{\nu_\mu} < 2.6 \cdot 10^{-7}$

GeV cm^{-2} sr^{-1} s^{-1} . Figure 7 shows several different obtained limits for diffuse cosmic neutrinos (sum of all flavours). These are based on both ν_μ and cascade event analysis. The single neutrino flavour limits have been multiplied by a factor of three for neutrino oscillations to obtain the all flavour flux at Earth. The lines labelled 7 and 8 in figure 7 are not limits but expected sensitivity since the data were still not unblinded. The sensitivity for detecting a diffuse cosmic neutrino flux has increased by an order of magnitude in less than 10 years.

3.3 Point source searches

The most sensitive search for neutrino point sources on the northern sky has been performed by the AMANDA collaboration using 3329 ν_μ events (purity about 95 %) obtained during the years 2000-2003[23]. Figure 8 shows the directions in the northern sky (in celestial coordinates) and the corresponding sig-

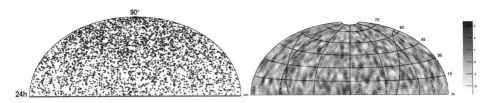

Figure 8. Left: AMANDA northern sky-plot (in celestial coordinates) for 3329 up-going neutrino candidates. Right: The corresponding significance map for a search for point sources showing the deviation from a uniform background. The scale corresponds to -3σ (white) to +3σ (black). No statistical evidence for cosmic neutrino point sources is observed

nificance map for the AMANDA events. No evidence for any extraterrestrial point source is found. The highest excess corresponds to a significance of about 3.4σ. The probability to observe this or a higher excess, taking into account the trial factor, is 92 %.

A search for 33 preselected neutrino candidate sources including galactic and extra-galactic sources has been done by the AMANDA collaboration[23] based on the four years data. The result is shown in Table 1. The highest excess is found in the direction of the Crab Nebula, with 10 observed events compared to an average of 5.4 expected background events (about 1.7σ). The probability that a background fluctuation produces this or a larger deviation in any of the 33 search bins is 64 %, taking into account the trial factor (due to the multiplicity of the directions examined and the correlation between overlapping search bins).

3.4 Gamma-ray bursts

Gamma-ray bursts are candidate sources for the ultra high energy cosmic rays. They are the most energetic objects in the universe and observing neutrinos in coincidence with the bursts would give evidence for acceleration of hadrons. The search for neutrinos in coincidence in direction and time with a GRB is almost free of background from atmospheric neutrinos. AMANDA and Super Kamiokande[12] have searched for neutrinos from GRB without any observed signal.

AMANDA has presented the most sensitive limits[13,14,15].

3.5 Searches for neutralino dark matter

The underground neutrino detectors Baksan and Super-Kamiokande have searched for neutrinos from neutralino annihilation in the Earth and the Sun for many years. The high energy neutrino telescopes Baikal and AMANDA have started to give limits. The most sensitive limits on muon flux due to neutrinos from neutralino annihilation in the centre of the Earth has been given by AMANDA using data from 1997-1999 (AMANDA-B10)[24]. The best corresponding limit for the Sun is given by the Super-Kamiokande collaboration[25]. AMANDA has presented a limit based on one year of data (2000)[26] and will, when all available data are analyzed, be comparable with the best direct search limit by the CDMS-II collaboration[27] (see figure 9). The km^3 telescopes will be able to probe MSSM models not accessible for direct searches. The direct and indirect methods are complementary since they probe different parts of the neutralino velocity distribution.

3.6 Magnetic monopole search

Magnetic monopoles with a magnetic Dirac charge and moving above the Cherenkov threshold in water ($\beta > 0.75$) would emit Cherenkov light 8300 times more intense than

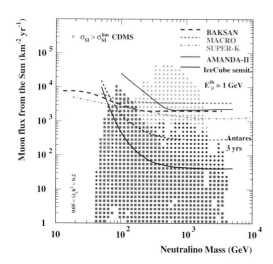

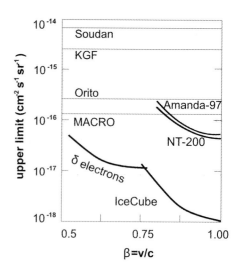

Figure 10. Upper limits, at 90 % CL, for magnetic monopoles as function of monopole velocity.

Figure 9. Limits on the muon flux due to neutrinos from neutralino annihilations in the Sun as function of neutralino mass. The points correspond to predictions from different MSSM models[28]. The dots correspond to models excluded by CDMS-II [27]. The plus signs (+) correspond to models testable to a tenfold improved sensitive for direct searches. The crosses (x) are models demanding more than a factor of 10 in increased sensitivity for direct searches.

4 Conclusions

Ultra high energy neutrino astronomy has now become a reality but so far no extraterrestrial neutrinos have been observed. The sensitivity is increasing every year and the new km^3 telescopes like IceCube will strongly improve existing limits or observe extraterrestrial neutrinos in the coming years.

References

1. P. Gorham, 1st International Workshop on the Saltdome Shower Array (SalSA), SLAC, Feb 2005
2. T. Gaisser 2005. Private communication.
3. J. Bahcall and E. Waxman, *Phys. Rev.* **D64**, 023002 (2001)
4. K. Mannheim, R. J. Protheroe and J. R. Rachen, *Phys. Rev.* **D63**, 023003 (2001); T. Gaisser 2005. Private communication.
5. M. Markov, Proc. 1960 Int. Conf. on High Energy Physics, Rochester 1960, 578
6. G. Askaryan, *Zh. Eksp. Fiz.* **41**, 616 (1961) [*Soviet Physics JETP* **14**, 441

a low energy muon. In high energy neutrino telescope this is a very strong signal and is easy to observe. Figure 10 shows the most recent flux limits obtained from different detectors. The km^3 telescopes like IceCube will be able to improve limits by more than an order of magnitude, or detect a signal. Detecting the δ-electrons emitted along the monopole path would make it possible to extend the search to lower β. In general large neutrino detectors are able to look for exotic particles like GUT magnetic monopoles catalyzing proton decay, Q-balls or nuclearites. The trigger in IceCube will make it possible to search for very slowly moving bright particles, which was not the case in AMANDA.

(1962)]; G. A. Akaryan, *Zh. Eksp. Fiz.* **48**, 988 (1965) [*Soviet Physics JETP* **21**, 658 (1965)]

7. G. Aggouras *et al.*, *Astropart. Phys.* Vol **23**, 377-392, (2005)

8. P. Miocinovic *et al.*, astro-ph/0503304.

9. K. Daum, W. Rhode, Frejus Collaboration, *Zeitschrift fur Physik* C**66** 177, (1995)

10. L. V. Volkowa, *Sov. J. Nucl. Phys.* **31** (1980); L. V. Volkowa, G.T. Zatsepin, *Sov. J. Nucl. Phys.* **37**, 212 (1980); M. Honda *et al*, *Phys. Rev.* D**52**, 4985 (1995)

11. K. Münich for the IceCube Collaboration, Proc. 29th Int. Cosmic Ray Conf., Pune, India (2005)

12. S. Fukuda *et al.*, *ApJ*, **578**, 317 (2002)

13. K. Kuehn for the IceCube Collaboration, Proc. 29th Int. Cosmic Ray Conf., Pune, India (2005)

14. M. Stamatikos for the IceCube Collaboration, Proc. 29th Int. Cosmic Ray Conf., Pune, India (2005), astro-ph/0510336

15. B. Hughey for the IceCube Collaboration, Proc. 29th Int. Cosmic Ray Conf., Pune, India (2005), astro-ph/0509570

16. M. Ambrosio *et al.*, *Astropart. Phys.* **19**, 1 (2003)

17. J. Ahrens *et al.*, *Phys. Rev. Lett.* **90**, 251101 (2003)

18. M. Ackermann *et al.*, *Astropart. Phys.* **22**, 339 (2005)

19. M. Ackermann *et al.*, *Astropart. Phys.* **22**, 127 (2004)

20. V. Aynutdinov *et al.*, for the Baikal Collaboration, Proc. 29th Int. Cosmic Ray Conf., Pune, India (2005)

21. L. Gerhardt for the IceCube Collaboration, Proc. 29th Int. Cosmic Ray Conf., Pune, India (2005)

22. J. Hodges for the IceCube Collaboration, Proc. 29th Int. Cosmic Ray Conf., Pune, India (2005)

23. M. Ackermann for the IceCube Collaboration, Proc. 29th Int. Cosmic Ray Conf., Pune, India (2005)

24. P. Ekström. PhD thesis, Stockholm University, 2004. ISBN 91-7265-886-X; D. Hubert, A. Davour and C. P. de los Heros for the IceCube Collaboration, Proc. 29th Int. Cosmic Ray Conf., Pune, India (2005)

25. S. Desai *et al.*, *Phys. Rev.* D**70**, 083523 (2004); erratum *ibid.*, **D70**, 109901 (2004)

26. M. Ackermann *et al.*, accepted for publication in Astropart. Phys. [astro-ph/0508518]

27. D. S. Akerib *et al.*, astro-ph/0509269

28. J. Edsjö, private communication.

DARK MATTER SEARCHES

LAURA BAUDIS

Physics Department, University of Florida, Gainesville, FL 32611, USA
E-mail: lbaudis@ufl.edu

More than 90% of matter in the Universe could be composed of heavy particles, which were non-relativistic, or 'cold', when they froze-out from the primordial soup. I will review current searches for these hypothetical particles, both via interactions with nuclei in deep underground detectors, and via the observation of their annihilation products in the Sun, galactic halo and galactic center.

1 Introduction

Seventy two years after Zwicky's first accounts of dark matter in galaxy clusters, and thirty five years after Rubin's measurements of rotational velocities of spirals, the case for non-baryonic dark matter remains convincing. Precision observations of the cosmic microwave background and of large scale structures confirm the picture in which more than 90% of the matter in the universe is revealed only by its gravitational interaction. The nature of this matter is not known. A class of generic candidates are weakly interacting massive particles (WIMPs) which could have been thermally produced in the very early universe. It is well known that if the mass and cross section of these particles is determined by the weak scale, the freeze-out relic density is around the observed value, $\Omega \sim 0.1$. The prototype WIMP candidate is the neutralino, or the lightest supersymmetric particle, which is stable in supersymmetric models where R-parity is conserved. Another recently discussed candidate is the lightest Kaluza-Klein excitation (LKP) in theories with universal extra dimensions. If a new discrete symmetry, called KK-parity is conserved, and if the KK particle masses are related to the weak scale, the LKP is stable and makes an excellent dark matter candidate. A vast experimental effort to detect WIMPs is underway. Cryogenic direct detection experiments are for the first time probing the parameter space predicted by SUSY theories for neutralinos, while indirect detection experiments may start to probe the distribution of dark matter in the halo and galactic center. In the following, I will give a brief overview of the main search techniques, focusing on most recent results.

2 Direct Detection

WIMPs can be detected directly, via their interactions with nuclei in ultra-low-background terrestrial targets[1]. Direct detection experiments attempt to measure the tiny energy deposition ($<50\,\mathrm{keV}$) when a WIMP scatters off a nucleus in the target material. Predicted event rates for neutralinos range from 10^{-6} to 10 events per kilogram detector material and day, assuming a typical halo density of $0.3\ \mathrm{GeV/cm^3}$. The nuclear recoil spectrum is featureless, but depends on the WIMP and target nucleus mass. Figure 1 shows differential spectra for Si, Ar, Ge and Xe, calculated for a WIMP mass of $100\,\mathrm{GeV}$, a WIMP-nucleon cross section of $\sigma = 10^{-43}$ cm^2 and using standard halo parameters.

Basic requirements for direct detection detectors are low energy thresholds, low backgrounds and high masses. The recoil energy of the scattered nucleus is transformed into a measurable signal, such as charge, scintillation light or phonons, and at least one of the above quantities is detected. Observing two signals simultaneously yields a powerful discrimination against background events, which are mostly interactions with electrons,

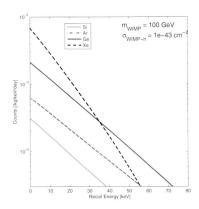

Figure 1. Differential WIMP recoil spectrum for a WIMP mass of 100 GeV and a WIMP-nucleon cross section $\sigma = 10^{-43} \text{cm}^2$. The spectrum was calculated for illustrative nuclei such as Si (light solid), Ar (light dot-dashed), Ge (dark solid), Xe (dark dashed).

as opposed to WIMPs and neutrons scattering off nuclei.

In order to convincingly detect a WIMP signal, a specific signature from a particle populating our galactic halo is important. The Earth's motion through the galaxy induces both a seasonal variation of the total event rate[2,3] and a forward-backward asymmetry in a directional signal[4,5]. The expected seasonal modulation effect is of the order of $\mathcal{O}(15\text{kms}^{-1}/220\text{kms}^{-1}) \approx 0.07$, requiring large masses and long counting times as well as an excellent long-term stability of the experiment. The forward-backward asymmetry yields a larger effect, of the order of $\mathcal{O}(v_\odot/220\text{km/s}) \approx 1$, and fewer events are needed to discover a WIMP signal[5]. The challenge is to build massive detectors capable of detecting the direction of the incoming WIMP.

2.1 Experiments

First limits on WIMP-nucleon cross sections were derived about twenty years ago, from at that time already existing germanium double beta decay experiments[6]. With low intrinsic backgrounds and already operating in underground laboratories, these detectors were essential in ruling out first WIMP candidates such as a heavy Dirac neutrino[7]. Present Ge ionization experiments dedicated to dark matter searches such as HDMS[8] are limited in their sensitivity by irreducible electromagnetic backgrounds close to the crystals or from production of radioactive isotopes in the crystals by cosmic ray induced spallation. Next generation projects based on high-purity germanium (HPGe) ionization detectors, such as the proposed GENIUS[9], GERDA[10], and Majorana[11] experiments, aim at an absolute background reduction by more than three orders in magnitude, compensating for their inability to differentiate between electron- and nuclear recoils on an event-by-event basis. Solid scintillators operated at room temperatures had soon caught up with HPGe experiments, despite their higher radioactive backgrounds. Being intrinsically fast, these experiments can discern on a statistical basis between electron and nuclear recoils, by using the timing parameters of the pulse shape of a signal. Typical examples are NaI experiments such as DAMA[12] and NAIAD[13], with DAMA reporting first evidence for a positive WIMP signal in 1997[14]. The DAMA results have not been confirmed by three different mK cryogenic experiments (CDMS[15], CRESST[16] and EDELWEISS[17]) and one liquid xenon experiment (ZEPLIN[18]), independent of the halo model assumed[19] or whether the WIMP-nucleon interaction is taken as purely spin-dependent[20,21]. The DAMA collaboration has installed a new, 250 kg NaI experiment (LIBRA) in the Gran Sasso Laboratory, and began taking data in March 2003. With lower backgrounds and increased statistics, LIBRA should soon be able to confirm the annual modulation signal. The Zaragosa group plans to operate a 107 kg NaI array (ANAIS) at the Canfranc Underground Laboratory (2450 mwe) in Spain[23], and deliver an independent check of the DAMA signal in NaI. Cryogenic experiments operated at sub-Kelvin temperatures are now leading the field with sensitivities of one order of magnitude above the best solid scintillator experiments. Specifically, the CDMS experiment can probe WIMP-nucleon cross sections as low as 10^{-43}cm^2 [24]. Liquid noble element detectors are rapidly evolving, and seem a very promising avenue towards the goal of constructing ton-scale or even multiton WIMP detectors. Many other interesting WIMP search techniques have been deployed, yet it is not the scope of this paper to deliver a full overwiev (for two recent reviews see[25,26]).

2.2 Cryogenic Detectors at mK Temperatures

Cryogenic calorimeters are meeting crucial characteristics of a successful WIMP detector: low energy threshold (<10 keV), excellent energy resolution (<1% at 10 keV) and the ability to differentiate nuclear from electron recoils on an event-by-event basis. Their development was driven by the exciting possibility of doing a calorimetric energy measurement down to very low energies with unsurpassed energy resolution. Because of the T^3 dependence of the heat capacity of a dielectric crystal, at low temperatures a small energy deposition can significantly change the temperature of the absorber. The change in temperature is measured either after the phonons (or lattice vibration quanta) reach equilibrium, or thermalize, or when they are still out of equilibrium, or athermal, the latter providing additional information about the location of an event.
CDMS: the Cold Dark Matter Search experiment operates low-temperature Ge and Si detectors at the Soudan Underground Laboratory in Minnesota (at a depth of 2080 m.w.e.). The high-purity Ge and Si crystals are 1 cm thick and 7.6 cm in diameter, and have a mass of 250 g and 100 g, respectively. Superconducting transition edge sensors photolitographically patterned onto one of the

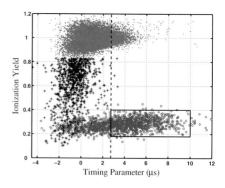

Figure 2. Ionization yield versus phonon timing parameter for ^{133}Ba gamma calibration events (dots and crosses) and ^{252}Cf neutron calibration events (circles). Low-yield ^{133}Ba events (crosses) have small values of the timing parameter, and the dashed vertical line indicates a timing cut, resulting in a high rate of nuclear recoil efficiency and a low rate of misidentified surface events.

crystal surfaces detect the athermal phonons from particle interactions. The phonon sensors are divided into 4 different channels, allowing to reconstruct the x-y position of an event with a resolution of ∼1 mm. If an event occurs close to the detector's surface, the phonon signal is faster than for events far from the surface, because of phonon interactions in the thin metallic films. The risetime of the phonon pulses, as well as the time difference between the charge and phonon signals allow to reject surface events caused by electron recoils. Figure 2 shows the ionization yield (ratio of ionization to recoil energy) versus the sum of above timing parameters for electron recoil events (collected with a ^{133}Ba source) and nuclear recoil events (collected with a ^{252}Cf source). Events below a yield around 0.75 typically occur within 0-30 μm of the surface, and can be effectively discriminated while preserving a large part of the nuclear recoil signal.

Charge electrodes are used for the ionization measurement. They are divided into an inner disk, covering 85% of the surface, and an outer ring, which is used to reject events near the edges of the crystal, where background interactions are more likely to occur. The discrimination against the electron recoil background is based on the fact that nuclear recoils (caused by WIMPs or neutrons) produce fewer charge pairs than electron recoils of the same energy. The ionization yield is about 0.3 in Ge, and 0.25 in Si for recoil energies above 20 keV. Electron recoils with complete charge collection show an ionization yield of ≈ 1. For recoil energies above 10 keV, bulk electron recoils are rejected with >99.9% efficiency, and surface events are rejected with >95% efficiency. The two different materials are used to distinguish between WIMP and neutron interactions by comparing the rate and the spectrum shape of nuclear recoil events.

A stack of six Ge or Si detectors together with the corresponding cold electronics is named a 'tower'. Five towers are currently installed in the 'cold volume' at Soudan, shielded by about 3 mm of Cu, 22.5 cm of Pb, 50 cm of polyethylene and by a 5 cm thick plastic scintillator detector which identifies interactions caused by cosmic rays penetrating the Soudan rock. In 2004, two towers were operated for 74.5 live days at Soudan, yielding an exposure of 34 kg d in Ge and 12 kg d in Si in the 10-100 keV nuclear recoil energy range. One candidate nuclear recoil event at 10.5 keV was observed in Ge, while no events were seen in the Si data[24]. This result was consistent with the expected background from surface events, and resulted in a new upper limit on spin-independent WIMP-nucleon cross sections in Ge of 1.6×10^{-43} cm^2 at the 90%CL at a WIMP mass of 60 GeV/c^2 (see Fig. 3).

The limits on spin-dependent WIMP interactions, shown in Fig. 4, are competitive with other experiments, in spite of the low abundance of ^{73}Ge (7.8%) in natural germanium. In particular, in the case of a pure neutron coupling, CDMS yields the most stringent limit obtained so far, thus strongly constraining interpretations of the DAMA signal region[21,27].

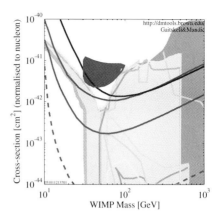

Figure 3. Experimental results and theoretical predictions for spin-independent WIMP nucleon cross sections versus WIMP mass. The data (from high to low cross sections) show the DAMA allowed region (red)[14], the latest EDELWEISS result (blue)[17], the ZEPLIN I result (green)[18] and the CDMS results from 2 towers at Soudan (red)[24]. Also shown is the expectation for 5 CDMS towers at Soudan (red dashed). The SUSY theory regions are shown as filled regions or contour lines, and are taken from [22].

EDELWEISS: the EDELWEISS experiment operates Ge bolometers at 17 mK in the Laboratoire Souterrain de Modane, at 4800 m.w.e. The detectors are further shielded by 30 cm of paraffin, 15 cm of Pb and 10 cm of Cu. They simultaneously detect the phonon and the ionization signals, allowing a discrimination against bulk electron recoils of better than 99.9% above 15 keV recoil energy. The charge signal is measured by Al electrodes sputtered on each side of the crystals, the phonon signal by a neutron transmutation doped (NTD) heat sensor glued onto one of the charge collection electrodes. The NTD sensors read out the thermal phonon signal on a time scale of about 100 ms.

Between 2000-2003, EDELWEISS performed four physics runs with five 320 g Ge crystals, accumulating a total exposure of 62 kg days[17]. Above an analysis threshold of 20 keV, a total of 23 events compatible with nuclear recoils have been observed. Figure 5 shows the ionization yield versus recoil energy for one EDELWEISS detector for an exposure

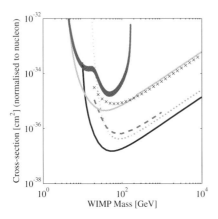

Figure 4. Experimental results for spin-dependent WIMP couplings (90% C.L. contours), for the case of a pure neutron coupling. The curves (from high to low cross sections) show the DAMA annual modulation signal (filled red region), the CDMS Soudan Si data (red crosses), the CDMS Stanford Si data (cyan), EDELWEISS (magenta dashed), DAMA/Xe (green dotted) and the CDMS Soudan Ge data (solid blue).

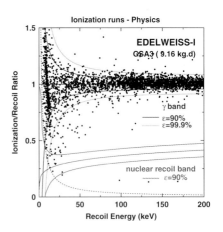

Figure 5. Ionization yield versus recoil energy for one EDELWEISS 320 g Ge detector with an exposure of 9.16 kg days. Also shown are the electron recoil (blue) and neutron recoil (red) bands. Figure taken from[17].

of 9.16 kg days. The derived upper limits on WIMP-nucleon couplings under the hypothesis that all above events are caused by WIMP interactions, and for a standard isothermal halo, are shown in Fig. 3 and Fig. 4.

The EDELWEISS experiment has ceased running in March 2004, in order to allow the upgrade to a second phase, with an aimed sensitivity of 10^{-44}cm^2. The new 50 liter low-radioactivity cryostat will be able to house up to 120 detectors. Because of the inability of slow thermal detectors to distinguish between low-yield surface events and nuclear recoils and the inherent radioactivity of NTD sensors, the collaboration has been developing a new design based on NbSi thin-film sensors. These films, besides providing a lower mass and radioactivity per sensor, show a strong difference in the pulse shape, depending on the interaction depth of an event[29]. The EDELWEISS collaboration plans to operate twenty-one 320 g Ge detectors equipped with NTD sensors, and seven 400 g Ge detectors with NbSi thin-films in the new cryostat starting in 2005.

CRESST: the CRESST collaboration has developed cryogenic detectors based on CaWO$_4$ crystals, which show a higher light yield at low temperatures compared to other scintillating materials. The detectors are also equipped with a separate, cryogenic light detector made of a $30\times30\times0.4$ mm^3 silicon wafer, which is mounted close to a flat surface of the CaWO$_4$ crystal. The temperature rise in both CaWO$_4$ and light detector is measured with tungsten superconducting phase transition thermometers, kept around 10 mK, in the middle of their transition between the superconducting and normal conducting state. A nuclear recoil in the 300 g CaWO$_4$ detector has a different scintillation light yield than an electron recoil of the same energy, allowing to discriminate between the two type of events when both the phonon and the light signals are observed. The advantage of CaWO$_4$ detectors is their low energy threshold in the phonon signal, and the fact that no light yield degradation for surface events has been detected so far. However, about 1% or less of the energy deposited in the CaWO$_4$ is seen as scintillation light[16]. Only a few tens of photons are emitted per keV electron recoil, a number which is further diminished for nuclear recoils, because of the involved quenching factor. The quenching factor of oxygen nuclear recoils for scintillation light is around 13.5% relative to electron recoils[16], leading to a rather high effective recoil energy threshold for the detection of the light signal. While neutrons will scatter predominantly on oxygen nuclei, it is expected that WIMPs will more likely scatter on the heavier calcium and tungsten.

The most recent CRESST results[16] were obtained by operating two 300 g $CaWO_4$ detectors at the Gran Sasso Underground Laboratory (3800 m.w.e) for two months at the beginning of 2004. The total exposure after cuts was 20.5 kg days. A total of 16 events were observed in the 12 keV - 40 keV recoil energy region, a number which seems consistent with the expected neutron background, since the experiment had no neutron shield at that time. No phonon-only events (as expected for WIMP recoils on tungsten) were observed between 12 keV - 40 keV in the module with better resolution in the light channel, yielding a limit on coherent WIMP interaction cross sections similar to the one obtained by EDELWEISS. CRESST has stopped taking data in March 2004, to upgrade with a neutron shield, an active muon veto, and a 66-channels SQUID read-out system. It will allow to operate 33 $CaWO_4$ detector modules, providing a total of 10 kg of target material and a final sensitivity of 10^{-44}cm^2.

2.3 Liquid Xenon Detectors

Liquid xenon (LXe) has excellent properties as a dark matter detector. It has a high density (3 g/cm^3) and high atomic number (Z=54, A=131.3), allowing experiments to be compact. The high mass of the Xe nucleus is favorable for WIMP scalar interactions provided that a low energy threshold can be achieved (Fig. 1 shows a comparison with other target nuclei). LXe is an intrinsic scintillator, having high scintillation ($\lambda =$ 178 nm) and ionization yields because of its low ionization potential (12.13 eV). Scintillation in LXe is produced by the formation of excimer states, which are bound states of ion-atom systems. If a high electric field ($\sim$1 kV/cm) is applied, ionization electrons can also be detected, either directly or through the secondary process of proportional scintillation. The elastic scattering of a WIMP produces a low-energy xenon recoil, which loses its energy through ionization and scintillation. Both signals are quenched when compared to an electron recoil of the same energy, but by different amounts, allowing to use the ratio for distinguishing between electron and nuclear recoils. The quenching factors depend on the drift field and on the energy of the recoil. At zero electric field, the relative scintillation efficiency of nuclear recoils in LXe was recently measured to be in the range of 0.13-0.23 for Xe recoil energies of 10 keV-56 keV [30].

ZEPLIN: the Boulby Dark Matter collaboration has been operating a single-phase LXe detector, ZEPLIN I, at the Boulby Mine ($\sim$3000 m.w.e.) during 2001-2002. ZEPLIN I had a fiducial mass of 3.2 kg of liquid xenon, viewed by 3 PMTs through silica windows and inclosed in a 0.93 ton active scintillator veto. A total exposure of 293 kg days had been accumulated. With a light yield of 1.5 electrons/keV, the energy threshold was at 2 keV electron recoil (corresponding to 10 keV nuclear recoil energy for a quenching factor of 20%). A discrimination between electron and nuclear recoils was applied by using the difference in the mean time of the corresponding pulses. Using this statistical discrimination method, a limit on spin-independent WIMP cross sections comparable to CRESST and EDELWEISS has been achieved (see Fig. 3).

The collaboration has developed two concepts for dual-phase detectors, ZEPLIN II and ZEPLIN III. ZEPLIN II will have a 30 kg fiducial target mass, observed by 7 PMTs.

356

ZEPLIN III will operate a lower target mass (6 kg LXe viewed by 31 PMTs) at a higher field (> 5 kV/cm). Both ZEPLIN II and ZEPLIN III are now being deployed at the Boulby Mine and are expected to take data by 2006.

XENON: the XENON collaboration, including groups from US, Italy and Portugal, will operate a 10 kg dual-phase detector in the Gran Sasso Underground Laboratory by 2005-2006. At present, a 3 kg prototype is under operation above ground, at the Columbia Nevis Laboratory. The detector is operated at a drift field of 1 kV/cm, and both primary and proportional light are detected by an array of seven 2 inch PMTs operating in the cold gas above the liquid. The performance of the chamber was tested with gamma (^{57}Co), alpha (^{210}Pb) and neutron (^{241}AmBe) sources. The depth of an event is reconstructed by looking at the separation in time between the primary and proportional scintillation signal. The x-y position is inferred with a resolution of 1 cm from the center of gravity of the proportional light emitted close to the seven PMTs. The measured ratio of proportional light (S2) to direct light (S1) for alpha recoils is 0.03 if the corresponding ratio for gamma events (electron recoils) is normalized to 1, providing a very clear separation between these type of events.

More interesting is the ratio S2/S1 for nuclear recoil events. It was established using a ^{241}AmBe neutron source, by selecting events which were tagged as neutron recoils in a separate neutron detector placed under a scattering angle of 130 deg. If the S2/S1 ratio for electron recoils (provided by a ^{137}Cs source) is normalized to 1, then S2/S1 for nuclear recoils was measured to be around 0.1, the leakage of electron recoils into the S2/S1 region for nuclear recoils being < 1% [31]. Figure 6 shows a histogram of S2/S1 for events taken with the ^{241}AmBe source, compared to the corresponding distribution of events from the ^{137}Cs gamma source.

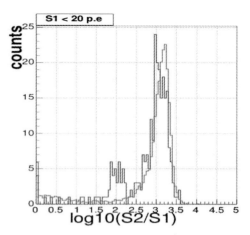

Figure 6. Histogram showing the S2/S1 distribution for AmBe events (blue) versus 662 keV gamma events from ^{137}Cs (red). Two distinct populations are visible in the AmBe data (from[31]).

The first XENON detector with a fiducial mass of 10 kg (XENON10) to be operated in Gran Sasso is currently under construction. Its goal is to achieve a sensitivity of a factor of 10 below the current CDMS results, thus probing WIMP cross sections around 2×10^{-44} cm^2.

2.4 The Future: Direct Detection

We live in suspenseful times for the field of direct detection: for the first time, a couple of experiments operating deep underground probe the most optimistic supersymmetric models. The best limits on WIMP-nucleon cross sections come from cryogenic experiments with ultra-low backgrounds and excellent event-by-event discrimination power. Although these experiments had started with target masses around 1 kg, upgrades to several kilograms have already taken place or are foreseen for the near future, ensuring (along with improved backgrounds) an increase in sensitivity by a factor of 10-100. Other techniques, using liquid noble elements such as Xe and Ar, may soon catch up and probe similar parameter spaces to low-temperature cryogenic detectors. It is worth emphasizing here that given the importance of the endeavor and the challenge in unequivocally identifying and measuring the properties of a dark matter particle, it is essential that more than one technique will move forward.

In supersymmetry, WIMP-nucleon cross sections as low as 10^{-48}cm^2 are likely[32]. Likewise, in theories with universal extra dimensions, it is predicted that the lightest Kaluza Klein particle would have a scattering cross section with nucleons in the range of 10^{-46} - 10^{-48}cm^2 [33]. Thus, to observe a signal of a few events per year, ton or multi-ton experiments are inevitable. There are several proposals to build larger and improved dark matter detectors. The selection presented below is likely biased, but based on technologies which seem the most promising to date.

The SuperCDMS project[28] is a three-phase proposal to utilize CDMS-style detectors with target masses growing from 27 kg to 145 kg and up to 1100 kg, with the aim of reaching a final sensitivity of 3×10^{-46}cm^2 by mid 2015. This goal will be realized by developing improved detectors (for a more precise event reconstruction) and analysis techniques, and at the same time by strongly reducing the intrinsic surface contamination of the crystals. A possible site is the recently approved SNO-Lab Deep-site facility in Canada (at 6000 m.w.e.), where the neutron background would be reduced by more than two orders of magnitude compared to the Soudan Mine, thus ensuring the mandatory conditions to build a zero-background experiment. In Europe, a similar project to develop a 100 kg-1 ton cryogenic experiment, EURECA (European Underground Rare Event search with Calorimeter Array) [25] is underway. The XENON collaboration is designing a 100 kg scale dual-phase xenon detector (XENON100), towards a modular one tonne experiment[31]. ZEPLIN MAX, a R&D project of the Boulby Dark Matter collaboration, is a further proposal to build a ton scale liquid xenon experiment. The design will be based on the experience and re-

sults with ZEPLIN II/III at the Boulby Mine. WARP[34] and ArDM[35] are two proposals to build ton-scale dark matter detectors based on the detection of nuclear recoils in liquid argon. The physics and design concepts are similar to the discussed dual-phase xenon detectors. The Boulby Collaboration is developing a large directional sensitive detector based on the experience with DRIFT[36], a time projection chamber with a total active mass of $\sim$170 g of CS$_2$.

3 Indirect Detection

WIMPs can be detected by observing the radiation produced when they annihilate. The flux of annihilation products is proportional to $(\rho_{WIMP}/m_{WIMP})^2$, thus regions of interest are those expected to have a relatively high WIMP concentration. Possible signatures for dark matter annihilation are high energy neutrinos from the Sun's core and from the galactic center, gamma-rays from the galactic center and halo and antiprotons and positrons from the galactic halo. The predicted fluxes depend on the particle physics model delivering the WIMP candidate and on astrophysical quantities such as the dark matter halo profile, the presence of sub-structure and the galactic cosmic ray diffusion model.

3.1 Gamma Rays

WIMP annihilation can result in a continuum of gamma rays (via hadronization and decay of π_0's) or in a monochromatic flux (from direct annihilation into $\gamma\gamma$ or Zγ), in which case the energy of the gamma line gives direct information on the WIMP mass. While the predicted fluxes for the gamma continuum are higher, the energies are lower and the signature wouldn't be as clear as in the monochromatic case. In both cases, the expected fluxes are strongly dependent on the halo density profile. Direct observation of gamma rays in the energy range of interest to dark matter searches (GeV–TeV) can only occur in space, as the gamma will interact with matter via e^+e^--pair production, with an interaction length much shorter than the thickness of the atmosphere. However, high-energy gamma rays can be detected on the ground with air shower detectors. Of these, the atmospheric cerenkov telescopes (ACTs) detect the Cerenkov light produced by the cascade of secondary particles in the atmosphere. The background comes from cosmic ray induced showers, and imaging ACTs for instance can distinguish between gamma and cosmic ray events based on the light distribution in the Cerenkov cone. Examples of ACTs either taking data or in construction are HESS[37], MAGIC[38], CANGAROO[39] and VERITAS[40]. Their sensitivity typically is in the range 10 GeV - 10 TeV. EGRET[41], a space-based telescope on the Compton Gamma Ray Observatory, took data from 1991-2000 in the energy range 30 MeV–30 GeV. The next telescope to be launched in 2007 is GLAST[42] , which will observe the gamma sky up to ∼100 GeV. In general, space-based telescopes are complementary to ground-based ones, as their range of energies is lower and their field of view and duty cycle larger.

Recently an excess of high-energy (10^{12} TeV) gammas rays from the galactic center has been detected. This region had been observed by the VERITAS and CANGAROO groups, but the angular resolution was greatly increased by the HESS four-telescope array. The HESS 2004 data confirmed the excess and is consistent with the position of Sgr A* and a point-like source within the angular resolution of the detector (5.8') [43]. This signal has been interpreted as due to WIMP annihilation, with a WIMP mass around 19 TeV providing the best fit to the data[43]. Apart from the high WIMP mass, the observed signal would require large WIMP annihilation cross sections and a cuspy halo profile.

EGRET provided an all-sky gamma-ray survey, with about 60% of the sources yet to be identified. A reanalysis of EGRET data, which is publicly available, revealed that the diffuse component shows an excess by a factor of two above the background expected from π^o's produced in nuclear interactions, inverse Compton scattering and bremsstrahlung. This excess, which is observed in all sky directions, has been interpreted as due to WIMP annihilation, with a best fit WIMP mass around 60 GeV [44]. The relative contributions of the galactic background have been estimated with the GALPROP code[45], while the extragalactic background was obtained by subtracting the galactic 'foregrounds' (as given by GALPROP) from the EGRET data. The predicted cross section for elastic scatters on nucleons is in the 5×10^{-8}-2×10^{-7}pb range, and thus testable by the CDMS experiment.

An analysis of the EGRET extragalactic background revealed two components, a steep spectrum power law with index α=-2.33 and a strong bump at a few GeV [46]. Such a multi-GeV bump is difficult to explain in conventional astrophysical models, with contributions from faint blazars, radiogalaxies, gamma-ray bursts and large scale structures, and has been interpreted as evidence for WIMP annihilation[46]. The best fit is provided by a WIMP mass around 500 GeV and an annihilation cross section of $\langle\sigma v\rangle \approx 3 \times 10^{-25} cm^3 s^{-1}$. A typical candidate from supersymmetry would have cross sections on nucleons $<10^{-7}$pb, thus below the present sensitivity of direct detection experiments at these masses. However, while a >500 GeV WIMP would be difficult to detect at the LHC (and even at the ILC), it is within the reach of next generation direct detection experiments.

3.2 Neutrinos from the Sun

WIMPs with orbits passing through the Sun can scatter from nuclei and lose kinetic energy. If their final velocity is smaller than the escape velocity, they will be gravitationally trapped and will settle to the Sun's core. Over the age of the solar system, a sufficiently large number of WIMP can accumulate and efficiently annihilate, whereby only neutrinos are able to escape and be observed in terrestrial detectors. Typical neutrino energies are 1/3–1/2 of the WIMP mass, thus well above the solar neutrino background. Observation of high energy neutrinos from the direction of the Sun would thus provide a clear signature for dark matter in the halo. The annihilation rate is set by the capture rate, which scales with $(\mathrm{m}_{WIMP})^{-1}$ for a given halo density. Thus, annihilation and direct detection rates have the same scaling with the WIMP mass. However, the probability of detecting a neutrino by searching for muons produced in charge-current interactions scales with E_ν^2, making these searches more sensitive at high WIMP masses when compared to direct detection experiments. The best technique to detect high-energy neutrinos is to observe the upward-going muons produced in charged-current interactions in the rock below the detector. To distinguish neutrinos coming from the Sun's core from backgrounds induced by atmospheric neutrinos, directional information is needed. The direction of the upward-going muon and the primary neutrino direction are correlated, the rms angle scaling roughly with $\sim 20(E_\nu/10 GeV)^{-1/2}$.

Two types of detectors are used to search for high-energy neutrino signals, with no excess above the atmospheric neutrino background reported so far. In the first category are large underground detectors, such as MACRO[47] and SuperKamiokande [48], while the second type are dedicated neutrino telescopes, employing large arrays of PMTs deep in glacier ice, in the ocean or in a lake, such as AMANDA[49], BAIKAL[50], NESTOR[51] and ANTARES[52]. These experiments detect the Cerenkov light emitted when muons move with speeds larger than the velocity of light in

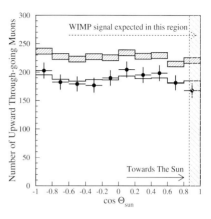

Figure 7. Angular distribution of upward through-going muons with respect to the Sun in Super-Kamiokande. Data are black circles, while the hatched region and the solid line show the expected atmospheric neutrino background before and after taking into account ν-oscillations. Figure from[48].

water/ice, with ~ 1 ns timing resolution. The PMT hit pattern and relative arrival times of the photons are used to reconstruct the direction of the incoming particle, which is correlated with the direction of the neutrino. Neutrino telescopes have higher energy thresholds (in the range 50-100 GeV), but their effective area is much larger, thus compensating for the lower fluxes predicted for heavy WIMPs.

The strongest limits on high-energy neutrinos coming from the Sun are placed by Super-Kamiokande[48]. Figure 7 shows the angular distribution of upward through-going muons with respect to the Sun, and the expected atmospheric neutrino background in the Super-Kamiokande detector. While the limits on scalar WIMP-nucleon interactions are not competitive to direct detection experiments, Super-Kamiokande gives the most stringent limit on spin-dependent WIMP-nucleon cross sections for pure proton couplings above a WIMP mass of ~ 20 GeV (see Fig. 8).

360

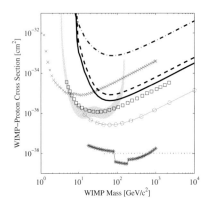

Figure 8. Upper limits on WIMP-nucleon cross sections for pure proton coupling. The lowest curve (green asterisks) is the SuperKamiokande limit, the filled region shows the 3-σ DAMA region. For details, see[27].

3.3 Positrons and antiprotons

Cosmic positrons and antiprotons produced in WIMP annihilations in the galactic halo can be observed with balloon or space-based experiments. For antiprotons, the background spectrum produced by spallation of primary cosmic rays on H atoms in the interstellar medium is expected to fall steeply at energies below 1 GeV. Thus, observation of low-energy cosmic-ray antiprotons could provide evidence for WIMPs in the halo. The BESS collaboration has measured the flux of antiprotons in several balloon flights between 1193-1997 [53]. No convincing excess above the cosmic ray background has been observed. For positrons, the background flux is expected to decrease slowly as a function of energy. The HEAT experiment has measured a positron excess at energies ~8 GeV [54], which can been interpreted as coming from WIMP annihilation in the halo[55]. The spectra can be fit by a WIMP mass of 200-300 GeV, but the fit is far from perfect. The signal requires a boost of a factor of ~30 in the WIMP density, for instance from the presence of dark matter clumps in the halo.

3.4 Future: Indirect Detection

Several existing observations have been interpreted as signatures for dark matter annihilation in our galaxy, or in extragalactic dark matter halos. There is no single WIMP capable of explaining all the data: WIMP masses from 60 GeV to 18 TeV are required, with large boost factors in the halo density and a cuspy inner halo. There is a clear demand for more data. Existing and future ACTs, such as HESS, MAGIC, CANGAROO and VERITAS will map out the galactic center with improved position, energy and timing resolution, and will likely reveal the source of high-energy gammas. Planned space-based detectors such as GLAST will map the gamma-ray sky with unsurpassed sensitivity, while cubic kilometer detectors in ice (IceCube) or water (Km3Net) will considerably increase existing sensitivities to high-energy neutrinos coming from the Sun. Finally, PAMELA, to be launched in 2007, and AMS-2, to be operated on the ISS starting in 2007 will greatly improve the sensitivity to antimatter from WIMP annihilation in the galactic halo.

In looking back over the fantastic progress made in the last couple of years, and extrapolating into the future, it seems probable that these, and other proposed projects, will have a fair chance to discover a WIMP signature within the present decade. In conjunction with direct WIMP searches and accelerator production of new particles at the weak scale, they will allow to reveal the detailed properties of WIMPs, such as their mass, spin and couplings to ordinary matter, and shed light on the density profile of dark matter in the halo.

References

1. M. W. Goodman and E. Witten, *Phys. Rev. D* **31**, 3059 (1985).
2. A. Drukier, K. Freese and D. Spergel, *Phys. Rev. D* **33**, 3495 (1986).
3. K. Freese, J. Frieman and A. Gould, *Phys. Rev. D* **37**, 3388 (1988).

4. D. N. Spergel, *Phys. Rev.* D **37**, 1353 (1988).

5. C. J. Copi, J. Heo and L. M. Krauss, *Phys. Lett.* B **461**, 43 (1999).

6. S. P. Ahlen *et al.*, *Phys. Lett.* B **195**, 603 (1987); D.O. Caldwell *et al.*, *Phys. Rev. Lett.* 61, 510 (1988).

7. M. Beck *et al.*, *Phys. Lett.* B **336**, 141 (1994).

8. L. Baudis *et al.* Phys. Rev. D **63**, 022001 (2001)

9. L. Baudis *et al.*, *Nucl. Instr. Meth. in Phys. Res.* A **426**, 425 (1999).

10. http://www.mpihd.mpg.de/ge76

11. R. Gaitskell *et al.*, nucl-ex/0311013.

12. R. Bernabei *et al.*, *Riv. Nuo. Cim.* **26**, 1 (2003).

13. B. Ahmed *et al.*, *Astropart. Phys.* **19**, 691 (2003).

14. R. Bernabei *et al.*, *Phys. Lett.* B **424**, 195 (1998).

15. D. Akerib *et al.* (CDMS Collaboration), *Phys. Rev. Lett.* **93**, 211301 (2004).

16. C. Angloher *et al.*, *Astropart. Phys.* **23**, 325 (2005).

17. V. Sanglard *et al.* (EDELWEISS Collaboration), *Phys.Rev.* D **71**, 122002 (2005).

18. G. J. Alner *et al.* (UKDM Collaboration), *Astropart. Phys.* **23**, 444 (2005).

19. C. J. Copi and L. M. Krauss, *Phys. Rev. D* **67**, 103507 (2003).

20. P. Ullio, M. Kamionkowski and P. Vogel, *JHEP* **0107**, 044 (2001).

21. C. Savage, P. Gondolo and K. Freese, *Phys. Rev. D* **70**, 123513 (2004).

22. H. Baer *et al.*, *JCAP* **0309**, 007 (2003), E. Baltz and P. Gondolo, *Phys. Rev. D* **67**, 063503 (2003), A. Bottino *et al.*, *Phys. Rev. D* **69**, 037302 (2004), U. Chattopadhyay, A. Corsetti and P. Nath, *Phys.Atom.Nucl.* **67**, 1188 (2004), M. Battaglia *et al.*, *Eur. Phys. J. C* **33**, 273 (2004), Y. G. Kim *et al.*, *JHEP* **0212**, 034 (2002), A. B. Lahanas and D. V. Nanopoulos, *Phys. Lett.* B **568**, 55 (2003).

23. S. Cebrian *et al.*, *Nucl. Phys. Proc. Suppl.* **114**, 111 (2003).

24. D. Akerib *et al.* (CDMS Collaboration), astro-ph/0509259, submitted to PRL.

25. G. Chardin, astro-ph/0411503.

26. R. J. Gaitskell, *Ann. Rev. Nucl. Part. Sci.* **54**, 315 (2004).

27. D. Akerib *et al.* (CDMS Collaboration), astro-ph/0509269, submitted to PRL.

28. P. Brink *et al.*, *Texas Symposium on Relativistic Astrophysics*, Stanford (2004).

29. N. Mirabolfati *et al.*, *AIP conference proceedings*, Volume **605**, 517 (2001).

30. E. Aprile *et al.*, astro-ph/050362, accepted in *Phys. Rev.* D (2005).

31. E. Aprile (XENON Collaboration), astro-ph/0502279 (2005).

32. J. Ellis, K. A. Olive, Y. Santoso and V. C. Spanos, *Phys. Rev. D* **71**, 095007 (2005).

33. G. Servant and T. M. Tait, *New J. Phys.* **4**, 99 (2002).

34. R. Brunetti *et al.*, astro-ph/0411491, Proceedings of the fifth international workshop on the Identification of Dark Matter, editors N.J.C. Spooner and V. Kudryavtsev, World Scientific (2004).

35. http://neutrino.ethz.ch/ArDM

36. D. P. Snowden-Ifft *et al.*, *Nucl. Instrum. and Meth. in Phys. Res.* A **498**, 155 (2003).

37. http://www.mpi-hd.mpg.de/hfm/HESS

38. http://magic.mppmu.mpg.de

39. http://icrhp9.icrr.u-tokyo.ac.jp

40. http://veritas.sao.arizona.edu

41. P. L. Nolan *et al.*, *IEEE Transactions Nucl.Sci.* **39**, 993 (1992).

42. http://www-glast.stanford.edu

43. D. Horns, *Phys.Lett.* B **607**, 225 (2005).

44. W. de Boer *et al.*, astro-ph/0508617, accepted by A&A.

45. A. W. Strong, I.V. Moskalenko, O. Reimer, *Astroph. J.* **613**, 956 (2004).

46. D. Elsaesser, K. Mannheim, *Phys. Rev. Lett.* **94**, 171302 (2005).

47. M. Ambrosio *et al.* *Nucl. Instr. Meth. Phys. Res.* A **486**, 663 (2002).

48. S. Desai *et al.* (Super-Kamiokande Collaboration), *Phys. Rev. D* **70**, 083523 (2004).

49. M. Ackermann *et al.* (AMANDA Collaboration), astro-ph/0508518. accepted by *Astropart. Phys.*

50. L. Kuzmichev (Baikal Collaboration), astro-ph/0507709.

51. http://www.nestor.org.gr

52. I. Sokalski (ANTARES Collaboration), hep-ex/0501003.

53. J. W. Mitchell *et al.*, *Advances in Space Research* in press (2005).

54. J. J. Beatty *et al.* Phys. Rev. Lett. **93**, 241102 (2004).

55. E. A. Baltz, J. Edsjö, K. Freese and P. Gondolo, astro-ph/0211239.

362

DISCUSSION

Guy Wormser (LAL Orsay):

Can you translate the limits on results obtained by indirect experiments like HESS on to the parameter space of direct searches?

Laura Baudis: HESS observes a signal, which can be interpreted as due to WIMP annihilation in the galactic center, with a WIMP mass higher than 12 TeV. In general, direct searches have reduced sensitivity at such high WIMP masses, because of the lower fluxes for a given halo density. The WIMP-nucleon scattering cross section depends on the particle physics candidate. For instance, for a scalar cross section larger than $\approx 4 \times 10^{-42} \text{cm}^2$, a 20 TeV particle would be ruled out by the CDMS experiment.

Peter Schuber (DESY):

You did not mention the recent publication in Phys. Rev. Lett. from the University of Wurzburg that reported a plausible evidence for neutralinos from satellite based high energy photon spectra. In fact they indicated even a high neutralino mass of about 500 GeV.

Laura Baudis: I misunderstood this question. I did not know about above publication, but have included it in these proceedings. A 500 GeV neutralino, although likely beyond the reach of planned accelerators, can be probed by future direct detection experiments.

Bennie Ward (Baylor University):

On the slide wherein you exhibited your candidate event as one of many background events presumably associated with a detector mis-function, what happened to the other events that wear near it?

Laura Baudis: The other events did not survive the standard CDMS cuts (such as data quality, fiducial volume, ionization yield and timing cuts).

EXPLORING THE UNIVERSE WITH DIFFERENT MESSENGERS

ELI WAXMANN

Department of Physics, Weizmann Institute of Science, 76100 Rehovot, Israel
E-mail: waxman@wicc.weizmann.ac.il

NO CONTRIBUTION RECEIVED

DARK ENERGY AND DARK MATTER FROM COSMOLOGICAL OBSERVATIONS

STEEN HANNESTAD

Department of Physics and Astronomy, University of Aarhus, Ny Munkegade, DK-8000 Aarhus C, Denmark, E-mail: sth@phys.au.dk

The present status of our knowledge about the dark matter and dark energy is reviewed. Particular emphasis is put on the. Bounds on the content of cold and hot dark matter from cosmological observations are discussed in some detail. I also review current bounds on the physical properties of dark energy, mainly its equation of state and effective speed of sound.

1 Introduction

The introduction of new observational techniques has in the past few years moved cosmology into the era of precision science. With the advent of precision measurements of the cosmic microwave background (CMB), large scale structure (LSS) of galaxies, and distant type Ia supernovae, a new paradigm of cosmology has been established. In this new standard model, the geometry is flat so that $\Omega_{\text{total}} = 1$, and the total energy density is made up of matter ($\Omega_m \sim 0.3$) [comprised of baryons ($\Omega_b \sim 0.05$) and cold dark matter ($\Omega_{\text{CDM}} \sim 0.25$)], and dark energy ($\Omega_X \sim 0.7$). With only a few free parameters this model provides an excellent fit to all current observations [1,2,4,7]. However, cosmology is currently very much a field driven by experiment, not theory. While all current data can be described by a relatively small number of fitting parameters the understanding of the underlying physics is still limited.

Here, I review the present knowledge about the observable cosmological parameters related to dark matter and dark energy, and relate them to the possible underlying particle physics models. I also discuss the new generation of experiments currently being planned and built, particularly those designed to measure weak gravitational lensing on large scales. These instruments are likely to bring answers to at least some of the fundamental questions about dark matter and dark energy.

2 Cosmological data

2.1 Large Scale Structure (LSS).

At present there are two large galaxy surveys of comparable size, the Sloan Digital Sky Survey (SDSS)[7,6] and the 2dFGRS (2 degree Field Galaxy Redshift Survey)[5]. Once the SDSS is completed in December 2005 it will be significantly larger and more accurate than the 2dFGRS, measuring in total about 10^6 galaxies.

Both surveys measure angular positions and distances of galaxies, producing a fully three dimensional map of the local Universe. From this map various statistical properties of the large scale matter distribution can be inferred.

The most commonly used is the power spectrum $P(k,\tau)$, defined as

$$P(k,\tau) = |\delta_k|^2(\tau), \qquad (1)$$

where k is the Fourier wave number and τ is conformal time. δ is the k'th Fourier mode of the density contrast, $\delta\rho/\rho$.

The power spectrum can be decomposed into a primordial part, $P_0(k)$, generated by some mechanism (presumably inflation) in the early universe, and a transfer function $T(k,\tau)$,

$$P(k,\tau) = P_0(k)T(k,\tau). \qquad (2)$$

The transfer function at a particular time is

found by solving the Boltzmann equation for $\delta(\tau)$ [125].

As long as fluctuations are Gaussian, the power spectrum contains all statistical information about the galaxy distribution. On fairly large scales $k \leq 0.1\,h/\mathrm{Mpc}$ this is the case, and for that reason the power spectrum is the form in which the observational data is normally presented.

2.2 Cosmic Microwave Background.

The CMB temperature fluctuations are conveniently described in terms of the spherical harmonics power spectrum $C_l^{TT} \equiv \langle |a_{lm}|^2 \rangle$, where $\frac{\Delta T}{T}(\theta, \phi) = \sum_{lm} a_{lm} Y_{lm}(\theta, \phi)$. Since Thomson scattering polarizes light, there are also power spectra coming from the polarization. The polarization can be divided into a curl-free ((E)) and a curl ((B)) component, much in the same way as $\vec{E}$ and $\vec{B}$ in electrodynamics can be derived from the gradient of a scalar field and the curl of a vector field respectively (see for instance[133] for a very detailed treatment). The polarization introduced a sequence of new power spectra, but because of different parity some of them are explicitly zero. Altogether there are four independent power spectra: C_l^{TT}, C_l^{EE}, C_l^{BB}, and the T-E cross-correlation C_l^{TE}.

The WMAP experiment has reported data only on C_l^{TT} and C_l^{TE} as described in Refs. [3,4]. Other experiments, while less precise in the measurement of the temperature anisotropy and not providing full-sky coverage, are much more sensitive to small scale anisotropies and to CMB polarization. Particularly the ground based CBI[58], DASI [59], and ACBAR[57] experiments, as well as the BOOMERANG balloon experiment [60,61,62] have provided useful data.

2.3 Type Ia supernovae

Observations of distant supernovae have been carried out on a large scale for about a decade. In 1998 two different projects almost simultaneously published measurements of about 50 distant type Ia supernovae, out to a redshift or about 0.8 [1,2]. These measurements were instrumental for the measurement of the late time expansion rate of the universe.

Since then a, new supernovae have continuously been added to the sample, with the Riess et al.[63] "gold" data set of 157 distant supernovae being the most recent. This includes several supernovae measured by the Hubble Space Telescope out to a redshift of 1.7.

3 Cosmological parameters

Based on the present cosmological data, many different groups have performed likelihood analyses based on various versions of the standard Friedmann-Robertson-Walker cosmology (see for instance [7,64] for recent analyses). A surprisingly good fit is provided by a simple, geometrically flat universe, in which 30% of the energy density is in the form of non-relativistic matter and 70% in the form of a new, unknown dark energy component with strongly negative pressure. In its most basic form, the dark energy is in the form of a cosmological constant where $w \equiv P/\rho = -1$. The only free parameters in this model are: Ω_m, the total matter density, Ω_b, the density in baryons, and H_0, the Hubble parameter. In addition to these there are parameters related to the spectrum of primordial fluctuations, presumably generated by inflation. Observations indicate that the fluctuations are Gaussian and with an almost scale invariant power spectrum. More generally, the primordial spectrum is usually parameterized by two parameters: A, the amplitude, and n_s the spectral tilt of the power spectrum. Finally, there is the parameter τ which is related to the redshift of reionization of the Universe. Altogether, standard cosmology is describable by only 6 parameters (5 if the spectrum is assumed to be scale invariant [a].

Adding other parameters to the fit does not significantly alter the determination of the 6 fundamental parameters, although in some cases the estimated error bars can increase substantially.

4 Dark matter

The current cosmological data provides a very precise bound on the physical dark matter density[7]

$$\Omega_m h^2 = 0.138 \pm 0.012, \qquad (3)$$

although this bound is somewhat model dependent. It also provides a very precise measurement of the cosmological density in baryons [7]

$$\Omega_b h^2 = 0.0230^{+0.0013}_{-0.0012}. \qquad (4)$$

This value is entirely consistent with the estimate from Big Bang nucleosynthesis, based on measurements of deuterium in high redshift absorption systems, $\Omega_b h^2 = 0.020 \pm 0.002$ [117,118].

The remaining matter density consists of dark matter with the density[7]

$$\Omega_{\mathrm{dm}} h^2 = 0.115 \pm 0.012. \qquad (5)$$

The bound on the dark matter density in turn provides strong input on any particle physics model for dark matter. Space limitations allow only for a very brief review of the cosmological constraints on dark matter. Very detailed reviews can be found in [126,127].

4.1 WIMPs

The simplest model for cold dark matter consists of WIMPs - weakly interacting massive particles. Generic WIMPs were once in thermal equilibrium, but decoupled while strongly non-relativistic. For typical models with TeV scale SUSY breaking where

---

[a]See [123,124] for a discussion about how to estimate the number of cosmological parameters needed to fit the data.

neutralinos are the LSPs, one finds that $T_D/m \sim 0.05$. SUSY WIMPs are currently the favoured candidate for cold dark matter (see [127]). The reason is that for massive particles coupled to the standard model via a coupling which is suppressed by 1/TeV and with a mass of order 100 GeV to 1 TeV a present density of $\Omega_m h^2 \sim 0.1$ comes out fairly naturally. SUSY WIMPs furthermore have the merit of being detectable. One possibility is that they can be detected directly when they deposit energy in a detector by elastically scattering (see the contribution by Laura Baudis to these proceedings). Another is that WIMPs annihilate and produce high energy photons and neutrinos which can subsequently be detected.

4.2 CDM Axions

WIMPs are by no means the only possibility for having cold dark matter. Another possibility is that CDM is in the form of axions, in which case the mass needed to produce the correct energy density is of order 10^{-3} eV. In this case the axions would be produced coherently in a condensate, effectively acting as CDM even though their mass is very low (see for instance [65] for a recent overview).

4.3 Exotica

Another interesting possibility is that dark matter consists of very heavy particles. A particle species which was once in thermal equilibrium cannot possible be the dark matter if its mass is heavier than about 350 TeV [66]. The reason is that its annihilation cross section cannot satisfy the unitarity bound. Therefore, heavy dark matter would have to be produced out of thermal equilibrium, typically by non-perturbative processes at preheating towards the end of inflation (see for instance [67]). These models have the problem of being exceedingly hard to verify or rule out experimentally.

4.4 Hot dark matter

In fact the only dark matter particle which is known to exist from experiment is the neutrino. From measurements of tritium decay, standard model neutrinos are known to be light. The current upper bound on the effective electron neutrino mass is 2.3 eV at 95% C.L. [68] (see also the contribution by Christian Weinheimer to these proceedings). Such neutrinos decouple from thermal equilibrium in the early universe while still relativistic. Subsequently they free-stream until the epoch around recombination where they become non-relativistic and begin to cluster. The free-streaming effect erases all neutrino perturbations on scales smaller than the free-streaming scale. For this reason neutrinos and other similar, light particles are generically known as hot dark matter. Models where all dark matter is hot are ruled out completely by present observations, and in fact the current data is so precise that an upper bound of order 1 eV can be put on the sum of all light neutrino masses [69,70,71,72,73,74,75,76,77,78,79]. This is one of the first examples where cosmology provides a much stronger constraint on particle physics parameters than direct measurements. The robustness of the neutrino mass bound has been a topic many papers over the past two years. While some derived mass bounds, as low as 0.5 eV are almost certainly too optimistic to consider robust at present, it is very hard to relax the upper bound to much more than 1.5 eV [79]. The reason for the difference in estimated precision lies both in the assumptions about cosmological parameters, and in the data sets used.

In the future, a much more stringent constraint will be possible, especially using data from weak lensing (see section 6).

4.5 General thermal relics

The arguments pertaining to neutrinos can be carried over to any thermal relic which decoupled while relativistic. As long as the mass is in the eV regime or lower the free streaming scale is large than the smallest scales in the linear regime probed by LSS surveys. This has for instance been used for particles such as axions[80]. It should of course be noted that these axions are in a completely different different mass range than the axions which could make up the CDM. At such high masses, the axions would be in thermal equilibrium in the early universe until after the QCD phase transition at $T \sim 100$ MeV and therefore behave very similarly to neutrinos.

However, for relics which decouples very early, the mass can be in the keV regime. In that case it is possible to derive mass bounds using data from the Lyman-α forest which is at much higher redshifts and therefore still in the semi-linear regime, even at sub-galactic scales. Using this data it has for instance been possible to set constraints on the mass of a warm dark matter particle which makes up all the dark matter[81].

4.6 Telling fermions from bosons

There is a fundamental difference between hot dark matter of fermionic and of bosonic nature. First of all, the number and energy densities are different. For equal values of Ωh^2 this leads to different particle masses and therefore also different free-streaming behaviour. The differences are at the few percent level, and although not visible with present data, should be clearly visible in the future[128]. The difference between the matter power spectra of two different models, both with $\Omega_{\mathrm{HDM}} = 0.02$, can be seen in Fig. 1. Even more interesting, in the central parts of dark matter halos, the density of a bosonic hot dark matter component can be several times higher than that of a fermionic component with the same mass, purely because of quantum statistics [128]. The reason is that the distribution function, $f = 1/(e^{E/T} + 1)$, for a non-degenerate fermion in thermal equilib-

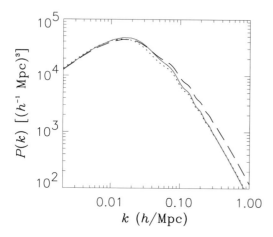

Figure 1. Linear power spectra for two different ΛHCDM models. The blue (dotted) line shows a model with three massless neutrinos and one massive Majorana fermion, contributing $\Omega = 0.02$. The red (solid) line shows the same, but with a massive scalar instead. The black (dashed) line is the standard ΛCDM model with no HDM. Note that these spectra have been normalized to have the same amplitude on large scales. [From [128]].

rium has a maximum at $p = 0$ where $f = 1/2$. This bound also applies to the species after decoupling, and provides an upper bound on the physical density of such particles in dark matter halos. This is known as the Tremaine-Gunn bound [129,130,131,132]. Because there is no such limit for non-degenerate bosons, their density in dark matter halos can be many times higher than that of fermions. Unfortunately the effect is most pronounced in the central parts of dark matter halos where the density is dominated by cold dark matter and baryons, and therefore it might not be observable[128].

5   Dark energy

From the present supernova data alone, the universe is known to accelerate. In terms of the deceleration parameter q_0, the bound is

$$q_0 = -\frac{\ddot{a}a}{\dot{a}^2} < -0.3 \qquad (6)$$

at 99% C.L. [63]. Such a behaviour can be explained by the presence of a component of the energy density with strongly negative pressure, which can be seen from the acceleration equation

$$\frac{\ddot{a}}{a} = -\frac{4\pi G \sum_i (\rho_i + 3P_i)}{3}. \qquad (7)$$

The cosmological constant is the simplest (from an observational point of view) version of dark energy, with $w \equiv P/\rho = -1$. However, there are many other possible models which produce cosmic acceleration.

However, since the cosmological constant has a value completely different from theoretical expectations one is naturally led to consider other explanations for the dark energy.

5.1   The equation of state

If the dark energy is a fluid, perfect or non-perfect, it can be described by an equation of state w which in principle is constrainable from observations. Secondly, this dark energy fluid must have an effective speed of sound c_s which in some cases can be important.

A light scalar field rolling in a very flat potential would for instance have a strongly negative equation of state, and would in the limit of a completely flat potential lead to $w = -1$ [86,87,88]. Such models are generically known as quintessence models. The scalar field is usually assumed to be minimally coupled to matter, but very interesting effects can occur if this assumption is relaxed (see for instance[89]).

In general such models would also require fine tuning in order to achieve $\Omega_X \sim \Omega_m$, where Ω_X and Ω_m are the dark energy and matter densities at present. However, by coupling quintessence to matter and radiation it is possible to achieve a tracking behavior of the scalar field so that $\Omega_X \sim \Omega_m$ comes out naturally of the evolution equation for the scalar field[8,9,10,11,12,13,14,15,16].

Many other possibilities have been considered, like k-essence, which is essentially a scalar field with a non-standard kinetic term [17,18,19,20,21,22,23]. It is also possible, although not without problems, to construct models which have $w < -1$, the so-called phantom energy models [24,28,25,26,27,29,30,31,32,33,34,35,37,36,38,39].

From an observational perspective there are numerous studies in which the effective equation of state of the dark energy has been constrained.

The simplest parametrization is $w = $ constant, for which constraints based on observational data have been calculated many times [82,83,84,85]. The bound on the equation of state, w, assuming that it is constant is roughly (see[41,90,40,64])

$$-1.2 \leq w \leq -0.8 \qquad (8)$$

at 95% C.L. Very interestingly, however, there is a very strong degeneracy between measurements of w and the neutrino mass $\sum m_\nu$. When the neutrino mass is included in fits of w the lower bound becomes much weaker and the allowed range is

$$-2.0 \leq w \leq -0.8 \qquad (9)$$

at 95% C.L. [79]. The result of a likelihood analysis taking both parameters to be free can be seen in Fig. 2.

Even though a constant equation of state is the simplest possibility, as the precision of observational data is increasing is it becoming feasible to search for time variation in w.

At present there is no indication that w is varying. Even though the present Type Ia supernova data seem to favour a rapid evolution of w, this indication vanishes if all available cosmological data is analyzed [41,90,40,64] (for other discussions of a time-varying w, see for instance [42,43,44,45,46,47,48,49,50,51,52,53,54,55,56,91,92,93].

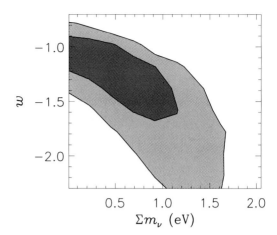

Figure 2. The 68% (dark) and 95% (light) likelihood contours for m_ν and w for WMAP, SDSS, and SNI-a data. [From [79]]

5.2 The sound speed of dark energy

In general the dark energy speed of sound is given by

$$c_s^2 = \frac{\delta P}{\delta \rho}, \qquad (10)$$

if it can be described as a fluid. The perturbation equations depend on the speed of sound in all components, including dark energy, and therefore c_s^2 can in principle be measured [119,120,121,122].

For a generic component with constant w, the density scales as $a^{-3(1+w)}$, where a is the scale factor. Therefore, the ratio of the energy density to that in CDM is given by $\rho/\rho_{CDM} \propto a^{-3w}$. If w is close to zero this means that dark energy can be important at early times and affect linear structure formation. If, on the other hand, w is very negative, dark energy will be unimportant during structure formation. This also means that since $w \leq -0.8$ there is effectively no present constraint on the dark energy equation of state. In Fig. 3 we show current constraints in w and c_s^2.

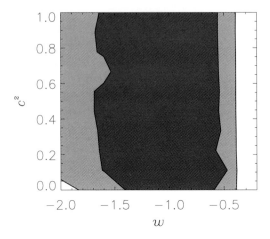

Figure 3. The 68% (dark) and 95% (light) likelihood contours for w and c_s^2 for WMAP, SDSS, and SNI-a data. [From [122]].

5.3 Dark energy or modified gravity?

A potentially very interesting possibility is that what we perceive as dark energy is in fact a modification of gravity on very large scales. General relativity has been tested to work in the weak field regime up to super-galactic scales. However, it is possible that at scales close to the Hubble horizon there might be modifications.

One possible scenario is that there are extra spatial dimensions into which gravity can propagate. For instance in the Dvali-Gabadadze-Porrati model[94], the standard model is confined to a 3+1 dimensional brane in a 4+1 dimensional bulk where gravity can propagate. On small scales, gravity can be made to look effectively four dimensional by an appropriate tuning of the model parameters, whereas on large scales gravity becomes weaker. This leads to an effect very similar to that of dark energy. Based on this idea, other authors have taken a more observational approach, adding extra terms to the Friedmann equation[95,96,98].

In this case the dark energy has no meaningful speed of sound since it is a change in gravity. However, exactly since it affects gravity it also affects the way in which structure grows in the universe. In[97] it was found that, unless the cross-over scale has very specific and fine tuned values, models with modified large scale gravity are almost impossible to reconcile with present observations.

6 Future observations

6.1 Cosmic microwave background

In the coming years, the present CMB experiments will be superseded by the Planck Surveyor satellite[99], due to be launched in 2007. It will carry instrumentation similar to that on the latest BOOMERANG flight, but will carry out observations from space, and for several years. The expectation is that the project will measure the CMB spectrum precisely up to $l \sim 2500$, being essentially limited only by foreground in this range. This experiment will be particularly important for the study of inflation because it will be able to measure the primordial spectrum of fluctuations extremely precisely.

On a longer timescale there will be dedicated experiments measuring small scales, such as the Atacama Cosmology Telescope [100]. Small scale observations will be instrumental in understanding non-linear effects on the CMB, arising from sources such as the Sunyaev-Zeldovich effect and weak gravitational lensing.

6.2 Type Ia supernovae

There are several ongoing programs dedicated to measuring high redshift supernovae. For instance the Supernova Legacy Survey is currently being carried out at the CFHT[101]. ESSENCE[102] is another project dedicated to improving the current measurement of w. The future Dark Energy Survey[103] is expected to find about 2000 Type Ia supernovae, and the Supernova Acceleration Probe (SNAP) satellite mission (one of the contenders for the NASA Dark Energy Probe program) will find several thousand supernovae out to redshifts of order 2 [104].

6.3 Weak lensing

Perhaps the most interesting future probe of cosmology is weak gravitational lensing on large scales. The shape of distant galaxies will be distorted by the matter distribution along the line of sight, and this effect allows for a direct probe of the large scale distribution of the gravitational potential (see for instance[105] for a review). Just as for the CMB the data can be converted into an angular power spectrum, in this case of the lensing convergence [105,106,107,108]. Several upcoming surveys aim at measuring this spectrum on a large scale. The first to become operational is the Pan-STARRS [109] project which will have first light in 2006. In the more distant future, the Large Synoptic Survey Telescope [110] will provide an even more detailed measurement of lensing distortions across large fractions of the sky.

6.4 The impact on cosmological parameters

Many of the cosmological parameters will be measured much more precisely with future data. For the standard cosmological parameters, a detailed discussion and analysis can be found in [111]. As an example, the bound on the physical matter density could be improved from the present ± 0.012 to ± 0.0022, at least an improvement by a factor 5.

With regards to hot dark matter, the neutrino mass could be constrainable to a precision of $\sigma(\sum m_\nu) \sim 0.1$ eV or better[112,113,114,115,116,79], perhaps allowing for a positive detection of a non-zero mass.

The equation of state of the dark energy could be measurable to a precision of about 5%, depending on whether it varies with time [40].

7 Discussion

We are currently in the middle of an immensely exciting period for cosmology. We now have estimates of most basic cosmological parameters at the percent level, something which was almost unthinkable a decade ago. Cosmology is now at the stage where it can contribute significant new information of relevance to particle physics. One notable example is the density of cold dark matter, which is relevant for SUSY parameter space exploration. Another is the bound on the mass of light neutrinos which is presently significantly stronger than the corresponding laboratory bound.

The precision with which most of the cosmological parameters can be measured is set to increase by a factor of 5-10 over the next ten years, given a whole range of new experiments. For the foreseeable future, cosmology will be an extremely interesting field, and its relevance to particle physics is set to increase with time.

References

1. A. G. Riess *et al.* [Supernova Search Team Collaboration], *Astron. J.* **116**, 1009, (1998), [arXiv:astro-ph/9805201].
2. S. Perlmutter *et al.* [Supernova Cosmology Project Collaboration], *Astrophys. J.* **517**, 565, (1999) [arXiv:astro-ph/9812133].
3. C. L. Bennett *et al.*, *Astrophys. J. Suppl.* **148**,1, (2003), [astro-ph/0302207].
4. D. N. Spergel *et al.* [WMAP Collaboration], *Astrophys. J. Suppl.* **148**, 175, (2003) [arXiv:astro-ph/0302209].
5. M. Colless *et al.*, astro-ph/0306581.
6. M. Tegmark *et al.* [SDSS Collaboration], astro-ph/0310725.
7. M. Tegmark *et al.* [SDSS Collaboration], *Phys. Rev.* D**69**, 103501, (2004), [arXiv:astro-ph/0310723].
8. I. Zlatev, L. M. Wang and P. J. Stein-

372

hardt, *Phys. Rev. Lett.* **82**, 896, (1999) [arXiv:astro-ph/9807002].

9. L. M. Wang, R. R. Caldwell, J. P. Ostriker and P. J. Steinhardt, *Astrophys. J.* **530**, 17, (2000), [arXiv:astro-ph/9901388].

10. P. J. Steinhardt, L. M. Wang and I. Zlatev, *Phys. Rev.* **D59**, 123504, (1999), [arXiv:astro-ph/9812313].

11. F. Perrotta, C. Baccigalupi and S. Matarrese, *Phys. Rev.* **D61**, 023507, (2000), [arXiv:astro-ph/9906066].

12. L. Amendola, *Phys. Rev.* **D62**, 043511, (2000), [arXiv:astro-ph/9908023].

13. T. Barreiro, E. J. Copeland and N. J. Nunes, *Phys. Rev.* **D61**, 127301, (2000), [arXiv:astro-ph/9910214].

14. O. Bertolami and P. J. Martins, *Phys. Rev.* **D61**, 064007, (2000), [arXiv:gr-qc/9910056].

15. C. Baccigalupi, A. Balbi, S. Matarrese, F. Perrotta and N. Vittorio, *Phys. Rev.* **D65**, 063520, (2002), [arXiv:astro-ph/0109097].

16. R. R. Caldwell, M. Doran, C. M. Mueller, G. Schaefer and C. Wetterich, *Astrophys. J.* **591**, L75, (2003), [arXiv:astro-ph/0302505].

17. C. Armendariz-Picon, T. Damour and V. Mukhanov, *Phys. Lett.* **B458**, 209, (1999), [arXiv:hep-th/9904075].

18. T. Chiba, T. Okabe and M. Yamaguchi, *Phys. Rev.* **D62**, 023511, (2000), [arXiv:astro-ph/9912463].

19. C. Armendariz-Picon, V. Mukhanov and P. J. Steinhardt, *Phys. Rev.* **D63**, 103510, (2001), [arXiv:astro-ph/0006373].

20. L. P. Chimento, *Phys. Rev.* **D69**, 123517, (2004), [arXiv:astro-ph/0311613].

21. P. F. Gonzalez-Diaz, *Phys. Lett.* **B586**, 1, (2004), [arXiv:astro-ph/0312579].

22. R. J. Scherrer, arXiv:astro-ph/0402316.

23. J. M. Aguirregabiria, L. P. Chimento and R. Lazkoz, arXiv:astro-ph/0403157.

24. R. R. Caldwell, *Phys. Lett.* **B545**, 23, (2002), [arXiv:astro-ph/9908168].

25. S. M. Carroll, M. Hoffman and M. Trodden, *Phys. Rev.* **D68**, 023509, (2003), [arXiv:astro-ph/0301273].

26. G. W. Gibbons, arXiv:hep-th/0302199.

27. R. R. Caldwell, M. Kamionkowski and N. N. Weinberg, *Phys. Rev. Lett.* **91**, 071301, (2003) [arXiv:astro-ph/0302506].

28. A. E. Schulz and M. J. White, *Phys. Rev.* **D64**, 043514, (2001), [arXiv:astro-ph/0104112].

29. S. Nojiri and S. D. Odintsov, *Phys. Lett.* **B562**, 147, (2003), [arXiv:hep-th/0303117].

30. P. Singh, M. Sami and N. Dadhich, *Phys. Rev.* **D68**, 023522, (2003), [arXiv:hep-th/0305110].

31. M. P. Dabrowski, T. Stachowiak and M. Szydlowski, *Phys. Rev.* **D68**, 103519, (2003), [arXiv:hep-th/0307128].

32. J. G. Hao and X. z. Li, arXiv:astro-ph/0309746.

33. H. Stefancic, *Phys. Lett.* **B586**, 5, (2004), [arXiv:astro-ph/0310904].

34. J. M. Cline, S. y. Jeon and G. D. Moore, arXiv:hep-ph/0311312.

35. M. G. Brown, K. Freese and W. H. Kinney, arXiv:astro-ph/0405353.

36. V. K. Onemli and R. P. Woodard, arXiv:gr-qc/0406098.

37. V. K. Onemli and R. P. Woodard, *Class. Quant. Grav.* **19**, 4607, (2002) [arXiv:gr-qc/0204065].

38. A. Vikman, arXiv:astro-ph/0407107.

39. B. Boisseau *et al.*, *Phys. Rev. Lett.* **85**, 2236, (2000).

40. A. Upadhye, M. Ishak and P.J. Steinhardt, arXiv:astro-ph/0411803.

41. Y. Wang and M. Tegmark, *Phys. Rev. Lett.* **92**, 241302 (2004) [arXiv:astro-ph/0403292].

42. P. S. Corasaniti, M. Kunz, D. Parkinson, E. J. Copeland, B. A. Bassett, arXiv:astro-ph/0406608

43. Gong, Y., arXiv:astro-ph/0405446

44. Gong, Y., arXiv:astro-ph/0401207

45. S. Nesseris and L. Perivolaropoulos, arXiv:astro-ph/0401556.

46. B. Feng, X. L. Wang and X. M. Zhang, arXiv:astro-ph/0404224.

47. U. Alam, V. Sahni, A. A. Starobinsky, arXiv:astro-ph/0406672

48. U. Alam, V. Sahni, A. A. Starobinsky, *JCAP* **6**, 8, (2004)

49. P. S. Corasaniti and E. J. Copeland, arXiv:astro-ph/0205544

50. H. K. Jassal, J. S. Bagla and T. Padmanabhan, arXiv:astro-ph/0404378.

51. T. R. Choudhury and T. Padmanabhan, arXiv:astro-ph/0311622.

52. D. Huterer and A. Cooray, arXiv:astro-ph/0404062

53. R. A. Daly and S. G. Djorgovski, *Astrophys. J.* **597**, 9, (2003)

54. Y. Wang and M. Tegmark, arXiv:astro-ph/0403292

55. Y. Wang and K. Freese, arXiv:astro-ph/0402208

56. Y. Wang and P. Mukherjee, *Astrophys. J.* **606**, 654, (2004)

57. C. l. Kuo *et al.* [ACBAR collaboration], *Astrophys. J.* **600**, 32, (2004) [arXiv:astro-ph/0212289].

58. T. J. Pearson *et al.*, *Astrophys. J.* **591**, 556, (2003) [arXiv:astro-ph/0205388].

59. J. Kovac, E. M. Leitch, C. Pryke, J. E. Carlstrom, N. W. Halverson and W. L. Holzapfel, *Nature* **420**, 772, (2002) [arXiv:astro-ph/0209478].

60. W. C. Jones *et al.*, arXiv:astro-ph/0507494.

61. F. Piacentini *et al.*, arXiv:astro-ph/0507507.

62. T. E. Montroy *et al.*, arXiv:astro-ph/0507514.

63. A. G. Riess *et al.* *Astrophys. J.* **607**, 665, (2004)

64. U. Seljak *et al.*, *Phys. Rev.* D**71**, 103515, (2005) [arXiv:astro-ph/0407372].

65. G. G. Raffelt, arXiv:hep-ph/0504152.

66. K. Griest and M. Kamionkowski, *Phys. Rev. Lett.* **64**, 615, (1990).

67. D. J. H. Chung, E. W. Kolb and A. Riotto, *Phys. Rev. Lett.* **81**, 4048, (1998) [arXiv:hep-ph/9805473].

68. C. Kraus *et al.* European Physical Journal C (2003), proceedings of the EPS 2003 - High Energy Physics (HEP) conference.

69. S. Hannestad, *JCAP* **0305**, 004, (2003)

70. O. Elgaroy and O. Lahav, *JCAP* **0304**, 004, (2003)

71. V. Barger, D. Marfatia and A. Tregre, hep-ph/0312065

72. S. Hannestad and G. Raffelt, *JCAP* **0404**, 008, (2004)

73. P. Crotty, J. Lesgourgues and S. Pastor, *Phys. Rev.* D**69**, 123007, (2004)

74. S. Hannestad, hep-ph/0404239;

75. U. Seljak *et al.*, astro-ph/0407372

76. G. L. Fogli, E. Lisi, A. Marrone, A. Melchiorri, A. Palazzo, P. Serra and J. Silk, *Phys. Rev.* D**70**, 113003, (2004)

77. S. Hannestad, hep-ph/0409108

78. M. Tegmark, hep-ph/0503257

79. S. Hannestad, astro-ph/0505551.

80. S. Hannestad, A. Mirizzi and G. Raffelt, *JCAP* **0507**, 002, (2005) [arXiv:hep-ph/0504059].

81. M. Viel, J. Lesgourgues, M. G. Haehnelt, S. Matarrese and A. Riotto, *Phys. Rev.* D**71**, 063534, (2005) [arXiv:astro-ph/0501562].

82. P. S. Corasaniti and E. J. Copeland, *Phys. Rev.* D**65**, 043004, (2002), [arXiv:astro-ph/0107378].

83. R. Bean and A. Melchiorri, *Phys. Rev.* D**65**, 041302, (2002), [arXiv:astro-ph/0110472].

84. S. Hannestad and E. Mortsell, *Phys. Rev.* D**66**, 063508, (2002), [arXiv:astro-ph/0205096].

85. A. Melchiorri, L. Mersini, C. J. Odman and M. Trodden, *Phys. Rev.* D**68**, 043509, (2003), [arXiv:astro-ph/0211522].

374

86. C. Wetterich, *Nucl. Phys.* B**302**, 668, (1988).

87. P. J. E. Peebles and B. Ratra, *Astrophys. J.* **325**, L17, (1988).

88. B. Ratra and P. J. E. Peebles, *Phys. Rev.* D**37**, 3406, (1988).

89. D. F. Mota and C. van de Bruck, arXiv:astro-ph/0401504.

90. S. Hannestad and E. Mortsell, *JCAP* **0409**, 001, (2004), [arXiv:astro-ph/0407259]

91. J. Jonsson, A. Goobar, R. Amanullah and L. Bergstrom, arXiv:astro-ph/0404468

92. J. Weller and A. M. Lewis, *Mon. Not. Roy. Astron. Soc.* **346**, 987, (2003) [arXiv:astro-ph/0307104].

93. E. V. Linder, *Phys. Rev. Lett.* **90**, 091301, 2003

94. G. R. Dvali, G. Gabadadze, M. Porrati, *Phys. Lett.* B**485**, 208, (2000) [arXiv:hep-th/0005016].

95. G. Dvali and M. S. Turner, arXiv:astro-ph/0301510.

96. O. Elgaroy and T. Multamaki, *Mon. Not. Roy. Astron. Soc.* **356**, 475, (2005) [arXiv:astro-ph/0404402].

97. S. Hannestad and L. Mersini-Houghton, *Phys. Rev.* D**71**, 123504, (2005) [arXiv:hep-ph/0405218].

98. M. Ishak, A. Upadhye and D. N. Spergel, arXiv:astro-ph/0507184.

99. http://astro.estec.esa.nl/Planck/

100. A. Kosowsky, *New Astron. Rev.* **47**, 939, (2003) [arXiv:astro-ph/0402234].

101. R. Pain [SNLS Collaboration], eConf **C041213**, 1413, (2004).

102. T. Matheson *et al.*, arXiv:astro-ph/0411357.

103. B. Flaugher [the Dark Energy Survey Collaboration], *Int. J. Mod. Phys.* A**20**, 3121, (2005).

104. http://snap.lbl.gov/

105. M. Bartelmann and P. Schneider, *Phys. Rept.* **340**, 291, (2001) [arXiv:astro-ph/9912508].

106. N. Kaiser, *Astrophys. J.* **388**, 271, (1992)

107. N. Kaiser, *Astrophys. J.* **498**, 26, (1998) [arXiv:astro-ph/9610120].

108. B. Jain and U. Seljak, *Astrophys. J.* **484**, 560, (1997) [arXiv:astro-ph/9611077].

109. http://pan-starrs.ifa.hawaii.edu/public/index.html

110. http://www.lsst.org/lsst_home.shtml

111. M. Ishak, C. M. Hirata, P. McDonald and U. Seljak, *Phys. Rev.* D**69**, 083514, (2004) [arXiv:astro-ph/0308446].

112. S. Hannestad, *Phys. Rev.* D **67**, 085017, (2003).

113. J. Lesgourgues, S. Pastor and L. Perotto, *Phys. Rev.* D**70**, 045016, (2004);

114. S. Wang, Z. Haiman, W. Hu, J. Khoury and M. May, astro-ph/0505390

115. K. N. Abazajian and S. Dodelson, *Phys. Rev. Lett.* **91**, 041301, (2003);

116. M. Kaplinghat, L. Knox and Y. S. Song, *Phys. Rev. Lett.* **91**, 241301, (2003)

117. S. Burles, K. M. Nollett and M. S. Turner, *Astrophys. J.* **552**, L1, (2001) [arXiv:astro-ph/0010171].

118. R. H. Cyburt, B. D. Fields and K. A. Olive, *Phys. Lett.* B**567**, 227, (2003) [arXiv:astro-ph/0302431].

119. W. Hu, D. J. Eisenstein, M. Tegmark and M. J. White, *Phys. Rev.* D**59**, 023512, (1999) [arXiv:astro-ph/9806362].

120. J. K. Erickson, R. R. Caldwell, P. J. Steinhardt, C. Armendariz-Picon and V. Mukhanov, *Phys. Rev. Lett.* **88**, 121301, (2002) [arXiv:astro-ph/0112438].

121. R. Bean and O. Dore, *Phys. Rev.* D**69**, 083503, (2004) [arXiv:astro-ph/0307100].

122. S. Hannestad, *Phys. Rev.* D**71**, 103519, (2005) [arXiv:astro-ph/0504017].

123. A. R. Liddle, *Mon. Not. Roy. Astron. Soc.* **351**, L49, (2004) [arXiv:astro-ph/0401198].

124. P. Mukherjee, D. Parkinson and A. R. Liddle, arXiv:astro-ph/0508461.

125. C. P. Ma and E. Bertschinger, *Astrophys. J.* **455**, 7, (1995) [arXiv:astro-ph/9506072].

126. L. Bergstrom, *Rept. Prog. Phys.* **63**, 793, (2000) [arXiv:hep-ph/0002126].

127. G. Bertone, D. Hooper and J. Silk, *Phys. Rept.* **405**, 279, (2005) [arXiv:hep-ph/0404175].

128. S. Hannestad, A. Ringwald, H. Tu and Y. Y. Y. Wong, arXiv:astro-ph/0507544.

129. S. Tremaine and J. E. Gunn, *Phys. Rev. Lett.* **42**, 407, (1979)

130. A. Kull, R. A. Treumann and H. Böhringer, *Astrophys. J.* **466**, L1, (1996) [arXiv:astro-ph/9606057].

131. J. Madsen, *Phys. Rev. Lett.* **64**, 2744, (1990)

132. J. Madsen, *Phys. Rev.* D**44**, 999, (1991)

133. M. Kamionkowski, A. Kosowsky and A. Stebbins, *Phys. Rev.* D**55**, 7368, (1997) [arXiv:astro-ph/9611125].

ULTRA–HIGH-ENERGY COSMIC RAYS

STEFAN WESTERHOFF

Columbia University, Department of Physics, New York, NY 10027, USA
E-mail: westerhoff@nevis.columbia.edu

One of the most striking astrophysical phenomena today is the existence of cosmic ray particles with energies in excess of 10^{20} eV. While their presence has been confirmed by a number of experiments, it is not clear where and how these particles are accelerated to these energies and how they travel astronomical distances without substantial energy loss. We are entering an exciting new era in cosmic ray physics, with instruments now producing data of unprecedented quality and quantity to tackle the many open questions. This paper reviews the current experimental status of cosmic ray physics and summarizes recent results on the energy spectrum and arrival directions of ultra-high-energy cosmic rays.

1 Introduction

Cosmic ray particles were discovered almost one hundred years ago, and yet very little is known about the origin of the most energetic particles above and around 10^{18} eV, traditionally referred to as "ultra–high-energy cosmic rays." The measured spectrum of cosmic rays extends beyond 10^{20} eV, 11 orders of magnitude greater than the equivalent rest mass of the proton. While the presence of particles at these energies has been confirmed by a number of experiments, it is not clear where and how these particles are accelerated to these energies and how they travel astronomical distances without substantial energy loss. Some indication comes from the energy spectrum itself, which roughly follows a power law $E^{-2.8}$ and is therefore "non-thermal." The power law behavior and the "universal" spectral index indicate that the underlying acceleration mechanism could be Fermi acceleration[1] at shock fronts, e.g. in Supernova Remnants (SNRs) and in the jets of Active Galactic Nuclei (AGNs). Regardless of the actual acceleration process, it is clear that thermal emission processes cannot generate such energies.

Several quantities accessible to experiment can help to reveal the sources of ultra–high-energy cosmic rays: the *flux* of cosmic rays; their *chemical composition*; and their *arrival direction*. Charged cosmic ray primaries are subject to deflection in Galactic and intergalactic magnetic fields and do not necessarily point back to their sources. The strength and orientation of these fields is poorly known and estimates vary[2,3], but their impact should decrease at the highest energies, above several times 10^{19} eV; here, cosmic ray astronomy might be possible.

Since the cosmic ray flux drops almost three orders of magnitude for each energy decade, the flux at the highest energies is very low, about one particle per km^2 per year above 5×10^{18} eV. Low statistics have historically plagued the field; the total published number of events above 4.0×10^{19} eV is still less than 100, and drawing conclusions on the basis of such a small data set has proved rather perilous. However, the experimental situation is finally improving; new instruments are now collecting data of unprecedented quality and quantity. Since 1999, the High Resolution Fly's Eye (HiRes) air fluorescence stereo detector in Utah has accumulated data with excellent angular resolution. The HiRes data set has been used extensively over the last two years to search for small-scale anisotropies in the arrival directions of cosmic rays, correlations of cosmic rays with known astrophysical sources, and to study the composition of the primary cosmic ray flux.

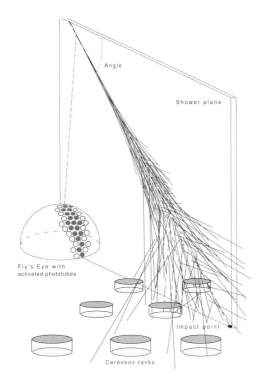

Figure 1. Scheme of the different cosmic ray detection techniques (Auger Collaboration).

In the southern hemisphere, the world's largest detector for cosmic radiation, the Pierre Auger Observatory in Argentina, is nearing completion, and first results on the energy spectrum and the arrival direction distribution of cosmic rays have recently been published.

In this paper, I will review some recent developments in ultra–high-energy cosmic ray physics. The paper is organized as follows: Section 2 gives a short review of the experimental techniques and the current major instruments in the field. I will then discuss new results from the Auger and HiRes experiments on the energy spectrum (Section 3) and the arrival direction distribution (Section 4). Concluding remarks follow in Section 5.

2 Experimental Techniques

Because the flux at ultra–high energies is small, experiments need a large detector volume. Consequently, detectors have to be earth-bound, and the primary cosmic ray particles can not be observed directly. Primaries interact in the upper atmosphere and induce extensive air showers with on the order of 10^{10} particles for a 10^{19} eV primary. The properties of the original cosmic ray particle, such as arrival direction and energy, have to be inferred from the observed properties of the extensive air shower.

There are two different techniques to study cosmic ray air showers at ultra–high-energies. Both are shown schematically in Fig. 1. *Ground arrays* sample the shower front arriving on the Earth's surface with an array of particle detectors, for example scintillation counters or water Cherenkov detectors. The arrival direction of the air shower and the primary cosmic ray particle is reconstructed from the differences in trigger times for individual detectors as the narrow shower front passes. The advantage of ground arrays is their near 100 % duty cycle and the robustness of the detectors. A disadvantage is that ground arrays sample the shower at one altitude only and do not record the development of the shower in the atmosphere. Moreover, the sampling density is typically very small.

The classic example of a pure ground array is the AGASA (Akeno Giant Air Shower Array) experiment, which operated in Japan from 1984 to 2003. In its final stage, the array consisted of 111 scintillation counters of $2.2\,\mathrm{m}^2$ area each on a 1 km spacing, leading to a total area of about $100\,\mathrm{km}^2$.

Apart from the cascade of secondary particles, air showers also produce Cherenkov light and fluorescence light. The latter is produced when the particles in the air shower cascade excite air molecules, which fluoresce in the UV. *Air fluorescence detectors* measure this light with photomultiplier cameras

Table 1. Comparison of Instruments. *Comments:* [1] above 10^{20} eV, 10 % duty cycle, [2] January 2004 – June 2004, [3] error bars are strongly asymmetric

Experiment	Operation	Aperture [km^2 sr]	Exposure [km^2 sr yr]	Angular Resolution (68 %)	$\Delta E/E$ [%]
AGASA	1984-2003	$\simeq 250$	1620	2.5°	25
HiRes mono	1997 -	10000[1]	5000	> 2.5°,[3]	25
HiRes stereo	2000 -	10000[1]	3400	0.6°	15
Auger (under construction)	2004 -	7400	1750 (SD)[2]	2.0° - 0.9° (SD) 0.6° (FD)	10

that observe the night sky. The shower is observed by a succession of tubes and reconstructed using the photomultiplier timing and pulseheight information. Air fluorescence detectors can only operate on clear, moonless nights with good atmospheric conditions, so the duty cycle is about 10 %; however, they observe the shower development in the atmosphere and provide us with a nearly calorimetric energy estimate. In addition, the instantaneous detector volume is rather large, of order $10000 \, \text{km}^2$ sr at 10^{20} eV.

The High Resolution Fly's Eye (HiRes) experiment in Utah is a stereo air fluorescence detector with 2 sites roughly 13 km apart. Each site is made up of several telescope units monitoring different parts of the night sky. With 22 (42) telescopes at the first (second) site, the full detector covers about 360° (336°) in azimuth and $3° - 16.5°$ ($3° - 30°$) in elevation above horizon. Each telescope consists of a mirror with an area of about $5 \, \text{m}^2$ for light collection and a cluster of 256 hexagonal photomultiplier tubes in the focal plane.

The HiRes air fluorescence detector can operate in "monocular mode," with air showers only observed from one site, or in "stereoscopic mode," with the same shower observed by both detectors simultaneously. The monocular operation suffers from poor angular and energy resolution. With only one "eye," the shower-detector-plane (*i.e.* the plane that contains the shower and the de-

tector (see Fig. 1)) can be reconstructed with high accuracy. Unfortunately, the position of the shower within that plane, determined using the photomultiplier trigger times, is ambiguous. Stereo viewing of the shower with two detectors breaks the ambiguity and leads to an excellent angular resolution of order 0.5°. HiRes has been taking data in monocular mode since 1996 and in stereo mode since December 1999.

In summary, stereo air fluorescence detectors have excellent angular resolution and give a nearly calorimetric energy determination, while ground arrays have the advantage of a relatively straightforward determination of the detector aperture. Obviously, the best detector is a detector that combines the two techniques. The Pierre Auger Observatory, currently under construction in Malargue, Argentina, and scheduled to be completed in 2006, is such a *hybrid detector*, combining both a ground array and fluorescence detectors.

The Auger surface detector array[4] will eventually comprise 1600 detector stations with 1500 m separation, spread over a total area of $3000 \, \text{km}^2$. One year of Auger datataking will therefore correspond to about 30 AGASA years. Each detector station is a light-tight 11 000 liter tank filled with pure water. Three 9-inch photomultipliers in the tank measure the Cherenkov light from shower particles crossing the tank. The stations are self-contained and work on solar power. The surface detector array is complemented by 4 fluorescence stations[5] with 6 telescopes each. The field of view of a single telescope covers 30° in azimuth and 28.6° in elevation, adding up to a total field of view of 180° × 28.6° for each site. Each telescope consists of a spherical mirror of area 3.5 m × 3.5 m and a focal surface with 440 hexagonal (45 mm diameter) photomultipliers. Because of the large field of view of each telescope, a Schmidt optics is used.

While still under construction, the Auger experiment has recorded data with a growing detector since January 2004. The total exposure of the surface detector array already exceeds the total AGASA exposure, making Auger a competitive experiment.

Table 1 gives an overview of experimental parameters for AGASA, Auger, HiRes monocular and HiRes stereo, including the aperture of the instruments and the exposure used in publications of recent results.

3   Results

Until a few years ago, the world data set of ultra–high-energy cosmic rays was dominated by data recorded with the AGASA air shower array. In a number of important publications, the AGASA group has described several exciting and controversial results, including the shape of the energy spectrum above 10^{18} eV, studies of possible anisotropies in the arrival directions above 4.0×10^{19} eV, and an excess of cosmic ray flux from the Galactic center region around 10^{18} eV. Due to the small number of events, some of these results have a small statistical significance. For several years, statistically independent data sets to support or refute these findings were not available. Only recently, with the start of HiRes stereo data-taking in 1999 and the beginning of operations at the Pierre Auger Observatory in January 2004, are we reaching the point where we can study these topics with new data.

3.1   Energy Spectrum

Proton primaries above 5.0×10^{19} eV interact with the 2.7 K microwave background via photo-pion production, losing energy in each interaction until they eventually fall below the energy threshold. This so-called GZK effect, postulated by Greisen[6] and independently by Zatsepin and Kuzmin[7] shortly after the discovery of the microwave background, is expected to lead to a rapid fall-off of the cosmic ray energy spectrum above this energy, but it has not been experimentally confirmed at this point.

The AGASA group claimed[8,9] that the GZK suppression is not observed, raising questions as to the nature of the primary particles or even the particle physics involved. More recently, the HiRes monocular spectrum[10,11] has been interpreted as being in agreement with a GZK suppression. While the disagreement between the AGASA and the HiRes mono result has received a lot of attention, it has also been pointed out that the two spectra actually agree reasonable well if systematic errors are taken seriously[12,13].

Fig. 2 depicts the situation. It shows the differential energy spectrum as reported by AGASA[9] and HiRes[10,11] in monocular mode (the HiRes collaboration has not yet published a stereo spectrum). While only statistical errors are shown, systematic errors of order 25 % (roughly what is reported by the experiments) would manifest themselves simply by a shift of one bin in this plot. Keeping this in mind, both spectra agree quite well below $10^{19.8}$ eV. The disagreement at the highest energies has been estimated to be of order 2.0σ [12].

Reversing the argument, it is actually quite surprising how well the spectra agree, given that ground arrays and fluorescence detectors have entirely different systematic errors in their estimate of the shower energy. AGASA has given a detailed account of their energy resolution[9]. Ground arrays determine

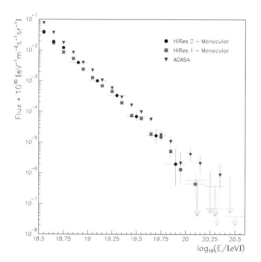

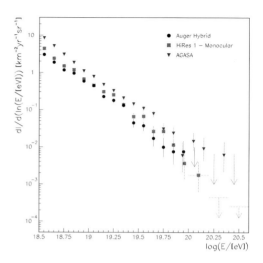

Figure 2. Ultra–high-energy cosmic ray flux as a function of energy as measured by the AGASA experiment[9] and the HiRes detectors in monocular mode, HiRes 1[10] and HiRes 2[11].

Figure 3. Intensity of the ultra–high-energy cosmic ray flux as a function of energy measured by the AGASA experiment[9] and HiRes 1 in monocular mode[10], compared to a first estimate by the Pierre Auger Observatory[14].

the energy of a shower from the measured particle density S at some (fixed) distance from the shower core[15]. The optimal distance is the one where the spread in particle density is minimal, and this optimal distance depends on the detector geometry (600 m for AGASA[9] and 1000 m for Auger[16]). The correlation between the particle density and the energy is then established using simulated data. This of course leads to a strong mass and model dependence.

Air fluorescence detectors image the shower development in the atmosphere and obtain a nearly calorimetric signal. However, the Earth's atmosphere is a tricky calorimeter. It is not only highly inhomogeneous, but also changes on short time scales. The atmosphere needs to be monitored continuously in order to be able to correctly account for Rayleigh and Mie scattering. In addition, there are systematic errors from the fluorescence yield, the absolute calibration, and the accounting for "missing energy," *i.e.* energy that goes into particles that do not contribute to air fluorescence.

The hybrid nature of the Pierre Auger Observatory will enable us to understand whether there are intrinsic difficulties with one of the two methods that contribute to the disagreement in the two spectra. At this time, the Auger experiment has not yet accumulated the necessary statistics to make a definitive statement on the shape of the energy spectrum at the highest energies. The collaboration has, however, already published a first estimate of the spectrum. Fig. 3 compares the Auger spectrum to both the AGASA and HiRes 1 monocular measurement. The spectrum is based on 3525 events with energies above 3×10^{18} eV data taken with the ground array between January 2004 and June 2005. At these energies, the ground array is fully efficient and the total exposure can be inferred from the total running time (where the growing size of the detector is accounted for) and the geometrical aperture of the ground array. The energy of the showers is established using the subset of *hybrid* events in the data sample, *i.e.* the subset of events that are also detected in one of the fluorescence detectors. These events are used to derive a relationship between shower energy (as determined by the fluorescence detector) and the ground parameter S (as measured by the ground array)[14].

By using the statistics and the exposure of the ground array and the nearly calorimetric energy estimate of the fluorescence detectors, it is ensured that the Auger spectrum does not rely strongly on either simulations or assumptions about the chemical composition of the cosmic ray flux. Unlike the AGASA and HiRes spectra, it is therefore nearly model-independent.

As shown in Fig. 3, the Auger spectrum agrees quite well with the HiRes measurement at this point. Considering the large systematic uncertainties, there is also little disagreement with AGASA. For this first estimate of the spectrum, systematic errors range from about 30 % at 3×10^{18} eV to 50 % at 10^{20} eV, of which about 25 % stem from total systematic uncertainties in the fluorescence detector energy measurements. Errors will soon become considerably smaller with larger statistics and the application of more sophisticated reconstruction methods.

Two additional aspects of the Auger spectrum should be stressed.

(1) While the spectrum does not show any events above 10^{20} eV, the Auger experiment has detected an event above that energy[17]. The event does not enter the energy spectrum because its core falls outside the surface detector array, but the event is well-reconstructed and passes all other quality cuts. This means that all major cosmic ray experiments have now confirmed the existence of particles above 10^{20} eV.

(2) As described above, the energy scale of the spectrum is normalized to the fluorescence detector. When simulations are used to determine the shower energy from surface detector data alone, the energies are systematically higher by at least 25 %. Energies determined in this way depend on the shower simulation code, the hadronic model, and the assumed composition of the cosmic ray flux. The difference shows that many systematics need to be addressed. Auger is in a unique position to study these systematic uncertainties and will measure the spectrum in the southern hemisphere accurately in coming years. There is no reason to believe that the energy spectra in the northern sky and the southern sky are identical, although differences are probably subtle and will require a larger instrument in the northern hemisphere.

3.2 Search for Small-Scale Clustering and Point Sources

A direct way to search for the sources of ultra–high energy cosmic rays is to study their arrival direction distribution. Astronomy with charged particles of course faces a serious problem. At low energies, Galactic and intergalactic magnetic fields will bend the particle's trajectories sufficiently to render the direction information useless. However, the Larmor radius increases with energy, and cosmic ray astronomy may be possible above some energy threshold. But little is known about intergalactic magnetic fields, so it is not straightforward to determine an optimal energy cut for arrival direction studies. Choosing an energy threshold too low means that deflections destroy any correlation, but too high a threshold also weakens the statistical power of the data set. Given these uncertainties, an *a priori* optimal choice for the energy threshold and the angular separation for clustering searches does not exist.

AGASA claimed a significant amount of clustering in the arrival direction distribution as early as 1996[18] and has updated these results frequently[19,20,21,22]. The current claim is that the 5 "doublets" and 1 "triplet" observed in the set of 57 events with energies above 4.0×10^{19} eV have a probability of less

382

than 0.1 % to occur by chance in an AGASA data set of this size. This is an extremely important result, but its validity has been questioned on statistical grounds. The problems stem from the fact that the data set used to evaluate the chance probability *includes* the data used for formulating the clustering hypothesis in the first place; parameters like the energy threshold and the angular scale that defines a "cluster" of arrival directions were not defined *a priori*. The probability of 0.1 % has therefore little meaning, and analyses that test the hypothesis only with the part of the data set that was recorded after the hypothesis was formulated find a much higher chance probability of around 8 % [23], making the result insignificant.

The AGASA clustering claim has been tested with the statistically independent HiRes stereo data set. The HiRes collaboration has published several papers describing searches for deviation from isotropy, including the calculation of the standard two-point correlation function[24], a two-point correlation scan[25], and a search for point sources using an unbinned maximum likelihood ratio test[26,27]. None of these searches have uncovered any deviation from isotropy, and no evidence for statistically significant point sources in the HiRes or the combined HiRes/AGASA data sets above 4.0×10^{19} eV or 1.0×10^{19} eV were found[26,27] (although a point source has been claimed by other authors[28,29]). With an overall exposure that exceeds the AGASA exposure and an angular resolution that is 4 to 10 times sharper than AGASA's, HiRes stereo finds no small-scale anisotropy at any energy threshold above 10^{19} eV and any angular scale out to 5°.

3.3 Correlation with BL Lac Objects

There has recently been marginal evidence that a small fraction of the ultra–high-energy cosmic ray flux originates from BL Lacertae Objects[30,31]. BL Lacs are a subclass of blazars, which are active galaxies in which the jet axis happens to point almost directly along the line of sight. The EGRET instrument on board the Compton Gamma Ray Observatory (CGRO) has firmly established blazars as sources of high energy γ-rays above 100 MeV [32], and several BL Lac objects have been observed at TeV energies with ground-based air Cherenkov telescopes[33]. High energy γ-rays could be by-products of electromagnetic cascades from energy losses associated with the acceleration of ultra–high-energy cosmic rays and their propagation in intergalactic space[34,35].

The history of claims for a correlation between BL Lacs and ultra–high-energy cosmic rays is rather convoluted. Correlation claims based on data recorded with AGASA and the Yakutsk array have been published since 2001[36,37,38], but a problematic aspect of the claims is the procedure used to establish correlations and evaluate their statistical significance. The authors explicitly tuned their selection criteria to assemble catalogs showing a maximum correlation with arrival directions of cosmic rays above some energy. The statistical significance quoted is therefore meaningless, and claims of BL Lac correlations have been criticized on these grounds[39,40]. In some cases it has been shown that statistically independent data sets do not confirm the correlations[41].

Recently, the HiRes collaboration published its own analysis of the data[31], using an unbinned maximum likelihood ratio test which accounts for the individual point spread function of each event. The HiRes stereo data does not confirm any of the previous claims based on AGASA and Yakutsk data, in spite of its larger statistical power. It does, however, verify a recent analysis[30] of correlations between published HiRes stereo events above 10^{19} eV and a subset of confirmed BL Lacs from the 10th Veron Catalog[42]. This subset[36] contains 157 confirmed BL Lacs from the catalog with optical magnitude $m < 18$. Since the cuts[30] used to isolate this signal are not *a priori*, the correlation needs to be confirmed with statistically independent data before any claims can be made.

The correlations are strongest at energies around 10^{19} eV, where magnetic fields are expected to sufficiently scramble the arrival directions of charged particles, yet the correlations are on the scale of the detector angular resolution. This would suggest neutral cosmic ray primaries for these events, or at least that the primaries were neutral during significant portions of their journey through Galactic and extragalactic magnetic fields. If verified with future data, the correlation with BL Lac objects would be the first evidence for an extragalactic origin of the highest energy cosmic rays, and a first indication that at least a fraction of this flux originates in known astrophysical objects.

3.4 Galactic Center

One of the most interesting regions of the sky to search for an excess of ultra–high-energy cosmic rays is around the Galactic center. This region is a natural site for cosmic ray acceleration. It harbors a black hole of mass 2.6×10^6 solar masses whose position is consistent with the radio source Sagittarius A*. Hour-scale X-ray and rapid IR flaring indicate the presence of an active nucleus with low bolometric luminosity. In addition, this crowded part of the sky contains a dense cluster of stars, stellar remnants, and the supernova remnant Sgr A East. The Galactic center is now also established as a source of TeV γ-radiation[43,44,45].

The AGASA experiment reported an excess with a statistical significance of order $4\,\sigma$ near the Galactic center[46,47]. The excess is observed in a narrow energy band from $10^{17.9}$ eV to $10^{18.3}$ eV only. Due to its location in the northern hemisphere and a zenith angle cut of $60°$, AGASA's field of view cuts off roughly $5°$ north of the Galactic center, so the center itself, at right ascension $\alpha = 266.4°$ and declination $\delta = -28.9°$, is outside the field of view. The AGASA excess is found roughly around $\alpha = 280°$, $\delta = -16°$, so it is offset considerably from the location of the Galactic center. To produce this excess, the event density is integrated over a circle with a radius of $20°$. Several other "beam sizes" were tried, but this integration radius was found to maximize the signal.

The chance probability in the AGASA publication is *a posteriori*. Since the analysis has several "tunable" parameters like the energy bin and the integration radius, the true chance probability cannot be derived from the AGASA data set itself. The AGASA result was, however, supported by a re-analysis of archival data taken with the SUGAR array[48], a cosmic ray experiment that operated in Australia between 1968 and 1979. Unlike AGASA, SUGAR operated from a location with good visibility of the Galactic center region. The SUGAR data showed an excess of about $2.9\,\sigma$ in the energy bin $10^{17.9}$ eV to $10^{18.5}$ eV, roughly the same energy bin as AGASA, but at $\alpha = 274°$ and $\delta = -22°$, so offset both from the location of the AGASA excess and the Galactic center. Furthermore, the SUGAR excess was consistent with a point source, indicating neutral primaries.

In such a context, it is interesting to point out that *neutron* primaries are a viable hypothesis, as they could travel undeflected. The neutron hypothesis would also explain the narrow energy range of the signal. In an amazing coincidence, the neutron decay length at 10^{18} eV roughly corresponds to the distance between us and the Galactic center, about 8.5 kpc.

Like SUGAR, the Auger detector has good visibility of the Galactic center region. The Auger coverage map in the vicinity of the Galactic center is smooth and shows no strong variations (see upper left part of

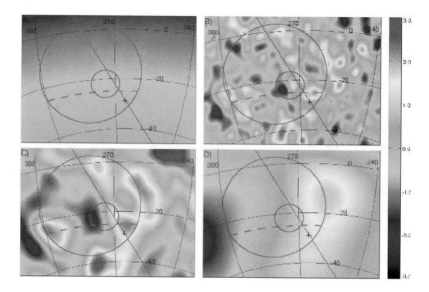

Figure 4. Auger skymap in the vicinity of the Galactic center[49], using surface detector data. Shown are the coverage map (upper left) and the significance maps for the data smoothed using three different resolutions: the Auger resolution (upper right), the large integration radius of used by the AGASA collaboration for their claim of a Galactic center source (lower right), and the resolution of the SUGAR experiment (lower left). The Galactic center is indicated by a cross, the regions of excess for AGASA and SUGAR are indicated by the red circles. The dashed line indicates the limit of the AGASA field of view.

Fig. 4). The Auger group has searched for an excess in the Galactic center region in the data set taken from 1 January 2004 through 5 June 2005[49], both in the surface detector data (angular resolution 1.5°) and hybrid data (0.6°). The event statistics for the surface detector data are already larger than those of the two previous experiments (3 times AGASA, 10 times SUGAR), but the AGASA excess occurs in an energy range where Auger is not fully efficient ($> 30\%$ for protons, $> 50\%$ for iron). However, even in the worst case (a proton signal on an iron-dominated background) a $5.2\,\sigma$ excess is expected if the AGASA source is real.

Fig. 4 shows the results of an analysis using surface detector data taken between January 2004 and June 2005. The figure shows the map smoothed using different angular resolutions, corresponding to the Auger resolution (upper right), the AGASA signal (lower right), and the SUGAR resolution (lower left). No significant excess is found in all cases. The Auger group also studied neighboring energy bins to account for a possible systematic difference in the energy calibration, but no excess is found in any scenario. For the Galactic center itself, a 95% confidence level upper limit for the flux from a point source is derived[49] which excludes the neutron source scenario suggested to explain the previous claims[47,50].

4 Concluding Remarks

The most important and most fascinating result at this point is that *cosmic ray particles with energies above $10^{20}\ eV$ exist* – their existence has been confirmed by all experiments, regardless of the experimental technique, and regardless of whether the GZK suppression is present or absent in the data.

With current statistics, the cosmic ray sky is remarkably isotropic. One possible reason for this could be magnetic smearing. This effect should be smallest at the highest energies, above 10^{20} eV, but until now the number of events has been too small to draw any conclusions. In another possible scenario, we could be dealing with many sources, each currently contributing at most one or two events, and again, more data will help to eventually resolve the strongest ones.

The Pierre Auger Observatory will dramatically increase the high-energy sample size over the next several years. Auger is still under construction (to be completed in 2006), but the collaboration has already published first results, among them a first estimate of the cosmic ray energy spectrum. With its location in the southern hemisphere, its hybrid design and unprecedented size, Auger is in an excellent position to answer definitively the questions left open by the previous generation of cosmic ray experiments, and to make the discoveries that will challenge the next one.

Acknowledgments

It is a pleasure to thank the organizers of Lepton-Photon 2005 for a delightful meeting in Uppsala. The author is supported by the National Science Foundation under contract numbers NSF-PHY-9321949 and NSF-PHY-0134007.

References

1. E. Fermi, *Phys. Rev.* **75**, 1169 (1949).
2. G. Sigl, F. Miniati, and T. Ensslin, *Phys. Rev.* D **68**, 044008 (2003).
3. K. Dolag, D. Grasso, V. Springel, and I.I. Tkachev, *JETP Lett.* **79**, 583 (2004).
4. Pierre Auger Collaboration, Proc. 29th Int. Cosmic Ray Conf., Pune, India (2005) (arXiv: astro-ph/0508466).
5. Pierre Auger Collaboration, Proc. 29th Int. Cosmic Ray Conf., Pune, India (2005) (arXiv: astro-ph/0508389).
6. K. Greisen, *Phys. Rev. Lett.* **16**, 748 (1966).
7. G.T. Zatsepin and V.A. Kuzmin, *Zh. Eksp. Teor. Fiz* **4**, 114 (1966).
8. M. Takeda *et al.*, *Phys. Rev. Lett.* **81**,1163 (1998).
9. M. Takeda *et al.*, *Astropart. Phys.* **19**, 447 (2003).
10. R.U. Abbasi *et al.*, *Phys. Rev. Lett.* **92**, 151101 (2004).
11. R.U. Abbasi *et al.*, *Phys. Lett.* B **619**, 271 (2005).
12. D. De Marco, P. Blasi, and A.V. Olinto, *Astroparticle Phys.* **20**, 53 (2003).
13. J.N. Bahcall and E. Waxman, *Phys.Lett.* B **556**, 1 (2003).
14. Pierre Auger Collaboration, Proc. 29th Int. Cosmic Ray Conf., Pune, India (2005) (arXiv: astro-ph/0507150).
15. A.M. Hillas, D.J. Marsden, J.D. Hollows, and H.W. Hunter, Proc. 12th Int. Cosmic Ray Conf., Hobart, Australia, **3**, 1001 (1971),
16. Pierre Auger Collaboration, Proc. 29th Int. Cosmic Ray Conf., Pune, India (2005) (arXiv: astro-ph/0507029).
17. Pierre Auger Collaboration, Proc. 29th Int. Cosmic Ray Conf., Pune, India (2005).
18. N. Hayashida *et al.*, *Phys. Rev. Lett.* **77**, 1000 (1996).
19. M. Takeda *et al.*, *Astrophys. J.* **522**, 225 (1999).
20. N. Hayashida *et al.*, astro-ph/0008102 (2000).
21. M. Takeda *et al.*, Proceedings 27th Int. Cosmic Ray Conference (ICRC), Hamburg, 345 (2001).
22. M. Teshima *et al.*, Proceedings 28th Int. Cosmic Ray Conference (ICRC), Tsukuba, 437 (2003).
23. C.B. Finley and S. Westerhoff, *Astroparticle Physics* **21**, 359 (2004).
24. C.B. Finley *et al.*, *Int. J. Mod. Phys.* A **20**, 3147 (2005).
25. R.U. Abbasi *et al.*, *Astrophys. J.* **610**,

L73 (2004).

26. R.U. Abbasi *et al.*, *Astrophys. J.* **623**, 164 (2005).

27. HiRes Collaboration, Proc. 29th Int. Cosmic Ray Conf., Pune, India (2005) (arXiv: astro-ph/0507574).

28. G.R. Farrar, arXiv: astro-ph/0501388 (2005).

29. G.R. Farrar, A.A. Berlind, and D.W. Hogg, arXiv: astro-ph/0507657 (2005).

30. D.S. Gorbunov, P.G. Tinyakov, I.I. Tkachev, and S.V. Troitsky, *JETP Lett.* **80**, 145 (2004).

31. R.U. Abbasi *et al.*, *Astrophys. J.*, in press (2005) (arXiv: astro-ph/0507120).

32. R.C. Hartman *et al.*, *Astrophys. J. Suppl.* **123**, 79 (1999).

33. D. Horan and T.C. Weekes, *New Astron. Rev.* **48**, 527 (2004).

34. V.S. Berezinskii, S.V. Bulanov, V.A. Dogiel, and V.S. Ptuskin, *Astrophysics of cosmic rays*, Amsterdam: North-Holland, edited by V.L. Ginzburg (1990).

35. P.S. Coppi and F.A. Aharonian, *Astrophys. J.* **487**, L9 (1997).

36. P.G. Tinyakov and I.I. Tkachev, *JETP Lett.* **74**, 445 (2001).

37. P.G. Tinyakov and I.I. Tkachev, *Astroparticle Phys.* **18**, 165 (2002).

38. D.S. Gorbunov, P.G. Tinyakov, I.I. Tkachev, and S.V. Troitsky, *Astrophys. J.* **577**, L93 (2002).

39. N.W. Evans, F. Ferrer, and S. Sarkar, *Phys. Rev. D* **67**, 103005 (2003).

40. B.E. Stern and J. Poutanen, *Astrophys. J.* **623**, L33 (2005).

41. D.F. Torres, S. Reucroft, O. Reimer, and L.A. Anchordoqui, *Astrophys. J.* **595** L13 (2003).

42. M.P. Veron-Cetty and P. Veron, *Astronomy and Astrophysics* **374**, 92 (2001).

43. K. Tsuchiya *et al.*, *Astrophys. J.* **606**, L115 (2004).

44. K. Kosack *et al.*, *Astrophys. J.* **608**,

L97 (2004).

45. F. Aharonian *et al.*, *Astronomy and Astrophysics* **425**, L13 (2004).

46. N. Hayashida *et al.*, *Astroparticle Phys.* **10**, 303 (1999).

47. N. Hayashida *et al.*, Proc. 26th Int. Cosmic Ray Conf., Salt Lake City, USA (1999) (arXiv:astro-ph/9906056).

48. J.A. Bellido *et al.*, *Astroparticle Phys.* **15**, 167 (2001).

49. Pierre Auger Collaboration, Proc. 29th Int. Cosmic Ray Conf., Pune, India (2005) (arXiv: astro-ph/0507331).

50. M. Bossa *et al.*, *J. Phys. G* **29**, 1409 (2003).

51. L.A. Anchordoqui, H. Goldberg, F. Halzen, T.J. Weiler, *Phys. Lett. B* **593**, 42 (2004).

52. R.U. Abbasi *et al.*, *Astrophys. J.* **622**, 910 (2005).

DISCUSSION

Vincenzo Cavasinni (Universita di Pisa/INFN):

If the cosmic ray excess from the Galactic center is due to neutrons, would you expect also a neutrino flux coming from the neutron decays which could be measured by large volume neutrino detectors?

Stefan Westerhoff : Yes. Neutrons with energy less than EeV will decay in flight and produce an antineutrino flux above TeV that could be detected by the next generation neutrino telescopes such as IceCube. The expected event rate per year above 1 TeV for IceCube was estimated to be about 20. See reference [51] for details.

Tim Greenshaw (Liverpool University):

Does the simultaneous observation of the fluorescence and surface signals at Auger give additional information on the composition of UHE cosmic rays and, if so, do you have any preliminary results?

Stefan Westerhoff : There is a wealth of information in the combined measurement, but studies are ongoing and there are no published results from Auger yet. The HiRes collaboration has published results based on the stereo data. See reference [52] for details.

Jaques Lefrancois (LAL, Orsay): You showed a plot with the Auger and AGASA energy spectra. Can the different rate of almost a factor of three be explained by energy calibration?

Stefan Westerhoff : The rate differences between AGASA, HiRes mono, and Auger indeed suggest problems in the energy calibration. Further evidence comes from the first analysis of the Auger spectrum. Here, it has already been shown that there is a discrepancy in energies if the spectrum has its energy scale normalized to the fluorescence detector or if simulations are used to determine the shower energy from surface detector data alone. This might indicate problems with the shower simulation code, the hadronic model, and the assumed composition of the cosmic ray flux. The difference shows that many systematics need to be addressed.

FUTURE PERSPECTIVES AND FACILITIES

Previous page:
Above, the anatomical theater, built in 1662 with the idea to improve the practical training of medical students.

The Ångstrom Laboratory, housing all the physics departments in Uppsala. (Foto: Pereric Öberg)

FUTURE FACILITIES IN ASTROPARTICLE PHYSICS AND COSMOLOGY

RENE ONG

Department of Physics and Astronomy University of California, Los Angeles, CA 90095-1562, USA.
E-mail: rene@astro.ucla.edu

NO CONTRIBUTION RECEIVED

PARTICLE DETECTOR R&D

MICHAEL V.DANILOV

Institute for Theoretical and Experimental Physics, B.Cheremushkinskaya 25, 117218 Moscow, RUSSIA

E-mail: danilov@itep.ru

Recent results on the particle detector R&D for new accelerators are reviewed. Different approaches for the muon systems, hadronic and electromagnetic calorimeters, particle identification devices, and central trackers are discussed. Main emphasis is made on the detectors for the International Linear Collider and Super B-factory. A detailed description of a novel photodetector, a so called Silicon Photomultiplier, and its applications in scintillator detectors is presented.

1 Introduction

Particle detector R&D is a very active field. Impressive results of the long term R&D for the Large Hadron Collider (LHC) are being summarized now by the four LHC detector collaborations. A worldwide effort is shifted to the detector development for the future International Linear Collider (ILC) and for the Super B-factory. The detector development for the FAIR facility has already started. Several groups perform R&D studies on detectors for the next generation of the hadron colliders.

This review is devoted mainly to the detector development for the ILC and Super B-factory. The vertex detectors are not discussed here in order to provide more details in other fields. R&D on the vertex detectors is very active and deserves a separate review. This review is organized following the radius of a typical collider detector from outside to inside.

2 Muon Detectors

Muon detectors cover very large areas. Therefore they should be robust and inexpensive. Resistive Plate Chambers (RPC) are often used in the present detectors, for example at the B-factories. In the streamer mode RPCs provide large signals. Hence it is possible to use very simple electronics. Another advantage is a possibility to have different shapes of read out electrodes that match best the physics requirements. For example the BELLE RPCs have ring and sector shaped readout electrodes in the end cap regions.

The European CaPiRe Collaboration developed a reliable industrial technique for the glass RPC production[1]. The production rate of more than 1000 square meters per day is possible. The RPC efficiency is larger than 95% up to the counting rates of $1 Hz/cm^2$. This is reasonably adequate for the ILC detector but at the Super-B factory one expects by far larger rates. The RPCs in the proportional mode can stand about hundred times higher counting rates.

Scintillator strip detectors can work at even higher rates. A very attractive possibility is to use scintillator strips with Wave Length Shifting (WLS) fibers read out by so called Silicon Photo Multipliers (SiPM).

SiPM is a novel photo detector developed in Russia[2,3,4]. It will be mentioned many times in this review. Therefore we shall discuss its properties in detail[a]. SiPM is a matrix of $1024 = 32 \times 32$ independent silicon photodiodes[b] covering the area of $1 \times 1 \, mm^2$.

---

[a] Three groups developed such devices and produce them. They use different names for their products. We will use a generic name SiPM for all types of multipixel Si diodes working in the Geiger mode. New types of SiPMs are being developed by several groups including Hamamatsu

[b] SiPMs can be produced with different number of pixels in the range 500-5000. We describe here the SiPM used for the hadronic calorimeter prototype for the ILC[5]

Each diode has its own quenching polysilicon resistor of the order of a few hundred kΩ. All diode-resistor pairs, called pixels later on, are connected in parallel. A common reverse bias voltage V_{bias} is applied across them. Its magnitude of the order of $40-60V$ is high enough to start the Geiger discharge if any free charge carrier appears in the $p-n$ junction depletion region. The diode discharge current causes a voltage drop across the resistor. This reduces the voltage across the diode below the breakdown voltage $V_{breakdown}$ and the avalanche dies out. One diode signal is $Q_{pixel} = C_{pixel}(V_{bias} - V_{breakdown})$ where C_{pixel} is the pixel capacitance. Typically $C_{pixel} \sim 50$ fF and $\Delta V = V_{bias} - V_{breakdown} \sim 3V$ yielding $Q_{pixel} \sim 10^6$ electrons. Such an amplification is similar to the one of a typical photomultiplier and 3–4 orders of magnitude larger than the amplification of an Avalanche Photo Diode (APD) working in the proportional mode. Q_{pixel} does not depend on the number of primary carriers which start the Geiger discharge. Thus each diode detects the carriers created e.g. by a photon, a charged particle or by a thermal noise with the same response signal of $\sim 10^6$ electrons. Moreover the characteristics of different diodes inside the SiPM are also very similar. When fired, they produce approximately the same signals. This is illustrated in Fig. 1a. It shows the SiPM response spectrum when it is illuminated by weak flashes of a Light Emitting Diode (LED). First peak in this figure is the pedestal. The second one is the SiPM response when it detects exactly one photon. It is not known which diode inside the SiPM produces the signal since all of them are connected to the same output. However since the responses of all pixels are similar, the peak width is small. If several pixels in the SiPM are fired, the net charge signal is the sum of all charges. The third, forth and so on peaks in Fig. 1a correspond to 2, 3, ... fired pixels.

The SiPM photodetection efficiency de-

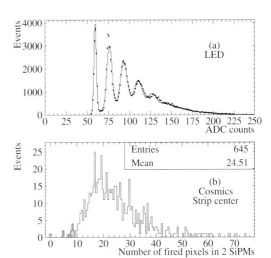

Figure 1. (a) SiPM response to short weak LED flashes. The fit curve is a simple model of the SiPM response. (b) Number of fired pixels in two SiPMs by a cosmic particle at the strip center. Few entries around zero belong to the pedestal.

pends on the light wave length and the overvoltage ΔV. A typical value is about 10-15% for the green light. It includes geometrical inefficiency due to dead regions in the SiPM between the pixels. Thus SiPM and traditional photomultipliers have similar gain and efficiency. However, SiPM is approximately twice cheaper than one channel in the multianode photomultiplier and further cost reductions are expected in case of mass production. SiPM can work in the magnetic field, so there is no need in the light transportation out of the magnetic field. SiPM is so tiny that it can be mounted directly on the detector. This minimizes the light losses because of a shorter fiber length. SiPM has a quite high noise rate of about 2 MHz at 0.1 photoelectron threshold. However the noise rate drops fast with increasing threshold.

Fig. 1b shows the pulse height spectrum for cosmic particles obtained with the scintillator strip detector read out by two SiPMs[6]. The detector consists of a $200 \times 2.5 \times 1\,cm^3$ plastic scintillator strip and a wavelength shifting fiber read out by two SiPMs installed at the strip ends. The strip is extruded from the granulated polystyrene with two dyes at the "Uniplast" enterprise in Vladimir, Russia. The Kuraray multicladding WLS fiber Y11 (200) with 1 mm diameter is put in the 2.5 mm deep groove in the middle of the strip. No gluing is used to attach the WLS fiber to the SiPM or to the strip. There is about 200 μm air gap between the fiber end and the SiPM. To improve the light collection efficiency, the strip is wrapped in the Superradiant VN2000 foil produced by the 3M company.

In the worst case when the particle passes through the strip center there are 13.7 detected photons. The Minimum Ionizing Particle (MIP) signal in Fig. 1b is well separated from the pedestal. The detector efficiency averaged over the strip length is as high as $99.3 \pm 0.3\%$ at the 8 pixel threshold. Such a threshold is sufficient to reduce the SiPM noise rate to $5kHz$.

The ITEP group has also studied $100 \times 4 \times 1\,cm^3$ strips[7] with a SiPM[3] at one end of the WLS fiber and a 3M Superradiant foil mirror at the other end. The strips were produced by the extrusion technique in Kharkov. The strip surface was covered by a Ti oxide reflector co-extruded together with the strip. The Kuraray Y11, 1mm diameter fiber was glued into the 3mm deep groove with an optical glue. SiPM was also glued to the fiber. More than 13 photoelectrons per MIP were detected at the strip end far from the SiPM. With such a large number of photoelectrons the efficiency of more than 99% for MIP was obtained with the threshold of 7 photoelectrons. The detector can work at counting rates above $1kHz/cm^2$. This is sufficient for the Super B-factory. Therefore the Belle Collaboration plans to use this technique for the K_L and muon system upgrade in the end cap region[8].

A scintillator tile structure can be used for even higher rates. Sixteen $10 \times 10 \times 1\,cm^3$

tiles read out by two SiPM[3] each were tested at the KEK B-factory[8]. They demonstrated a stable performance adequate for the Super B-factory. An eight square meter cosmic test system for ALICE TOF RPC chambers is constructed at ITEP[9]. It consists of $15 \times 15 \times 1\,cm^3$ tiles read out by two SiPMs[3] each. The counters have an intrinsic noise rate below $0.01\,Hz$, the time resolution of $1.2\,nsec$, and the rate capabilities up to $10\,kHz/cm^2$.

3 Hadronic Calorimeters

The precision physics program at the future International Linear Collider (ILC) requires to reconstruct heavy bosons (W,Z,H) in hadronic final states in multijet events. In order to do this a jet energy resolution of better than $30\%/\sqrt{E}$ is required[10]. The energy E is measured in GeV in this expression and in similar expressions for the energy resolution below. Monte Carlo (MC) simulations demonstrate that such a resolution can be achieved using a novel "particle flow" (PF) approach in which each particle in a jet is measured individually[11]. Momenta of charged particles are determined using tracker information. Photons are measured in the electromagnetic calorimeter (ECAL). Only neutrons and K_L should be measured in the Hadronic calorimeter (HCAL). They carry on average only about 12% of the jet energy. Therefore the HCAL can have modest energy resolution. The major problem is to reconstruct showers produced by charged tracks and to remove the corresponding energy from the calorimetric measurements. This requirement makes the pattern recognition ability to be a major optimization parameter of HCAL.

The CALICE Collaboration investigates two approaches for the HCAL. In the digital approach only one bit yes/no information is recorded for each cell. Extremely high granularity of about $1\,cm^2$/cell is required in this case. In the analog approach the pulse height information is recorded for each cell. However a very high granularity of about $5 \times 5\,cm^2$/cell is still required[12]. Such a granularity practically can not be achieved with a conventional readout approach with WLS fiber and a multianode photomultiplier (MAPM). The use of tiny SiPMs makes such a granularity achievable.

3.1 Analog Hadronic Calorimeters

A small 108 channel hadronic calorimeter prototype has been built in order to gain experience with this novel technique[5]. The calorimeter active modules have been made at ITEP and MEPhI. Scintillator tiles are made of a cheap Russian scintillator using a molding technique. A Kuraray Y11 1mm diameter double clad WLS fiber is inserted into a 2mm deep circular groove without gluing. The SiPM is placed directly on the tile and occupies less than 0.5% of a sensitive area. There is an air gap of about $100\mu m$ between the fiber and SiPM. Signals from SiPMs are sent directly to LeCroy 2249A ADCs via 25 meter long 50 Ω cables.

A lot of R&D has been performed in order to increase the light yield and the uniformity of the response. For better light collection the surface of the tiles is covered with 3M Superradiant foil. The tile edges are chemically treated in order to provide diffuse light reflection and separation between tiles. A light yield of more than 20 photoelectrons per MIP has been achieved for $5 \times 5 \times 0.5\,cm^3$ tiles. Fig. 2 shows LED and β-source (^{90}Sr) signals from such a tile. Peaks with different number of photoelectrons are clearly seen. Signals from the β-source are very similar to MIP signals.

The HCAL prototype was successfully operated at the DESY electron test beam. Fig. 3 shows the linearity of the calorimeter response measured with SiPM (circles) and MAPM (squares). The agreement between

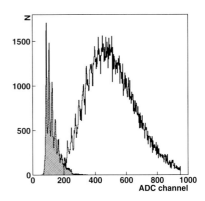

Figure 2. Pulse height spectrum from a tile with SiPM for low intensity LED light (hatched histogram) and for MIP signals from a β-source.

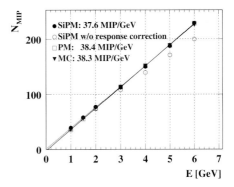

Figure 3. Calorimeter response normalized to number of MIPs versus beam energy; solid points (open circles) show SiPM data with (without) response function correction, squares are MAPM data and triangles are MC predictions .

two measurements is better than 2%. The linear behavior of the SiPM result (better than 2%) demonstrates that the applied saturation correction due to limited number of pixels in the SiPM is reliable. The obtained energy resolution agrees well with MC expectations and with a resolution obtained using conventional MAPMs as well as APDs[13]. The obtained resolution of about $21\%/\sqrt{E}$ is modest since this is a hadron calorimeter.

The 8000 channel HCAL prototype with the SiPM readout is being constructed by a subgroup of the CALICE Collaboration[14]. The $3 \times 3\,cm^2$ tiles are used in the central part of the calorimeter in order to test a semidigital approach. MC studies predict that the calorimeter with $3 \times 3\,cm^2$ cells and the 3 threshold measurement of the energy deposited in the tile should provide as good performance as a digital (yes/no) calorimeter with the $1 \times 1\,cm^2$ granularity[15].

3.2 Digital Hadronic Calorimeters

The RPC based digital HCAL is developed by a subgroup of the CALICE collaboration[16]. They studied several RPC geometries and gases in order to optimize the efficiency and to reduce the cross-talk between pads. Fig. 4 shows the efficiency and pad multiplicity due to cross-talk obtained with the developed RPC prototype. The prototype consists of two sheets of floating glass with the resistive paint layer (1Mohm/square) and the gas gap of 1.2 mm. In works in the proportional mode and has the efficiency above 90% up to the rates of $50\,Hz/cm^2$. The pad multiplicity is about 1.5. Much smaller pad multiplicity is observed in the RPC in which the readout electrode defines the gas sensitive volume instead of the glass sheet (see Fig. 4). It will be interesting to study further the properties of this promising RPC.

The GEM based digital HCAL is studied by another subgroup of the CALICE Collaboration[17]. They developed a procedure for the large area double GEM chamber production. A small prototype demonstrates 95% efficiency at 40mV threshold and pad multiplicity of 1.27. The 3M company plans to produce already in 2005 very long GEM foils of about $30\,cm$ width.

The number of channels in the digital HCAL is enormous. Therefore cheap and reliable electronics is the key issue for this approach. The RPC and GEM digital HCAL teams develop jointly the electronics suitable for both techniques.

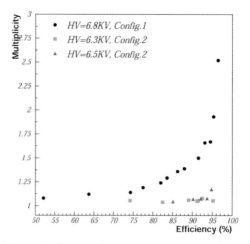

Figure 4. Pad multiplicity dependence on the efficiency for two types of RPC: full circles - standard RPC with two glass sheets; triangles and squares - the RPC with one glass sheet.

3.3 The DREAM Calorimeter

Usually calorimeters have different response to electromagnetic and hadronic showers of the same energy. Therefore the fluctuations of the electromagnetic energy fraction in the hadron shower is one of the main reasons for the deterioration of the energy resolution.

In the Dual Readout Module (DREAM) calorimeter[18] the electromagnetic energy in the hadronic shower is measured independently using quartz fibers sensitive only to the Cherenkov light produced dominantly by electrons. The visible energy is measured by scintillation fibers. The electromagnetic energy fraction in the shower can be determined by the comparison of the two measurements. This allows to correct for the different calorimeter response to the electromagnetic showers and to improve the energy resolution. A very similar response to electrons, hadrons, and jets was obtained in the DREAM calorimeter prototype after this correction. The ultimate energy resolution of the DREAM calorimeter is expected to be better than $30\%/\sqrt{E}$. Unfortunately the shower leakage and insufficient amount of Cherenkov light limited the measured prototype calorimeter resolution to $64\%/\sqrt{E}$ only.

The fluctuations of the visible energy because of the nuclear energy loss can be corrected for by adding to the DREAM structure a third type of fibers sensitive to neutrons. In this case the ultimate energy resolution of $15\%/\sqrt{E}$ is expected[18]. There are many nice ideas how to separate different mechanisms in the hadronic shower and to improve the energy resolution[18]. However they should be first demonstrated experimentally.

4 Electromagnetic Calorimeters

4.1 Electromagnetic Calorimeters for ILC

The requirement of a high granularity for the ILC detectors leads to the choice of very dense electromagnetic calorimeters with a small Mollier radius (R_M). Silicon/tungsten, scintillator/tungsten and scintillator/lead sandwich options are developed. The price for the high granularity is a modest energy resolution of the proposed calorimeters.

The CALICE collaboration constructs the Si/W prototype with about 10 thousand channels[19]. The pad size is as small as $1 \times 1\,cm^2$. The Si thickness is $500\mu m$. The tungsten plate thickness is $1.4\,mm$, $2.8\,mm$, and $4.2\,mm$ in the front, middle, and rare parts of the calorimeter. One third of the prototype has already been tested at the DESY electron beam and demonstrated a stable behavior. The signal to noise ratio of 8.5 was obtained for MIP. The tests of the whole calorimeter will start this Winter. The combined tests with the analog hadronic calorimeter are planned in 2006 as well.

The detector and readout plane thickness is $3.4\,mm$ in the present prototype. It will be reduced to $1.75\,mm$ including the readout chip in the final design resulting in $R_M = 1.4\,cm$.

The US groups (SLAC, UO, BNL) develop even more aggressive design of the Si/W calorimeter[20] for the small radius Si based ILC detector (ILC SiD). The detector and readout plane thickness is $1\,mm$ only which results in the $R_M = 1.4\,cm$. Together with HPK they developed a Si detector consisting of 1024 hexagonal pads with $5\,mm$ inner diameter. The detector is read out by a specially developed electronic chip[21]. The measured MIP signal in this detector is 26k electrons while the pedestal width is 780 electrons. The Si/W calorimeter for ILC is also developed in Korea[22].

A hybrid scintillator/lead calorimeter prototype with three Si layers has been built and tested by the INFN groups[23]. The $5 \times 5 \times 0.3\,cm^3$ scintillator tiles are combined into 4 longitudinal sections. Three layers of $9 \times 9mm^2$ Si pads are placed between the sections at 2, 6, and $12X_0$. The prototype demonstrated a good energy resolution of $11.1\%/\sqrt{E}$. It has the impressive spatial resolution of $2mm$ at $30GeV$ and e/π rejection below 10^{-3}. However it is not clear whether the granularity is sufficient for the PF method. Also the light transportation in the real detector will be extremely difficult. The use of SiPMs can solve the last problem.

The Japan-Korea-Russia Collaboration develops a scintillator/lead calorimeter with the SiPM readout[24]. The active layer consists of two orthogonal planes of $200 \times 10 \times 2mm^3$ scintillator strips and a plane of $40 \times 40 \times 2mm^3$ tiles with WLS fibers. The fibers are readout by SiPMs developed at Dubna[4]. Even shorter strips of $40 \times 10 \times 2mm^3$ are considered as an alternative. The signal of 5p.e./MIP was obtained with the $200 \times 10 \times 2mm^3$ strips.

4.2 Electromagnetic Calorimeters for the Super B-Factory

Electromagnetic calorimeters for the Super B-factory should have a very good energy resolution and a fast response. They should be radiation hard up to about 10 kRad in the endcap region. The present CsI(Tl) calorimeters at the KEKB and SLAC B-factories can not stand the planned increase of the luminosity above $20\,ab^{-1}$. The CsI(Tl) light yield decreases to about 60% already at $10\,ab^{-1}$. There is also a large increase of PIN diode dark current. Finally the long decay time of about $1\,\mu sec$ leads to the pile up noise and fake clusters.

The BELLE Collaboration proposes to use pure CsI crystals with a phototelectrode readout and a waveform analysis in the end cap region[25]. The shaping time is reduced from $1\,\mu sec$ to $30\,nsec$. The time resolution of better than $1\,nsec$ is achieved for energies above $25\,MeV$. The electronic noise is similar to the present CsI(Tl) calorimeters. The pure CsI crystals keep more than 90% of the light output after the irradiation of $7\,kRad$.

The BaBar Collaboration considers more radiation hard options of LSO or LYSO crystals and a liquid Xe calorimeter with the light readout[26]. The LSO and LYSO crystals are radiation hard, fast, and dense (see Table 1). They meet perfectly the requirements of the Super B-factory but their cost is prohibitively high at the moment. Liquid Xe is also an attractive option as it is seen in Table 1. The challenge here is the UV light collection. BaBar proposes to use WLS fibers and WLS cell coating for an immediate shift of the light wave length into a region with smaller absorption.

There is a good experience with very large liquid noble gas calorimeters. For example the $11\,m^3$ LiKr calorimeter at VEPP-4 has an excellent spatial ($\sim 1\,mm$) and energy ($\sim 3\%/\sqrt{E}$) resolution[27].

4.3 The CMS Lead Tungstate Calorimeter

The CMS collaboration summarized at this conference their more than 10 year long R&D on the lead tungstate ($PbWO_4$)

Table 1. Properties of different scintillators.

Scintillator	CsI(Tl)	LSO	LiXe
Density (g/cc)	4.53	7.40	2.95
X_0 (cm)	1.85	1.14	2.87
R_M (cm)	3.8	2.3	5.7
λ scint.(nm)	550	420	175
τ scint.(ns)	680	47	4.2,
	3340		22, 45
Photons/MeV	56k	27k	75k
Radiation hardness(Mrad)	0.01	100	-
cost($/cc)	3.2	~ 50	2.5

calorimeter[28]. The choice of $PbWO_4$(Y/Nb) is driven by its small $X_0 = 0.89\,cm$, small $R_M = 2.19\,cm$, fast decay time of $\tau \sim 10\,nsec$, and a very high radiation hardness above $200\,kGy$. More than 37.000 crystal have already been produced at the Bogoroditsk (Russia). In spite of a small light yield of $\sim 8\,p.e./MeV$ the excellent energy resolution of 0.51% has been achieved at $120\,GeV$. Intensive R&D together with Hamamatsu resulted in excellent APD operated at a gain of 50. All 120.000 APDs passed a very strict acceptance test which included a $500\,krad$ irradiation and accelerated aging. Vacuum phototriodes (RIA, St.Petersburg) will be used in the endcaps because they are more radiation hard. The main challenge for CMS is to finish the production of crystals and to maintain the advantages of this approach in the big calorimeter.

5 Particle Identification

5.1 Cherenkov Counters

A novel type of proximity focusing RICH counter with a multiple refractive index (n) aerogel radiator has been developed for the

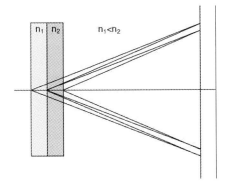

Figure 5. Principle of the dual radiator Ring Imaging Cherenkov counter .

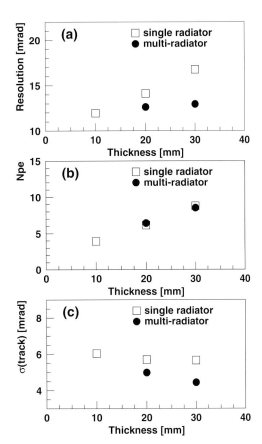

Figure 7. Single photon resolution (a), number of detected photons (b), and single track Cherenkov angle resolution for single and multiple focusing radiators for $4GeV$ pions (c).

BELLE detector upgrade[29]. The multiple radiator allows to increase the radiator thickness and hence the Cherenkov photon yield without degradation in single photon angular resolution. With the refractive index of the consecutive layers suitably increasing in the downstream direction (focusing combination) one can achieve overlapping of Cherenkov rings from all layers (see Fig. 5). With the decreasing n (defocusing combination) one can obtain well separated rings from different layers (see Fig. 6).

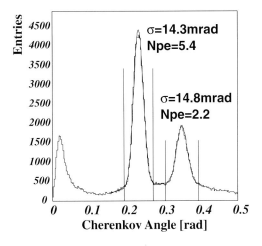

Figure 6. Distribution of the Cherenkov photon angles from $4GeV$ pions for a defocusing dual radiator with $n_1 = 1.057$ and $n_2 = 1.027$

Fig. 7 shows the performance of the de-

tector with a single and multiple layer radiators. The number of detected photons is similar in two approaches but the single photon resolution is much better in the multiple layer configuration. The Cherenkov angle resolution of $4.5\,mrad$ per track was achieved with the triple layer radiator. This corresponds to the 5.1σ K/π separation at $4GeV$.

The radiators with different refraction index layers attached directly at the molecular level have been produced at Novosibirsk[30] and in Japan[29].

The BaBar DIRC detector demonstrated an excellent performance. It is natural to consider the improved version of this technique for the Super B-factory. The SLAC

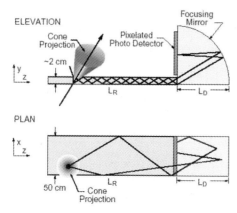

Figure 8. The principle of FDIRC operation.

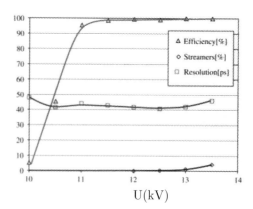

$U(kV)$

Figure 9. Efficiency (triangles,%), time resolution (squares, $nsec$), and streamer probability (circles,%) of MRPC versus applied voltage across 5 gaps (kV).

and Cincinnati groups develop the Fast Focusing DIRC (FDIRC) detector[26]. The idea of this detector is illustrated in Fig. 8. With the accurate time measurement one gets a 3D image of the Cherenkov cone. In FDIRC the photon detection part is by far smaller than in DIRC. The development of the pixelated photodetectors with better than $100\,nsec$ time resolution is a challenging task. The detail studies of Hamamatsu MAPM and Burley MCP PM at SLAC give very promising results. The FDIRC prototype is ready for tests at SLAC.

In the Time of Propagation (TOP) counter the Cherenkov cone image is reconstructed from the coordinate at the quartz bar end and the TOP[31]. The MCP PM SL10 is developed for TOP together with HPK. SL10 has 5 mm pitch and a single photon sensitivity in the 1.5T magnetic field. A time resolution of $30\,psec$ has been achieved however the cross-talk is still a problem. The Ga/As photocathodes developed by HPK and Novosibirsk provide enough light for the 4σ π/K separation at $4\,GeV$. However the cathode life time is not sufficient yet. It looses 40% of quantum efficiency after collecting $350\,mC/cm^2$ which corresponds to 6 month operation at the Super B-factory.

5.2 TOF systems

A multilayer RPC (MRPC) with the excellent time resolution of better than $50\,psec$ (see Fig.9) has been developed for the ALICE TOF system[32]. It has an efficiency of about 99% at counting rates as high as few hundred Hz/cm^2. The MRPC has ten $220\,\mu m$ gaps. It would be interesting to investigate a possibility to use MRPC for K_L momentum measurements in the muon system at the Super B-factory.

A time resolution of $48\,psec$ was obtained with a $3 \times 3 \times 40\,mm^3$ Bicron-418 scintillator read out directly by a $3 \times 3\,mm^2$ SiPM without preamplifier[33]. The MIPs were crossing $40\,mm$ in the scintillator. Therefore the signal was as big as 2700 pixels in the SiPM with 5625 pixels. The threshold was at 100 pixels. This approach is very promising for a super high granularity TOF capable to work in a very high intensity beams for example at FAIR.

6 Tracking

The Time Projection Chamber (TPC) is a natural choice for the ILC detector central tracker. This approach is developed by a large world wide collaboration[34]. TPC provides continues tracking through a large vol-

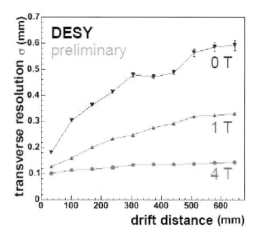

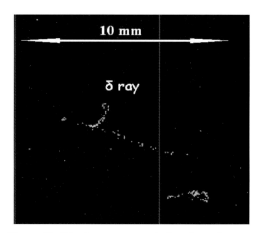

Figure 10. The transverse resolution dependence on the drift distance for three values of the magnetic field obtained in the TPC with a GEM readout.

Figure 11. The transverse resolution dependence on the drift distance for three values of the magnetic field obtained in the TPC with a GEM readout.

ume with a very small amount of material in front of the ECAL ($X_0 \sim 3\%$ in the barrel region). The dE/dx resolution of better than 5% helps in particle identification.

The thrust of the R&D is in the development of novel micro-pattern gas detectors which promise to have a better point and two track resolution than the traditional wire chambers. These detectors have smaller ion feedback into the TPC volume. Micromegas meshes and GEM foils are considered as main candidates. The spatial resolution of $\sim 100\,\mu m$ was already achieved with GEM after $65\,cm$ drift in a $4\,T$ field (see Fig. 10). Tests at smaller fields demonstrate that a similar resolution can be achieved with Micromegas as well. The double track resolution of $\sim 2mm$ has been already demonstrated in small prototypes. By pitting a resistive foil above the readout pads it is possible to spread the signal over several pads. As a result the resolution improves up to the diffusion limit.

A very exciting approach is a direct TPC readout with the MediPix2 chip[34]. This CMOS chip contains a square matrix of 256×256 pixels of $55 \times 55\,\mu m^2$. Each pixel is equipped with a low noise preamplifier, dis-

criminator, threshold DAC and communication logic. The extremely high granularity allows to distinguish individual clusters in a track. Thus the ultimate spatial and dE/dx resolution can be achieved. Unfortunately the diffusion will severely limit both measurements. Nice tracks have been recorded by a prototype chamber equipped with Micromegas and MediPix2 (see Fig. 11). The number of observed clusters ($0.52/mm$ in a He/Isobutane 80/20 mixture) agrees within 15% with the expectations. The next step is to integrate the chip and Micromegas at the postprocessing step and to add the (drift) time measurement. Tracks were observed also with a GEM/MediPix2 prototype[36]. A compact all Si tracker is vigorously developed by the US groups[35] for the small radius Si Detector for ILC. With small detector modules it is possible to reach a very good S/N ratio of about 20, to have a simple low risk assembly and relatively small amount of material of $\sim 0.8\% X_0$ per layer including a support structure. The pattern recognition is a serious issue for the Si tracker especially for tracks not coming from the main vertex.

The choice of the central tracker for the Super B-factory depends crucially on the expected background which depends on the interaction region design.

In the BELLE study[25] the background is expected to increase by a factor 20 from the present values. In this case the drift chamber with $13.3 \times 16 mm^2$ cells is still adequate for the radius above $12.8\,cm$. Small $5.4 \times 5.0 mm^2$ cells are foreseen for the radius between $10.2\,cm$ and $11.6\,cm$.

In the BaBar study[37] the luminosity term in the background extrapolation dominates. Therefore the background estimates are much higher than in the BELLE case. A drift chamber can not work in such environment. Therefore it is proposed to use an all Si tracker up to $R = 60\,cm$. A relatively large amount of material in the Si sensors and support structures leads to multiple scattering and considerable deterioration of the momentum and mass resolution. For example the mass resolution in the $B \to \pi^+\pi^-$ decay mode deteriorates from $23 MeV$ in case of the drift chamber to $35 MeV$ in case of a conservative Si tracker design. Serious R&D efforts are required to make the Si tracker thinner. It should be also demonstrated that the pattern recognition in the Si tracker is good enough.

May be it is possible to develop an alternative solution to the Si tracker. Using the controlled etching the BINP-CERN group reduced the Cu thickness in GEM foils from 5 to $1 \mu m$[38]. This allows to build the light triple GEM chamber with less than $0.15\% X_0$ including the readout electrode. The light double GEM chamber has even smaller thickness. The double and triple light GEM chambers were constructed and demonstrated identical performance with the standard GEM chambers. The light GEM chambers have a potential to provide the granularity and spatial resolution comparable to the Si tracker but with considerably smaller amount of material. However it is not clear so far how thick support structure is needed. A lot of R&D studies are required to demonstrate the feasibility of this approach.

7   Conclusions

The ongoing R&D should be sufficient to demonstrate the feasibility of detectors for the ILC and the Super B-factory. However there are many promising new ideas which have a potential to improve considerably the performance of the detectors and to exploit fully the physics potential of these colliders. The technologies for practically all detector subsystems are still to be selected on the basis of the R&D results. It is very important to strengthen and to focus the detector R&D especially for the ILC as it was done for the LHC collider.

8   Aknowledgments

This review would be impossible without many fruitful discussions with physicists working on the detector R&D for the LHC, ILC, and Super B-factory. In particular we are grateful to A. Bondar, J. Brau, B. Dolgoshein, J. Haba, E. Popova, F. Sefkow, R. Settles, A. Smirnitsky. This work was supported in part by the Russian grants SS551722.2003.2 and RFBR0402/17307a.

References

1. M. Piccolo, Proc. LCWS2005, SLAC (2004).

2. G. Bondarenko et al., Nucl. Phys. Proc. Suppl. **61B** 347 (1998).
   G. Bondarenko et al., Nucl. Instr. Meth. **A442** 187 (2000).
   P. Buzhan et al., ICFA Intstr.Bull. **23** 28 (2001).
   P. Buzhan et al., Nucl. Instr. Meth. **A504** 48 (2003).

3. A.Akindinov et al., Nucl. Instr. Meth. **A387** 231 (1997).

4. Z. Sadygov et al., arXiv:hep-ex/9909017 and references therein.

5. V. Andreev et al., Nucl. Instr. Meth. **A540** 368 (2005).

6. V. Balagura *et al.* Paper 241 contributed to this Symposium. V. Balagura *et al.* arXiv: Physics/0504194.

7. V. Balagura *et al.*, To be published in *Nucl. Instr. Meth.*.

8. M. Danilov, Talk at the 6th Workshop on Higher Luminosity B-Factory, KEK (2004), http://belle.kek.jp/superb /workshop/2004/HL6/.

9. A. Akindinov *et al.*, Submited to *Nucl. Instr. Meth.*

10. See e.g. F. Sefkow, Proc. Calor2004, Perugia (2004).

11. V. Morgunov, Proc. Calor2002, CALTECH (2002). H. Videau and J. C. Brient, Proc. Calor2002, CALTECH (2002).

12. M. Danilov, Proc. LCWS04, Paris (2004).

13. V. Andreev *et al.*, To be published in *Nucl. Instr. Meth.*

14. F. Sefkow, Proc. LCWS2005, SLAC (2005), http://www-conf.slac. stanford.edu /lcws05/.

15. V. Zutshi, Proc. LCWS04, Paris (2004).

16. J. Repond, Proc. LCWS05, SLAC (2005).

17. A. White, Proc. LCWS05, SLAC (2005).

18. R. Wigmans, Proc. LCWS05, SLAC (2005).

19. J-C. Brient, Proc. LCWS05, SLAC (2005).

20. D. Strom, Proc. LCWS05, SLAC (2005).

21. M. Breidenbach, Proc. LCWS05, SLAC (2005).

22. S. Nam, Proc. LCWS05, SLAC (2005).

23. P. Checchia, Proc. LCWS05, SLAC (2005).

24. D. H. Kim, Proc. LCWS05, SLAC (2005).

25. K. Abe *et al.* SuperKEKB LoI, KEK Report 04-4 (2004).

26. D. Hitlin, Talk at the Super B-Factory Workshop, Hawaii (2004) http://www. phys.hawaii.edu/ superb04/.

27. V. A. Aulchenko *et al.*, *Nucl. Instr. Meth.* **A419** 602 (1998); Yu. Tikhonov, Private communication.

28. M. Lethuillier (CMS), Paper 131 contributed to this Symposium.

29. T. Iijima *et al.*, arXiv: Physics/0504220.

30. A. Yu. Barnyakov *et al.*, Proc. RICH2005, to be published in *Nucl. Instr. Meth.*

31. K. Inami, Talk at the Super B-Factory Workshop, Hawaii (2004).

32. A. Akindinov *et al.*, *Nucl. Instr. Meth.* **A456** 16 (2000).

33. A. Karakash, Talk at the 4th Conference on New Developments in Photodetection Beaune (2005).

34. R. Settles, Paper 222 contributed to this Symposium.

35. T. Nelson, Proc. LCWS05, SLAC (2005).

36. M. Titov, Private communication; Paper submited to IEEE Nuclear Science symposium, Puerto Rico (2005).

37. G. Calderini, Talk at the Super B-Factory Workshop, Hawaii (2004).

38. A. Bondar, Private communication.

PHYSICS AT FUTURE LINEAR COLLIDERS

KLAUS MÖNIG

DESY, Zeuthen, Germany and LAL, Orsay France
E-mail: klaus.moenig@desy.de

This article summarizes the physics at future linear colliders. It will be shown that in all studied physics scenarios a 1 TeV linear collider in addition to the LHC will enhance our knowledge significantly and helps to reconstruct the model of new physics nature has chosen.

1 Introduction

Most physicists agree that the International Linear Collider, ILC, should be the next large scale project in high energy physics[1]. The ILC is an e^+e^- linear collider with a center of mass energy of $\sqrt{s} \leq 500\,\text{GeV}$ in the first phase, upgradable to about $1\,\text{TeV}$[2]. The luminosity will be $\mathcal{L} \approx 2 - 5 \cdot 10^{34}\text{cm}^{-2}\text{s}^{-1}$ corresponding to $200 - 500\ \text{fb}^{-1}/\text{year}$. The electron beam will be polarisable with a polarisation of $\mathcal{P} = 80 - 90\%$.

In addition to this baseline mode there are a couple of options whose realization depends on the physics needs. With relatively little effort also the positron beam can be polarized with a polarization of $40 - 60\%$. The machine can be run on the Z resonance producing $> 10^9$ hadronically decaying Z bosons in less than a year or at the W-pair production threshold to measure the W-mass to a precision around $6\,\text{MeV}$ (GigaZ). The ILC can also be operated as an e^-e^- collider. With much more effort one or both beams can be brought into collision with a high power laser a few mm in front of the interaction point realizing a $\gamma\gamma$ or $e\gamma$ collider with a photon energy of up to 80% of the beam energy.

At a later stage one may need an e^+e^- collider with $\sqrt{s} = 3 - 5\,\text{TeV}$. Such a collider may be realized in a two-beam acceleration scheme (CLIC). Extensive R&D for such a machine is currently going on[3].

ILC will run after LHC[4,5] has taken already several years of data. However the two machines are to a large extend complementary. The LHC reaches a center of mass energy of $\sqrt{s} = 14\,\text{TeV}$ leading to a very high discovery range. However not the full $\sqrt{s}$ is available due to parton distributions inside the proton ($\sqrt{s}_{\text{eff}} \sim 3\,\text{TeV}$). The initial state is unknown and the proton remnants disappear in the beam pipe so that energy-momentum conservation cannot be employed in the analyses. There is a huge QCD background and thus not all processes are visible.

ILC has with its $\sqrt{s} \leq 1\,\text{TeV}$ a lower reach for direct discoveries. However the full $\sqrt{s}$ is available for the primary interaction and the initial state is well defined, including its helicity. The full final state is visible in the detector so that energy-momentum conservation also allows reconstruction of invisible particles. Since the background is small, basically all processes are visible at the ILC.

The LHC is mainly the "discovery machine" that can find new particles up to the highest available energy and should show the direction nature has taken. On the contrary ILC is the "precision machine" that can reconstruct the underlying laws of nature. Only a combination of the LHC reach with the ILC precision is thus able to solve our present questions in particle physics.

Better measurement precision can not only improve existing knowledge but allows to reconstruct completely new effects. For example Cobe discovered the inhomogeneities of the cosmic microwave background but only the precision of WMAP allowed to conclude that the universe is flat. As another example, from the electroweak precision measurements before LEP and SLD one could verify that the lepton couplings to the Z were consistent with the Standard Model prediction but only the high precision of LEP and SLD could predict the Higgs mass within this model.

The ILC has a chance to answer several of the most important questions in particle physics. Roughly ordered in the chances of the ILC to find some answers they are:

- How is the electroweak symmetry broken? The ILC can either perform a precision study of the Higgs system or see first signs of strong electroweak symmetry breaking.

- What is the matter from which our universe is made off? ILC has a high chance to see supersymmetric dark matter, also some other solutions like Kaluza Klein dark matter might give visible signals.

- Is there a common origin of forces? Inside supersymmetric theories the unification of couplings as well as of the SUSY breaking parameters can be checked with high precision.

- Why is there a surplus of matter in the universe? Some SUSY models of baryogenesis make testable predictions for the ILC. Also CP violation in the Higgs system should be visible.

- How can gravity be quantized? The ILC is sensitive to extra dimensions up scales of a few TeV and tests of unification in SUSY may give a hint towards quantum gravity at the GUT scale.

2 The Top Quark Mass and why we need it

ILC can measure the top mass precisely from a scan of the $t\bar{t}$ threshold. With the appropriate mass definition the cross section near threshold is well under control[6] (see fig. 1).

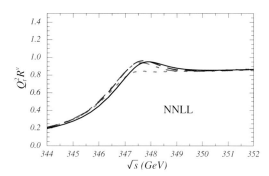

Figure 1. Top pair cross section using the NNLL pole mass for different values of the top velocity parameter[6].

With a ten-point scan an experimental precision of $\Delta m_{\mathrm{t}} = 34\,\mathrm{MeV}$ and $\Delta\Gamma_{\mathrm{t}} = 42\,\mathrm{MeV}$ is possible[7], so that, including theoretical uncertainties $\Delta m_{\mathrm{t}}(\overline{\mathrm{MS}}) \approx 100\,\mathrm{MeV}$ can be reached.

A precise top mass measurement is needed in many applications. The interpretation of the electroweak precision data after GigaZ needs a top mass precision better than $2\,\mathrm{GeV}$ (fig. 2 left) and the interpretation of the MSSM Higgs system even needs a top mass precision of about the same size as the uncertainty on the Higgs mass (fig. 2 right)[8]. Also the interpretation of the WMAP cosmic microwave data in terms so the MSSM needs a precise top mass in some regions of the parameter space[9].

3 Higgs Physics and Electroweak Symmetry Breaking

If a roughly Standard Model like Higgs exists, it will be found by the LHC. However the ILC has still a lot to do to figure out the exact model and to measure its parameters. If only one Higgs exists it can be the Standard Model, a little Higgs model or the Higgs can be mixed with a Radion from extra dimensions. If two Higgs doublets exist it can be a general two Higgs doublet model or the MSSM. However the Higgs structure may be even more complicated like in the

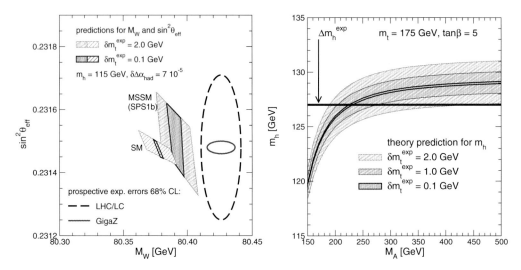

Figure 2. Required top mass precision for the interpretation of the electroweak precision data (left) and for the MSSM Higgs system (right).

NMSSM with an additional Higgs singlet or the top quark can play a special role as in little Higgs or top-colour models. In all cases there maybe only one Higgs visible at LHC that looks Standard-Model like, but the precision at ILC can distinguish between the models.

The Higgs can be identified independent from its decay mode using the $\mu^+\mu^-$ recoil mass in the process $e^+e^- \rightarrow HZ$ with $Z \rightarrow \mu^+\mu^-$ (see fig. 3)[10]. The cross section of this process is a direct measurement of the HZZ coupling and it gives a bias free normalization for the Higgs branching ratio measurements. Together with the cross section of the WW fusion channel ($e^+e^- \rightarrow \nu\nu H$) this allows for a model independent determination of the Higgs width and its couplings to W, Z, b-quarks, τ-leptons, c-quarks and gluons on the $1 - 5\%$ level[11].

At higher energies the $t\bar{t}H$ Yukawa coupling can be measured from the process $e^+e^- \rightarrow t\bar{t}H$ where the Higgs is radiated off a t-quark. At low Higgs masses, using $H \rightarrow b\bar{b}$, a precision around 5% can be reached. For higher Higgs masses, using $H \rightarrow WW$, 10% accuracy will be possible (see fig. 4)[12].

If the Higgs is not too heavy the

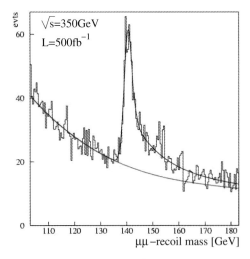

Figure 3. Measurement of $e^+e^- \rightarrow HZ$ from the $\mu^+\mu^-$ recoil mass.

triple Higgs self-coupling can be measured to around 10% using the double-Higgs production channels $e^+e^- \rightarrow ZHH$ and $e^+e^- \rightarrow \nu\bar{\nu}HH$[13]. As shown in fig. 5 all these Higgs coupling measurements allow to show, that the Higgs really couples to the mass of the particles[13].

These measurements present a powerful tool to test the model from which the Standard-Model-like Higgs arises. Figure 6[13]

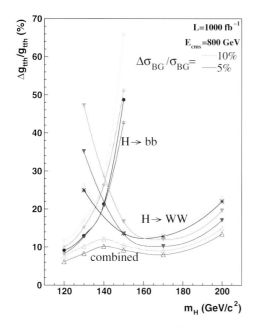

Figure 4. Expected precision of the $t\bar{t}H$ Yukawa coupling as a function of the Higgs mass.

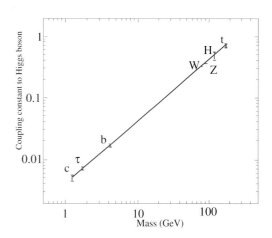

Figure 5. Higgs-particle coupling and expected uncertainty as a function of the particle mass.

shows possible deviations of the Higgs couplings from the the Standard Model prediction together with the expected uncertainties for a two Higgs doublet model, a model with Higgs-Radion mixing and a model incorporating baryogenesis[14]. In all cases the ILC allows to separate clearly between the Standard Model and the considered one.

Further information can be obtained from loop decays of the Higgs, namely $H \to gg$ and $H \to \gamma\gamma$. Loop decays probe the Higgs coupling to all particles, also to those that are too heavy to be produced directly. The Higgs-decay into gluons probes the coupling to all coloured particles which is completely dominated by the top-quark in the Standard Model. The one to photons is sensitive to all charged particles, dominantly the top quark and the W-boson in the SM. The partial width $\Gamma(H \to gg)$ can be measured on the 5% level from Higgs decays in e^+e^-. The photonic coupling of the Higgs can be obtained from the Higgs production cross section at a photon collider (see fig. 7)[15,16]. The

loop decays of the Higgs are sensitive to the model-parameters in many models. As an example figure 8 shows the expected range of couplings within a little Higgs model[17].

3.1 Heavy SUSY-Higgses

In the relevant parameter range of the MSSM the heavy scalar, H, the pseudoscalar, A, and the charged Higgses H$^{\pm}$ are almost degenerate in mass and the coupling ZZH vanishes or gets at least very small. At the ILC they are thus pair-produced, either as HA or H$^+$H$^-$ and the cross section depends only very little on the model parameters. All states are therefore visible basically up to the kinematic limit $m(H) < \sqrt{s}/2$. As shown in figure 9[13] at least one of the heavy states should be visible in another channel in most of the parameter space. The additional channels serve as redundancy and can be used to measure model parameters.

In addition to the direct searches the precision branching ratio measurements of the light Higgs can give indications of the H and A mass. Figure 10 shows the ratio of branching ratios $BR(h \to b\bar{b})/BR(h \to WW)$ of the MSSM relative to the Standard Model as a function of m_H[18]. The width of the band gives the uncertainty from the measurement

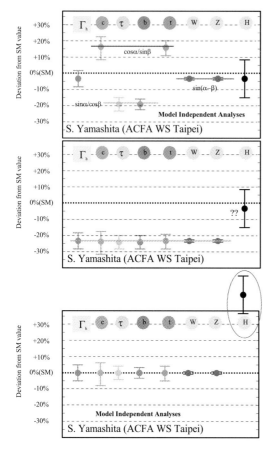

S. Yamashita (ACFA WS Taipei)

S. Yamashita (ACFA WS Taipei)

S. Yamashita (ACFA WS Taipei)

Figure 6. Deviation of the Higgs couplings from the Standard Model together with the expected ILC precision for a two Higgs doublet model (upper), a model with Higgs-Radion mixing (middle) and a model incorporating baryogenesis[14] (lower).

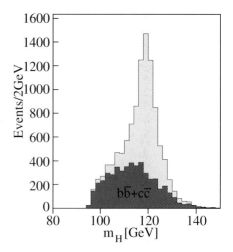

Figure 7. $b\bar{b}$ mass spectrum in the $\gamma\gamma \to H$ analysis after all cuts[15].

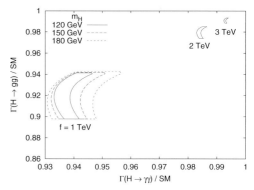

Figure 8. Possible deviations of Higgs loop decays from the Standard Model prediction in little Higgs models.

of the MSSM parameters. Up to A masses of a few hundred GeV one can get a good indication of m_A.

Another possibility to find the heavy SUSY Higgses is the photon collider. Since Higgses are produced in the s-channel the maximum reach is twice the beam energy corresponding to $0.8\sqrt{s_{ee}}$. Figure 11 shows the expected sensitivity in one year of running for $m_A = 350\,\text{GeV}$, $\sqrt{s_{ee}} = 500\,\text{GeV}$ and different SUSY parameters[19]. In general H and A are clearly visible, however due to the loop coupling of the γ to the Higgs the sensitivity becomes model dependent.

4 Supersymmetry and Dark Matter

Supersymmetry (SUSY) is the best motivated extension of the Standard Model. Up to now all data are consistent with SUSY, however also with the pure Standard Model. Contrary to the SM, SUSY allows the unification of couplings at the GUT scale and, if R-parity is conserved, SUSY offers a perfect dark matter candidate. If some superpartners are visible at the ILC they will be discovered by the LHC in most part of the parameter space. However many tasks are left for the ILC in this case. First the ILC has

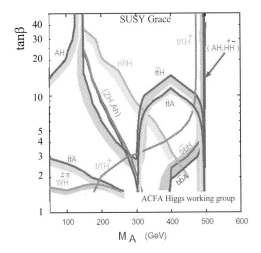

Figure 9. Visibility of heavy SUSY Higgses at ILC ($\sqrt{s} = 1$ TeV).

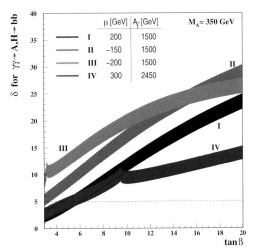

Figure 11. Sensitivity of the $\gamma\gamma$ collider to heavy MSSM Higgses. ($m_A = 350$ GeV, $\sqrt{s}_{ee} = 500$ GeV, $M_2 = 200$ GeV, $M_{\tilde{f}} = 1000$ GeV)

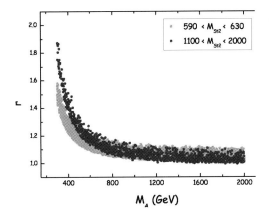

Figure 10. $BR(h \to b\bar{b})/BR(h \to WW)$ MSSM/SM within the SPS1a scenario as a function of m_A.

to confirm that the discovered new states are really superpartners of the Standard Model particles. Then it has to measure as many of the > 100 free parameters as possible in a model independent way which allows to check if grand unification works and to get an idea by which mechanism Supersymmetry is broken. If Supersymmetric particles are a source of dark matter the ILC has to measure their properties.

Within the minimal supergravity model (mSUGRA) the parameter space can be strongly restricted requiring that the abun-

dance of the lightest neutralino, which is stable in this model, is consistent with the dark matter density measured by WMAP. Figure 12 shows the allowed region in a pictorial way[20]. In the so called "bulk region" all superpartners are light and many are visible at the LHC and the ILC. In the "coannihilation region" the mass difference between the lightest neutralino, $\tilde{\chi}_1^0$, and the lighter stau, $\tilde{\tau}_1$, is very small so that the $\tilde{\tau}_1$-decay particles that are visible by the detector have only a very small momentum. In the "focus point region" the $\tilde{\chi}_1^0$ gets a significant Higgsino component enhancing its annihilation cross section. This leads to relatively heavy scalars, probably invisible at the ILC and the LHC. Other regions, like the "rapid annihilation funnel" are characterized by special resonance conditions, like $2m(\tilde{\chi}_1^0) \approx m_A$, increasing the annihilation rate. All these special regions tend to be challenging for both machines.

After new states consistent with SUSY have been discovered at the LHC, the ILC can check, if it is really Supersymmetry. As an example fig. 13 shows the threshold behaviour of smuon production and the pro-

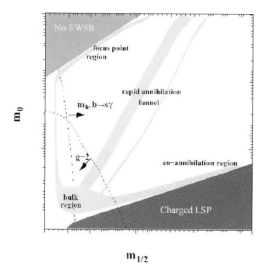

Figure 12. Dark matter allowed regions of mSUGRA.

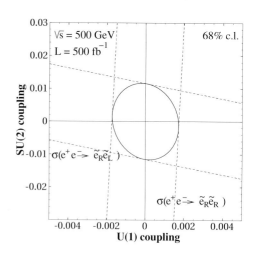

Figure 14. Measurement of the SU(2) and U(1) coupling of the selectron at the ILC.

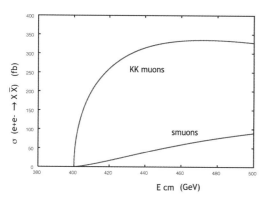

Figure 13. Threshold behaviour of smuon production and the production of Kaluza Klein excitations of the muon.

duction of Kaluza Klein excitations of the muon[21]. There is no problem for the ILC to distinguish the two possibilities. Figure 14 shows the expected precision of the measurement of the SU(2) and U(1) coupling of the selectron[22]. The agreement with the couplings of the electron can be tested to the percent to per mille level.

4.1 SUSY in the bulk region

An often studied benchmark point in the bulk region is the SPS1a scenario[23]. In this scenario all sleptons, neutralinos and charginos are visible at ILC and and in addition squarks and gluinos at the LHC. The LHC can measure mass differences pretty accurately, but has difficulties to measure absolute masses. The ILC, however can measure absolute masses with good precision, including the one of the $\tilde{\chi}_1^0$. Table 1 shows the expected precision for the mass measurements for the LHC and ILC alone and for the combination[24]. In many cases the combination is significantly better than the LHC or even the ILC alone. As an example figure 15 shows the correlation between the squark mass and the $\tilde{\chi}_1^0$ mass from LHC together with $m(\tilde{\chi}_1^0)$ from ILC[24]. The improvement in $m(\tilde{q})$ is evident.

With these inputs it is then possible to fit many of the low energy SUSY breaking parameters in a model independent way. Figure 16 shows the result of this fit to the combined ILC and LHC results for the SPS1a scenario[25]. Most parameters can be measured on the percent level.

These parameters can then be extrapolated to high scales using the renormalization group equations to check grand unification[26]. Figure 17 shows the expected precision for the gaugino and slepton mass parameters and for the coupling constants.

	m_{SPS1a}	LHC	ILC	LHC+ILC		m_{SPS1a}	LHC	ILC	LHC+ILC
h	111.6	0.25	0.05	0.05	H	399.6		1.5	1.5
A	399.1		1.5	1.5	$H+$	407.1		1.5	1.5
χ_1^0	97.03	4.8	0.05	0.05	χ_2^0	182.9	4.7	1.2	0.08
χ_3^0	349.2		4.0	4.0	χ_4^0	370.3	5.1	4.0	2.3
$\chi_1^\pm$	182.3		0.55	0.55	$\chi_2^\pm$	370.6		3.0	3.0
$\tilde{g}$	615.7	8.0		6.5					
$\tilde{t}_1$	411.8		2.0	2.0					
$\tilde{b}_1$	520.8	7.5		5.7	$\tilde{b}_2$	550.4	7.9		6.2
$\tilde{u}_1, \tilde{c}_1$	551.0	19.0		16.0	$\tilde{u}_2, \tilde{c}_2$	570.8	17.4		9.8
$\tilde{d}_1, \tilde{s}_1$	549.9	19.0		16.0	$\tilde{d}_2, \tilde{s}_2$	576.4	17.4		9.8
$\tilde{e}_1$	144.9	4.8	.05	0.05	$\tilde{e}_2$	204.2	5.0	0.2	0.2
$\tilde{\mu}_1$	144.9	4.8	0.2	0.2	$\tilde{\mu}_2$	204.2	5.0	0.5	0.5
$\tilde{\tau}_1$	135.5	6.5	0.3	0.3	$\tilde{\tau}_2$	207.9		1.1	1.1
$\tilde{\nu}_e$	188.2		1.2	1.2					

Table 1. Expected precision of mass measurements at LHC and ILC in the SPS1a scenario.

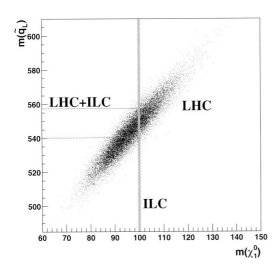

Figure 15. Correlation between $m(\tilde{\chi}_1^0)$ and $m(\tilde{q})$ measurements at LHC together with $m(\tilde{\chi}_1^0)$ from ILC.

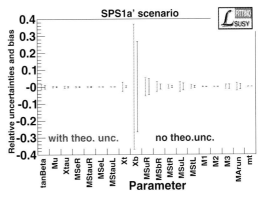

Figure 16. Low energy SUSY breaking parameters from a fit to the LHC and ILC results.

4.2 Reconstruction of dark matter

As already mentioned the lightest neutralino is a good candidate for the dark matter particle. To calculate its density in the universe, the properties of all particles contributing to the annihilation have to be reconstructed with good precision. In any case the mixing angles and mass of the $\tilde{\chi}_1^0$ need to be known.

However also the properties of other particles can be important. For example in the $\tilde{\chi}_1^0 - \tilde{\tau}_1$ coannihilation region the $\tilde{\chi}_1^0 - \tilde{\tau}_1$ mass difference is essential. Figure 18 shows the possible precision with which the dark matter density and neutralino mass can be reconstructed from the LHC and the ILC measurements[27]. ILC matches nicely the expected precision of the Planck satellite, allowing a stringent test whether Supersymmetry can account for all dark matter in the universe.

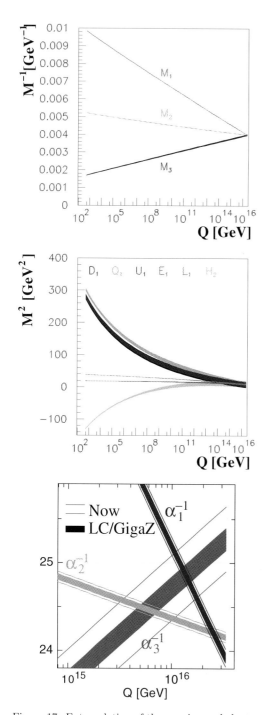

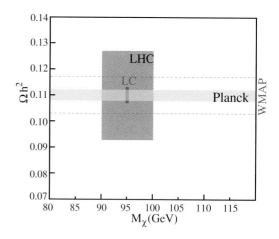

Figure 18. Projected precision of the dark matter density in the coannihilation region from WMAP, Planck, LHC and ILC.

5 Models without a Higgs

Without a Higgs WW scattering becomes strong at high energy, finally violating unitarity at 1.2 TeV. One can thus expect new physics the latest at this scale. At the moment there are mainly two classes of models that explain electroweak symmetry breaking without a Higgs boson. In Technicolour like models[28] new strong interactions are introduced at the TeV scale. In Higgsless models the unitarity violation is postponed to higher energy by new gauge bosons, typically KK excitations of the Standard Model gauge bosons. Both classes should give visible signals at the ILC. The accessible channels are W-pair production, where the exchanged γ or Z may fluctuate into a new state, vector boson scattering, where the new states can be exchanged in the s- or t-channel of the scattering process and three gauge boson production where the new states can appear in the decay of the primary γ or Z.

5.1 Strong electroweak symmetry breaking

As already said, in technicolour like models one expects new strong interactions, includ-

Figure 17. Extrapolation of the gaugino and slepton mass parameters and of the coupling constants to the GUT scale.

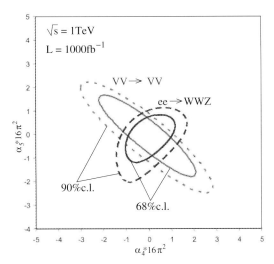

Figure 19. Expected sensitivity on α_4 and α_5 from vector boson scattering and three vector boson production.

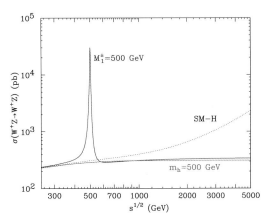

Figure 20. Cross section $\sigma(WZ \to WZ)$ in a Higgsless model and in the Standard Model with and without a Higgs[32].

ing resonances, at the TeV scale. To analyze these models in a model independent way, the triple and quartic couplings can be parameterized by an effective Lagrangian in a dimensional expansion[29]. For the interpretation the effects of resonances on these couplings can then be calculated. Figure 19 shows the possible sensitivity to α_4 and α_5 at $\sqrt{s} = 1$ TeV from vector-boson scattering and three vector boson production[30]. Typical sensitivities are $\mathcal{O}(0.1/16\pi^2)$ for triple and $\mathcal{O}(1/16\pi^2)$ for quartic couplings. This corresponds to mass limits around 3 TeV for maximally coupled resonances. The different processes can then distinguish between the different resonances. For example W-pair production is only sensitive to vector resonances.

5.2 Higgsless models

Higgsless models predict new gauge bosons at higher energies. Especially also charged states are predicted that cannot be confused with a heavy Higgs. Figure 20 shows the cross section for the process $WZ \to WZ$ in a Higgsless model, the Standard Model without a Higgs and the SM where unitarity is restored by a 600 GeV Higgs[31,32]. Detailed

studies show that these states can be seen at LHC, however it is out of question that such a state would also give a signal at ILC in $WZ \to WZ$ and in WWZ production so that its properties could be measured in detail.

6 Extra Gauge Bosons

The ILC is sensitive to new gauge bosons in $e^+e^- \to f\bar{f}$ via the interference with the Standard Model amplitude far beyond $\sqrt{s}$. The sensitivity is typically even larger than at the LHC. If the LHC measures the mass of a new Z' a precise coupling measurement is possible at the ILC. In addition angular distributions are sensitive to the spin of the new state and can thus distinguish for example between a Z' and KK graviton towers. A review of the sensitivity can be found in[11].

An interesting possibility is the reconstruction of the 2nd excitation of the Z and γ in universal extra dimensions. In this models an excitation quantum number may be defined that is conserved and makes the lightest excitation stable and thus a good dark matter candidate[33]. The second excitations couple to Standard Model particles only loop suppressed and thus weakly[34]. Cosmology suggests $\frac{1}{R} \approx m(\gamma') < 1$ TeV corresponding

414

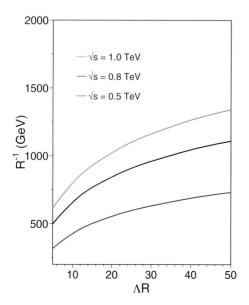

Figure 21. ILC reach for the Z'', γ'' expressed in $1/R$ as a function of the cutoff parameter ΛR[35]. A ΛR value around 30 is suggested theoretically[34].

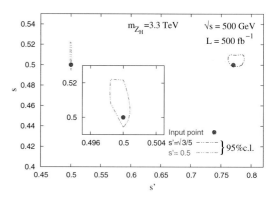

Figure 22. Measurement of the Z_H mixing angles at the ILC.

to $m(\gamma'') < 2\,\mathrm{TeV}$[33]. The LHC can see the γ' in pair production up to about this energy. The ILC is sensitive to the Z'' and γ'' up to $2\sqrt{s}$ which corresponds to the same $1/R$ reach for $\sqrt{s} = 1\,\mathrm{TeV}$ (see fig. 21)[35], helping enormously in the interpretation of a possible LHC signal as KK excitation.

Little Higgs models explain the "little hierarchy problem" by a new gauge structure and a new top-like quark[36]. The new gauge structure also predicts new vector bosons (Z_H, A_H, W_H) at masses of a few TeV. Figure 22 shows the precision with which the mixing angles of the Z_H can be measured at $\sqrt{s} = 500\,\mathrm{GeV}$ once its mass (3.3 TeV in this example) in measured at the LHC[37].

7 Conclusions

Independent of which physics scenario nature has chosen, the ILC will be needed in addition to the LHC. If there is a Higgs and SUSY the ILC has to reconstruct as many of the SUSY-breaking parameters as possible, extrapolate them to the GUT scale to get

some understanding of the breaking mechanism and measure the properties of the dark matter particle.

If there is a Higgs without Supersymmetry the precision measurements of the Higgs boson guide the way to the model of electroweak symmetry breaking. In addition several models, like some extra dimension models or little Higgs models have extra gauge bosons that are visible via their indirect effects.

If the LHC doesn't find any Higgs boson, the ILC can fill some loopholes that still exist, can see signals of strong electroweak symmetry breaking and is sensitive to a new gauge sector.

In any case we know that the top quark is accessible to the ILC and that its properties can be measured with great precision.

Acknowledgements

I would like to thank everybody who helped me in the preparation of this talk, especially Klaus Desch, Jonathan Feng, Fabiola Gianotti, Francois Richard, Sabine Riemann and Satoru Yamashita. Special thanks also to Tord Ekelöf and the organizing committee for the splendid organization and hospitality during the conference!

References

1. *Understanding Matter, Energy, Space and Time: the Case for the Linear Collider*,
http://sbhep1.physics.sunysb.edu/
~grannis/lc_consensus.html

2. J. Dorfan, these proceedings.

3. F. Zimmermann, these proceedings.

4. G. Rolandi, these proceedings.

5. F. Gianotti, these proceedings.

6. A. H. Hoang *et al.*, *Phys. Rev. D* **65** 014014 (2002) [arXiv:hep-ph/0107144].

7. M. Martinez and R. Miquel, *Eur. Phys. J. C* **27** 49 (2003) [arXiv:hep-ph/0207315].

8. S. Heinemeyer, S. Kraml, W. Porod and G. Weiglein, *JHEP* **0309** 075 (2003) [arXiv:hep-ph/0306181].

9. J. R. Ellis, S. Heinemeyer, K. A. Olive and G. Weiglein, *JHEP* **0502** 013 (2005) [arXiv:hep-ph/0411216].

10. J.C.Brient, private communication

11. J. A. Aguilar-Saavedra *et al.*, *TESLA Technical Design Report Part III: Physics at an e^+e^- Linear Collider*, DESY-01-011C.

12. A.Gay, talk presented at LCWS04, Paris, April 2004.

13. S.Yamashita talk presented at the 7th ACFA workshop, Taipei, November 2004.

14. S. Kanemura, Y. Okada and E. Senaha, arXiv:hep-ph/0507259.

15. K. Mönig and A. Rosca, arXiv:hep-ph/0506271.

16. P. Niezurawski, A. F. Zarnecki and M. Krawczyk, arXiv:hep-ph/0307183.

17. T. Han, H. E. Logan, B. McElrath and L. T. Wang, *Phys. Lett. B* **563** 191 (2003) [Erratum-ibid. B **603** 257 (2004)] [arXiv:hep-ph/0302188].

18. K. Desch, E. Gross, S. Heinemeyer, G. Weiglein and L. Zivkovic, *JHEP* **0409** 062 (2004) [arXiv:hep-ph/0406322].

19. P. Niezurawski, A. F. Zarnecki and M. Krawczyk, arXiv:hep-ph/0507006.

20. J. L. Feng, arXiv:hep-ph/0509309.

21. M. Peskin, talk at the ALPG meeting, Victoria, Canada, July 2004.

22. A. Freitas, A. von Manteuffel and P. M. Zerwas, *Eur. Phys. J. C* **34** 487 (2004) [arXiv:hep-ph/0310182].

23. B. C. Allanach *et al.*, *Eur. Phys. J.* **C25** 113 (2002).

24. G. Weiglein *et al.* arXiv:hep-ph/0410364.

25. P. Bechtle, K. Desch and P. Wienemann, arXiv:hep-ph/0506244.

26. B. C. Allanach *et al.*, *Nucl. Phys. Proc. Suppl.* **135** 107 (2004) [arXiv:hep-ph/0407067].

27. M. Berggren, F. Richard and Z. Zhang, LAL 05-104, arXiv:hep-ph/0510088.

28. C. T. Hill and E. H. Simmons, *Phys. Rept.* **381** 235 (2003) [Erratum-ibid. **390** 553 (2004)] [arXiv:hep-ph/0203079].

29. W. Kilian and J. Reuter, arXiv:hep-ph/0507099.

30. P. Krstonosic *et al.*, arXiv:hep-ph/0508179.

31. A. Birkedal, K. Matchev and M. Perelstein, *Phys. Rev. Lett.* **94** 191803 (2005) [arXiv:hep-ph/0412278].

32. A. Birkedal, K. T. Matchev and M. Perelstein, arXiv:hep-ph/0508185.

33. G. Servant and T. M. P. Tait, *Nucl. Phys. B* **650**, 391 (2003) [arXiv:hep-ph/0206071].

34. H. C. P. Cheng, K. T. Matchev and M. Schmaltz, *Phys. Rev. D* **66**, 036005 (2002) [arXiv:hep-ph/0204342].

35. S. Riemann, arXiv:hep-ph/0508136.

36. N. Arkani-Hamed, A. G. Cohen and H. Georgi, *Phys. Lett. B* **513** 232 (2001) [arXiv:hep-ph/0105239].

37. J. A. Conley, J. Hewett and M. P. Le, arXiv:hep-ph/0507198.

DISCUSSION

Bernd Jantzen (Univ. of Karlsruhe):
Can anything be said about if we need
CLIC and what we would like to explore
with it before the data of LHC and/or
ILC has been analyzed?

Klaus Mönig: The detailed physics case for
CLIC can only be made once we know
the scenario realized in nature. For
example if relatively light SUSY exists
CLIC can extend the ILC precision measurements to the coloured part of the
spectrum. However, it may also be possible, that a hadron collider at very high
energy may be the better next machine
at the energy frontier.

R&D FOR FUTURE ACCELERATORS

FRANK ZIMMERMANN

CERN, 1211 Geneva, Switzerland
E-mail: frank.zimmermann@cern.ch

Research & development for future accelerators are reviewed. First, I discuss colliding hadron beams, in particular upgrades to the Large Hadron Collider (LHC). This is followed by an overview of new concepts and technologies for lepton ring colliders, with examples taken from VEPP-2000, DAFNE-2, and Super-KEKB. I then turn to recent progress and studies for the multi-TeV Compact Linear Collider (CLIC). Some generic linear-collider research, centered at the KEK Accelerator Test Facility, is described next. Subsequently, I survey the neutrino factory R&D performed in the framework of the US feasibility study IIa, and I also comment on a novel scheme for producing monochromatic neutrinos from an electron-capture beta beam. Finally, I present innovative ideas for a high-energy muon collider and I consider recent experimental progress on laser and plasma acceleration.

1 Introduction

The past 40 years have seen a remarkable increase in the energy of colliding particle beams, which is illustrated in Fig. 1. This progress in acceleration technology has paved the way for many fundamental discoveries in particle physics. With the advent of the Large Hadron Collider (LHC) in 2007, the steep improvement is set to continue. Upgrades of the LHC are already considered, first, around 2014, in luminosity and after 2020 also in energy. Later perhaps a Very Large Hadron Collider (VLHC) could become the ultimate hadron collider. In parallel, the lepton colliders have followed a similar trend. The final incarnation of the Large Electron Positron Ring (LEP-2) presumably marks the maximum energy which will ever be reached by storage-ring electron-positron colliders. For the future, three novel types of lepton colliders are, therefore, proposed, which in different ways overcome the fundamental obstacle of synchrotron radiation. Two of them are linear colliders, with the design extrapolated over many orders of magnitude from the only linear collider ever in operation, namely the Stanford Linear Collider (SLC). The International Linear Collider (ILC) uses superconducting accelerating structures and is limited to a maximum

beam energy of 500 GeV. The Compact Linear Collider (CLIC) is based on two-beam acceleration with an ultimate beam energy of 2.5 TeV. In the longer-distant future, a muon ring collider may conceivably reach beam energies of 10 TeV. The dates indicated in Fig. 1 for the various future colliders represent subjective guesses based on the current state of planning and development, optimistcally assuming that all proposed projects will be realized. Beyond the muon collider, several evolving advanced acceleration techniques, e.g., ones utilizing lasers or plasmas, offer the exciting prospect of pushing the high-energy frontier even further.

2 Hadron Colliders

Presently two hadron colliders are in operation — the Tevatron and RHIC. A third, the LHC, is scheduled to start in 2007. The R&D at the Tevatron is limited to electron cooling in the recycler ring and to beam-beam compensation using a pair of electron lenses. At the time of writing, a first successful cooling of unbunched antiprotons in the recycler ring has been reported, with an electron beam energy of 4.3 MeV, which constitutes a world record[1]. The beam-beam compensation is on a promising track as well, with a successful demonstration of beam-lifetime improvement

beam energy [GeV]

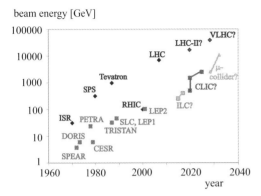

Figure 1. Schematic of collider energy as a function of year. Shown are the beam energies of hadron ring colliders (blue rhombi), including an intrepid extrapolation into the future, those of electron-positron colliders (pink squares) as well as hypothetical points for future electron-positron linear colliders (red squares) and a high-energy muon collider (green triangles). Only data points before 2005 are certain.

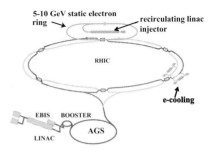

Figure 2. Schematic of the upgraded RHIC complex, with high-energy electron cooling and additional ion-electron collisions (bottom right and top, in green).

in 2004, by a single lens[2]. RHIC pursues an ambitious extension plan. Studies for an upgrade of the LHC have begun.

2.1 RHIC Upgrade

The expanded RHIC complex is illustrated in Fig. 2. It consists of two parts. First, the RHIC luminosity will be increased by a factor of 10, primarily by a pioneering scheme of high-energy bunched-beam electron cooling, which requires a 54-MeV electron beam[3]. Second, an additional electron beam accelerated in a recirculating linac and stored in a 5–10 GeV static ring will collide with the ion beams stored in RHIC. The upgraded RHIC will explore QCD at high energy and high temperature (strongly coupled quark gluon plasma), as well as QCD at low x (colour glass condensate), and the origin of nuclear spin.

2.2 LHC Upgrade

In 2001 two CERN task forces investigated the physics potential[4] and accelerator aspects[5] of an LHC upgrade. Presently,

the European CARE-HHH Network[6] and the US DOE US LARP programme[7] are jointly studying ways of increasing the LHC luminosity by a factor of 10, from the nominal value of 10^{34} cm^{-2}s^{-1} to 10^{35} cm^{-2}s^{-1} at two interaction points (IPs) by about 2014[8,9]. A factor two increase in luminosity can be achieved by a reduction of the IP beta functions $\beta^*_{x,y}$. The remaining factor of five may be obtained by raising the beam current: The baseline upgrade scheme foresees decreasing the bunch spacing from 25 ns to either 15 or 10 ns, at the ultimate bunch population of $N_b \approx 1.7 \times 10^{11}$ protons[8]. An alternative scheme would collide fewer, but longer and more intense bunches ($N_b \approx 6 \times 10^{11}$), and operate with a large Piwinski angle $\phi \equiv \theta_c \sigma_z/(2\sigma^*) \approx 3$, where σ_z denotes the bunch length, θ_c the full crossing angle, and σ^* the rms beam size[8]. An upgrade of the LHC injectors for higher energy and reduced LHC turnaround time is under consideration, and for a later stage the increase of the LHC beam energy itself.

Two reasons call for an upgrade of the LHC after about 6 years of operation. They are illustrated in Fig. 3, due to J. Strait[10]. The figure illustrates two possible evolutions of peak luminosity as a function of the year, as well as the corresponding integrated luminosity and the run-time needed to halve the statistical error of experimental measure-

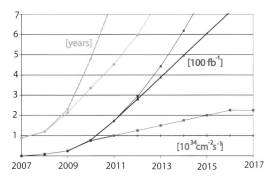

Figure 3. Time to halve the statistical error (green curves), integrated luminosity (blue curves), and peak luminosity (red curves) for two different scenarios compatible with the baseline LHC: (1) the luminosity is raised to the nominal value by 2011 and then stays constant, (2) it continues to increase linearly, reaching the ultimate value by 2016. The assumed radiation damage limit of the IR magnets is 700 fb^{-1}[10] [J. Strait].

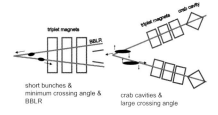

Figure 4. IR 'baseline' schemes for the LHC upgrade with minimum crossing angle and possibly long-range beam-beam compensation (left) or with large crossing angle and possibly crab cavities (right); see also[12].

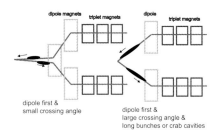

Figure 5. Alternative IR schemes for the LHC upgrade with dipole first for small (left) or large crossing angle (right). The right layout needs to either operate with large Piwinski angle or employ crab cavities; see also[12].

ments. Both scenarios assume an LHC start up in 2007, and that 10% of the design luminosity is reached in 2008, and 100% in 2011[11]. In one case, the luminosity is taken to be constant from then on, in the other it continues to increase linearly until the so-called ultimate luminosity (2.3 times the nominal) would be reached by 2016. The radiation damage limit of the LHC low-β quadrupoles is estimated at an integrated luminosity of 600–700 fb^{-1}[13]. As the figure shows, this value would be exceeded in 2014 or 2016 depending on the scenario. The additional runtime required to halve the statistical error rises more steeply. It would exceed 7 years by 2011 or 2013, respectively. In view of the life expectancy of the interaction-region magnets and the forecast for the statistical-error halving time, it is reasonable to plan a machine luminosity upgrade based on new low-β IR magnets before about 2014.

Figures 4 and 5 illustrate various options for the IR configuration of an upgraded LHC[11,12]. Either quadrupoles or dipoles are the first magnetic elements closest to the main collision points. Placing the quadrupoles in front (Fig. 4) reduces the chromaticity, while "dipoles first" (Fig. 5) leads to a rapid separation or the two beams and minimizes the number of long-range collision points. In the dipoles-first scheme, collision debris is swept into the magnet, and for this reason the US-LARP program studies the design of open-midplane s.c. dipoles. In addition, the two LHC beams could be collided either with a small crossing angle, necessitating long-range beam-beam compensation, e.g., using current-carrying wires, or with a large crossing angle, requiring crab cavities, much shorter bunches, or operation in a regime of large Piwinski angle.

A possible injector upgrade consists of refurbishing the PS, and replacing the SPS with two new rings, a first ring using normal conducting or super-ferric magnets and accel-

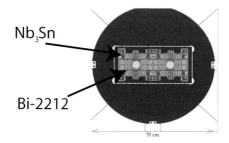

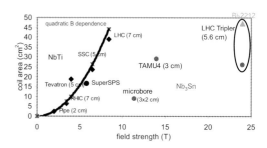

Figure 6. Schematic of a dual-pipe 24-T block dipole magnet with Bi-2122 in inner high field windings (green) and Nb$_3$Sn in outer low field windings (red)[17].

Figure 7. Magnet coil area vs. field strength for different s.c. dipoles, showing a reduced size for the proposed block-dipole magnets of the LHC energy tripler[17].

erating to 150 GeV and a second fast cycling superconducting ring, raising the LHC injection energy to 1 TeV[14]. The higher injection energy will alleviate dynamic effects in the LHC and should significantly reduce the LHC turnaround time. In addition, beams with larger normalized emittance and higher intensity can be injected at the higher energy. Finally, the increased injection energy can be the first step of an energy upgrade for the LHC.

The LHC beam energy is determined by the main dipole field, which nominally is 8.39 T corresponding to 7 TeV beam energy. A proof-of-principle magnet based on Nb$_3$Sn s.c. material at LBNL has reached 16 T a few years ago, with a 10-mm aperture[15]. The European NED activity[16] aims to develop a large-aperture (up to 88 mm), 15-T dipole-magnet model. A 24-T block-coil dipole for an LHC energy tripler is also being developed by Texas A&M University[17]. It employs high-Tc superconductor (Bi-2212) in the inner high-field windings and Ti$_3$Sn for the outer low-field windings. The magnet layout is illustrated in Fig. 6 and its small coil area is emphasized in Fig. 7.

3 Electron-Positron Ring Colliders

Among the e^+e^- colliders, several high-luminosity factories will explore innovative beam-manipulation techniques. Though

BEPC-II, CESR-c and PEP-II are contributing to this effort, below we can only present a few selected highlights from VEPP-2000, DAFNE, and Super-KEKB.

3.1 VEPP-2000

VEPP-2000 will fill the missing-data gap for hadron production in the range of 1.4–2.0 GeV c.m. energies. The start of operation is foreseen for 2007. VEPP-2000 will collide round beams focused by s.c. 13-T solenoids. The round beam collisions offer two distinctive advantages: (1) for the same particle density (tune shift), the luminosity doubles, (2) the transverse particle motion becomes 1-dimensional, introducing an additional integral of motion, which, according to simulations, should allow about 3 times higher tune shifts than flat beams[18]. The VEPP-2000 luminosity scales with ξ^2. Its design value is 10^{32} cm^{-2}s^{-1}.

3.2 DAFNE-2

The upgrade of DAFNE, DAFNE-2, will study discrete symmetries in the neutral kaon system, and perform the most sensitive test of CPT invariance. For DAFNE-2, a new concept of "strong rf focusing" has been proposed, where the bunch length and momentum spread vary around the ring[19]. At the collision point, the bunch length assumes a minimum, so that the beta function can be squeezed to a small value. Over the rest of the ring a long bunch length is desirable, since this improves the beam lifetime and mitigates impedance effects. The strong rf focusing is realized by a higher rf voltage and by tayloring the local momentum compaction around the ring. It has recently been shown that such a scheme can be realized without a synchrotron tune near the half integer. An experiment of strong rf focusing at DAFNE is foreseen for 2007, with the installation of a 10-MV 1.3-GHz multi-cell s.c. rf cavity[20].

3.3 Super-KEKB

KEKB and its upgrade, Super-KEKB, represent the luminosity frontier. Super-KEKB will provide definitive answers on new physics beyond the standard model in the heavy-flavor sector. With a peak luminosity of 1.58×10^{34} cm^{-2}s^{-1}, KEKB presently holds the record luminosity of any collider. The goal of Super-KEKB is a luminosity of 4×10^{35} cm^{-2}s^{-1}, which would make it a truly unique facility, as illustrated in Fig. 8. The luminosity can be expressed as

$$L = \frac{\gamma_\pm}{2er_e}\left(1+\frac{\sigma_y^*}{\sigma_x^*}\right)\frac{I_\pm\xi_\pm}{\beta_y^*}\left(\frac{R_L}{R_y}\right), \quad (1)$$

where R_L and R_y denote geometric reduction factors, due to crossing angle and hourglass effect, r_e is the classical electron radius, and γ the Lorentz factor. Three improvements will increase the luminosity by more than a factor 20 compared with the present KEKB: (1) The stored beam currents $I_\pm$ are raised from 1.27/1.7 A to 4.1/9.4 A, (2) the vertical beta function at the collision point is squeezed from about 6 mm to 3 mm, and (3) the beam-beam tune-shift parameter $\xi_\pm$ will be enlarged by a factor of two or three by using crab cavities[21]. These are transversely deflecting dipole-mode cavities, which rotate the bunches at the collision point, so that the particle dynamics becomes equivalent to the one of the head-on case, even though the

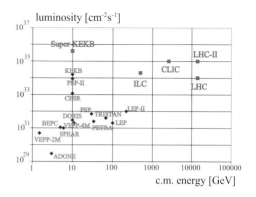

Figure 8. Peak luminosity of various past and proposed colliders vs. centre-of-mass energy.

bunch centroids cross with an angle. In early 2006, single crab cavities will be installed in both KEKB rings. Simulations predict a resulting large increase in the achievable beam-beam tune-shift parameter.

A schematic of Super-KEKB is shown in Fig. 9. All elements in color are new. The upgrade notably comprises a new interaction region with a new detector, additional rf stations, new vacuum chambers for both rings including antechambers and novel low-impedance bellows, as well as an exchange of the electron and positron beams between the two rings. Positrons will be stored in the high-energy ring, where they both have a lower current and are more rigid, which is expected to combat the "electron-cloud effect", i.e., a beam-size blow up due to photo-electrons and secondary electrons. To accelerate the positrons to the higher injection energy (8 GeV instead of 3.5 GeV), their beam size is reduced in a new damping ring, before they are accelerated by a new C-band linac, which replaces the second half of the existing lower-gradient S-band linac.

4 Linear Colliders

Reaching the high-energy frontier in electron-positron collisions requires a linear collider. From initially five proposals (VLEPP, NLC, JLC/GLC, TESLA and CLIC), only two

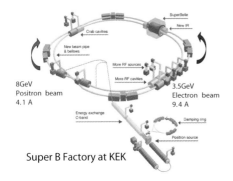

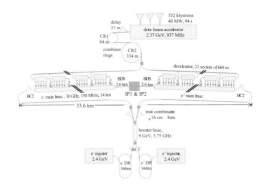

Figure 9. Schematic of Super-KEKB, with existing components shown in grey, and new ones in color.

Figure 10. Layout of CLIC for 3 TeV c.m. energy.

have survived, namely the ILC (based on TESLA technology) and CLIC. Since ILC is covered by a separate talk[22], we discuss CLIC. Afterwards we present some generic linear-collider R&D of interest to both ILC and CLIC.

4.1 Compact Linear Collider (CLIC)

The physics motivation of CLIC is to probe beyond the standard model and to address, in particular, the origin of mass, the unification of forces, and the origin of flavors. With its centre-of-mass energy reach of 3–5 TeV, CLIC is fully complementary to the LHC. Its physics potential is described in[23,24,25,26].

The key features of CLIC are[27]: (1) a high accelerating gradient, 150 MV/m, which — in view of various limits due to dark-current capture, surface heating, and rf breakdown — requires a high rf frequency, namely 30 GHz; (2) two-beam acceleration, where the energy is stored in a drive beam, which can be transported over long distances with small losses, and the rf power is generated locally from the drive beam, where it is required; and (3) a central injector producing the drive beam, which comprises a fully loaded normal conducting linac with 96% rf-to-beam power-transfer efficiency followed by rf multiplication and power compression. The overall layout of the CLIC complex for

3-TeV c.m. energy is presented in Fig. 10.

The high accelerating gradient of CLIC allows constructing a multi-TeV electron-positron collider with an acceptable length. High-gradient tests of 30-GHz rf structures with molybdenum and tungsten irises exceeded the CLIC design goal of 150 MV/m accelerating gradient and reached a world record of 190 MV/m[28] (Fig. 11), but with an rf pulse length of 16 ns only, while the nominal CLIC value is 70 ns. Tests with the nominal pulse length are foreseen at CTF-3 (Fig. 12), presently under construction and commissioning, in the second half of 2005. CTF-3 has already demonstrated two other aspects of the CLIC scheme, namely drive-beam acceleration with full beam loading, shown in Fig. 13, and frequency multiplication[29], in Fig. 14.

The two beam acceleration scheme ensures that the main-linac tunnel is simple, without any active elements, which is illustrated in Fig. 15. An additional advantage is that the construction can be staged by simply adding additional structures and corresponding drive beams, each of which accelerates by about 75 GeV/c.

CLIC aims at colliding nanometre-size beams, for which ground motion and support vibrations are a concern. Using commercially available active stabilization systems, CLIC prototype quadrupoles were stabilized to less than 0.5-nm above 4 Hz, on the CERN site

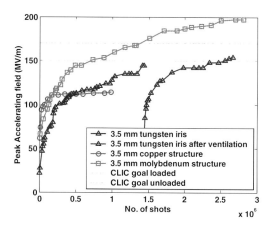

Figure 11. Peak accelerating gradients vs. conditioning time for structures with copper, molybdenum and tungsten irises.

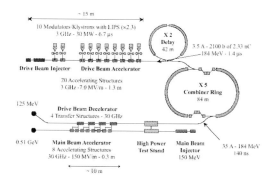

Figure 12. CLIC Test Facility (CTF) 3.

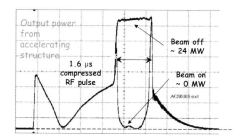

Figure 13. Fully loaded operation of the CTF-3 linac, with higher-order mode damping and short fill time.

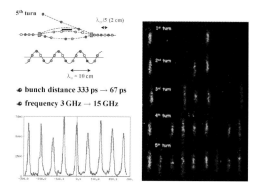

Figure 14. Bunch frequency multiplication by factors of 2–5 demonstrated in the CTF-3 preliminary phase. Left top plot shows a schematic of the injection into the combiner ring using rf delectors; the right and bottom picture show streak-camera measurements during and after bunch multiplication.

next to a busy street, which is a factor 2 better than the requirement for the CLIC linac and represents a factor 10 suppression of the ground motion[30].

The design optics of the CLIC damping ring provides the exceedingly small target emittances of $\gamma\epsilon_x = 550$ nm, $\gamma\epsilon_y = 3.3$ nm, $(\sigma_z\ \Delta E_{\rm rms}) = 4725$ eV-m, taking into account the strong effect of intrabeam scattering which far outweighs the quantum excitation from synchrotron radiation[31]. A collaboration with BINP is investigating wiggler-design options for the damping ring. A possible prototype is the permanent wiggler built for the PETRA-3 light source.

In 2005, the CLIC parameters have been re-optimized[32]. As a result, the bunch-train length was reduced by almost a factor of 2 to 58 ns, with the same luminosity above 99% of the nominal c.m. energy, which is 3.3×10^{34} cm^{-2}s^1 including the effect of beamstrahlung, and lower detector background.

The CLIC schedule anticipates a complete demonstration of the CLIC feasibility by the end of 2009, at a time when first results from the LHC may offer a clearer picture of the physics ahead.

5 Generic Linear Collider R&D

We briefly describe three items of generical linear-collider R&D centered at the KEK Accelerator Test Facility (ATF).

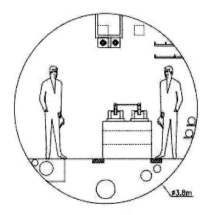

Figure 15. CLIC tunnel cross section.

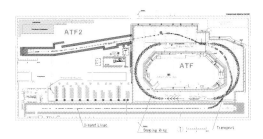

Figure 16. Layout of ATF/ATF-2 facility with the existing ATF damping ring on the right, and the planned ATF-2 final focus on the top left.

5.1 ATF Damping Ring

The ATF is the world's largest linear-collider test facility. It accommodates several sources, an S-band injector linac, a prototype damping ring and an extraction line. The 1.3-GeV ATF ring produces the world's smallest-emittance beams. It has demonstrated single-bunch emittances of $\gamma\epsilon_x \approx 3.5 - 4.3$ μm and $\gamma\epsilon_y \approx 13 - 18$ nm, at $N_b \approx 8 \times 10^9$ electrons per bunch[33]; the emittances are highly current dependent due to the strong effect of intrabeam scattering (for comparison, the CLIC design values are $\gamma\epsilon_x \approx 0.45$ μm and $\gamma\epsilon_y \approx 3$ nm, at $N_b \approx 2.6 \times 10^9$). Numerous new beam-diagnostic instruments were developed at the ATF, for monitoring the unprecedentedly small beam size in a storage ring, among which are laser wires, non-invasive diffraction radiation diagnostics, X-ray optics, and interferometry. Damping-ring tuning and correction algorithms were developed, and a young generation of accelerator physicists was trained.

5.2 ATF-2 Final Focus

The ATF-2 is a proposed extension of the ATF extraction line, accommodating a prototype compact final focus[34]. The ATF-2 final focus is a scaled down version of the NLC, ILC or CLIC final-focus optics, based on a de-sign principle which has never been put into practice. The ATF-2 final focus will squeeze the extremely low-emittance beam from the ATF damping ring to a spot size of 30 nm with less than 5% orbit jitter. The ATF-2 will also test nm-resolution beam position monitors, various advanced spot-size monitors, as well as feedback and stabilization schemes. Figure 16 presents the ATF layout including the ATF-2 extension.

5.3 Polarized Positron Source

A possible scheme of producing highly polarized positron or electron beams is laser Compton back-scatterng of a circularly polarized laser off an unpolarized electron beam. The resulting X-ray photons are polarized and their polarization is inherited by the higher-energy component of pairs created when the X-rays impinge on a thin tungsten target. In pioneering experiments at the ATF, about 10^4 positrons were produced per pulse with a measured average polarization of about 75%[35]. A recent improvement of this scheme, for ILC or CLIC, is the proposal to accumulate the positrons in a dedicated accumulator ring, or in the main damping ring, and, e.g., to combine many small bunches from the Compton-ring source to form one nominal bunch[36]. The accumulation greatly relaxes the requirements on the laser pulse energy.

6 Neutrino Beams

Neutrino beams can be produced by smashing proton superbeams on a target, by muon decay in a future muon storage ring (ν factory), and by β decay or inverse *beta* decay of unstable isotopes. We discuss the latter two options.

6.1 Neutrino Factory

The physics goals of the neutrino factory are the measurement of the θ_{13} mixing angle, determining the neutrino mass hierarchy, and detecting the CP violating phase δ in the neutrino sector. The target intensity of neutrinos required is a few 10^{20} per year. All neutrino-factory designs so far are based on existing sites with appropriately adapted accelerator infrastructure, like CERN (3.5 GeV s.c. proton linac), FNAL (6 GeV s.c. proton linac), BNL (AGS upgrade), J-PARC (50 GeV RCS), or RAL. In 1999-2001 two US feasibility studies were conducted for the FNAL and BNL sites. As part of the 2003 APS Study on the Physics of Neutrinos, the design of the second feasibility study was re-optimized for cost reduction. The outcome is known as feasibility study IIa and was presented at this conference[37].

The ν factory proposed here consists of a high-power proton source (24-GeV BNL AGS ugrade), a target (mercury jet in 20-T solenoid), first a bunching and then a phase-rotation section, cooling (solid LiH absorbers, closed cavity apertures), fast acceleration (comprising a s.c. linac, a recirculating linac, 'RLA', and two non-scaling fixed-field alternating gradient synchrotrons or 'FFAGs'), and a storage decay ring. Figure 17 illustrates the muon acceleration.

Crucial demonstration experiments relate to the three areas of targetry, muon cooling, and acceleration: (1) A mercury jet target with 20 m/s speed will be tested in a 15-T solenoid at the CERN nTOF beamline (experiment "nTOF11"). The instantaneous

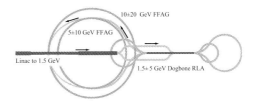

Figure 17. Layout of neutrino-factory muon acceleration for feasibility[37].

power deposition of 180 J/g in this experiment corresponds to that from a 4-MW proton driver. (2) The first phase of on ionization cooling experiment, "MICE"[38], has been approved for construction at RAL. It comprises the assembly of two solenoid tracking spectrometers, which will be cross-calibrated using a muon beam. In a second phase it is foreseen to install one lattice cell of a cooling channel between these spectrometers. The expected emittance reduction is of order 10%. At MICE various absorbers and lattice optics can be tested and compared. (3) The "non-scaling" FFAG is a novel approach, entailing unconventional beam dynamics; building a scaled-down electron-beam model of a non-scaling FFAG is under discussion.

6.2 Beta and Electron-Capture Beams

The physics goals of β or electron-capture beam facilities are similar to those of the ν factories. The main difference is that the neutrinos are produced by the beta decay or inverse beta decay of unstable nuclei instead of muon decay. This type of neutrino beam has recently has become more attractive thanks to the discovery of nuclei that decay fast through atomic electron capture, like ^{150}Dy or ^{148}Gd [39], which provides the fascinating possibility to create mono-energetic neutrino beams. The energy of the emitted neutrinos is Lorentz boosted, e.g., in the forward direction $E_\nu = 2\gamma E_0$, where E_0 is the emission energy in the rest frame. It is assumed that 10^{18} neutrinos can be obtained per year, for example, at EURISOL. Figure 18 shows the

426

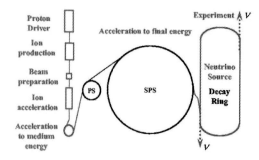

Figure 18. CERN part of a "CERN to Frejus" electron-capture neutrino beam facility[39].

schematic of a proposed "CERN to Frejus" (130 km) electron-capture neutrino beam facility. Such a project would greatly profit from an LHC injector upgrade, e.g., from a Super-SPS.

7 Muon Collider

A collaboration of Muons Inc., IIT, FNAL and JLAB is developing innovative schemes for a muon collider. The new ideas include[40] (1) high-gradient rf cavities filled with pressurized H_2, which simultaneously capture, bunch rotate and cool the muon beam [no strong interactions, no showers], (2) continuous absorbers for emittance exchange, (3) helical cooling channels consisting of solenoid plus transverse helical dipole and quadrupole fields, (4) parametric-resonance ionization cooling, and (5) reverse emittance exchange. These novel concepts are expected to reduce the 6D emittance by about 8 orders or magnitude, which would produce high luminosity, e.g., 10^{35} cm^{-2}s^{-1} at 5 TeV c.m. energy, with 10 bunches of 10^{11} muons each per beam.

8 Advanced Acceleration Schemes

Going to even higher energies, beyond several or tens of TeV, will require radically novel concepts. Candidate schemes include laser and plasma acceleration[41].

8.1 Plasma Acceleration

Plasmas can sustain high accelerating gradients of 10–100 GV/m. To produce these fields the plasma must be excited either by a laser of by a drive beam. The resulting plasma wake may be used to accelerate a witness bunch.

One long-standing problem has been the extremely poor beam quality and the large energy spread of electrons accelerated by a laser in a plasma. Recently, three experimental groups accelerated more than 10^9 electrons in a few-mm long plasma interacting with an intense laser pulse to 100-200 MeV with less than 10–20% full energy spread[42]. Planned as the next step is the development of a 1 GeV compact module utilizing a 100 TW laser[43]. Numerous issues still need to be addressed, however, before a laser-driven plasma can be employed as the primary accelerator in a high-energy collider. Examples are the limited efficiencies of the laser and the plasma, both directly proportional to the collider luminosity.

Closer to a practical application in high-energy physics appears the beam-driven plasma acceleration, which has produced spectacular results in the Final Focus Test Beam (FFTB) line at the end of the SLAC linac. Here, 28.5-GeV bunches of $N_b \approx 2 \times 10^{10}$ electrons, or positrons, with an rms length σ_z of 12–20 μm are sent through a 10-cm Li plasma gas cell of maximum density of 10^{18} cm^{-3} [44,45]. In passing the plasma, the bunch excites a large amplitude plasma wave which focuses and accelerates beam particles. For a single bunch, particles in the head of the bunch lose energy, while those at the end of the bunch are accelerated. For the parameter regime of this experiment, the accelerating field scales as $E_z \propto N_b/\sigma_z^2$. Figure 19 shows the energy distribution of a bunch measured by an energy spectrometer without a plasma and when passing through a 10-cm long plasma of 2.8×10^{17} cm^{-3}. The

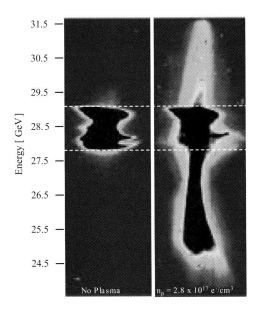

Figure 19. Single-bunch energy spectra at the end of the SLAC FFTB line without plasma (left), and after traversing a 10-cm long 2.8×10^{17} cm^{-3} lithium plasma (right)[44,45].

figure demonstrates that some particles gain more than 2.7 GeV in energy, which corresponds to an accelerating gradient above 27 GV/m. The large energy spread is an artifact of the single-bunch experiment. Future studies will aim at accelerating a second witness bunch travelling at a variable distance behind a short drive bunch.

8.2 Production of Ultra-Short Bunches

Laser-driven accelerators require the injected bunches to be of femtosecond length. Such bunches can be produced by an inverse free electron laser (IFEL) inducing an energy modulation which is followed by a magnetic chicane for microbunching. In an IFEL, the electron beam travelling through an undulator absorbs energy from a co-propagating laser beam. Experiments by a SLAC-Stanford collaboration have recently demonstrated IFEL operation for a 30-MeV beam at 800 nm laser wavelength over a length of about 6 cm[46]. A peak-to-peak energy mod-

ulation of 50 keV was observed. Adjusting the undulator gap in situ revealed multiple resonances of the IFEL interaction.

8.3 Laser Acceleration

Laser-driven particle acceleration in a structure-loaded vacuum instead of a plasma is also under study. It has some similarities to conventional microwave acceleration. A proof-of-principle experiment has accelerated electrons in a semi-infinite vacuum using 800-nm laser radiation, and it achieved a 40 MV/m peak gradient over a distance of 1000 λ[47]. For net laser acceleration to occur a boundary was proven to be necessary. In the experiment an 8-μm gold-coated Kapton tape was employed. Future applications would use more efficient structures, e.g., ones resembling miniature versions of disk-loaded rf waveguides coupled to a laser.

9 Summary and Conclusion

Ongoing R&D activities for future accelerators are multiple. Exciting results achieved so far and ambitious plans for the future hold great promise for the next decades.

Acknowledgements

I would like to thank R. Assmann, H. Braun, H. Burkhardt, B. Cros, W. Fischer, S. Gilardoni, M. Hogan, R. Johnson, D. Kaplan, E. Levichev, M. Lindroos, P. McIntyre, P. Muggli, K. Oide, S. Peggs, T. Plettner, F. Ruggiero, W. Scandale, C. Sears, V. Shiltsev, R. Siemann, J. Urakawa, I. Wilson, M. Zisman and M. Zobov for helpful discussions and providing much of the material.

References

1. S. Nagaitsev, FNAL All Experimenters Meeting, 18 July (2005).
2. V. Shiltsev et al, *AIP Conf. Proc.* **693**, 256 (2004).

428

3. M. Farkondeh and V. Ptitsyn (eds.), "eRHIC Zeroth Order Design Report," BNL C-A/AP/142 (2004).

4. F. Gianotti, M. Mangano, T. Virdee, CERN-TH-2002-078 (2002).

5. O. Bruning et al., LHC-Project-Report-626 (2002).

6. http://care-hhh.web.cern/-CARE-HHH/

7. http://www-td.fnal.gov/LHC/-USLARP.html

8. F. Ruggiero, F. Zimmermann, in *CARE-HHH-APD workshop HHH-2004*, Geneva, CERN-2005-006 (2005).

9. W. Scandale, F. Zimmermann, CERN Courier 45, no. 3, April 2005.

10. J. Strait, private communication, 12 April 2003 (2003).

11. J. Strait, N. Mokhov, T. Sen, in Ref.[8].

12. J. Strait et al., PAC2003 Portland (2003).

13. N. Mokhov, I.L. Rakhno, J.S. Kerby, J.B. Strait, FERMILB-FN-0732 (2003).

14. F. Ruggiero, W. Scandale, private communication (2005).

15. LBNL Superconducting Magnet Program Newsletter, Issue no. 2 (2003).

16. http://lt.tnw.utwente.nl/-project.php?projectid=9

17. P. McIntyre, A. Sattorov, in *PAC'05*, Knoxville (2005).

18. I. Nesterenko, D. Shatilov, E. Simonov, in *'Round Beams and Related Concepts in Beam Dynamics,'* FNAL Dec. (1996).

19. A. Gallo, 30th Meeting of LNF Scientific Committee May 23, 2005 (2005).

20. D. Alesini et al., LNF-05/04 (IR) (2005)

21. K. Ohmi et al., PRST-AB 7, 104401 (2004).

22. J. Dorfan, this symposium.

23. J. Ellis and I. Wilson, *Nature* 409, 431 (2001).

24. http://clicphysics.web.cern.ch/-CLICphysics

25. M. Battaglia et al, "Physics at the CLIC Multi-TeV Linear Collider," CERN-2004-005 (2004).

26. J. Ellis, in *Nanobeam'02*, Lausanne, CERN-Proceedings-2003-001 (2003).

27. The CLIC Study Team, "A 3-TeV e^+e^- Linear Collider based on CLIC Technology," CERN 2000-008 (2000).

28. C. Achard et al., CERN-AB-2003-048, CLIC Note 569 (2003).

29. R. Corsini et al., PRST-AB 7, 040101 (2004).

30. R. Assmann et al., "The CLIC Stability Study on the Feasibility of Colliding High Energy Nanobeams," in Ref.[26].

31. M. Korostelev, F. Zimmermann, "A Lattice Design for the CLIC Damping Ring," in Ref.[26].

32. H. Braun et al., "Updated CLIC Parameters 2005," CLIC Note 627 (2005).

33. Y. Honda et al., PRST-AB 92, 5 (2004).

34. ATF2 Collaboration, CLIC N. 636, KEK Report-2005-2, SLAC-R-771 (2005).

35. T. Omori et al., KEK Preprint 2005-56 (2005).

36. K. Moenig, T. Omori, J. Urakawa et al., Proposal subm. to *Snowmass 2005*.

37. D. Kaplan, Paper-450, this symposium.

38. http://mice.iit.edu/

39. J. Bernabeu, et al., "Monochromatic Neutrino Beams," (2005).

40. R. Johnson et al., in Ref.[17].

41. T. Tajima and J.M. Dawson, *Phys. Rev. Lett.* 43, 267 (1979).

42. S. Mangles et al., C. Geddes et al., J. Faure et al., *Nature* 431, 30 September 2004, pages 535, 538 and 541 (2004).

43. W. Leemans, at HEEAUP, Paris, June (2005).

44. P. Muggli, ibd.

45. M.J. Hogan et al., *Phys. Rev. Lett.* 95, 054802 (2005).

46. C. Sears et al., 'High Harmonic Inverse Free Electron Laser Interaction at 800 nm,' subm. to *Phys. Rev. Lett.* (2005).

47. T. Plettner et al., 'Visible-Laser Acceleration of Relativistic Electrons in a Semi-Infinite Vacuum,' submitted to *Phys. Rev. Lett.* (2005).

DISCUSSION

Gerhard Brandt (Uni. of Heidelberg):
Do you know attempts to use high temperature superconductors, like these ceramics, to build dipoles?

Frank Zimmermann: Yes. For example, the 24-T block-coil dipole, which is being developed by P. McIntyre at Texas A&M University for an LHC "energy tripler", combines Nb_3Sn superconductor in the outer low-field coil windings with Bi-2212 in the inner high-field windings. I have shown a cross section of this magnet towards the beginning of my presentation. The cuprate ceramic Bi-2212 (bismuth strontium calcium copper oxide) is a typical high-temperature superconductor.

Anna Lipniacka (University of Bergen):
What are the currents achieved in these plasma accelerators so far?

Frank Zimmermann: Bunches of $1-2 \times 10^{10}$ electrons were used in the beam-driven plasma wake-field experiments at SLAC. The peak current at the bunch center was increased up to 30 kA by compressing the bunch length to less than 100 fs. Since the same bunch acts as drive and as witness, only 5–10% of its electrons are accelerated, however, i.e., about 2×10^9 electrons per pulse, at a linac repetition rate of 1 or 10 Hz. For the complementary approach of laser-driven plasma wake fields, the maximum number of electrons which were recently accelerated in the so-called "bubble" regime was of comparable magnitude. Earlier, in the mid-90's a different laser-plasma scheme was pursued, where pairs of less-intense lasers at two different wavelengths were used to generate "plasma beat waves" for electron acceleration. At that time and with this alternative scheme, the typically accelerated charge was much lower, namely of order 10^5 electrons, spread over about 100 ps.

STATUS OF THE INTERNATIONAL LINEAR COLLIDER PROJECT

JONATHAN DORFAN

SLAC, 2575 Sand Hill Road, Menlo Park, CA 94025, USA.
E-mail: Jonathan.Dorfan@slac.stanford.edu

NO CONTRIBUTION RECEIVED

SUMMARY TALK

LEPTON PHOTON SYMPOSIUM 2005: SUMMARY AND OUTLOOK

FRANCIS HALZEN

Physics Department, University of Wisconsin, Madison, WI 53706, USA
Email: halzen@pheno.physics.wisc.edu

Lepton Photon 2005 told the saga of the Standard Model which is still exhilarating because it leaves all questions of consequence unanswered.

Over the last decade the biennial gathering discussing leptons and photons has broadened its horizons to reflect the excursions particle physics techniques have made into astronomy and cosmology. It was in the grandest of particle physics traditions however that five days of talks in the historic aula of one of Europe's oldest universities, the home of Linnaeus, Manne and Kai Siegbahn and Dag Hammarskjold, erected the impressive edifice that is called the Standard Model. Experimental ingenuity has not been able to pierce the Model's armor and I cannot help thinking of the prophetic words of Leon Lederman at the Rochester meeting held in Madison twenty five years ago: "the experimentalists do not have enough money and the theorists are overconfident". Where experimentalists are concerned, nobody could have anticipated that today we would be studying the proton structure to one thousandth its size and would have established the Standard Model as a gauge theory with a precision of one in a thousand, pushing any interference of possible new physics to energy scales beyond 10 TeV. The theorists can modestly claim that they have taken revenge for Leon's remark. Because all the big questions remain unanswered, there is no feeling though that we are now dotting the i's and crossing the t's of a mature theory. Worse, the theory has its own demise built into its radiative corrections.

The most evident of unanswered questions is why the weak interactions are weak. Though unified with electromagnetism, electromagnetism is apparent in daily life while the weak interactions are not. Already

in 1934 Fermi provided an answer with a theory[1] that prescribed a quantitative relation between the fine-structure constant and the weak coupling $G \sim \alpha/m_W^2$. Although Fermi adjusted m_W to accommodate the strength and range of nuclear radioactive decays, one can readily obtain a value of m_W of 40 GeV from the observed decay rate of the muon for which the proportionality factor is $\pi/\sqrt{2}$. The answer is off by a factor of 2 because the discovery of parity violation and neutral currents was in the future and introduces an additional factor $1 - m_W^2/m_Z^2$:

$$G_\mu = \left[\frac{\pi\alpha}{\sqrt{2}m_W^2} \right] \left[\frac{1}{1 - m_W^2/m_Z^2} \right] (1 + \Delta r). \tag{1}$$

Fermi could certainly not have anticipated that we now have a renormalizable gauge theory that allows us to calculate the radiative corrections Δr to his formula. Besides regular higher order diagrams, loops associated with the top quark and the Higgs boson contribute; they have been observed[2,3,4].

I once heard one of my favorite physicists refer to the Higgs as the "ugly" particle, but this is nowadays politically incorrect. Indeed, scalar particles are unnatural. If one calculates the radiative corrections to the mass m appearing in the Higgs potential, the same gauge theory that withstood the onslaught of precision experiments at LEP/SLC and the Tevatron yields a result that grows quadratically:

$$\delta m^2 = \frac{3}{16\pi^2 v^2}(2m_W^2 + m_Z^2 + m_H^2 - 4m_t^2)\Lambda^2, \tag{2}$$

where $m_H^2 = 2\lambda v^2$, λ is the quartic Higgs coupling, $v = 246$ GeV and Λ a cutoff. Upon minimization of the potential, this translates into a dangerous contribution to the Higgs vacuum expectation value which destabilizes the electroweak scale[5]. The Standard Model works amazingly well by fixing Λ at the electroweak scale. It is generally assumed that this indicates the existence of new physics beyond the Standard Model; following Weinberg

$$\mathcal{L}(m_W) = \frac{1}{2}m^2 H^\dagger H + \frac{1}{4}\lambda(H^\dagger H)^2 + \mathcal{L}_{SM}^{gauge}$$
$$+ \mathcal{L}_{SM}^{Yukawa} + \frac{1}{\Lambda}\mathcal{L}^5 + \frac{1}{\Lambda^2}\mathcal{L}^6 + ... \quad (3)$$

The operators of higher dimension parametrize physics beyond the Standard Model. The optimistic interpretation of all this is that, just like Fermi anticipated particle physics at 100 GeV in 1934, the electroweak gauge theory requires new physics to tame the divergences associated with the Higgs potential. By the most conservative estimates this new physics is within our reach. Avoiding fine-tuning requires $\Lambda \lesssim 2{\sim}3$ TeV to be revealed by the LHC, possibly by the Tevatron. For instance, for $m_H = 115{-}200$ GeV

$$\left|\frac{\delta m^2}{m^2}\right| = \left|\frac{\delta v^2}{v^2}\right| \leq 10 \ \Rightarrow \ \Lambda \lesssim 2{-}3 \text{ TeV}. \quad (4)$$

Dark clouds have built up around this sunny horizon because some electroweak precision measurements match the Standard Model predictions with too high precision, pushing Λ to 10 TeV. The data pushes some of the higher order dimensional operators in Weinberg's effective Lagrangian to scales beyond 10 TeV. Some theorists have panicked by proposing that the factor multiplying the unruly quadratic correction $(2m_W^2 + m_Z^2 + m_H^2 - 4m_t^2)$ must vanish; exactly! This has been dubbed the Veltman condition. The problem is now "solved" because scales as large as 10 TeV, possibly even higher, can be accommodated by the observations once one eliminates the dominant contribution. One

can even make this stick to all orders and for $\Lambda \leq 10$ TeV, this requires that $m_H \sim 210{-}225$ GeV [5].

Let's contemplate the possibilities. The Veltman condition happens to be satisfied and this would leave particle physics with an ugly fine tuning problem reminiscent of the cosmological constant. This is very unlikely; LHC must reveal the Higgs physics already observed via radiative correction, or at least discover the physics that implements the Veltman condition[6]. It must appear at 2~3 TeV, even though higher scales can be rationalized when accommodating selected experiments[2]. Minimal supersymmetry is a textbook example. Even though it elegantly controls the quadratic divergence by the cancellation of boson and fermion contributions, it is already fine-tuned at a scale of $2 \sim 3$ TeV. There has been an explosion of creativity to resolve the challenge in other ways; the good news is that all involve new physics in the form of scalars, new gauge bosons, non-standard interactions... Alternatively, it is possible that we may be guessing the future while holding too small a deck of cards and LHC will open a new world that we did not anticipate. Particle physics would return to its early traditions where experiment leads theory, as it should be, and where innovative techniques introduce new accelerators and detection methods that allow us to observe with an open mind and without a plan, leading us to unexpected discoveries.

There is good news from Fermilab[7]. The Tevatron experiments are within an order of magnitude of the sensitivity where they may discover the Higgs; see Fig. 1. The integrated luminosity of $8\,\text{fb}^{-1}$, expected by extrapolating present collider performance, bridges that gap. Discovery will require an additional boost in sensitivity from improved detector performance which is actually expected for lepton identification and jet mass resolution. The performance of the detectors has been nothing short of spectacular as illustrated by

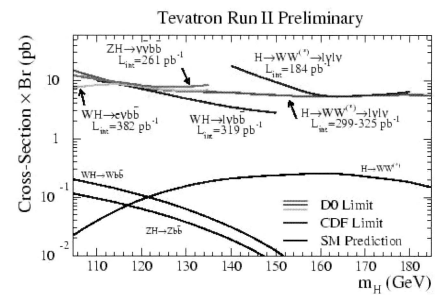

Figure 1. The Tevatron roadmap to the Higgs by increased luminosity and improved detector performance.

the identification of the top quark in 6-jet events[3].

Baryogenesis is another one of the grand issues left unresolved by the Standard Model. We know that at some early time in the evolution of the Universe quarks and anti-quarks annihilated into light, except for just one quark in 10^{10} that failed to find a partner and became us. We are here because baryogenesis managed to accommodate the three Zakharov conditions; one of them dictates CP-violation. Evidence for the indirect violation of CP-invariance was first revealed in 1964 in the mixing of neutral kaons. Direct CP-violation, not mixing-assisted, was not discovered until 1999. Today, precision data on neutral kaons have been accumulated over 40 years; the measurements can, without exception, be accommodated by the Standard Model with three families[8]. History has repeated itself for B mesons, but in three years only, thanks to the magnificent performance of the B-meson factories Belle and BaBar[9]. Direct CP-violation has been established in the decay $B_d \to K\pi$ with a significance in excess of 5 sigma. Unfortunately, this re-

sult, as well as a wealth of data contributed by CLEO, BES and Dafne, fails to reveal evidence for new physics[10]. Whenever the experimental precision increases, the higher precision measurements invariably collapse onto the Standard Model values; see Fig. 2. Given the rapid progress and the better theoretical understanding of the Standard Model expectations relative to the K system[11], the hope is that at this point the glass is half full and that improved data will pierce the Standard Model's resistant armor[12]. Where theory is concerned, it is noteworthy that lattice techniques have reached the maturity to perform computer experiments that are confirmed by experiment.

The rise and fall of theories, or at least of their popularity, can be easily assessed by consulting the citation index. The number of citations to Wolfenstein's seminal paper on neutrino oscillations in the presence of matter has, after a steady increase from 1978–2000, dropped by almost a factor of two since. Progress in neutrino physics has been led by a string of fundamental experimental measurements summarized by the simple vacuum re-

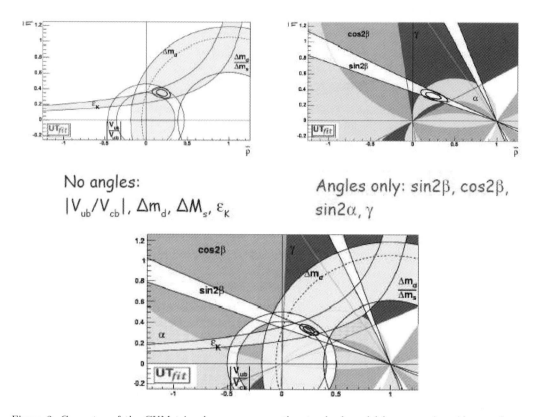

No angles:

$|V_{ub}/V_{cb}|$, Δm_d, ΔM_s, ε_K

Angles only: $\sin 2\beta$, $\cos 2\beta$, $\sin 2\alpha$, γ

Figure 2. Geometry of the CKM triangle converges on the standard model by measuring sides, angles, or both.

lations between the neutrino states produced by the weak interactions in e, mu and tau flavors and propagating as mixed states ν_1, ν_2 and ν_3:

$$\nu_1 = -\cos\theta\,\nu_e + \sin\theta\left(\frac{\nu_\mu - \nu_\tau}{\sqrt{2}}\right),$$

$$\nu_2 = \sin\theta\,\nu_e + \cos\theta\left(\frac{\nu_\mu - \nu_\tau}{\sqrt{2}}\right),$$

$$\nu_3 = \left(\frac{\nu_\mu + \nu_\tau}{\sqrt{2}}\right). \qquad (5)$$

Here θ is the solar mixing angle. Discovery of neutrino oscillations in solar and atmospheric beams has been confirmed by supporting evidence from reactor and accelerator beams[13].

As usual, next-generation experiments are a lot more challenging and the boom times of neutrino physics are probably over as reflected by Wolfenstein's citations. Also,

high-precision data from the pioneering experiments trickle in at a slower pace, although new evidence for the oscillatory behavior in L/E of the muon-neutrinos in the atmospheric neutrino beam has become very convincing. The new results included first data from the reborn SuperKamiokande experiment[14]. The future of neutrino physics is undoubtedly bright. Construction of the KATRIN spectrometer measuring neutrino mass to 0.02 eV by studying the kinematics of tritium decay is in progress and a wealth of ideas on double beta decay and long-baseline experiments is approaching reality[15]. These experiments will have to answer the great "known-unknowns" of neutrino physics: their absolute mass and hierarchy, the precise value of the second and third small mixing angle and its associated CP-violating phase and whether neutrinos are really Majorana particles. In Eq. (5) we assumed that the mixing of mu and tau neutrinos is maximal, with no admixture of electron neutrinos. Observing otherwise will most likely require next-generation experiments.

Among these, discovery of neutrinoless double beta decay would be especially rewarding[16]. Its observation would confirm the theoretical bias that neutrinos are their own antiparticles, yield critical information on the absolute mass scale and, possibly, resolve the hierarchy problem. In the meantime we will keep wondering whether small neutrino masses are our first glimpse at grand unified theories via the see-saw mechanism, or represent a new Yukawa scale tantalizingly connected to lepton conservation and, possibly, the cosmological constant.

The cosmological constant represents a thorny issue for the Standard Model[17]. New physics is also required to control the Standard Model calculation of the vacuum energy, also known as the cosmological constant, which diverges as

$$\int^{\Lambda} \frac{1}{2}\hbar\omega = \int^{\Lambda} \frac{1}{2}\hbar\sqrt{k^2 + m^2}\, d^2k \ \sim \ \Lambda^4 \,.$$
(6)

It has not escaped attention that the cutoff energy required to accommodate its now "observed" value happens to be $\Lambda = 10^{-3}$ eV, of the order of the neutrino mass.

Information on neutrino mass has emerged from an unexpected direction: cosmology[18]. The structure of the Universe is dictated by the physics of cold dark matter and the galaxies we see today are the remnants of relatively small over-densities in the nearly uniform distribution of matter in the very early Universe. Overdensity means overpressure that drives an acoustic wave into the other components making up the Universe: the hot gas of nuclei and photons and the neutrinos. These acoustic waves are seen today in the temperature fluctuations of the microwave background as well as in the distribution of galaxies on the sky. With a contribution to the Universe's matter balance similar to that of light, neutrinos play a secondary role. The role is however identifiable — neutrinos, because of their large mean-free paths, prevent the smaller structures in the cold dark matter from fully developing and this is visible in the observed distribution of galaxies; see Fig. 3. Simulations of structure formation with varying amounts of matter in the neutrino component, i.e. varying neutrino mass, can be matched to a variety of observations of today's sky, including measurements of galaxy-galaxy correlations and temperature fluctuations on the surface of last scattering. The results suggest a neutrino mass of at most 1 eV, summed over the 3 neutrino flavors, a range compatible with the one deduced from oscillations.

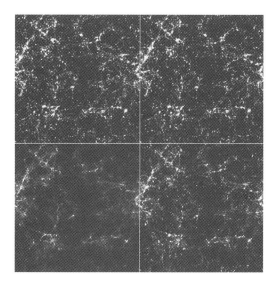

Figure 3. Simulations of structure formation with varying amounts of matter in the neutrino component, i.e. varying neutrino mass: (top left) $m_\nu = 0$ eV; (top right) $m_\nu = 1$ eV; (bottom right) $m_\nu = 4$ GeV; (bottom left) $m_\nu = 7$ eV.

Cosmology, in association with the discovery of neutrino mass, has also been responsible for renewed interest in deciphering baryogenesis — a tally of the rapidly increasing number of citations to the 1986 paper

438

by Fukugita and Yanagida underscores the point. The problem is more clearly framed than ever before. The imprint on the surface of last scattering of the acoustic waves driven into the hot gas of nuclei and photons reveals a relative abundance of baryons to photons of $6.5^{+0.4}_{-0.3} \times 10^{-10}$ (WMAP observation). Gamov realized that a Universe born as hot plasma must consist mostly of hydrogen and helium, with small amounts of deuterium and lithium added. The detailed balance depends on basic nuclear physics as well as the same relative abundance of baryons to photons; the state of the art result of this exercise yields $4.7^{+1.0}_{-0.8} \times 10^{-10}$. The agreement of the two observations is stunning, not just because of the precision, but because of the concordance of two results derived by totally unrelated ways to probe the early Universe.

Physics at the high energy frontier is the physics of partons. For instance, at the LHC gluons produce the Higgs boson and the highest energy neutrinos interact with sea-quarks in the detector. We master this physics with unforeseen precision because of a decade of steadily improving HERA measurements of the nucleon structure[19]. These now include experiments using targets of polarized protons and neutrons. HERA is our nucleon microscope, tunable by the wavelength and the fluctuation time of the virtual photon exchanged in the electron proton collision. The wavelength of the virtual photons probing the nucleon is reduced with increased momentum transfer Q. The proton has now been probed to distances of one thousandth of its size of 1 fm. In the interaction the fluctuations of the virtual photons survive over distances $ct \sim 1/x$, where x is the relative momentum of the parton. HERA now studies the production of chains of gluons as long as 10 fm, an order of magnitude larger than and probably totally insensitive to the proton target. These are novel QCD structures, the understanding of which has been challenging[20]. We should not forget however that theorists ana-

lyze HERA data with calculations performed to next-to-next to leading order in the strong coupling. In fact, beyond this precision one has to include the photon as a parton inside the proton[21]. These electromagnetic structure functions violate isospin and differentiate a u-quark in a proton from a d-quark in a neutron because of the different electric charge of the quark. Interestingly, their inclusion in the structure functions modifies the extraction of the Weinberg angle from NuTeV data, bridging roughly half of its discrepancy with the particle data book value. Added to already anticipated intrinsic isospin violations associated with sea-quarks, the NuTeV anomaly may be on its way out.

Recalling Lederman, whatever the actual funding, the experimenters managed to deliver most highlights of this conference. And where history has proven that theorists had the right to be confident in 1980, they have not faded into the background, and provided some highlights of their own. Developing QCD calculations to the level that the photon structure of the proton becomes a factor is a tour de force and, there were others at this meeting. Progress in higher order QCD computations of hard processes is mind boggling — progress useful, sometimes essential, for the interpretation of LHC experiments[22]. Discussions of strings, supersymmetry and additional dimensions were very much focused on the capability of experiments to confirm or debunk these concepts[23].

Theory and experiment joined forces in the ongoing attempts to read the information supplied by the data on heavy ion collisions from Brookhaven. Rather than the anticipated quark gluon plasma, the data suggests the formation of a strongly interacting fluid with very low viscosity for its entropy[24]. Similar fluids of cold ^{6}Li atoms have been created in atomic traps. Interestingly, theorists are exploiting the Maldacena connection between four dimensional gauge theory and ten dimensional string theory to model such a thermodynamic system[25]. The model is that of a 10-D rotating black hole with Hawking-Beckenstein entropy. It accommodates the low viscosities observed. This should put us on notice that very high energy collisions of nuclei may be more interesting than anticipated from QCD-inspired logarithmic extrapolations of accelerator data. This is relevant to the analysis of cosmic ray experiments. Enter particle astrophysics.

Conventional astronomy spans 60 octaves in photon frequency, from 10^4 cm radiowaves to 10^{-14} cm photons of GeV energy. This is an amazing expansion of the power of our eyes that scan the sky over less than a single octave just above 10^{-5} cm wavelength. Recently detection and data handling techniques of particle physics[26] are reborn in instrumentation to probe the Universe at new wavelengths, smaller than 10^{-14} cm, or photon energies larger than 10 GeV. Besides gamma rays, gravitational waves and neutrinos as well as very high-energy protons that are only weakly deflected by the magnetic field of our galaxy, have become astronomical messengers from the Universe[27]. As exemplified time and again, the development of novel ways of looking into space invariably results in the discovery of unanticipated phenomena. For particle physicists the sexiest astrophysics problem is undoubtedly how Nature manages to impart an energy of more than 10^8 TeV to a single elementary particle.

Although cosmic rays were discovered almost a century ago, we do not know how and where they are accelerated[28]. This may be the oldest mystery in astronomy and solving it is challenging as can be seen by the following argument. It is sensible to assume that, in order to accelerate a proton to energy E in a magnetic field B, the size R of the accelerator must encompass the gyroradius of the particle: $R > R_{gyro} = E/B$, i.e. the accelerating magnetic field must contain the particle orbit. This condition yields a maximum energy $E < \Gamma B R$ by dimensional analysis and noth-

ing more. The factor Γ has been included to allow for the possibility that we may not be at rest in the frame of the cosmic accelerator resulting in the observation of boosted particle energies. Opportunity for particle acceleration to the highest energies is limited to dense regions where exceptional gravitational forces create relativistic particle flows: the dense cores of exploding stars, inflows on supermassive black holes at the centers of active galaxies, annihilating black holes or neutron stars? All speculations involve collapsed objects and we can therefore replace R by the Schwartzschild radius $R \sim GM/c^2$ to obtain $E < \Gamma B M$.

The above speculations are reinforced by the fact that the sources listed happen to also be the sources of the highest energy gamma rays observed. At this point a reality check is in order. Note that the above dimensional analysis applies to the Fermilab accelerator: kGauss fields over several kilometers (covered with a repetition rate of 10^5 revolutions per second) yield 1 TeV. The argument holds because, with optimized design and perfect alignment of magnets, the accelerator reaches efficiencies close to the dimensional limit. It is highly questionable that Nature can achieve this feat. Theorists can imagine acceleration in shocks with efficiency of perhaps 1–10%.

Given the microgauss magnetic field of our galaxy, no structures seem large or massive enough to reach the energies of the highest energy cosmic rays. Dimensional analysis therefore limits their sources to extragalactic objects. A common speculation is that they may be relatively nearby active galactic nuclei powered by a billion solar mass black holes. With kilo-Gauss fields we reach 100 EeV, or 10^{20} eV. The jets (blazars) emitted by the central black hole could reach similar energies in accelerating substructures boosted in our direction by a Γ-factor of 10, possibly higher. The neutron star or black hole remnant of a collapsing supermassive star could support magnetic fields of 10^{12} Gauss, possibly larger. Shocks with $\Gamma > 10^2$ emanating from the collapsed black hole could be the origin of gamma ray bursts and, possibly, the source of the highest energy cosmic rays.

The astrophysics problem is so daunting that many believe that cosmic rays are not the beam of cosmic accelerators but the decay products of remnants from the early Universe, for instance topological defects associated with a grand unified GUT phase transition near 10^{24} eV. A topological defect will suffer a chain decay into GUT particles X,Y, that subsequently decay to familiar weak bosons, leptons and quark- or gluon jets. Cosmic rays are the fragmentation products of these jets. HERA again revealed to us the composition of these jets that count relatively few protons, *i.e.* cosmic rays, among their fragmentation products and this is increasingly becoming a problem when one confronts this idea with data.

We conclude that, where the highest energy cosmic rays are concerned, both the accelerator mechanism and the particle physics are enigmatic. There is a realistic hope that the oldest problem in astronomy will be resolved soon by ambitious experimentation: air shower arrays of 10^4 km^2 area (Auger), arrays of air Cerenkov detectors (H.E.S.S. and Veritas, as well as the Magic 17 m mirror telescope) and kilometer-scale neutrino observatories (IceCube and NEMO). Some of these instruments have other missions; all are likely to have a major impact on cosmic ray physics. While no breakthroughs were reported, preliminary data forecast rapid progress and imminent results in all three areas[27].

The Auger air shower array is confronting the low statistics problem at the highest energies by instrumenting a huge collection area covering 3000 square kilometers on an elevated plane in Western Argentina. The instrumentation consists of 1600 water Cherenkov detectors spaced by 1.5 km. For calibration, showers occurring at night, about 10 percent of them, are also viewed by four fluorescence detectors. The detector will observe several thousand events per year above 10 EeV and tens above 100 EeV, with the exact numbers depending on the detailed shape of the observed spectrum. The end of the cosmic ray spectrum is a matter of speculation given the somewhat conflicting results from existing experiments, most notably the HiRes fluorescence detector and the AGASA scintillator array; see Fig. 4.

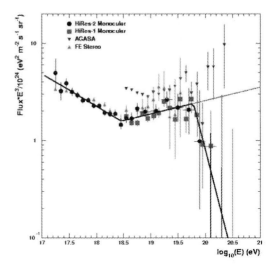

Figure 4. Extragalactic cosmic ray spectrum before Auger and HiRes stereo measurements.

Above a threshold of 50 EeV the cosmic rays interact with cosmic microwave photons and lose energy to pions before reaching our detectors. This is the Greissen-Zatsepin-Kuzmin cutoff that limits the sources to our supercluster of galaxies. The feature in the spectrum is claimed at the 5 sigma level in the latest HiRes data. It is totally absent in the AGASA data, a fact that would require some radical departure from established particle physics or astrophysics. At this meeting Auger presented the first results from the partially deployed array[28]. The exposure is similar to that of the final AGASA data. The data confirms the existence of super EeV events. There is no evidence, either in the latest HiRes or the Auger data, however, for anisotropies in the arrival directions of the cosmic rays claimed mostly on the basis of the AGASA data. Importantly, Auger observes a discrepancy between the energy measurements of showers obtained from the fluorescence and particle array techniques. The discrepancy suggests that very high energy air showers do not develop as fast as modeled by the particle physics simulations used to analyze previous experiments, *i.e.* the data necessitate deeper penetration of the primary, less inelasticity and more energy in fewer leading particles than anticipated. Auger data definitely indicate that the experiment is likely to qualitatively improve existing observations of the highest energy cosmic rays in the near future.

Cosmic accelerators are also cosmic beam dumps producing secondary photon and neutrino beams[29]. Particles accelerated near black holes pass through intense radiation fields or dense clouds of gas leading to production of secondary photons and neutrinos that accompany the primary cosmic-ray beam. The target material, whether a gas or photons, is likely to be sufficiently tenuous so that the primary beam and the photon beam are only partially attenuated; see Fig. 5.

Although gamma ray and neutrino telescopes have multiple interdisciplinary science missions, in the case of neutrinos the real challenge has been to develop a reliable, expandable and affordable detector technology[30]. The South Pole AMANDA neutrino telescope, now in its fifth year of operation, has improved its sensitivity by more than an order of magnitude since reporting its first results in 2000. It has now reached a sensitivity close to the neutrino flux anticipated to accompany the highest energy cosmic rays, dubbed the Waxman-Bahcall bound. Expansion into the Ice-Cube kilometer-scale neutrino observatory,

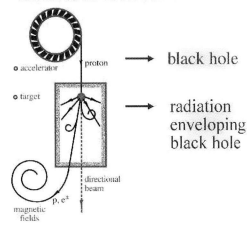

NEUTRINO BEAMS: HEAVEN & EARTH

Figure 5. Cosmic ray accelerators are also cosmic beamdumps producing fluxes of neutrinos and TeV photons accompanying the cosmic rays.

required to be sensitive to the best estimates of potential cosmic neutrino fluxes, is in progress. Companion experiments in the deep Mediterranean are moving from the R&D into the construction phase with the goal to eventually build an IceCube size detector. With the sun and SN87 neutrino observations as proofs of concepts, next-generation neutrino experiments will also scrutinize their data for new particle physics, from the signatures of dark matter to the evidence for additional dimensions of space.

It is however the H.E.S.S. array of four air Cherenkov gamma ray telescopes deployed under the southern sky of Namibia that delivered the highlights in the particle astrophysics corner[27]. For the first time an instrument is capable of imaging astronomical sources in TeV gamma rays. Its images of young galactic supernova remnants shows filament structures of high magnetic fields that are capable of accelerating protons to the energies, and with the energy balance, required to explain the galactic cosmic rays; Fig. 6. Although the smoking gun for cosmic ray acceleration is still missing, the evidence is tantalizingly close.

442

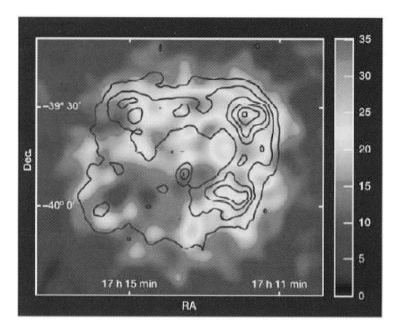

Figure 6. TeV gamma ray image of a young supernova remnant.

The big event of the next biennium is the commissioning of the LHC. With dark matter and energy[18,31], astronomers have raised physics problems that seem as daunting as the problem of the lifetime of the sun over one century ago. Evolution and geology required a sun that was older than several tens of millions of years. Chemistry established its lifetime at 3000 years. Neither chemistry nor astronomy solved the puzzle, Rutherford did. May history repeat itself with the solution revealed by the accelerators in our future, LHC and a linear collider[32]!

Rendez vous in 2007 in Daegu, Korea.

Acknowledgments

I am grateful for the superb hospitality of Professor Tord Ekelöf, my Uppsala IceCube colleagues and other organizers of the meeting. This talk is not only inspired by the excellent talks at this meeting, but also by a series of stimulating and provocative talks at meetings I attended prior to this conference, by Alan Martin, K. Kinoshita and C. W. Chiang at Pheno 05 in Madison and by Pilar Hernández, Hitoshi Murayama, Carlos Peña-Garay and G. Raffelt at a neutrino workshop in Madrid. I thank Vernon Barger, Hooman Davoudiasl, Concha González-García, Tao Han, and Patrick Huber for reading the manuscript. This research was supported in part by the National Science Foundation under Grant No. OPP-0236449, in part by the U.S. Department of Energy under Grant No. DE-FG02-95ER40896, and in part by the University of Wisconsin Research Committee with funds granted by the Wisconsin Alumni Research Foundation.

References

1. E. Fermi, *Z. Phys.* **88**, 161 (1934); *Nuovo Cim.* **11**, 1 (1934).
2. S. Dawson, these proceedings.
3. A. Juste, these proceedings.
4. D. Diaconnu, these proceedings.
5. For a nice update, see J.A. Casas, J.R. Espinosa and I. Hidalgo, arXiv:hep-ph/0410298, *JHEP* **0411**, 057 (2004).

6. F. Gianotti and L. Rolandi, these proceedings.
7. S. Lammel, these proceedings.
8. E. Blucher, these proceedings.
9. K. Abe and F. Forti, these proceedings.
10. M. Artuso, R. Jesic, X. Shen and V. Burkert, these proceedings.
11. U. Nierste and I. Stewart, these proceedings.
12. L. Silvestrini, these proceedings.
13. A. Poon, E. Zimmerman and S. Goswami, these proceedings.
14. Y. Suzuki, these proceedings.
15. C. Weinheimer, these proceedings.
16. O. Cremonesi, these proceedings.
17. E. Witten, Proceedings of the 2003 Lepton-Photon Conference, Fermilab, Chicago (2003)
18. S. Hannestad, these proceedings.
19. J. Butterworth and J.-M. Le Goff, these proceedings.
20. G. Ingelman, these proceedings.
21. M. Gluck, P. Jimenez-Delgado and E. Reya, arXiv:hep-ph/0503103, *Phys. Rev. Lett.* **95**, 022002 (2005); A.D. Martin, R.G. Roberts, W.J. Stirling and R.S. Thorne, arXiv:hep-ph/0411040, *Eur. Phys. J.* **C39**, 155 (2005).
22. G. Salam, these proceedings.
23. I. Antoniadis, these proceedings.
24. J. Stachel, these proceedings.
25. I. Klebanov, these proceedings.
26. M. Danilov, these proceedings.
27. R. Ong, talk at this conference.
28. S. Westerhoff, these proceedings.
29. E. Waxman, these proceedings.
30. P. Hulth, these proceedings.
31. L. Baudis, these proceedings.
32. K. Moenig, F. Zimmerman and J. Dorfan, these proceedings.

PUBLIC TALK

THE UNIVERSE IS A STRANGE PLACE

FRANK WILCZEK

Center for Theoretical Physics, Department of Physics
Massachusetts Institute of Technology
Cambridge, Massachusetts 02139, USA
E-mail: wilczek@mit.edu

Our understanding of ordinary matter is remarkably accurate and complete, but it is based on principles that are very strange and unfamiliar. As I'll explain, we've come to understand matter to be a Music of the Void, in a remarkably literal sense. Just as we physicists finalized that wonderful understanding, towards the end of the twentieth century, astronomers gave us back our humility, by informing us that ordinary matter – what we, and chemists and biologists, and astronomers themselves, have been studying all these centuries constitutes only about 5% of the mass of the universe as a whole. I'll describe some of our promising attempts to rise to this challenge by improving, rather than merely complicating, our description of the world.

In a lecture of one hour I will not be able to do justice to all the ways in which the Universe is a strange place. But I'll share a few highlights with you.

1 Interior Strangeness

First I'd like to talk about how strange I am. Oh, and you too, of course. That is, I'd like to describe how strange and far removed from everyday experience is the accurate picture of ordinary matter, the stuff that we're made of, that comes from modern physics.

I think that 10,000 years from now, our descendants will look back on the twentieth century as a very special time. It was the time when humankind first came to understood how matter works, indeed what it *is*.

By 1900, physicists had a great deal of Newtonian-level knowledge. The paradigm for this level of understanding is Newton's account of the motion of planets, moons, and comets in the solar system. Newtonian-level knowledge takes the form: if bodies are arranged with given positions and velocities at time t_0, then the laws of physics tell you what their positions and velocities will be at any other time. Newton's theory is extremely accurate and fruitful. It led, for example, to the discovery of a new planet – Neptune – whose gravity was necessary to account for anomalies in motion of the known planets. (We're facing similar problems today, as I'll describe later. Neptune was the "dark matter" of its day.)

But nothing in Newton's theory prescribes how many planets there are. Nothing predicts their masses, or their distances from the Sun. For planetary systems all that freedom is a good thing, we know now, because astronomers are discovering other systems of planets around other stars, and these solar systems are quite different from ours, but they still obey Newton's laws.

But no such freedom is observed for the building blocks of ordinary matter. Those building blocks come in only a few kinds, that can only fit together in very restricted ways. Otherwise there could not be such a subject as chemistry, because each sample of matter would be different from every other sample. Pre-modern physics could not account for that fact, and therefore it could not even begin to account for the specific chemical, electrical, and mechanical properties of matter.

The turning point came in 1913, with Bohr's model of the hydrogen atom. Bohr's model brought quantum ideas into the description of matter. It pictured the hydrogen atom as analogous to a simplified solar system (just one planet!) held together by electric rather than gravitational forces, with the proton playing the role of the sun and the electron the role of the planet. The crucial new idea in Bohr's model is that only certain orbits are allowed to the electron: namely, oribts for which the angular momentum is a whole-number multiple of Planck's constant. When the atom exchanges energy with the outside world, by emitting or absorbing light, the electron jumps from one orbit to another, decreasing its energy by emitting a photon or increasing its energy by absorbing one. The frequency of the photon, which tells us what color of light it conveys, is proportional to its energy according to the Planck-Einstein relation $E = h\nu$, where E is the energy, ν is the frequency, and h is Planck's constant. Because the allowed orbits form a discrete set, rather than a continuum, we find discrete spectral lines corresponding to the allowed energy changes.

Bohr's model predicts the colors of the spectral lines remarkably accurately, though not perfectly. When Einstein learned of Bohr's model, he called it "the highest form of musicality in the sphere of thought". I suppose Einstein was alluding here to the ancient idea of "Music of the Spheres", which has enchanted mathematically inclined mystics from Pythagoras to Kepler. According to that idea the planets produce a kind of music through their stately periodic motions, as the strings of a violin do through their periodic vibrations. Whether the "Music of the Spheres" was intended to be actual sounds or something less tangible, a harmonious state induced in the appreciative mind, I'm not sure. But in Bohr's model the connection is close to being tangible: the electron really does signal its motion to us in a sensory form, as light, and the frequencies in the line spectrum are the tonal palette of the atomic instrument.

So that's one way in which Bohr's model

is musical. Another is in its power to suggest meanings far beyond its actual content. Einstein sensed immediately that Bohr's introduction of discreteness into the description of matter, with its constraint on the possible motions, intimated that the profound limitations of pre-modern physics which I just described would be overcome – even though, in itself, Bohr's model only supplied an approximate description of the simplest sort of atom.

In one respect, however, Einstein was wrong. Bohr's model is definitely *not* the highest form of musicality in the sphere of thought. The theory that replaced Bohr's model, modern quantum mechanics, outdoes it by far.

In modern quantum mechanics, an electron is no longer described as a particle in orbit. Rather, it is described by a vibrating wave pattern in all space. The equation that describes the electron's wave function, Schrödinger's equation, is very similar to the equations that describe the vibrations of musical instruments. In Schrödinger's account light is emitted or absorbed when the electron's vibrations set the electromagnetic field – aether, if you like – in motion, by the same sort of sympathetic vibration that leads to the emission of sound by musical instruments, when their vibrations set air in motion. These regular, continuous processes replace the mysterious "quantum jumps" from one orbit to another that were assumed, but not explained, in Bohr's model.

A major reason that physicists were able to make rapid progress in atomic physics, once Schrödinger found his equation, is that they were able to borrow techniques that had already been used to analyze problems in sound production and music. Ironically, despite his well-known love of music, Einstein himself never accepted modern quantum mechanics.

After the consolidation of atomic physics in the early 1930s, the inner boundary of physics shrank by a factor of a hundred thousand. The challenge was to understand the tiny atomic nuclei, wherein are concentrated almost all the mass of matter. The nuclei could only held together by some new force, which came to be called the strong force, since gravity is much too feeble and electrical forces are both too feeble and of the wrong sign to do the job (being repulsive, not attractive). Experimenters found that it was useful to think of nuclei as being built up from protons and neutrons, and so the program initially was to study the forces between protons and neutrons. You would do this by shooting protons and neutrons at each other, and studying how they swerved. But when the experiments were done, what emerged was not just existing particles swerving around in informative directions, but a whole new world containing many many new particles, a Greek and Roman alphabet soup containing among others $\pi, \rho, K, \omega, \phi$ mesons and $\Delta, \Lambda, \Sigma, \Omega$ baryons and their antiparticles, that are unstable, but otherwise bear a strong family resemblance to protons and neutrons. So the notion of using protons and neutrons as elementary building blocks, bound together by forces you would just go ahead and measure, became untenable.

I'll skip over the complicated story of how physicists struggled to recover from this confusing shock, and just proceed to the answer, as it began to emerge in the 1970s, and was firmly established during the 1990s. The building blocks of nuclei are quarks and gluons. Add electrons and photons, and you have all the building blocks you need for atoms, molecules, ions, and indeed all ordinary matter.

Quarks once had a bad reputation, because for many years attempts to produce them in the laboratory, or to find them anywhere, failed. People went so far as to search with great care through cosmic ray tracks, and even in moon rocks. And gluons had no reputation at all. But now we know that quarks and gluons are very real. You can see

them, quite literally — once you know what to look for! Here (Fig. 1) is a picture of a

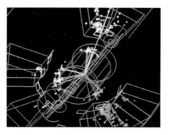

Figure 1. A three-jet event at LEP. Courtesy L3 collaboration.

quark, an antiquark, and a gluon. What is shown here is the result of colliding electrons and positrons at very high energy. It was taken at the Large Electron Positron collider (LEP) at CERN, near Geneva, which operated through the 1990s. You see that the particles are emerging in three groups, which we call jets. I want to convince you that one of these jets represents a quark, one an antiquark, and one a gluon.

The key idea for justifying this interpretation is *asymptotic freedom*, for which the Nobel Prize was awarded in 2004. Asymptotic freedom says that an energetic quark (or antiquark or gluon) will frequently emit soft radiation, which does not significantly change the overall flow of energy and momentum; but only rarely emit hard radiation, which does produce changes in the flow. Here's what asymptotic freedom means, concretely, in the LEP situation. Right after the electron and positron annihilate, their energy is deposited into a quark-antiquark pair moving rapidly in opposite directions. Usually the quark and antiquark will only emit soft radiation. In that case, which occurs about 90% of the time, we will have two jets of particles, moving in opposite directions, whose total energy and momentum reflect those of the original quark and antiquark. More rarely, about 10% of the time, there will be a hard radiation, where a gluon is emitted. That gluon will then initiate a third jet. And 10% of the 10%, that is 1%, of the time a second hard radiation will occur, and we'll have four jets, and so forth. Having invested hundreds of millions of Euros to construct the LEP machine, physicists exploited it intensely, producing many millions of collisions, working up to about one bang per buck. So it was possible to study the properties of these multi-jet events, to see how likely it was for them to emerge at different angles and to share the available energy in various ways, and thereby to test precisely whether they behave in the way that our basic theory of quarks and gluons, namely quantum chromodynamics (QCD), predicts. It works. That's why I can tell you, with complete confidence, that what you're seeing in the picture is a quark, an antiquark, and a gluon. We see them not as particles in the conventional sense, but through the flows they imprint on the visible particles.

There's another aspect of this business that I think is extremely profound, though it is so deeply ingrained among physicists that they tend to take it for granted. It is this: These observations provide a wonderfully direct illustration of the probabilistic nature of quantum theory. At LEP experimentalists just did one thing over and over again, that is collide electrons and positrons. We know from many experiments that electrons and positrons have no significant internal structure, so there's no question that when we make these collisions we really are doing the same thing over and over again. If the world were governed by deterministic equations, then the final result of every collision would be the same, and the hundreds of millions of Euros would have had a very meagre payoff. But according to the principles of quantum theory many outcomes are possible, and our task is to compute the probabilities. That task, computing the probabilities, is exactly what QCD accomplishes so successfully.

By the way, one consequence of the probabilistic nature of our predictions is that while I can tell you that you're seeing a quark, an antiquark and a gluon, I can't say for sure which is which!

So we know by very direct observations that quarks and gluons are fundamental constituents of matter. QCD proposes that gluons and quarks are all we need to make protons, neutrons, and all the other strongly interacting particles. That's an amazing claim, because there's a big disconnect between the properties of quarks and gluons and the properties of the things they are supposed to make.

Most notably, gluons have strictly zero mass, and the relevant quarks have practically zero mass, but together, it's claimed, they make protons and neutrons, which provide overwhelmingly most of the mass of ordinary matter. That claim flies in the face of the "conservation of mass" principle that Newton used as the basis of classical mechanics and Lavoisier used as the foundation of quantitative chemistry. Indeed, before 1905, this idea of getting mass from no mass would have been inconceivable. But then Einstein discovered his second law.

My friend and mentor Sam Treiman liked to relate his experience of how, during World War II, the U.S. Army responded to the challenge of training a large number of radio engineers starting with very different levels of preparation, ranging down to near zero. They designed a crash course for it, which Sam took, and a training manual, which Sam loved, and showed me . In that training manual, the first chapter was devoted to Ohm's three laws. The Army designed Ohm's first law is $V = IR$. Ohm's second law is $I = V/R$. I'll leave it to you to reconstruct Ohm's third law.

Similarly, as a companion to Einstein's famous equation $E = mc^2$ we have his second law, $m = E/c^2$.

All this isn't quite as silly as it may seem, because different forms of the same equation can suggest very different things. The great theoretical physicist Paul Dirac described his method for making discoveries as "playing with equations".

The usual way of writing the equation, $E = mc^2$, suggests the possibility of obtaining large amounts of energy by converting small amounts of mass. It brings to mind the possibilities of nuclear reactors, or bombs. In the alternative form $m = E/c^2$, Einstein's law suggests the possibility of explaining mass in terms of energy. That is a good thing to do, because in modern physics energy is a more basic concept than mass. It is energy that is strictly conserved, energy that appears in the laws of thermodynamics, energy that appears in Schrödinger's equation. Mass, by contrast, is a rather special, technical concept – it labels irreducible representations of the Poincaré group (I won't elaborate on that.)

Actually, Einstein's original paper does not contain the equation $E = mc^2$, but rather $m = E/c^2$. So maybe I should have called $m = E/c^2$ the zeroth law, but I thought that might be confusing. The title of the original paper is a question: "Does the Inertia of a Body Depend Upon its Energy Content?" From the beginning, Einstein was thinking about the foundations of physics, not about making bombs. I think he would have been delighted to learn that our answer to his question is a resounding "Yes!" Not only does the inertia of bodies depend on its energy content; for ordinary matter most of inertia *is* the energy associated with moving quarks and gluons, themselves essentially massless, following $m = E/c^2$. Who knows, it might even have inspired him to accept modern quantum mechanics.

To solve the equations of QCD, and identify the different ways in which quarks and gluons can organize themselves into the particles we observe, physicists have pushed the envelope of modern computing. They sculpt upwards of 10^{30} protons and neutrons into a massively parallel computer, which runs at

teraflop speeds – that is, a thousand billion, or 10^{12} multiplications of big numbers *per second* for months, that is 10^7 seconds. All to calculate what every single proton does in 10^{-24} seconds, that is figure out how to arrange its quarks and gluons efficiently, to get the minimum energy.

Evidently there's room for improvement in our methods of calculation. But already the results are most remarkable. They are displayed in Fig. 2. I think what you see in

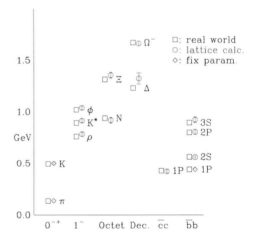

Figure 2. Hadron masses as computed numerically in QCD. Courtesy D. Toussaint.

that modest-looking plot is one of the greatest scientific achievements of all time. We start with a very specific and mathematically tight theory, QCD. An objective sign of how tight the theory is, is that just a very few parameters have to be taken from experiment. Then everything else is fixed, and must be computed by algorithms the theory supplies, with no room for maneuver or fudge factors. Here three parameters were fixed, indicated by the diamonds, by matching the masses of the π and K mesons and a splitting among heavy quark mesons (I won't enter the technicalities) and then all the other calculated masses are displayed as circles, with line intervals indicating the remaining uncertainty in calculation (due to computer limitations).

452

As you see, they agree quite well with the measured values, indicated as squares.

What makes this *tour de force* not only impressive, but also historic, is that one of the entries, N, means "nucleon", that is, proton or neutron. So QCD really does account for the mass of protons and neutrons, and therefore of ordinary matter, starting from ideally simple elementary objects, the quarks and gluons, which themselves have essentially zero mass. In this way, QCD fulfills the promise of Einstein's second law.

Another important aspect of Figure 2 is what you *don't* see. The computations do not produce particles that have the properties of individual quarks or gluons. Those objects, while they are the building block, are calculated never to occur as distinct individual particles. They are always confined within more complex particles – or, as we've seen, reconstructed from jets. This confinement property, which historically made quarks and gluons difficult to conceive and even more difficult to accept, is now a calculated consequence of our equations for their behavior.

Thus our theory corresponds to reality, in considerable detail, wherever we can check it. Therefore we can use it with some confidence to explore domains of reality that are extremely interesting, but difficult to access directly by experiment.

I think that for the future of physics, and certainly for the future of this lecture, the most profound and surprising result to emerge from late twentieth-century physics may be the realization that what we perceive as empty space is in reality a highly structured and vibrant dynamical medium. Our eyes were not evolved to see that structure, but we can use our theories to calculate what empty space might look like if we had eyes that could resolve down to distances of order 10^{-14} centimeters, and times of order 10^{-24} seconds. Derek Leinweber in particular has put a lot of effort into producing visualizations of the behavior of quark and gluon fields, and I highly recommend his website www.physics.adelaide.edu.au/theory/staff/leinweber/VisualQCD/QCDvacuum/welcome.html as a source of enlightenment. Figure 3 shows gluon fields as they fluctuate in "empty" space. I want to emphasize that this is not a free fantasy, but part of the calculation that leads to Fig. 2 (for experts: what is shown is a smoothed distribution of topological charge).

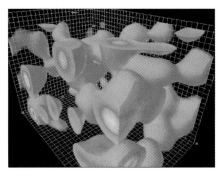

Figure 3. Gluon fields fluctuating in Void. Courtesy D. Leinweber.

The different particles we observe are the vibration patterns that are set up in the medium of "empty" space, let's call it Void, when it is disturbed in different ways. Stable particles such as protons correspond to stable vibration patterns; unstable particles correspond to vibration patterns that hold together for a while, then break apart. This is not a metaphor, it is our most profound understanding. Rather it is more familiar and conventional ideas about matter that are metaphors, for this deeper reality.

Indeed, the way Figure 2 was produced, and more generally the way we compute the properties of matter from first principles, is to introduce some disturbance in Void, let it settle down for a while, and observe what kind of stable or long-lived patterns emerge. An example is shown in Fig. 4. Here a quark and an antiquark are inserted on the left, the medium responds, and a stable vibration pattern emerges (the plane shows a slice of time).

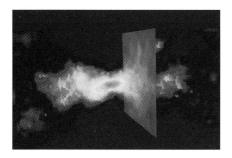

Figure 4. Disturbance in Void from injecting quark-antiquark pair. Courtesy G. Kilcup.

This picture is obtained by averaging over the fluctuations that would occur in the absence of the quark and antiquark, so the stuff in Figure 3 has been subtracted; we're interested in the net disturbance. This is how we produce a π meson for numerical study. The picture for a proton would look similar; you'd get it by studying the disturbance when you introduce three quarks, instead of a quark and antiquark.

We go from these vibration patterns to masses by combining Einstein's second law $m = E/c^2$ with the Einstein-Planck relation $E = h\nu$ between the energy of a state and the frequency at which its wave function vibrates. Thus

$$m \;=\; E/c^2 \;=\; h\nu/c^2$$

or alternatively

$$\nu \;=\; mc^2/h.$$

Thus masses of particles correspond in a very direct and literal way to the frequencies at which the Void vibrates, when it is disturbed. That is how we calculate them. The ancient "Music of the Spheres" was an inspiring concept, but it never corresponded to ideas that are very precise or impressive. Now we have a Music of the Void, which I trust you'll agree is all three of these things.

2 Our Strange Surroundings

Just as we physicists were finally consolidating this powerful understanding of ordinary matter, astronomers made some amazing new discoveries, to help us maintain our humility. They discovered that the sort of matter we've been familiar with, made from electrons, photons, quarks, and gluons — the stuff we're made of, the stuff of chemistry and biology, the stuff of stars and nebulae — makes up only 5% of the Universe by mass. The remainder consists of at least new substances. There's 25% in something we call dark matter, and 70% in what we call dark energy.

Very little is known about dark matter, and even less about dark energy. One thing we do know is that neither of them is really dark. They're transparent. They neither emit nor absorb light to any significant extent — if they did, we'd have discovered them a long time ago. In fact, dark matter and dark energy seem to interact very feebly not only with photons, but with ordinary matter altogether. They've only been detected through their gravitational influence on the (ordinary) kind of matter we do observe.

Other things we know: Dark matter forms clumps, but not such dense clumps as ordinary matter. Around every visible galaxy that's been carefully studied we find an extended halo of dark matter, whose density falls off much more slowly than that of ordinary matter as you recede from the center. It's because it is more diffusely distributed that averaged over the Universe as a whole the dark matter has more total mass than ordinary matter, even though ordinary matter tends to dominate in the regions where it is found.

Dark energy doesn't seem to clump at all. It is equally dense everywhere, as far as we can tell, as if it is an intrinsic property of space-time itself. Most strange of all, dark energy exerts negative pressure, causing the expansion of the Universe to accelerate.

With that, I've basically told you everything we know about dark matter and dark energy. It's not very satisfying. We'd like to know, for example, if the dark matter is

454

made of particles. If so, what do those particles weigh? Do they really not interact with matter at all, except by gravitation, or just a little more feebly than we've been sensitive to so far? Do dark matter particles interact strongly with each other? Can they collide with one another and annihilate?

How do you find answers to questions like those? One way, of course, is to do experiments. But the experiments have not borne fruit so far, as I mentioned, and as a general rule it's difficult to find something if you don't know what you're looking for. There's another way to proceed, which has a great history in physics. That is, you can improve the equations of fundamental physics. You can try to make them more consistent, or to improve their mathematical beauty.

For example, in the middle of the nineteenth century James Clerk Maxwell studied the equations of electricity and magnetism as they were then known, and discovered that they contained a mathematical inconsistency. At the same time Michael Faraday, a self-taught genius of experimental physics who did not have great skill in mathematics, had developed a picture of electric and magnetic phenomena that suggested to Maxwell how he might fix the inconsistency, by adding another term to the equations. When Maxwell added this term, he found that the new equations had solutions where changing electric fields induce magnetic fields, and vice versa, so that you could have self-supporting waves of electromagnetic disturbance traveling through space at the speed of light. Maxwell proposed that his electromagnetic disturbances in fact *are* light, and in this way produced one of the great unifications in the history of physics. As a bonus, he predicted that there could be electromagnetic waves of different wavelength and frequency, which would in effect be new forms of "light", not visible to human eyes. Waves of this sort were finally produced and detected by Heinrich Hertz in 1888; today of course we

call them radio waves, and also microwaves, gamma rays, and so on. Another example came around 1930, when Paul Dirac worked to improve Erwin Schrödinger's equation for the quantum mechanical wave function of electrons. Schrödinger's equation, as I mentioned before, made a big logical improvement on Bohr's model and is quite successful in giving a first account of atomic spectra. We still use it today. But Schrödinger's equation has a severe theoretical flaw: it is not consistent with special relativity. In 1928 Dirac invented an improved equation for electrons, that implements quantum dynamics and is also consistent with special relativity. Some of the solutions of Dirac's equation correspond closely to solutions of solutions of Schrödinger's equation, with small corrections. But Dirac's equation has additional solutions, that are completely new. At first it was quite unclear what these solutions meant physically, but after some struggles and false starts in 1932 Dirac put forward a convincing interpretation. The new solutions describe a new kind of matter, previously unsuspected. Dirac predicted the existence of antielectrons, or positrons. Within a few months, the experimentalist Carl Anderson found positrons, by studying cosmic rays. It was the first example of antimatter. Nowadays positrons are used for medical purposes (Positron Emission Tomography, or PET), and many other kinds of antimatter have been discovered.

Today we have several good new ideas for how to improve the equations of physics. I'll mention a few momentarily. But first let me make a preliminary observation: Because we understand so much about how matter behaves in extreme conditions, and because that behavior is remarkably simple, we can work out the cosmological consequences of changes in our fundamental equations. If our suggestion for improving the equations predicts new kinds of particles, we can predict the abundance with which those particles would be produced during the big bang. If any of the particles are stable, we will predict their present cosmological density. If we're lucky, we might find that one of our new particles is produced in the right amount, and has the right properties, to supply the astronomers' dark matter.

Recently experimenters have been testing our theoretical understanding of the early universe in a most remarkable way. By colliding gold nuclei at extremely high energies they create, for very brief times and in a very small volume (around 10^{-20} seconds, and 10^{-12} centimeters), conditions of temperature and density in terrestrial laboratories similar to those that last occurred throughout the universe a hundredth of a second or so after the Big Bang. The ashes of these tiny fireballs are thousands of particles, as shown in Fig. 5. It looks very complicated, and

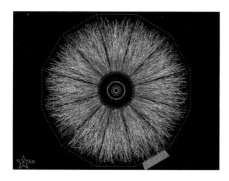

Figure 5. Particles emerging from a violent collision of gold nuclei, which reproduces physical conditions close to the big bang. Courtesy STAR collaboration.

in many ways it is, but our theories makes many predictions about the overall flow of energy and the properties of the most energetic particles that can be compared with observations, and those predictions work pretty well, so we're encouraged, and emboldened.

So, how do we go about improving our equations? Over the course of the twentieth century, symmetry has been immensely fruitful as a source of insight into Nature's basic operating principles. QCD, in particular, is constructed as the unique embodiment of a huge symmetry group, local $SU(3)$

color gauge symmetry (working together with special relativity, in the context of quantum field theory). As we try to discover new laws that improve on what we know, it seems good strategy to continue to use symmetry as our guide. This strategy has led physicists to several compelling suggestions. Let me very briefly mention four of them:

1. We can combine our theories of the strong, weak, and electromagnetic interactions into a single unified theory, by extending the symmetries that form the basis of these theories into a larger symmetry, that contains all of them (and more). This is known as grand unification. Grand unification predicts the existence of new particles and new phenomena, including mass for neutrinos (which has been observed) and instability of protons (not yet observed).

2. We can extend the space-time symmetry of special relativity to include mixing of space and time with additional quantum dimensions. This is known as supersymmetry. Supersymmetry predicts the existence of a whole new world of particles. Each currently known particle will have a heavier superpartner, with different spin.

3. We can enhance the equations of QCD by adding a symmetry that explains why the strong interactions exhibit no preferred arrow of time. This leads us to predict, by quite subtle and advanced arguments, the existence of a new kind of extremely light, extremely feebly interacting particle, the axion.

4. We can enhance the symmetry of our equations for the weak and electromagnetic interactions, and achieve a partial unification, by postulating the existence of a universal background field, the so-called Higgs condensate, that fills all space and time. There is already a lot

of indirect evidence for that idea, but we'd like to produce the new particles, the Higgs particles, that this field is supposed to be made of.

Each of these items leads into a beautiful story and an active area of research, and it's somewhat of a torture for me to restrain myself from saying a lot more about them, but time forbids. I'll just describe one particular line of ideas, that ties together the first two of these items with the dark matter problem, and that will soon be tested decisively.

Both QCD and the modern theory of electromagnetic and weak interactions are founded on similar mathematical ideas. The combination of theories gives a wonderfully economical and powerful account of an astonishing range of phenomena. It constitutes what we call the Standard Model. Just because it is so concrete and so successful, the Standard Model can and should be closely scrutinized for its aesthetic flaws and possibilities. In fact, the structure of the Standard Model gives powerful suggestions for its further fruitful development. They are a bit technical to describe, but I'll say a few words, which you can take as poetry if not as information.

The product structure $SU(3) \times SU(2) \times U(1)$ or the gauge symmetry of the Standard Model, the reducibility of the fermion representation (that is, the fact that the symmetry does not make connections linking all the fermions), and the peculiar values of the hypercharge quantum numbers assigned to the known particles all suggest the desirability of a larger symmetry. The devil is in the details, and it is not at all automatic that the superficially complex and messy observed pattern of matter will fit neatly into a simple mathematical structure. But, to a remarkable extent, it does.

There seems to be a big problem with implementing more perfect symmetry among the different interactions, however. The different interactions, as observed, do not have the same overall strength, as would be required by the extended symmetry. The strong interaction, mediated by gluons, really is observed to be much stronger than the electromagnetic interaction, mediated by photons. That makes it difficult, on the face of it, to interpret gluons and photons as different aspects of a common reality.

But now we should recall that empty space is a dynamical medium, aboil with quantum fluctuations. Gluons or photons see particles not in their pristine form, but rather through the distorting effects of this unavoidable medium. Could it be that when we correct for the distortion, the underlying equality of different interactions is revealed?

To try out that idea, we have to extend our theory to distances far smaller than, or equivalently energies far larger than, we have so far accessed experimentally. Fig. 6 gives a sense of what's involved. The left-most part of the graph, with the discs, represents our actual measurements. It's a logarithmic scale, so each tick on the horizontal axis means a factor of ten in energy. Building an accelerator capable of supplying one more factor of ten in energy will cost a few billion Euros. After that it gets *really* difficult. So the prospects for getting this information directly are not bright. Nevertheless, it's interesting to calculate, and if we do that we find some very intriguing results.

In correcting for the medium, we must include fluctuations due to all kinds of fields, including those that create and destroy particles we haven't yet discovered. So we have to make some hypothesis, about what kind of additional particles there might be. The simplest hypothesis is just that there are none, beyond those we know already. Assuming this hypothesis, we arrive at the calculation displayed in the top panel of Fig. 6. You see that the approach of coupling strengths to a unified value is suggested, but it is not quite accurately realized.

Physicists react to this near-success in

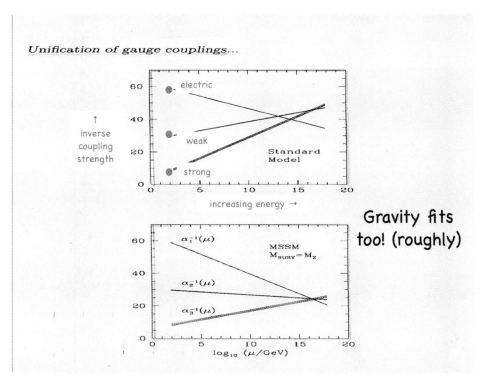

Figure 6. Unification of couplings using the currently known particles (upper panel) and with low-energy supersymmetry (lower panel).

different ways. One school says that near-success is still failure, and you should just give up this dream of unification, that the world is bound to be a lot more complicated. Another school says that there is some truth in the basic idea, but the straightforward extrapolation of physics as we know it (based on quantum field theory) to such extreme energies and distances is wrong. You might have to include the effect of extra dimensions, or of strings, for example. So you should be grateful that this calculation works as well as it does, and wait for revolutionary developments in physics to teach you how to improve it.

Either or both of these reactions might turn out to be right. But I've long advocated a more definite and (relatively) conservative proposal that still seems promising, and I'd like to mention it now, having warned you that not all my colleagues have

signed on to it, by any means. It is based on yet another way to improve the equations of physics, known as low-energy supersymmetry.

As the name suggests, supersymmetry involves expanding the symmetry of the basic equations of physics. This proposed expansion of symmetry goes in a different direction from the enlargement of gauge symmetry, which we've just been considering. Supersymmetry connects particles having the same color charges and different spins, whereas expanded gauge symmetry changes the color charges while leaving spin untouched. Supersymmetry expands the space-time symmetry of special relativity.

In order to implement low-energy supersymmetry, we must postulate the existence of a whole new world of heavy particles, none of which has yet been observed directly. There is, however, a most intriguing indirect hint that this idea may be on the right track. If we include the particles needed for low-energy supersymmetry, in their virtual form, into the calculation of how couplings evolve with energy, then accurate unification is achieved! This is shown in the bottom panel of Fig. 6.

Among the many new particles, one is particularly interesting. It is electrically neutral and has no color charge, so it interacts very weakly with ordinary matter. It lives a very long time – longer than the current lifetime of the Universe. Finally, if we calculate how much of it would be produced during the big bang, we find that it supplies roughly the right density to supply the dark matter. All this adds up to suggest that maybe this new particle is supplying the dark matter astronomers have discovered.

By ascending a tower of speculation, involving now both extended gauge symmetry and extended space-time symmetry, we seem to break though the clouds, into clarity and breathtaking vision. Is it an illusion, or reality? That question creates a most exciting situation for the Large Hadron Collider (LHC), due to begin operating at CERN in 2007. For that great accelerator will achieve the energies necessary to access the new world of heavy particles, if it exists. How the story will play out, only time will tell. In any case, I think it is fair to say that the pursuit of unified field theories, which in past (and many present) incarnations has been vague and barren of testable consequences, has in the circle of ideas I've been describing here attained entirely new levels of concreteness and fertility.

3 Three Great Lessons

Now I'm done with what I planned to tell you today about the strangeness of the Universe. I think it's appropriate to conclude by connecting these grand considerations about the nature of reality to human life. So I'll conclude by drawing what I feel are three great lessons — I'm not sure whether I should call them moral, philosophical, or spiritual lessons — from the scientific results I've described.

1. **The part of the world we understand is strange and beautiful.** We, and all the things we deal with in everyday life, are Music of the Void.

2. **If we work to understand, we can understand.** Using hands and minds evolved for quite other purposes, by diligent labor and honest thought, we have come to comprehend vastly alien realms of the infinite and the infinitesimal.

3. **We still have a lot to learn.**

Acknowledgments

The work of FW is supported in part by funds provided by the U.S. Department of Energy under cooperative research agreement DE-FC02-94ER40818.

DISCUSSION

Nobu Katayama (KEK):

Do you think you can calculate the mass of the proton without using computers?

Frank Wilczek: I don't think it will be possible to do really accurate calculations, at the few per cent level, without heavy use of computers. I hope I'm wrong, but I don't see any real prospect of it.

One aspect of the problem is that the proton mass has to be considered in its natural context, which is the entire spectrum of hadrons, and that spectrum is quite complicated. It's hard to get a complicated answer from a simple calculation with simple equations. We know lots of examples in cellular automata, fractals, and chaos where simple equations or algorithms can give you tremendously complicated answers, but these connections always emerge from gigantic, computer-intensive calculations.

That said, if we lower our standards a bit, I think there's lots of room for creativity in devising approaches to strong interaction problems that give rough answers, say at the ten or twenty per cent level, without using computers, or using only modest computer power. We already have examples of that sort of thing, including the bag model, the strong coupling expansion, and the Regge-Chew-Frautschi phenomenology, which can be reproduced using a flux tube or string model. Each of these approaches could be considerably improved, I believe, and to make such improvements is a noble goal.

Betsy Devine

(betsydevine.weblogger.com) If ordinary matter is built up of massless particles and if, as you claim, mass is not conserved — then why did Lavoisier claim mass is conserved?

Frank Wilczek: It's a beautiful example of an emergent law: that is, a law that makes no sense, or is drastically wrong, directly at the level of fundamental equations, but is an important fact about solutions of those equations. When physicists go down to the level of elementary particles, they don't see anything like the conservation of mass. It's just not true that if you add up the masses of everything that comes out you get the same result as the sum of the masses of what came in, even roughly. At LEP, for example, you always start with an electron and a positron, but the output of their annihilation often contains tens of particles, whose masses adds up to many thousand times the electron plus positron mass. And, as I emphasized in the lecture, we build protons and neutrons, which provide overwhelmingly most of the mass ordinary matter – the stuff Lavoisier was concerned with – from essentially massless building blocks, namely gluons and the u and d quarks. So at that level, Lavoisier couldn't have been more wrong.

Yet Lavoisier's experimental proof of the conservation of mass was a great milestone in science, and remains the foundation of modern quantitative chemistry. And his experiments were correct, of course. How could it be that a principle that is totally wrong as a fundamental principle is incredibly accurate in practice?

The explanation is quite interesting and calls on some deep physical ideas. It would take me a whole lecture to do them all justice. Here I'll just mention one profound idea, that plays a central role. That is the idea of an energy gap, which comes from quantum mechanics.

As I explained, the proton is a stable vibration-pattern of the Void, that

460

you can set in motion by injecting three quarks. If you attempt to disturb this vibration-pattern a little bit, you find that you can't do it – the Void will relax back to the same stable pattern. That means that the energy in this pattern is locked up. You can't change it. According to Einstein's second law, that means the mass associated with the vibration pattern doesn't change either.

On the other hand if you make a really big disturbance, so you excite qualitatively different vibration patterns, corresponding to new kinds of particles, and then the mass can change.

In chemistry – and in particular, in the kinds of experiments Lavoisier performed – the disturbances involved are too small to change a proton's vibration pattern, and so the proton's mass is locked up. You can't change it a little bit, so you can't change it at all. But at accelerators we do violent enough operations to change the vibration pattern, and then all bets about the mass are off.

Anna Lipniacka (University of Bergen): What is your opinion on using the anthropic principle to explain the Universe we ended up in?

Frank Wilczek: My opinion is complicated. I've written about it at length, and to get the full answer, you'll have to read those papers. But I'll say just a few words now.

From the point of view of fundamental physics and cosmology, life is very fragile. If the electron mass, or the up-down quark mass difference, were significantly different, then you could wind up easily wind up in situations where it was impossible to form complex nuclei, for example, and there'd be no interesting chemistry. It's difficult to imagine that intelligent entities would emerge in such

circumstances. Yet we have no good ideas about why the values of these parameters are what they are. Similarly, if the amount of dark energy were much different, or the amount of dark matter, or the amplitude of the initial fluctuations that drive structure formation, various disasters would ensue, and again it's difficult to imagine that intelligent entities would emerge. And again, we have no convincing ideas from fundamental physics why the values of these parameters are what they are.

In the context of our overwhelming success in explaining so many things in physics and cosmology using the ideas we love about symmetry and quantum dynamics, and our failure despite much clever work to explain those conspiracies, it becomes tempting to turn to another sort of argument, which has been available for a long time, but not very popular until recently. This sort of argument, the anthropic reasoning you mentioned, turns the problem of conspiracies on its head. It takes the emergence of intelligent observers as a primary explanatory principle. Then the same facts that were called *conspiracies* among the parameters get called *explanations* of their relationships.

One reason this kind of idea has not been very popular is the empirical fact that when we look out at the universe, we seem to observe the same laws everywhere. If that's the case, we should search for "universal" explanations of the laws, and not selection for intelligent observers. But inflationary cosmology, which has been quite successful in recent years, suggests that we see only a small part of the whole universe, and the very distance parts might be quite different, so that reason doesn't seem convincing anymore.

Another reason is more methodological: anthropic reasoning is hard to justify properly, and it seems not to give you the kind of depth and precision that the grand old methods of theoretical physics do. So it's like a recreational drug, that's fun to use in the short run but dangerous, because it can ruin your mind. It might lead you to be satisfied with suboptimal explanations, and to give up prematurely.

I think that we may be driven to anthropic reasoning, and we should explore it and take it seriously, but that it should not be seen as a substitute for more dynamical arguments, and that in every specific application we should only accept it as the final answer if we can convince ourselves that there's no alternative.

Frederick Harris (Hawaii):
What are your conjectures concerning the source of dark energy?

Frank Wilczek: There are some enormous positive contributions to the density of empty space arising from the quark-antiquark condensate that drives chiral symmetry breaking, the Higgs condensate that drives electroweak symmetry breaking, and presumably other condensates that drive breaking of unified gauge symmetry and supersymmetry. If gravity truly measures the total density, then there has to be a source of negative energy density, and that's something we're not familiar with. Maybe it can't be explained in terms of anything else, it's just what we call a bare cosmological term.

The real question is why all these positive terms and some negative term add up to something that gravity doesn't care about. One possibility, championed by Steven Weinberg, is that the explanation is anthropic: without a pretty accurate cancellation, the Universe would either expand or contract too fast, and condensed structures would not emerge. Another possibility, that would be very interesting and exciting, is that gravity is not quite what Einstein thought, that in the true theory of gravity there's principle that zeros out its response to uniform distributions over cosmological distances. I've played with ideas of this sort, but I don't have anything I want to show in public.

Tord Ekelöf (Uppsala University):
Dark energy represents an energy density in the vacuum. In particle physics we have another hypothesis of a field that represents an energy density, which you did not speak of — the Higgs field, only that there is some 120 orders of magnitude difference in energy density. That seems to me to be one of the big problems of unification of particle physics and cosmology.

Frank Wilczek: I agree, it's a very suspicious conspiracy.

SYMPOSIUM PROGRAMME

Thrusday 30th June 2005

09:15 -**Welcome and Opening**
Kerstin Eliasson, State Secretary of Science and Education

SESSION 1 - ELECTROWEAK PHYSICS AND BEYOND

Chair: Tord Ekelöf, Uppsala

09:35 -**Top quark measurements**,   Aurelio Juste, FNAL

10:10 -**Electroweak measurements**,   Cristinel Diaconu, CPPM, Marseille.

10:45-11:15: Coffee

Chair: Piermaria Oddone, Fermilab

11:15 -**Searches for Higgs and new phenomena at colliders**
Stephan Lammel, FNAL.

11:50 -**Electroweak symmetry breaking**,   Sally Dawson, BNL.

12:25-14:00 Lunch

Chair: Sau Lan Wu, Wisconsin

14:00 -**Probing the hierarchy problem with LHC**,   Fabiola Gianotti, CERN.

14:35 -**The LHC machine and experiments**,   Luigi Rolandi, CERN.

15:10-15:45 Coffee

Chair: Vera Luth, SLAC

15:45 -**Experimental signatures of strings and branes**,
Ignatios Antoniadis, CERN.

SESSION 2 - FLAVOUR PHYSICS

16:20 -**CP violation in B mesons**,   Kazuo Abe, KEK.

16:55 -**CKM parameters and rare B decays**,   Francesco Forti, INFN-Pisa.

17:30 -**New K dacay results**,   Ed Blucher, Chicago.

18:15 - Reception at the Chancellor rooms in the University building.

Friday 1st July 2005
(SESSION ON FLAVOUR PHYSICS. Cont)

Chair:Hesheng Chen, Beijing

    09:00 -**Charm decay measurements**,   Marina Artuso, Syracuse.

    09:35 -**Heavy flavour oscillations and life times**,   Rick Jesik, IC, London.

    10:10 -**Heavy flavour and quarkonia production and decay**,
        Xiaoyan Shen, IHEP, Beijing.

    10:45-11:15 Coffee

Chair: Jonathan Dorfan, SLAC

    11:15 -**Quark mixing and CP violation- the CKM matrix**,   Ulrich Nierste, FNAL.

    11:50 -**QCD effects in weak decays**,   Iain Stewart, MIT.

    12:25-14:00 Lunch

Chair: Ken Peach, Rutherford lab.

    14:00 -**Rare decays and CP violation beyond the Standard Model**,
        Luca Silvestrini, INFN, Rome.

SESSION 3 - QCD AND HADRON STRUCTURE

    14:35 -**Has the quark-gluon plasma been seen?**,   Johanna Stachel, Heidelberg.

    15:10-15:45 Coffee

    15:45-16:20 POSTER SESSION

Chair: Alberto Santoro, Rio de Janeiro

    16:20 -**Have pentaquarks been seen?**,   Volker Burkert, Jefferson Lab.

    16:55 -**European Grids and infrastructure**,   Ulf Dahlsten, Eropean Comission.

    17:40-19:00 **Grid session.** Moderator Tord Ekelöf.
        Massimo Lamanna, CERN, John Renner Hansen, NBI, Ruth Pordes, FNAL.

Saturday 2nd July 2005
(SESSION ON QCD AND HADRON STRUCTURE. Cont)

Chair: Gösta Gustafson, Lund

    09:00 -**Developments of perturbative QCD**, Gavin Salam, Paris VI/VII.

    09:35 -**QCD at colliders**, Jon Butterworth, UCL, London.

    10:10 -**Hard diffraction– 20 years later**, Gunnar Ingelman, Uppsala.

    10:45-11:15 Coffee

Chair: Guy Worsmer, Orsay

    11:15 -**Polarization in QCD**, Jean-Marc Le Goff, CEA Saclay.

    11:50 -**QCD and string theory**, Igor Klebanov, Princeton.

    12:25-14:00 Lunch

SESSION 4 - NEUTRINO PHYSICS

Chair: Yoji Totsuka, KEK

    14:00 -**Accelerator and atmospheric neutrinos**, Yoichiro Suzuki, Tokio.

    14:35 -**Solar and reactor neutrinos**, Alan Poon, LBL.

    15:10-15:45 Coffee and group photograph

    15:45-16:20 POSTER SESSION

Chair: Olga Botner, Uppsala

    16:20 -**Short baseline neutrino oscillations**, Eric Zimmerman, Colorado.

    17:00 - 18:00 -**The Universe is a strange place**
        Public lecture by Frank Wilczek, MIT.

    18:30 Concert in the cathedral.

<u>Monday 4th July 2005</u>
(SESSION ON NEUTRINO PHYSICS. Cont)

Chair: Sachio Komamiya, Tokyo

09:00 -**Direct neutrino mass measurements**, Christian Weinheimer, Munich.

09:35 -**Double beta decay experiments and theory**, Oliviero Cremonesi, Milano.

10:10 -**Neutrino oscillations and masses**,
    Srubabati Goswami, Harish-Chandra Institute.

10:45-11:15 Coffee

SESSION 5 - ASTROPARTICLE PHYSICS AND COSMOLOGY

Chair: Boris Kayser, Fermilab

11:15 -**Telescopes for ultra-high energy neutrinos**, Per Olof Hulth, Stockholm.

11:50 -**Dark matter searches**, Laura Baudis, Florida.

12:25-14:00 Lunch

Chair: Allan Hallgren, Uppsala

14:00 -**Exploring the universe with different messangers**,
    Eli Waxman, Weizmann Institute.

14:35 -**Dark energy and dark matter from cosmological observations**,
    Steen Hannestead, NORDITA.

15:10-15:45 Coffee

15:45-16:20 POSTER SESSION

Chair: Barbro Åsman, Stockholm

16:20 -**Cosmic rays**, Stefan Westerhoff, Columbia.

SESSION 6 - FUTURE PERSPECTIVES AND FACILITIES

16:55 -**Future facilities in astroparticle physics and cosmology**,
    René Ong, UCLA.

17:30 -**Particle detector R&D**, Michael Danilov, ITEP.

19:30 - Banquet at Uppsala castle.

Tuesday 5th July 2005

Chair: Albrecht Wagner, DESY

09:00 -**Physics at future linear colliders**, Klaus Mönig, DESY.

09:35 -**R&D for future accelerators**, Frank Zimmermann, CERN.

10:10 -**Status of the International Linear Collider project**,
    Jonathan Dorfan, SLAC.

10:45-11:15 Coffee

Chair: Alexei Sissakian

11:15 - 11:50 -**InterAction report**, Neil Calder, SLAC.
         -**C11 report**, Vera Luth, SLAC.

11:50 -**Summary and Outlook**, Francis Halzen, Wisconsin.

LIST OF CONTRIBUTED PAPERS

A paper might be listed more than once if it was cross-submitted to more than one session

Electroweak Physics

Probing the Gluino and Neutralino phases via squark pair production at hadron colliders
K. Cankoçak and D. A. Demir

Search for events with an isolated lepton and missing transverse momentum at HERA
H1 Collaboration

Multi-lepton events at HERA and a general search for new phenomena at large transverse momentum
H1 Collaboration

Search for lepton flavor violation in ep collisions at HERA
H1 Collaboration

Search for Doubly Charged Higgs Production at HERA
H1 Collaboration

Search for events with tau leptons in ep collisions at HERA
H1 Collaboration

A Direct Search for Stable Magnetic Monopoles Produced in Positron-Proton Collisions at HERA
H1 Collaboration

Search for Light Gravitinos in Events with Photons and Missing Transverse Momentum at HERA
H1 Collaboration

A general search for new phenomena in ep scattering at HERA
H1 Collaboration

Search for bosonic stop decays in R-parity violating supersymmetry in e^+ p collisions at HERA
H1 Collaboration

Search for Squark Production in R-Parity Violating Supersymmetry at HERA
H1 Collaboration

Search for Single Top Quark Production in ep Collisions at HERA
H1 Collaboration

Muon Pair Production in ep Collisions at HERA
H1 Collaboration

First Measurement of the Polarisation Dependence of the Total Charged Current Cross Sections
H1 Collaboration

Measurement of the Polarisation Dependence of the Total e^+p Charged Current Cross Section
H1 Collaboration

Determination of Electroweak Parameters at HERA
H1 Collaboration

The Virtual Correction to Bremsstrahlung in High-Energy e^+e^- Annihilation: Comparison of Exact Results
S. A. Yost, C. Glosser, S. Jadach and B. F. L. Ward

Comparisons of Fully Differential Exact Results for $O(\alpha)$ Virtual Corrections to Single Hard Bremsstrahlung in e^+e^- Annihilation at High Energies
C. Glosser, S. Jadach, B. F. L. Ward and S. A. Yost

Weak Effects in Strong Interactions
E. Maina, S. Moretti, M. R. Nolten and D. A Ross

Study of Bose-Einstein Correlations in $e^+e^- \to W^+W^-$ Events at LEP
OPAL Collaboration

Search for Neutral MSSM Higgs Bosons at LEP
ALEPH, DELPHI, L3 and OPAL Collaborations,

Constraints on Anomalous Quartic Gauge Boson Couplings from $\nu\bar{\nu}\gamma\gamma$ and $q\bar{q}\gamma\gamma$ Events at LEP2
OPAL Collaboration

W Boson Polarisation at LEP2
OPAL Collaboration

Measurement of Rb at LEP2
OPAL Collaboration

Measurement of the running of $\alpha_{QED}(t)$
OPAL Collaboration

Determination of the LEP Beam Energy using Radiative Fermion-pair Events
OPAL Collaboration

Measurement of the partial widths of the Z into up- and down-type quarks
OPAL Collaboration

Branching Fraction and Form Factors measurement of the $K^{+-} \to \pi^{+-}e^+e^-$ decay
NA48 Collaboration

Precise measurements of the Branching Ratios $K_L- \to \pi^+\pi^-$ and $K_L \to \pi^0\pi^0$.
NA48 Collaboration

New precise measurement of the ratio of leptonic decays of K+ and K- mesons.
NA48 Collaboration

Search for Chargino and Neutralino Production at $\sqrt{s}$=192-209 GeV at LEP
OPAL Collaboration

CP violation in semileptonic tau lepton decays
D. Delepine, G. López Castro and L.-T. López Lozano

Study of the $\tau^- \to 3h^-2h^+\nu_\tau$ Decay
BaBar Collaboration

Search for the Decay $\tau^- \to 4\pi^-3\pi^+(\pi^0)\nu_\tau$
BaBar Collaboration

Search for Lepton-Flavor and Lepton-Number Violation in the Decay $\tau^- \to \ell^\mp h^\pm h^-$
BaBar Collaboration

Search for Lepton Flavor Violation in the Decay $\tau^\pm \to \mu^\pm\gamma$
BaBar Collaboration

Four Leptons Production at Next Linear Colliders from 3-3-1 Model
E. Ramirez Barreto, Y. A. Coutinho and J. Sá Borges

Final results from DELPHI on neutral Higgs bosons in MSSM benchmark scenarios
DELPHI Collaboration

Study of $\tau-$lepton mass measurements at Belle
Belle Collaboration

Measurement of the $\pi^-\pi^0$ spectral function in the decay $\tau^- \to \pi^-\pi^0\nu_\tau$
Belle Collaboration

Search for lepton flavor violating decay $\tau \to lK_S^0$
Belle Collaboration

Search for the lepton flavor violating decay $\tau \to \mu\gamma$
Belle Collaboration

Transversely polarized beams and Z boson spin orientation in $e^+e^- \to Z\gamma$ with anomalous $ZZ\gamma$ and $Z\gamma\gamma$ couplings
I. Ots, H. Uibo, H. Liivat, R.-K. Loide and R. Saar

Electroweak corrections uncertainty on the W mass measurement at LEP
F. Cossutti

A Measurement of the Tau Hadronic Branching Ratios
F. Matorras (DELPHI Collaboration)

Flavour Independent Searches for Hadronically Decaying Neutral Higgs Bosons
DELPHI Collaboration

Measurement and Interpretation of Fermion-Pair Production at LEP Energies above the Z Resonance
DELPHI Collaboration

Z gamma* production in e^+e^- interactions at $\sqrt{s}$=183-209 GeV
DELPHI Collaboration

Search for excited leptons in e^+e^- collisions at sqrts=189-208 GeV
DELPHI Collaboration

Single Intermediate Vector Boson Production in e^+e^- Collisions at $\sqrt{s}$=183-209 GeV
DELPHI Collaboration

Hadronic contributions to the muon g -2 in the instanton liquid model
A. E. Dorokhov

Precision calculations of W and Z production at the LHC
S. A. Yost, M. Kalmykov, S. Majhi and B. F. L. Ward

Study of Trilinear Gauge Boson Couplings ZZZ , ZZγ and Z$\gamma\gamma$ at LEP
DELPHI Collaboration

A determination of the centre-of-mass energy at LEP2 using radiative
DELPHI Collaboration

Study of W boson polarisations and measurement of TGCs at LEP
DELPHI Collaboration

Electroweak Radiative Corrections
A. Aleksejevs, S. Barkanova and P. G. Blunden

Search for lepton-flavor violation at HERA
ZEUS Collaboration

Flavour Physics

A Challenge to the Mystery of the Charged Lepton Mass Formula
Y. Koide

A Study of Excited b-Hadron States with the DELPHI Detector at LEP
Z. Albrecht, G. Barker, M. Feindt, U. Kerzel, M. Moch and P. Kluit

Search for η_b in two-photon collisions at LEP II with the DELPHI detector
M. Chapkin, A. Sokolov and Ph. Gavillet

Determination of heavy quark non-perturbative parameters from spectral moments in semileptonic B decays
DELPHI Collaboration

Observation of anomalous hadron-hadron threshold structures at BESII
BES Collaboration

The observation of sigma and kappa at BESII
BES Collaboration

J/ψ and η_c branching ratio measurements at BESII
BES Collaboration

Measurement of the Λ_c^+ mass
BaBar Collaboration

Measurement of Ω_c Branching Ratios
BaBar Collaboration

Determination of $|V_{ub}|$ from measurements of the electron and neutrino momenta in inclusive semileptonic B decays
BaBar Collaboration

Determination of $|V_{ub}|$ in exclusive charmless B decays
BaBar Collaboration

Measurement of $B \to \pi\ell\nu$ decays in the recoil of $B \to D^{(*)}\ell\nu$ decays and determination of $|V_{ub}|$
BaBar Collaboration

Measurements of charmless semileptonic B decays and determination of $|V_{ub}|$
BaBar Collaboration

MEG, an experiment to measure the decay $\mu \to e\gamma$
C. Bemporad

Production of $'Xi_c^0$ and Ξ_b in Z decays and lifetime measurement of Ξ_b
DELPHI Collaboration

Measurements of $B \to K\ell^+\ell^-$ and $B \to K^*\ell^+\ell^-$ Decays
BaBar Collaboration

Measurements of $B \to X_s\gamma$ Decays from a Lepton-Tagged, Fully-Inclusive Photon Sample
BaBar Collaboration

Search for the Rare Decay $\bar{B}^0 \to D^{*0}\gamma$
BaBar Collaboration

Measurements of branching fractions and CP asymmetries for $B \to \omega K_S$
BaBar Collaboration

Improved Measurement of the CKM Angle α Using $B^0 \to \rho^+\rho^-$ Decays
BaBar Collaboration

Search for rare decays $B^+ \to D^{(*)+}K_S^0$
BaBar Collaboration

Measurement of the branching fraction and decay rate asymmetry of $B^- \to D_{\pi^+\pi^-\pi^0}K^-$
BaBar Collaboration

Direct Measurement of the Pseudoscalar Decay Constant f_{D^+}
BES Collaboration

Measurement of Time-Dependent CP asymmetries in $B^0 \to D^{*+}D^{*-}$ Decays
BaBar Collaboration

Direct Measurements of the Branching Fractions for $D^0 \to K^-e^+\nu_e$ and $D^0 \to \pi^-e^+\nu_e$ and Determinations of the Form Factors $f_+^K(0)$ and $f_+^\pi(0)$
BES Collaboration

Direct Measurement of the Branching Fraction for the Decay of $D^+ \to \overline{K}^0 e^+\nu_e$ and Determination of
$\Gamma(D^0 \to K^-e^+\nu_e)/\Gamma(D^+ \to \overline{K}^0 e^+\nu_e)$
BES Collaboration

Measurement of Time-Dependent CP asymmetries in $B^0 \to D^{*\pm}D^{\mp}$ and $B^0 \to D^+D^-$ Decays
BaBar Collaboration

Search for $\psi'' \to \rho\pi$ at BESII
BES Collaboration

Search for Factorization-Suppressed $B \to \chi_c K^{(*)}$ Decays
BaBar Collaboration

Production and Decay of Ξ_c^0 at BaBar
BaBar Collaboration

Evidence for the Decay $B^\pm \to K^{*\pm}\pi^0$
BaBar Collaboration

Measurement of CP Asymmetries in $B^0 \to \phi K^0$ and $B^0 \to K^+ K^- K_s^0$ Decays
BaBar Collaboration

Search for the Decay $B^+ \to K^+ \nu\bar{\nu}$
BaBar Collaboration

Measurement of the $B^0 \to K_2^*(1430)^0\gamma$ and $B^+ \to K_2^*(1430)^+\gamma$ branching fractions
BaBar Collaboration

Search for the Radiative Decay $B^0 \to \phi\gamma$
BaBar Collaboration

Measurements of B meson decays to $\omega K^\star$ and $\omega\rho$
BaBar Collaboration

Limit on the $B^0 \to \rho^0\rho^0$ Branching Fraction and Implications for the CKM Angle α
BaBar Collaboration

Measurements of Branching Fractions and Time-Dependent CP-Violating Asymmetries in $B \to \eta' K$ Decays
BaBar Collaboration

Measurement of branching fractions and charge asymmetries in B^+ decays to $\eta\pi^+$, ηK^+, $\eta\rho^+$, $\eta'\pi^+$, and search for B^0 decays to ηK^0 and $\eta\omega$
BaBar Collaboration

Measurement of Branching Fraction and Dalitz Distribution for $B^0 \to D^{(*)\pm} K^0 \pi^\mp$ Decays
BaBar Collaboration

Measurement of γ in $B^\pm \to D^{(*)} K^\pm$ decays with a Dalitz analysis of $D \to K_S \pi^- \pi^+$
BaBar Collaboration

Measurement of the Branching Fraction and the CP-Violating Asymmetry for the Decay $B^0 \to K_S^0 \pi^0$
BaBar Collaboration

Branching Fractions and CP Asymmetries in $B^0 \to \pi^0\pi^0$, $B^+ \to \pi^+\pi^0$ and $B^+ \to K^+\pi^0$ Decays and Isospin Analysis of the $B \to \pi\pi$ System
BaBar Collaboration

Improved Measurements of CP-Violating Asymmetry Amplitudes in $B^0 \to \pi^+\pi^-$ Decays
BaBar Collaboration

Measurement of the Ratio $\mathcal{B}(B^- \to D^{*0}K^-)/\mathcal{B}(B^- \to D^{*0}\pi^-)$ and of the CP Asymmetry of $B^- \to D_{CP+}^{*0}K^-$ Decays
BaBar Collaboration

Search for $b \to u$ transitions in $B^- \to D^0 K^-$ and $B^- \to D^{*0} K^-$
BaBar Collaboration

Impact of charm meson resonances on the form factors in D $\to$ P/V semileptonic decays
S. Fajfer and J. Kamenik

A Search for CP Violation and a Measurement of the Relative Branching Fraction in $D^+ \to K^- K^+ \pi^+$ Decays
BaBar Collaboration

Ambiguity-Free Measurement of $cos2\beta$: Time-Integrated and Time-Dependent Angular Analyses of $B \rightarrow J/\psi K\pi$
BaBar Collaboration

Measurement of the $B^0 \rightarrow D^{*-}D_s^{*+}$ and $D_s^+ \rightarrow \phi\pi^+$ Branching Fractions
BaBar Collaboration

Measurement of the Branching Fraction of $\Upsilon(4S) \rightarrow B^0\overline{B}^0$
BaBar Collaboration

Time-Dependent CP-Violating Asymmetries and Constraints on $\sin(2\beta + \gamma)$ with Partial Reconstruction of $B \rightarrow D^{*\mp}\pi^{\pm}$ Decays
BaBar Collaboration

Search for quark-annihilation type B-meson decays $B^- \rightarrow D_s^-\phi$ and $B^- \rightarrow D_s^{*-}\phi$
BaBar Collaboration

Measurement of CP asymmetry in $B^0 \rightarrow K^+K^-K_L^0$
BaBar Collaboration

Recent issues in charm spectroscopy
F. Fernández González

Study of CP violation with $B^- \rightarrow DK^{*-}$ Decays
BaBar Collaboration

Study of Double Charmonium Production in e^+e^- Annihilations around $\sqrt{s} = 10.58\,GeV$ with the BaBar Detector
BaBar Collaboration

Cosmological Family Asymmetry and CP violation
T. Fujihara, S. Kaneko, S.Kang, D. Kimura, M. Tanimoto

Measurement of F2ccbar and F2bbbar at High Q^2 using the H1 Vertex Detector at HERA
H1 Collaboration

Measurement of F2ccbar and F2bbbar at Low Q^2 using the H1 Vertex Detector at HERA
H1 Collaboration

Evidence for a Narrow Anti-Charmed Baryon State
H1 Collaboration

Diffractive D* Meson Production in DIS at HERA
H1 Collaboration

Acceptance corrected ratios of D*p(3100) and D* yields and differential cross sections of D*p(3100) production
H1 Collaboration

Inclusive Production of D$^+$, D^0, D$_s^+$ and D*+ Mesons in Deep Inelastic Scattering at HERA
H1 Collaboration

Measurement of Beauty Production at HERA Using Events with Muons and Jets
H1 Collaboration

Measurement of Charm and Beauty Photoproduction at HERA using D* mu Correlations
H1 Collaboration

Measurement of Charm and Beauty Photoproduction at HERA using Inclusive Lifetime Tagging
H1 Collaboration

Photoproduction of $D^{*\pm}$ Mesons Associated with a Jet at HERA
H1 Collaboration

The Charm Fragmentation Function in DIS
H1 Collaboration

The structure of Charm Jets in DIS
H1 Collaboration

Study of Jet Shapes in Charm Photoproduction at HERA
H1 Collaboration

Study of τ Decays to Four-Hadron Final States with Kaons
CLEO Collaboration

A New Measurement of the Masses and Widths of the Σ_c^{*++} and Σ_c^{*0} Charmed Baryons
CLEO Collaboration

Search for D^0–$\overline{D}^0$ Mixing in the Dalitz Plot Analysis of $D^0 to K_S^0 \pi^+ \pi^-$
CLEO Collaboration

Improved Measurement of $\mathcal{B}(D^+ to \mu^+ \nu)$ and the Pseudoscalar Decay Constant f_{D^+}
CLEO Collaboration

Measurements of the Branching Fractions for $D^0 \to X\ell^+\nu$ and $D^+ \to X\ell^+\nu$
CLEO Collaboration

Measurement of Absolute Hadronic Branching Fractions of D Mesons and $e^+e^- \to D\bar{D}$ Cross Sections at $E_{cm} = 3773$ MeV
CLEO Collaboration

Measurement of the Inclusive Decay Rates of D^0 and D^+ Mesons into η, η' and ϕ Mesons
CLEO Collaboration

Constraints on Weak Annihilation Contribution to the Inclusive $b \to u\ell\nu$ Decay Rate and Impact on $|V_{ub}|$
CLEO Collaboration

Search for $e^+e^- \to \lambda_{b^0} \bar{\Lambda}_{b^0}$ Near Threshold
CLEO Collaboration

Absolute Branching Fraction Measurements of Exclusive D^+ Semileptonic Decays
CLEO Collaboration

Absolute Branching Fraction Measurements of Exclusive D^0 Semileptonic Decays
CLEO Collaboration

Title of abstract: ¡b¿Direct measurement of the cross sections for $D^0\bar{D}^0$, D^+D^- and $D\bar{D}$ production by e^+e^- annihilation at the center-of-mass energy $\sqrt{s} = 3.773$ GeV
BES Collaboration

Measurements of branching fractions for the inclusive Cabibbo-Favored $\bar{K}^{*0}(892)$ and Cabibbo-Suppressed $K^{*0}(892)$ decay of neutral and charged D mesons
BES Collaboration

Measurements of Cabibbo Suppressed Hadronic Decay Fractions of Charmed D^0 and D^+ Mesons
BES Collaboration

A new resonance, X(1835), is observed at BESII
BES Collaboration

QCD and hadron structure

Effects of Soft Phases on Squark Pair Production at Hadron Colliders
K. Cankoçak, D. A. Demir

Annihilation contribution and $B \to a_0\pi, f_0K$ decays
D. Delepine, J. L. Lucio M. and Carlos A. Ramire

Lepton Flavour Violating Leptonic/Semileptonic Decays of Charged Leptons in the Minimal Supersymmetric Standard Model
T. Fukuyama, A. Ilakovac and T. Kikuchi

Quark helicity distributions and partonic orbital angular momentum
A. Mirjalili, S. Atashbar Tehrani and Ali N. Khorramian

Density perturbations of thermal fermions and the low CMB quadrupole
S. Mohanty

Study of deep inelastic inclusive and diffractive scattering with the ZEUS Forward Plug calorimeter
ZEUS Collaboration

Search for a narrow charmed baryonic state decaying to D*p in ep collisions at HERA
ZEUS Collaboration

Hadron annihilation into two photons and backward VCS in the scaling regime of QCD
B. Pire and L. Szymanowski

A QCD analysis of $\bar{p}N \to \gamma^*\pi$ at GSI-FAIR energies
B. Pire and L. Szymanowski

Forward jet production in deep inelastic ep scattering and low-x parton dynamics at HERA
ZEUS Collaboration

Search for pentaquarks decaying to $\Xi - \pi$ in deep inelastic scattering at HERA
ZEUS Collaboration

An NLO QCD analysis of inclusive cross-section and jet-production data from the ZEUS experiment
ZEUS Collaboration

Multijet production in neutral current deep inelastic scattering at HERA and determination of α_s
ZEUS Collaboration

Soft gluon resummation effects on parton distributions
G. Corcella and L. Magnea

Exclusive electroproduction of ϕ mesons at HERA
ZEUS Collaboration

Study of double-tagged gamma-gamma events at LEPII
DELPHI Collaboration

Di-jet production in gamma-gamma collisions at LEP
DELPHI Collaboration

Fermi-Dirac correlations in $Z \to \bar{p}p\bar{p}X$ events
DELPHI Collaboration

Measurement of the Electron Structure Function at LEP energies
DELPHI Collaboration

Determination of the b quark mass at the M_Z scale with the DELPHI detector at LEP
DELPHI Collaboration

Bose-Einstein Correlations in W^+W^- events at LEP2
DELPHI Collaboration

Search for η_b in two-photon collisions at LEP II with the DELPHI detector
DELPHI Collaboration

Gauge Covariance and Truncation of the Schwinger-Dyson Equations in Quantum Electrodynamics
A. Bashir and A. Raya

Hadronic contributions to the muon g− 2 in the instanton liquid model
A. Dorokhov

Single particle inclusive production in two-photon collisions at LEP II with the DELPHI detector
DELPHI Collaboration

Observation of $p\bar{p}\pi^0$ and $p\bar{p}\eta$ in ψ' decays
BES Collaboration

Measurement of diffractive D* photoproduction at HERA
ZEUS Collaboration

Measurement of beauty production from dimuon events at HERA
ZEUS Collaboration

Charm production in deep inelastic scattering using HERA II data
ZEUS Collaboration

Measurements of azimuthal asymmetries in neutral current deep inelastic scattering at HERA
ZEUS Collaboration

Neutral strange particle production in deep inelastic scattering at HERA
ZEUS Collaboration

Bose-Einstein correlations between neutral and charged kaons in deep inelastic scattering at HERA
ZEUS Collaboration

Search for new states decaying to strange particles in ep collisions at HERA
ZEUS Collaboration

Forward jet production in deep inelastic scattering at HERA
ZEUS Collaboration

Inclusive jet cross sections in deep inelastic scattering at HERA and determination of α_s
ZEUS Collaboration

Charged multiplicity distributions in deep inelastic scattering at HERA
ZEUS Collaboration

Study of colour dynamics in photoproduction at HERA
ZEUS Collaboration

Study of event shapes in deep inelastic scattering at HERA
ZEUS Collaboration

Cross section measurements of a narrow baryonic state decaying to $K_s^0 - p$ and $K_s^0 - \bar{p}$ in deep inelastic scattering at HERA
ZEUS Collaboration

Diffractive photoproduction of J/ψ mesons with large momentum transfer at HERA
ZEUS Collaboration

Diffractive photoproduction of dijets at HERA
ZEUS Collaboration

Dijet production in diffractive deep inelastic scattering at HERA
ZEUS Collaboration

Properties of leading neutrons in deep inelastic scattering and photoproduction at HERA
ZEUS Collaboration

The $e^+e^- \rightarrow \pi^+\pi^-\pi^+\pi^-$, $K^+K^-\pi^+\pi^-$, and $K^+K^-K^+K^-$ Cross Sections at Center-of-Mass Energies 0.5–4.5 GeV Measured with Initial-State Radiation
BaBar Collaboration

Search for Strange-Pentaquark Production in e^+e^- Annihilation at $\sqrt{s}$=10.58 GeV
BaBar Collaboration

Colour reconnection studies in $e^+e^- \rightarrow W^+W^-$ events at $\sqrt{s}$=190-208 GeV using particle flow
OPAL Collaboration

Scaling violations of quark and gluon jet fragmentation functions in e^+e^- annihilations at $\sqrt{s}$=91.2 and 183-209 GeV
OPAL Collaboration

Experimental studies of unbiased gluon jets from e^+e^- annihilations using the jet boost algorithm
OPAL Collaboration

Measurement of event shape distributions and moments in $e^+e^- \to$ hadrons at 91-209 GeV and a determination of α_s
OPAL Collaboration

Determination of α_s using jet rates at LEP
OPAL Collaboration

Impact of charm meson resonances on the form factors in D$\to$ P/V semileptonic decays
S. Fajfer and J. Kamenik

Measurement of the Strong Coupling Constant α_s from the Four-Jet Rate in e^+e^- Annihilation using JADE data
J. Shieck, S. Kluth, S. Bethke, P.A. Movilla Fernández and C. Pahl (for the JADE Collaboration)

Study of moments of event shapes in e^+e^- annihilation using JADE data
C. Pahl, S. Kluth, S. Bethke, P. A. Movilla Fernández and J. Shieck (for the JADE Collaboration)

Recent issues in charm spectroscopy
J. Vijande, A. Valcarce and F. Fernández

Search for the Θ^+ in high statistics photoproduction experiment on deuterium with CLAS
K. Hicks and S. Stepany (for the CLAS Collaboration)

Weak Effects in Strong Interactions
E. Maina, S. Moretti, M.R. Nolten and D. A. Ross

QED X QCD Threshold Corrections at the LHC
C.Glosser, S. Jadach, B.F.L. Ward and S.A. Yost

Threshold Corrections in Precision LHC Physics: QED X QCD
B.F.L. Ward, C. Glosser, S. Jadach, S. A. Yost

Measurement of Dijet Production at Low Q^2 at HERA
H1 Collaboration

Inclusive Dijet Production at Low Bjorken-x in Deep Inelastic Scattering
H1 Collaboration

Measurement of the Proton Structure Function F2 at Low Q2 in QED Compton Scattering at HERA
H1 Collaboration

Forward π^0 Production and Associated Transverse Energy Flow in Deep-Inelastic Scattering at HERA
H1 Collaboration

Measurement of Prompt Photon Cross Sections in Photoproduction at HERA
H1 Collaboration

Measurement of $F_2^{c\bar{c}}$ and $F_2^{b\bar{b}}$ at High Q^2 using the H1 Vertex Detector at HERA
H1 Collaboration

Measurements of Forward Jet Production at low x in DIS
H1 Collaboration

Measurement of $F_2^{c\bar{c}}$ and $F_2^{b\bar{b}}$ at Low Q^2 using the H1 Vertex Detector at HERA
H1 Collaboration

Determination of Electroweak Parameters at HERA
H1 Collaboration

Measurement of the Polarisation Dependence of the Total $e + p$ Charged Current Cross Section
H1 Collaboration

Measurement of the Polarisation Dependence of the Total $e + p$ Charged Current Cross Section
H1 Collaboration

Measurement of the Inclusive DIS Cross Section at Low Q^2 and High x using Events with Initial State Radiation
H1 Collaboration

Multi-jet production in high Q^2 neutral current deeply inelastic scattering at HERA and determination of α_s
H1 Collaboration

Evidence for a Narrow Anti-Charmed Baryon State
H1 Collaboration

Measurement of Anti-Deuteron Photoproduction and a Search for Heavy Stable Charged Particles at HERA
H1 Collaboration

Measurement of Dijet Cross Sections in ep Interactions with a Leading Neutron at HERA
H1 Collaboration

Measurement of DVCS at HERA
H1 Collaboration

Dijets in Diffractive Photoproduction and Deep-Inelastic Scattering at HERA
H1 Collaboration

Diffractive D^* Meson Production in DIS at HERA
H1 Collaboration

Measurement of the Diffractive Cross Section in Charged Current Interactions at HERA
H1 Collaboration

Diffractive Photoproduction of rho Mesons with Large Momentum Transfer at HERA
H1 Collaboration

H1 Search for a Narrow Baryonic Resonance Decaying to $K_s^0 p\bar{p}$
H1 Collaboration

Acceptance corrected ratios of $D^*p(3100)$ and D^* yields and differential cross sections of $D^*p(3100)$ production
H1 Collaboration

Inclusive Production of D^+, D^0, D_s^+ and D^{*+} Mesons in Deep Inelastic Scattering at HERA
H1 Collaboration

Measurement of Beauty Production at HERA Using Events with Muons and Jets
H1 Collaboration

Measurement of Charm and Beauty Photoproduction at HERA using $D^* \mu$ Correlations
H1 Collaboration

Measurement of Charm and Beauty Photoproduction at HERA using Inclusive Lifetime Tagging
H1 Collaboration

Photoproduction of $D^{*\pm}$ Mesons Associated with a Jet at HERA
H1 Collaboration

The Charm Fragmentation Function in DIS
H1 Collaboration

The structure of Charm Jets in DIS
H1 Collaboration

Study of Jet Shapes in Charm Photoproduction at HERA
H1 Collaboration

The Search for $\eta(1440) \rightarrow K_S^0 K^p m \pi^m p$ in Two-Photon Fusion at CLEO
CLEO Collaboration

Neutrino Physics

Cosmological Family Asymmetry and CP violation
T. Fujihara *et al.*

Recent Progress Towards a Cost-Effective Neutrino Factory Design
D. M. Kaplan

Astroparticle Physics and Cosmology

Low CMBR quadrupole from thermal fermions in the inflationary era
S. Mohanty

Cosmic ray propagation in the Galaxy and high energy neutrinos
J. Candia

Cosmological Family Asymmetry and CP violation
T. Fujihara *et al.*

Final State of Hawking Radiation in Quantum General Relativity
B. F. L. Ward

Massive Elementary Particles and Black Holes
B. F. L. Ward

Future Perspectives and Facilities

Development of a Monolithic Active Pixel Detector for a Super B Factory – the CAP series
H. Aihara *et al.*

The CMS Electromagnetic Calorimeter
M. Lethullier

Monte Carlo Studies of Moller and Bhabha Scattering in Iron Targets for Polarimetry in Linear Colliders
N. M. Shumeiko, P. M. Starovoitov and J.G. Suarez

TPC R& D for an ILC Detector
LC TPC groups

Study of Scintillator Strip with Wavelength Shifting Fiber and Silicon Photomultiplier
V. Balagura *et al.*

Distinguishing between MSSM and NMSSM by combined LHC and ILC analyses
G. Moortgat-Pick, S. Hesselbach, F. Franke and H. Fraas

Recent Progress Towards a Cost-Effective Neutrino Factory Design
D. M. Kaplan

LIST OF CONTRIBUTED POSTERS

Top Quark Spin Polarization and Wtb Couplings in ep Collision
S. Atağ and I. Şahin

Anomalous Quartic $W^+W^-\gamma\gamma$ Coupling in $e\gamma$ Collision With Polarization Effect
S. Atağ and I. Şahin

Gauge Covariance and Truncation of the Schwinger-Dyson Equations in Quantum Electrodynamics
A. Bashir and A. Raya

Determination of branching fractions and $|V_{ub}|$ in exclusive charmless B decays
M. Kelsey (BABAR Collaboration)

SHERPA, an Event Generator for the LHC
S. Schumann, T. Gleisberg, S. Höche, F. Krauss, A. Schälicke and J. Winter

Cosmic Ray Propagation in the Galaxy and High Energy Neutrinos
J. Candia

A Photon Beam Polarimeter Based on Nuclear e+e- Pair Production in an Amorphous Target
F. Adamyan, H. Hakobyan, R.Jones, Zh. Manukyan, A. Sirunyan and H. Vartapetian

Tritium Beta Decay in the Context from a Left-Right Symmetric Model
A. Gutiárrez-Rodríguez, M. A. Hernández-Ruíz and O. A. Sampayo

Pairs-Production of Higgs Bosons in Association with Quarks Pairs at e^+e^- Colliders
A. Gutiárrez-Rodríguez, M. A. Hernández-Ruíz and O. A. Sampayo

Phenomenological Analysis of Lepton and Quark Yukawa Couplings in SO(10) two Higgs Model
K. Matsuda

Electroweak Radiative Corrections
A. Aleksejevs, S. Barkanova and P. G. Blunden

Sudakov resummation effects for parton distributions
C. Corcella and L. Magnea

Lepton Flavour Violating Leptonic/Semileptonic Decays of Charged Leptons in the Minimal Supersymmetric Standard Model
T. Kikuchi

Effects of Soft Phases on Squark Pair Production at Hadron Colliders
K. Cankoçak and D. A. Demir

Higgs boson decays in the littlest Higss model
G. A. González-Sprinberg and J.-Alexis Rodríguez

PARTICIPANTS

Abbott, Brad	University of Oklahoma	USA
Abe, Kazuo	KEK	Japan
Aihara, Hiroaki	University of Tokyo	Japan
Aleksejevs, Aleksandrs	Saint Mary's University	Canada
Alexander, Gideon	Tel-Aviv University	Israel
Alexander, Jim	LEPP, Cornell University	USA
Alwall, Johan	Uppsala University	Sweden
Alviggi, Mariagrazia	Universitá Federic II, Napoli	Italy
Ambrosino, Fabio	Universitá Federico II '& INFN-Napoli	Italy
Antoniadis, Ignatios	CERN	Switzerland
Apollinari, Giorgio	Fermilab	USA
Artuso, Marina	Syracuse University	USA
Assaoui, Fatna	University Mohammed V Agdal	Morocco
Ayaz, Muhammad	University of Peshawar	Pakistan
Aziz, Tariq	Tata Institute of Fundamental Research	India
Babu, Kaladi	Oklahoma State University	USA
Badelek, Barbara	Uppsala University	Sweden
Barkanova, Svetlana	Acadia University	Canada
Barnes, Peter	Los Alamos National Laboratory	USA
Barnett, Michael	Lawrence Berkeley National Lab	USA
Bashir, Adnan	UMSNH	Mexico
Baudis, Laura	University of Florida	USA
Bay, Aurelio	EPFL	Switzerland
Bemporad, Carlo	INFN and University of Pisa	Italy
Benchouk, Chafik	CPPM–Marseille	France
Berger, Edmond	Argonne National Laboratory	USA
Berges, Elin	Stockholm University	Sweden
Berkelman, Karl	Cornell University	USA
Bernet, Colin	CERN	Switzerland
Bernhard, Ralf	University of Zurich	Switzerland
Bijnens, Johan	Lund University	Sweden
Bingefors, Nils	Uppsala University	Sweden
Blazey, Jerry	Northern Illinois University	USA
Blucher, Edward	The University of Chicago	USA
Boisvert, Veronique	University of Rochester	USA
Bolotov, Vladimir	Institute for Nuclear Research	Russia
Bondar, Alexander	Budker Institute of Nuclear Physics	Russia
Botner, Olga	Uppsala University	Sweden
Bouchta, Adam	Uppsala University	Sweden
Bowcock, Themis	University of Liverpool	UK
Brandt, Gerhard	University of Heidelberg	Germany
Brau, James	University of Oregon	USA
Brenner, Richard	Uppsala University	Sweden
Buchholz, Peter	University of Siegen	Germany
Burkert, Volker	Jefferson Lab	USA

Buschhorn, Gerd W.	Max Planck Inst. für Physik	Germany
Butler, John	Boston University	USA
Butterworth, Jonathan	University College London	UK
Calvetti, Mario	LNF-IFN	Italy
Calvi, Marta	University of Milano Bicocca	Italy
Campbell, Alan	DESY	Germany
Candia, Julián	The Abdus Salam ICTP	Italy
Cankocak, Kerem	University of Mugla	Turkey
Carlson, Per	KTH - Royal Institute of Technology	Sweden
Cavasinni, Vincenzo	Universitá di Pisa and INFN	Italy
Cavoto, Gianluca	LNF-INFN and Univ. Rome-La Sapinza	Italy
Chang, Ngee-Pong	City College of CUNY	USA
Chen, Hesheng	Inst. of High Energy Physics	China
Cheung, Harry W.K.	Fermilab	USA
Choi, Sookyung	Gyeongsang National Unversity	South Korea
Chou, Weiren	Fermilab	USA
Ciafaloni, Paolo	INFN Sezione di Lecce	Italy
Clare, Robert	University of California, Riverside	USA
Clark, Phil	University of Edinburgh	UK
Conrad, Janet	Columbia University	USA
Conta, Claudio	INFN and University di Pavia	Italy
Cremonesi, Oliviero	INFN/Milano Bicocca University	Italy
Crépé-Renaudin, Sabine	CNRS	France
Dahlsten, Ulf	European Commission	Belgium
Dallapiccola, Carlo	University of Massachusetts	USA
Danielsson, Ulf	Uppsala University	Sweden
Danilov, Mikhail	ITEP	Russia
Davies, Andrew	University of Glasgow	UK
Davour, Anna	Uppsala University	Sweden
Dawson, Sally	Brookhaven National Laboratory	USA
de los Heros, Carlos	Uppsala University	Sweden
Descotes-Genon, Sebastien	Universite Paris-Sud/CNRS	France
Deshpande, Nilendra	University of Oregon	USA
Di Ciaccio, Anna	Unviersity of Roma Tor Vergata	Italy
Di Giacomo, Adriano	INFN & Pisa University	Italy
Diaconu, Cristinel	CPPM	France
Dijkstra, Hans	CERN	Switzerland
Dittmann, Jay	Baylor University	USA
Dore, Ubaldo	INFN Sezione di Roma	Italy
Dorfan, Jonathan	Stanford Linear Accelerator Center	USA
Dornan, Peter	Imperial College	UK
Dosselli, Umberto	INFN-PADOVA	Italy
Du, Dong-Sheng	Institute of High Energy Physics	P.R. China
Dubois-Felsmann, Gregory	Caltech	USA
Eckmann, Reinhard	University of Texas at Austin	USA
Eduardo de Bernadini, Alex	Unicamp	Brazil
Eerola, Paula	Lund University	Sweden

Ehret, Claus	DESY	Germany
Ekelöf, Tord	Uppsala University	Sweden
Ellert, Mattias	Uppsala University	Sweden
Elsing, Markus	CERN	Switzerland
Emery, Sandrine	CEA-Saclay	France
Ericson, Inger	Uppsala University	Sweden
Eriksson, David	Uppsala University	Sweden
Fajfer, Svjetlana	University of Ljubljana	Slovenia
Fayyazuddin	National Centre for Physics	Pakistan
Feltesse, Joel	CEA-Saclay	France
Ferguson, Thomas	Carneige Mellon University	USA
Fernández, Enrique	IFAE/Univ. Autónoma Barcelona	Spain
Fernández, Francisco	University of Salamanca	Spain
Ferrari, Arnauld	Uppsala University	Sweden
Ferreira, Erasmo	Universidade Federal, Rio de Janeiro	Brazil
Fiore, Luigi	INFN-Bari	Italy
Focardi, Ettore	University of Florence	Italy
Ford, William T.	University of Colorado at Boulder	USA
Forti, Francesco	INFN-PISA	Italy
Foudas, Costas	Imperial College	UK
Gallo, Elisabetta	INFN-Firenze	Italy
Garisto, Robert	Brookhaven	USA
Gianotti, Fabiola	CERN	Switzerland
Glazov, Sasha	DESY	Germany
Glenzinski, Douglas	Fermilab	USA
Gollub, Nils	Uppsala University	Sweden
Golowich, Eugene	University of Massachusetts	USA
Golutvin, Igor	Joint Institute for Nuclear Research	Russia
González, Saul	US Department of Energy	USA
Goodman, Jordan	University of Maryland	USA
Goswami, Srubabati	Harish-Chandra Research Institute	India
Gotsman, Errol Asher	Tel Aviv University	Israel
Grab, Christoph	ETH Zurich	Switzerland
Grannis, Paul	Stony Brook University	USA
Grard, Fernand	Université de Mons-Hainaut	Belgium
Graziani, Enrico	INFN - ROMA 3	Italy
Green, Michael	Roal Holloway University of London	UK
Grindhammer, Guenter	Max Planck Institut für Physik	Germany
Gustafson, Gösta	Lund University	Sweden
Gutay, László J.	Purdue University	USA
Gutierrez, Gaston	Fermilab	USA
Gutiérrez-Rodríguez, Alejandro	Zacatecas	México
Güler, Ali Murat	Middle East Technical University	Turkey
Haba, Junji	KEK	Japan
Hallgren, Allan	Uppsala University	Sweden
Halzen, Francis	University of Wisconsin	USA
Hamadache, Clarisse	LAPC	France
Handoko, L.T.	Research Center for Physics	Indonesia

Hannestad, Steen	University of Aarhus	Denmark
Hanson, Gail G.	University of California, Riverside	USA
Hansson, Per	KTH, Royal Institute of Technology	Sweden
Harris, Frederick	University of Hawaii	USA
Heintz, Ulrich	Boston University	USA
Hernándes-Ruiz, María A.	Universidad de Zacatecas	Mxico
Herten, Gregor	University of Freiburg	Germany
Hertzbach, Stanley	University of Massachusetts	USA
Hesselbach, Stefan	Uppsala University	Sweden
Heusch, Clemens A.	University of California at Santa Cruz	USA
Hicheur, Adlene	Rutherford Appleton Laboratory	UK
Higashijima, Kiyoshi	Osaka University	Japan
Hioki, Zenro	University of Tokushima	Japan
Hirosky, Bob	The University of Virginia	USA
Honscheid, Klaus	Ohio State University	USA
Horejsi, Jiri	Charles University Prague	Czech Republic
Huang, Tao	Institute of High Energy Physics	China
Hulth, Per Olof	Stockholm University	Sweden
Hultqvist, Klas	Stockholm University	Sweden
Ianni, Aldo	INFN-LNGS	Italy
Iarocci, Enzo	LNF-INFN/Univ. Rome–La Sapienza	Italy
Ille, Bernard	IPNL, Lyon	France
Ingelman, Gunnar	Uppsala University	Sweden
Ishikawa, Kenzo	Hokkaido University	Japan
Jacewicz, Marek	Uppsala University	Sweden
Jakobs, Karl	Freiburg University	Gemany
Jantzen, Bernd	University of Karlsruhe	Germany
Jawahery, Abolhassan	University of Maryland	USA
Jesik, Rick	Imperial College London	UK
Jin, Shan	Institute of High Energy Physics	P.R.China
Jon-And, Kerstin	Stockholm University	Sweden
Jung, Chang Kee	Stony Brook University	USA
Juste, Aurelio	Fermilab	USA
Kaplan, Daniel	Illinois Institute of Technology	USA
Karlen, Dean	University of Victoria and TRIUMF	Canada
Karshon, Uri	Weizmann Institute of Science	Israel
Kashif, Lashkar	Harvard University	USA
Katayama, Nobuhiko	KEK	Japan
Kayser, Boris	Fermilab	USA
Kelsey, Michael H.	Stanford Linear Accelerator Center	USA
Khan, Harunor Rashid	KEK	Japan
Khorramian, Alinaghi	IPM	Iran
Kikuchi, Tatsuru	Ritsumeikan University	Japan
Kim, DongHee	Kyungpook National University	South Korea
Kim, Guinyun	Kyungpook National University	South Korea
Kim, Jae Yool	Chonnam National University	South Korea
Klebanov, Igor	Princeton University	USA

Klima, Boaz	Fermilab	USA
Ko, Pyungwon	KIAS	South Korea
Kobel, Michael	Bonn University	Germany
Koide, Yoshio	University of Shizuoka	Japan
Komamiya, Sachio	University of Tokyo	Japan
Kozlov, Gennady	Joint Institute for Nuclear Research	Russia
Kubota, Takahiro	Osaka University	Japan
Kuhn, Dietmar	Leoplold Franzens Universität Innsbruck	Austria
Kuno, Yoshitaka	Osaka University	Japan
Lager, Sara	Stockholm University	Sweden
Lammel, Stephan	Fermilab	USA
Lange, David	Lawrence Livermore National Laboratory	USA
Le Diberder, Francois	Universite Paris VII	France
Le Goff, Jean-Marc	CEA-Saclay	France
Lefrancois, Jaques	LAL ORSAY	France
Legendre, Marie	DAPNIA/SPP	France
Lehmann, Inti	Uppsala University	Sweden
Lethuillier, Morgan	IN2P3	France
Lewis, Jonathan	Fermilab	USA
Li, Ling-Fong	Carnegie Mellon University	USA
Li, Weiguo	Institute of High Energy Physics	China
Limentani, Silvia	Padova University and INFN	Italy
Lipniacka, Anna	University of Bergen	Norway
List, Jenny	University of Hamburg	Germany
Long, Hoang Ngoc	Vietnam Academy of Science and Technology	Vietnam
Loveless, Richard	University of Wisconsin	USA
Lu, Gongru	Henan Normal University	P.R. China
Lubrano, Pasquale	INFN-Perugia	Italy
Lundberg, Johan	Uppsala University	Sweden
Lundborg, Agneta	Uppsala University	Sweden
Luth, Vera	Stanford University	USA
Lutz, Pierre	CEA Saclay	France
Lyons, Louis	Oxford University	UK
MacFarlane, David	University of California at San Diego	USA
Magnea, Lorenzo	University of Torino	Italy
Mankel, Rainer	DESY	Germany
Marchiori, Giovanni	INFN–Pisa	Italy
Marlow, Daniel	Princeton University	USA
Martínez, Roberto	Universidad Nacional de Colombia	Colombia
Martyn, Hans-Ulrich	RWTH Aachen and DESY	Germany
Matsuda, Koichi	Osaka University	Japan
Matsuda, Satoshi	Kyoto University	Japan
McCarthy, Robert L.	State University of New York	USA
McCauley, Neil	University of Pennsylvania	USA
Mikuz, Marko	University of Ljubljana	Slovenia
Minard, Marie-Noelle	LAPP/CNRS	France

Mirjalili, Abolfazl	Yazd University	Iran
Mishra, Shekhar	Fermi National Accelerator Lab	USA
Moa, Torbjörn	University of Stockholm	Sweden
Moenig, Klaus	LAL/DESY	France
Mohanty, Subhendra	Physical Research Laboratory	India
Mohn, Bjarte	Uppsala University	Sweden
Morozumi, Takuya	Hiroshima University	Japan
Namkung, Won	Pohang University of Science anf Technology	South Koreaa
Nandi, Satyanarayan	Oklahoma State University	USA
Nappi, Aniello	University of Perugia and INFN	Italy
Narain, Meenakshi	Boston University	USA
Nasriddinov, Komiljon	Tashkent State Pedagogical University	Uzbekistan
Nauenberg, Uriel	University of Colorado at Boulder	USA
Newman, Harvey	California Institute of Technology	USA
Nierste, Ulrich	Femilab	USA
Nikolic-Audit, Irena	Université Pierre et Marie Curie	France
Novaes, Sergio F.	UNESP	Brazil
Oddone, Piermaria	Fermilab	USA
Oh, Sun Kun	Konkuk University	South Korea
Ohlsson, Tommy	Royal Institute of Technology (KTH)	Sweden
Olsen, Stephen L	University of Hawaii	USA
Ong, René A.	University of California	USA
Oren, Yona	Tel Aviv University	Israel
Ots, Ilmar	University of Tartu	Estonia
Oyanguren, Arantza	LAL-Orsay	FRANCE
Pac, Myoung Youl	Dongshin University	South Korea
Padilla, Cristóbal	CERN	Switzerland
Patel, Popat M.	McGill University	Canada
Peach, Ken	CCLRC-RAL	UK
Penev, Vladimir	JINR	Russia
Pérez, Miguel Angel	CINVESTAV	Mexico
Peroni, Cristiana	University of Torino	Italy
Perrino, Roberto	INFN–Lecce	Italy
Peruzzi, Piccolo Ida	Laboratori Nazionali di Frascati del'INFN	Italy
Pettersson, Henrik	Uppsala University	Sweden
Piilonen, Leo	Virginia Tech	USA
Polci, Francesco	Universitá di Roma La Sapienza & LAL Orsay	France
Poluektov, Anton	Budker Institute of Nuclear Physics	Russia
Poon, Alan	Lawrence Berkeley National Laboratory	USA
Predazzi, Enrico	University of Torino	Italy
Prepost, Richard	Unversity of Wisconsin	USA
Price, Lawrence	Argonne National Laboratory	USA
Quinn, Breese	University of Mississippi	USA
Rademacker, Jonas	University of Oxford	UK
Rathsman, Johan	Uppsala University	Sweden
Ray, Rashmi	American Physical Society	USA

Raya, Alfredo	National University of Mexico	Mexico
Regnault, Nicolas	IN2P3 CNRS Univ. Paris-VI/VII	France
Reidy, James	US Dept of Energy	USA
Reithler, Hans	RWTH Aachen, Germany	Germany
Ricciardi, Giulia	Universita' di Napoli Federico II	Italy
Riu, Imma	University of Geneva	Switzerland
Roberts, B. Lee	Boston University	USA
Rodejohann, Werner	TU Munich	Germany
Rolandi, Luigi	CERN	Switzerland
Rosen, Peter S.	U.S. Department of Energy	USA
Rosenberg, Eli	Iowa State University	USA
Rosenfeld, Carl	University of South Carolina	USA
Rosner, Jonathan	University of Chicago	USA
Rubinstein, Roy	Fermilab	USA
Ryd, Anders	Cornell University	USA
Sa Borges, José	Univ. do Estado do Rio de Janeiro	Brazil
Saar, Rein	University of Tartu	Estonia
Sahin, Banu	Ankara University	Turkey
Sahin, Inanc	Ankara University	Turkey
Sakai, Yoshihide	KEK	Japan
Salam, Gavin	CNRS and Univ. of Paris VI/VII	France
Sannino, Mario	INFN–Genova	Italy
Santoro, Alberto	UERJ/IF	Brazil
Schleper, Peter	University Hamburg	Germany
Schmelling, Michael	MPI for Nuclear Physics	Germany
Schmitz, Carsten	University of Zurich & DESY	Germany
Schneekloth, Uwe	DESY	Germany
Schoerner-Sadenius, Thomas	University of Hamburg	Germany
Schumann, Steffen	TU Dresden	Germany
Schwarz, Carl	Elsevier B.V.	The Netherlands
Schüler, Peter	DESY	Germany
Schönning, Karin	Uppsala University	Sweden
Serci, Sergio	INFN–Monserrato	Italy
Sharma, Vivek	University of California, San Diego	USA
Sheldon, Paul	Vanderbilt University	USA
Shen, Xiaoyan	Chinese Academy of Sciences	China
Shumeiko, Nikolai	National Center for HEP	Belarus
Sidorov, Alexander	Joint Institute for Nuclear Research	Russia
Silvestrini, Luca	INFN–Rome and TU-Muenchen	Italy
Sirunyan, Albert	Yerevan Physics Institute	Armenia
Sissakian, Alexey	Joint Institute for Nuclear Research	Russia
Skuja, Andris	University of Maryland	USA
Smith, John	S.U.N.Y.	USA
Snellman, Håkan	KTH, School of Engineering Sciences	Sweden
Snider, Frederick	Femilab	USA
Soa, Dang Van	Hanoi University of Education	Vietnam
Son, Dongchui	Kyungpook National University	South Korea

South, David	DESY	Germany
St. Denis, Richard	University of Glasgow	UK
Stachel, Johanna	University of Heidelberg	Germany
Staffin, Robin	U.S. Department of Energy	USA
Stewart, Iain	MIT	USA
Stone, Sheldon	Syracuse University	USA
Strandberg, Jonas	Stockholm University	Sweden
Sugimoto, Shojiro	KEK	Japan
Sun, Werner	Cornell University	USA
Suzuki, Yoichiro	University of Tokyo	Japan
Sznajder, Andre	UERJ	Brazil
Takasugi, Eiichi	Osaka University	Japan
Tauscher, Ludwig	ETH Zurich	Switzerland
Teixeira-Dias, Pedro	RHUL-Royal Holloway, Univ. of London	UK
Tekin, Bayram	Middle East Technical University	Turkey
Tengblad, Ulla	Uppsala University	Sweden
Tibell, Gunnar	Uppsala University	Sweden
Tikhonov, Yury	Budker Insitute of Nuclear Physics	Russia
Tipton, Paul	University of Rochester	USA
Tokushuku, Katsuo	KEK	Japan
Tortora, Ludovico	INFN–Roma III	Italy
Totsuka, Yoji	KEK	Japan
Trischuk, Wiliam	University of Toronto	Canada
Turala, Michal	Heneryk Niewodniczanski Institute	Poland
Turan, Gursevil	Middle East Technical University	Turkey
Uraltsev, Nikolai	INFN–Milano	Italy
Wagner, Albrecht	DESY	Germany
Walck, Christian	Stockholm University	Sweden
Waldi, Roland	Universität Rostock	Germany
Wali, Kameshwar C.	Syracuse University	USA
Wang, Qin	Radboud University Nijmegen	The Netherlands
Wang, Xue-Lei	Henan Normal University	P.R. China
Ward, Bennie	Baylor University	USA
Watkins, Peter	Birmingham University	UK
Waxman, Eli	Weizmann Institute	Israel
Weinheimer, Christian	Institute for Nuclear Physics	Germany
Weinstein, Alan	California Institute of Technology	USA
Westerhoff, Stefan	Columbia University	USA
Wiedner, Ulrich	Uppsala University	Sweden
Wilczek, Frank	MIT	USA
Wisniewski, William	Stanford Linear Accelerator Center	USA
Wojcicki, Stanley	Stanford University	USA
Wormser, Guy	LAL Orsay	France
Wright, Alison	Nature Physics	UK
Wright, Douglas	Lawrence Livermore National Laboratory	USA

Wu, Sau Lan	University of Wisconsin - Madison	USA
Yamada, Sakue	University Hamburg &Tokyo-KEK	Germany
Yamanaka, Taku	Osaka University	Japan
Yamauchi, Masanori	KEK	Japan
Yang, Ya-Dong	Henan Normal University	P. R. China
Yoshimura, Motohiko	Okayama University	Japan
Yost, Scott A.	Baylor University	USA
Zeppenfeld, Dieter	Universität Karlsruhe	Germany
Zimmerman, Eric D.	University of Colorado	USA
Zimmermann, Frank	CERN	Switzerland
Zumerle, Gianni	Universty of Padova/INFN	Italy
Åsman, Barbro	Stockholm University	Sweden

AUTHOR INDEX